Invasive Species and Global Climate Change

FSC
www.fsc.org
MIX
Paper from
responsible sources
FSC® C013604

CABI Invasive Series

Invasive species are plants, animals or microorganisms not native to an ecosystem, whose introduction has threatened biodiversity, food security, health or economic development. Many ecosystems are affected by invasive species and they pose one of the biggest threats to biodiversity worldwide. Globalization through increased trade, transport, travel and tourism will inevitably increase the intentional or accidental introduction of organisms to new environments, and it is widely predicted that climate change will further increase the threat posed by invasive species. To help control and mitigate the effects of invasive species, scientists need access to information that not only provides an overview of and background to the field, but also keeps them up to date with the latest research findings.

This series addresses all topics relating to invasive species, including biosecurity surveillance, mapping and modelling, economics of invasive species and species interactions in plant invasions. Aimed at researchers, upper-level students and policy makers, titles in the series provide international coverage of topics related to invasive species, including both a synthesis of facts and discussions of future research perspectives and possible solutions.

Titles Available

1. Invasive Alien Plants: An Ecological Appraisal for the Indian Subcontinent
   Edited by J.R. Bhatt, J.S. Singh, R.S. Tripathi, S.P. Singh and R.K. Kohli
2. Invasive Plant Ecology and Management: Linking Processes to Practice
   Edited by T.A. Monaco and R.L. Sheley
3. Potential Invasive Pests of Agricultural Crops
   Edited by J.E. Peña
4. Invasive Species and Global Climate Change
   Edited by L.H. Ziska and J.S. Dukes

Invasive Species and Global Climate Change

LEWIS H. ZISKA, PhD

Crop Systems and Global Change Laboratory, USDA-ARS, 10300 Baltimore Avenue, Beltsville, MD 20705

And

JEFFREY S. DUKES, PhD

Department of Forestry and Natural Resources & Department of Biological Sciences, Purdue University, 715 W. State Street, West Lafayette, IN 47907-2061

www.cabi.org

CABI is a trading name of CAB International

CABI
Nosworthy Way
Wallingford
Oxfordshire OX10 8DE
UK

Tel: +44 (0)1491 832111
Fax: +44 (0)1491 833508
E-mail: cabi@cabi.org
Website: www.cabi.org

CABI
38 Chauncy Street
Suite 1002
Boston, MA 02111
USA

Tel: +1 800 552 3083 (toll free)
E-mail: cabi-nao@cabi.org

A catalogue record for this book is available from the British Library, London, UK.

Library of Congress Cataloging-in-Publication Data

Invasive species and global climate change / [edited by] Lewis H. Ziska, PhD, and Jeffrey S. Dukes, PhD.
pages cm. -- (CABI invasives series ; 4)
Includes bibliographical references and index.
ISBN 978-1-78064-164-5 (hbk : alk. paper) 1. Introduced organisms. 2. Climatic changes. I. Ziska, Lewis H. II. Dukes, Jeffrey S. III. C.A.B. International. IV. Series: CABI invasive species series ; 4.

QH353.I586 2014
578.6'2--dc23

2014011558

ISBN-13: 978 1 78064 164 5

Commissioning editor: David Hemming
Editorial assistant: Emma McCann
Production editor: Simon Hill

Typeset by Columns Design XML Ltd, Reading, UK
Printed and bound in the UK by CPI Group (UK) Ltd, Croydon, CR0 4YY

Contents

Contributors

Charles T. Bargeron, Center for Invasive Species and Ecosystem Health, University of Georgia, Tifton, GA 31793, USA. E-mail: cbargero@uga.edu

Matthew A. Barnes, Department of Biological Sciences, University of Notre Dame, Notre Dame, IN 46556, USA. E-mail: mbarnes3@nd.edu

Dana M. Blumenthal, Rangeland Resources Research Unit, USDA-ARS, 1701 Center Avenue, Ft Collins, CO 80526, USA. E-mail: dana.blumenthal@ars.usda.gov

Bethany A. Bradley, Environmental Conservation, 318 Holdsworth Hall, University of Massachusetts, Amherst, MA 01003, USA. E-mail: bbradley@eco.umass.edu

Steven L. Chown, Centre for Invasion Biology, Department of Botany and Zoology, Stellenbosch University, Private Bag X1, Matieland 7602, South Africa. E-mail: sichown@sun.ac.za

David R. Clements, Biology Department, Trinity Western University, Langley, British Columbia, V2Y 1Y1 Canada. E-mail: clements@two.ca

Peter Convey, British Antarctic Survey, Natural Environment Research Council, High Cross, Madingley Road, Cambridge, CB3 0ET, UK. E-mail: pcon@bas.ac.uk

Sarah J. Davies, Centre for Invasion Biology, Department of Botany and Zoology, Stellenbosch University, Private Bag X1, Matieland 7602, South Africa. E-mail: sdavies@sun.ac.za

Áron J. Deák, Department of Physical Geography and Geoinformatics, University of Szeged, HU-6701 Szeged, POB 653, Hungary. E-mail: aron@geo.u-szeged.hu

Andrew M. Deines, Department of Biological Sciences, University of Notre Dame, Notre Dame, IN 46556, USA. E-mail: audiences@nd.edu

Hilda Diaz-Soltero, USDA, Office of the Secretary, Senior Invasive Species Coordinator, Washington, DC 20005, USA. E-mail: hdiazsoltero@fs.fed.us

Antonio DiTommaso, Department of Crop and Soil Sciences, Cornell University, Ithaca, NY 14853, USA. E-mail: ad97@cornell.edu

Jeffrey S. Dukes, Department of Forestry and Natural Resources and Department of Biological Sciences, Purdue University, 715 W. State Street, West Lafayette, IN 47907-2061, USA. E-mail: jsdukes@purdue.edu

Greg A. Forbes, International Potato Center, 12 Zhongguaneun South Street, Beijing, 100051, China. E-mail: g.forbes@cgiar.org

Karen A. Garrett, Department of Plant Pathology, 4024 Throckmorton, Kansas State University, Manhattan, KS 66502, USA, and Plant Biosecurity Cooperative Research Centre, GPO Box 5012, Bruce, ACT 2617, Australia. E-mail: kgarrett@ksu.edu

Rachel M. Gentile, Department of Biological Sciences, University of Notre Dame, Notre Dame, IN 46556, USA. E-mail: hesselink.1@nd.edu

Laura E. Grieneisen, Department of Biological Sciences, University of Notre Dame, Notre Dame, IN 46556, USA. E-mail: lgrienei@nd.edu

Andrew Paul Gutierrez, Division of Ecosystem Science, College of Natural Resources, University of California, 151 Hilgard Hall, Berkeley, CA 94720, USA. E-mail: casas_global@berkeley.edu

John Hernandez Nopsa, Department of Plant Pathology, 4024 Throckmorton, Kansas State University, Manhattan, KS 66502, USA, and Plant Biosecurity Cooperative Research Centre, GPO Box 5012, Bruce, ACT 2617, Australia. E-mail: nopsa@ksu.edu

Kevin A. Hughes, British Antarctic Survey, Natural Environment Research Council, High Cross, Madingley Road, Cambridge, CB3 0ET, UK. E-mail: kehu@bas.ac.uk

Ulrike M. Irlich, Centre for Invasion Biology, Department of Botany and Zoology, Stellenbosch University, Private Bag X1, Matieland 7602, South Africa. E-mail: Ulrike.Irlich@capetown.gov.za

Julie A. Kray, Rangeland Resources Research Unit, USDA-ARS, 1701 Center Avenue, Ft Collins, CO 80526, USA. E-mail: julie.kray@ars.usda.gov

Bo Li, Coastal Ecosystems Research Station of Yangtze River Estuary, Ministry of Education Key Laboratory for Biodiversity Science and Ecological Engineering, The Institute of Biodiversity Science, Fudan University, #220 Handan Road, Shanghai 200433, PR China. E-mail: bool@fudan.edu.cn

Hui Li, Coastal Ecosystems Research Station of Yangtze River Estuary, Ministry of Education Key Laboratory for Biodiversity Science and Ecological Engineering, The Institute of Biodiversity Science, Fudan University, #220 Handan Road, Shanghai 200433, PR China.

Meng Lu, Coastal Ecosystems Research Station of Yangtze River Estuary, Ministry of Education Key Laboratory for Biodiversity Science and Ecological Engineering, The Institute of Biodiversity Science, Fudan University, #220 Handan Road, Shanghai 200433, PR China.

László Makra, Department of Climatology and Landscape Ecology, University of Szeged, HU-6701 Szeged, POB 653, Hungary. E-mail: makra@geo.u-szeged.hu

Steven T. Manning, Invasive Plant Control, Inc, PO Box 50556, Nashville, TN 37205, USA. E-mail: steven.manning@intrstar.net

István Matyasovszky, Department of Meteorology, Eötvös Loránd University, HU-1117 Budapest, Pázmány Péter st. 1/A, Hungary. E-mail: matya@ludens.elte.hu

David Pimentel, Department of Entomology, Cornell University, Ithaca, NY 14853, USA.

Glenn E. Plumb, National Park Service, Biological Resource Management Division, 1201 Oakridge Drive, Fort Collins, CO 80525, USA. E-mail: glenn_plumb@nps.gov

Luigi Ponti, Laboratorio Gestione Sostenibile degli Agro-Ecosistemi (UTAGRI-ECO), Agenzia Nazionale per le Nuove Tecnologie, l'Energia e lo Sviluppo Economico Sostenibile (ENEA), Centro Ricerche Casaccia, Via Anguillarese 301, 00123 Rome, Italy. E-mail: quartese@gmail.com

Jessica R. Resnik, National Park Service, Biological Resource Management Division, 1201 Oakridge Drive, Fort Collins, CO 80525, USA. E-mail: Jessica_resnik@nps.gov

David M. Richardson, Centre for Invasion Biology, Department of Botany and Zoology, Stellenbosch University, Private Bag X1, Matieland 7602, South Africa. E-mail: rich@sun.ac.za

John K. Scott, CSIRO Ecosystem Sciences and Climate Adaptation Flagship, Private Bag 5, Wembley, WA 6913, Australia. E-mail: John.K.Scott@csiro.au

Peter R. Scott, CABI Head Office, Wallingford, OX10 8DE, UK. E-mail: p.scott@cabi.org

Cascade J.B. Sorte, Assistant Professor, Department of Ecology and Evolutionary Biology, 321 Steinhaus Hall, University of California, Irvine, CA 92697, USA. E-mail: cjsorte@ucdavis.edu

Thomas J. Stohlgren, US Geological Survey, Fort Collins Science Center, 2150 Centre Ave, Bldg C, Fort Collins, CO 80526, USA. E-mail: Thomas.Stohlgren@colostate.edu

Sara Thomas-Sharma, Department of Plant Pathology, 4024 Throckmorton, Kansas State University, Manhattan, KS 66502, USA. E-mail: sarathomas@ksu.edu

John Peter Thompson, Member US National Invasive Species Advisory Committee, Washington, DC, USA, and Consultant – Bioeconomic Policy, Prince George's County, MD 21324, USA. E-mail: ipetrus@msn.com

Rieks D. van Klinken, CSIRO Ecosystem Sciences, GPO Box 2583, Brisbane, QLD 4001, Australia. E-mail: Rieks.vanklinken@csiro.au

Rebekah D. Wallace, Center for Invasive Species and Ecosystem Health, University of Georgia, Tifton, GA 31793, USA. E-mail: bekahwal@uga.edu

John D. Waugh, Advisor, Integra, LLC, 1030 15th St NW, Ste 555W, Washington, DC 2005, USA. E-mail: jwaught@integrallc.com

Bruce L. Webber, School of Plant Biology, The University of Western Australia, 35 Stirling Highway, Crawley, WA 6009, Australia, and CSIRO Ecosystem Sciences and Climate Adaptation Flagship, Private Bag 5, Wembley, WA 6913, Australia. E-mail: Bruce.Webber@csiro.au

Shujuan Wei, Coastal Ecosystems Research Station of Yangtze River Estuary, Ministry of Education Key Laboratory for Biodiversity Science and Ecological Engineering, The Institute of Biodiversity Science, Fudan University, #220 Handan Road, Shanghai 200433, PR China.

Randy G. Westbrooks, Invasive Plant Control, Inc, 233 Border Belt Drive, Whiteville, NC 28472, USA. E-mail: rwestbrooks@intrstar.net

Qiang Yang, Coastal Ecosystems Research Station of Yangtze River Estuary, Ministry of Education Key Laboratory for Biodiversity Science and Ecological Engineering, The Institute of Biodiversity Science, Fudan University, #220 Handan Road, Shanghai 200433, PR China.

Qin Zhong, Department of Ecology, College of Agriculture, South China Agricultural University, Guangzhou 510642, China. E-mail: q_breeze@126.com

Lewis H. Ziska, Crop Systems and Global Change Laboratory, USDA-ARS, 10300 Baltimore Avenue, Beltsville, MD 20705, USA. E-mail: l.ziska@ars.usda.gov

Foreword

I am pleased and honoured to write the Foreword to this outstanding book by Dr Lewis Ziska of the USDA and Dr Jeff Dukes of Purdue University, which focuses on climate change, invasive species and, broadly, the environment. An estimated 50,000 species of plants, animals and microbes have been introduced into the USA. Of these species, 98% of our crops, including wheat, rice, maize and soybeans, and livestock, including cattle, hogs and chickens, have been introduced intentionally and are fundamental to our agricultural productivity. They provide us not only with our essential food base but also are valued at more than US$800 billion year^{-1}.

An estimated 20,000 species of plants, animals and microbes are pest species and cause approximately US$220 billion each year in damages, including diseases and pests of our crops and livestock, plus damage to native animals, plants and microbes. In addition, it is estimated that invasive pests cause most of the extinctions of native plants, animals and microbes.

Worldwide, invasive species contribute to the ongoing food insecurity of roughly 66% of the 7 billion people globally. It is extremely difficult to exterminate an invasive species once they become widespread. For example, in the USA, only three species (of the estimated 50,000 considered invasive) have been eliminated.

Climate change and the extensive burning of fossil fuels and forests appear to be increasing the level of CO_2 and other greenhouse gases in the atmosphere. Most meteorologists and physical scientists conclude that the continued increase in CO_2 will add to the warmth of the Earth and increase the impacts on crop and livestock production and the loss of native species.

The overall changes in temperature, rainfall, carbon dioxide, insect pests, plant pathogens and weeds associated with global climate change are projected to further reduce food production worldwide. Thus, climate change is likely to increase food insecurity and escalate the number of human deaths associated with malnutrition. Hence, there is need to reduce the number of invasive species being introduced worldwide, as well as to increase food production to help feed the world population.

David Pimentel
College of Agriculture and Life Sciences
Cornell University
Ithaca, New York

1 Introduction

Jeffrey S. Dukes[1] and Lewis H. Ziska[2]

[1]*Department of Forestry and Natural Resources & Department of Biological Sciences, Purdue University, West Lafayette, Indiana, USA;* [2]*Crop Systems and Global Change Laboratory, USDA-ARS, Beltsville, Maryland, USA*

As we write this, the global population has reached 7.1 billion. At present rates, approximately 5 million new individuals will be added each month, every month, for the foreseeable future. (www.census.gov/popclock).

Ultimately, it is our rapidly increasing population and our need to increase the production of food, feed, fibre and fuel from a finite set of natural resources that are driving the environmental issues in this book, and that give these issues urgency. We need to transition to a sustainable society if we are to provide for this population (or even a smaller one) into the future. Such sustainability is necessary if we are to preserve our planet's ecosystem services, maintain its capacity to produce food and protect its biodiversity.

However, at present, our population needs, and the unprecedented transportation of biota into new regions to achieve these needs, are occurring on a scale that threatens the planet's natural resource capacity. This book is a collective attempt from ecologists around the world to describe the interaction between two of the resulting consequences. Specifically, to examine the nexus of climate change and biological invasions, and the resulting impacts, and to identify means to reduce the vulnerability and increase the resiliency of managed and unmanaged ecosystems.

Such a complex global topic is best addressed from a variety of perspectives. We thank the many people who have contributed and commented on the chapters in this book. The individual chapter authors and the anonymous reviewers of those chapters are world experts, and very busy people. We appreciate their willingness to commit to this project and their faith that a contribution to this book would be a worthy use of their time. There is no question that their contributions have enabled this book to convey a detailed, globally relevant, and sometimes provocative, portrait of what is known and what is unknown regarding climate change and invasive species. In addition, their contributions present a valuable overview of strategies for managing natural and agricultural systems on a rapidly changing planet.

In examining a complex set of issues, it also helps to have common definitions. This book considers 'climate change' in a broad sense; that is, both the disruption of Earth's relatively stable recent climate and the ongoing increase in atmospheric CO_2 concentrations that are largely responsible for that disruption. The book also considers biological invasions broadly, including many taxa. We recognize that 'invasive species' can have a variety of meanings (and these, in turn, can be complicated by climate change, as noted by Webber and Scott, 2012). Biologists alternately refer to these species as 'biological invaders', 'alien species', 'exotic species' or simply as 'invasives'. Regardless of the term, biologists are characterizing species that have crossed a major biogeographic barrier (e.g. an ocean), usually with the assistance of humans, and whose introduction has, or will, result in significant negative economic or environmental impacts.

Given the complexity, we recognize that not all chapters will appeal to an individual reader; rather, the book is intended to be accessible to a range of interested parties, not only the academic specialist. We do hope the book can educate broadly and provide a means for understanding the consequences of invasive species and climate change, not in isolation (such efforts are already well documented) but in a *synergistic* context. Still, for most readers, to understand the synergism it is important to appreciate the components of the problem, and we attempt to provide some background here.

The Problem and Its Components

The desire for food and fuel has been endemic since the dawn of human civilization and the commencement of cultivated agriculture. As populations grew, and land/energy needs increased, the incorporation of fossil fuels, or energy captured from sunlight over millennia by plants, became an integral part of the Industrial Revolution, a revolution that, for billions of people, has provided ample food, water and an improved standard of living.

Use of natural resources to meet these human needs has also, since its inception, had some impact on climate. For example, by removing forests and native plants, early agriculture altered hydrologic cycles and changed surface albedo, with consequences for regional climate (Pielke *et al.*, 2007). In addition, the burning of fossil fuels has jolted Earth's atmosphere with a 40+% increase in carbon dioxide (CO_2) since the onset of the Industrial Revolution.

That CO_2 generated by the Industrial Revolution could influence climate is not a new concept. In the 19th century, two scientists, Fourier and Arrhenius, suggested that industrial pollutants, notably carbon dioxide, were building up in Earth's atmosphere and could, potentially, result in increased surface temperatures (Fourier, 1827; Arrhenius, 1896). Quantitative measurements by Keeling in the 1950s confirmed that CO_2 was, in fact, increasing globally (Revelle and Suess, 1957).

One of the properties of the CO_2 molecule is that it absorbs energy in the infrared portion of the electromagnetic spectrum (making it a 'greenhouse gas'). Adding carbon dioxide to the air causes the atmosphere to trap more of the heat radiated up from Earth's surface that would otherwise escape to space. The atmosphere warms up more, the rest of the planet heats up a bit to follow, and more water evaporates from the warmer seas into the warmer skies. Water vapour itself traps heat and further warms the planet, in what is known as a positive feedback loop.

Overall, model projections based on future emissions of greenhouse gases suggest a marked warming of Earth's surface and changes in precipitation patterns in many regions. Model projections also indicate clearly that the rate and degree of climate disruption over the coming decades will depend on how quickly we continue to release heat-trapping gases to the atmosphere (Solomon *et al.*, 2007). It is worth acknowledging that, given the lifespan and ongoing release of carbon dioxide and other greenhouse gases, there is sufficient momentum at present so that a significant change in Earth's climate is essentially guaranteed. Therefore, as we prepare for warmer, uncertain climatic conditions, it is important to consider the consequences of these conditions for the utility and health of managed and unmanaged ecosystems.

In considering the importance of carbon dioxide and climatic change on ecosystems, it is also important to consider carbon dioxide as an essential substrate in plant biology, providing a primary building block for photosynthesis. The recent rapid increase in atmospheric CO_2 has been felt directly by plants, some of which are growing faster, with less water consumption, in response (e.g. Keenan *et al.*, 2013). Reports on climate change in the media only infrequently discuss the direct effects of this CO_2 increase, which has been much larger, with much stronger effects on plants, than any changes in climate experienced to date (and indeed, this may remain the case for many decades).

From a human perspective, such direct effects of CO_2 may be of benefit, providing more food and faster fibre production, and potentially even helping to slow climate change by storing carbon more quickly. However, CO_2 is indiscriminate with respect to which plant species may be favoured. For example, each of us has demonstrated that unwanted plant species such as yellow starthistle and poison ivy can have very strong responses to rising CO_2 levels (Ziska, 2003; Mohan *et al.*, 2006; Dukes *et al.*, 2011). Clearly, how plant species and ecosystems respond, not only to climate but also to rising CO_2 directly, will have significant biological consequences. Several of the chapters in this book help to examine these consequences in the context of invasive species biology.

In addition to the build-up of greenhouse gases, other human activities associated with the need for increased feed and fuel have contributed to large-scale environmental perturbation. Especially relevant has been the transportation, on a massive scale, of organisms that had been restricted to certain biogeographic zones but which are now distributed globally. Many of these species, such as soybean, are important for human welfare and a strong economy, but forced reallocation of thousands of species outside of their native habitats can also result in the distribution of extraordinarily aggressive species, with severe economic and environmental consequences.

Invasive species come in many shapes and sizes; they can be hard to recognize since their only common feature is biological domination outside of their native range. This book includes discussion of invasive weeds, insects and pathogens in many disparate taxa, from the poles to the tropics. These species disrupt a wide variety of ecosystem processes (Dukes and Mooney, 2004; Vilá *et al.*, 2011), threaten biodiversity (Powell *et al.*, 2013), the provision of ecosystem services (Charles and Dukes, 2007; Pejchar and Mooney, 2009) and food (Oerke, 2006) and cause economic damage estimated to be around US$120 billion per year in the USA alone (Pimentel *et al.*, 2005).

Why This Book?

In addition to affecting the basic aspects of biology on a global scale, both climate change and invasive species pose existential threats to the basic ecosystem services necessary for human life. Furthermore, it should not be assumed that each threat acts independently of the other. The synergy between these issues is becoming increasingly evident. For example, changing climatic conditions (e.g. polar melting and the opening of new trade routes) will alter global commerce in the near future, with the subsequent introduction of unwanted species into new geographical regions (Hellmann *et al.*, 2008; Bradley *et al.*, 2012). Once they are introduced, climate change – either through changes in means or extremes – may then facilitate the establishment and spread of such species; or alternatively, may allow other species that are currently established to become invasive as environmental constraints (e.g. cold winters) are eased (Dukes and Mooney, 1999; Walther *et al.*, 2009, Bradley *et al.*, 2010; Diez *et al.*, 2012). Recent work also suggests that invasive species management, particularly chemical applications, may further exacerbate greenhouse gas emissions (Heimpel *et al.*, 2013).

While there have been many separate books documenting the impact of climate change or invasive species, only one has broadly linked these aspects of environmental transformation. In 2000, when Hal Mooney and Richard Hobbs published *Invasive Species in a Changing World*, very few researchers had thought about the combined implications of these two environmental changes (Mooney and Hobbs, 2000). Since then, the field has grown rapidly, but has not been reviewed comprehensively.

Here, we take a global look at what is currently known about the synergistic nature of these environmental changes.

Such synergism is explored by David Pimentel, among the world's foremost invasive species experts, in the Foreword and is exemplified across the book's four parts. These parts, in turn, provide an overview of the current state of understanding in this field, the tools available to manage the problem and the challenges for future research. The first part of the book outlines the dimensions of the problem. In Chapter 2, John Peter Thompson and Lewis H. Ziska present a brief overview of the science of climate change and invasion biology, but also examine how we can communicate the science more effectively to policy makers. The next three chapters lay out the science with respect to three classes of invasive species in the context of changing climate and carbon dioxide levels: Karen Garrett and colleagues discuss pathogens; Andrew Gutierrez and Luigi Ponti examine insects; and Dana Blumenthal and Julie Kray look at plants.

The second part of the book highlights the *global* synergy between climate change and invasive species with 'case studies' from around the world. We begin in Antarctica, where Kevin Hughes and Peter Convey provide an overview of climate and invasives; we segue to aquatic environments, where Cascade Sorte appraises how invasives respond; then to eastern Europe, for a more specific examination of the implications of changing CO_2 and temperature for ragweed by László Makra *et al.*, followed by a consideration of climate and invasives in South Africa by Ulrike Irlich and colleagues. Tom Stohlgren *et al.* next scrutinize invasives in national parks in the USA; Bruce Webber and colleagues examine how climate is affecting invasive species in Australia; and Bo Li *et al.* survey the current and future climate for invasives in China.

In Part III, we turn to the issue of managing new invasive threats in a changing climate. We begin by emphasizing that early detection, which has the best hope for allowing problem species to be stopped in their tracks, is critical, in a chapter by Rebekah Wallace and Chuck Bargeron focusing on tools for early detection and mapping. Hilda Diaz-Soltero and Peter Scott present information about the new CABI compendium on invasive species; Bethany Bradley examines approaches to modelling the current and future distributions of invasives; Toni DiTommaso *et al.* assess the implications of climate change for new invasive weeds in agriculture; and Andy Gutierrez and Luigi Ponti present a new approach for modelling the impact of climate change on invasive species.

Finally, we address the issue of what can be done when invasive species do show up on your doorstep. Lewis Ziska examines how chemical control of invasive weeds is likely to be impacted by climate and CO_2; Randy Westbrooks *et al.* examine whether the Early Detection and Rapid Response (EDRR) paradigm can be configured to cope with climate change. Finally, if all else fails, Matthew Barnes *et al.* ask whether invasive species can actually serve as an economic resource (i.e. it's not Asian carp, it's Kentucky tuna!).

What Do We Hope To Accomplish?

In the Twitter/Facebook/Instagram age, when visual overload can occur each time you stare at a flat screen, books may feel like an anachronism. But books – this one included – are not designed to provide you with information in 5-min increments. Rather, books function as a period in a long stream of text messaging – a chance to stop, re-read and reassess what we currently know.

And, as it turns out, we know quite a bit. We know the climate is changing, and that this change is due primarily to human activity. We also know that the extent of this change is likely to further alter the transport and biology of invasive species – species whose introduction, establishment and spread are likely to disrupt the world's ecosystems in unpredictable and undesirable ways. We know that this disruption, in turn, will almost certainly alter human welfare, with consequences that range from food security to ocean ecology to forest dynamics.

But while the general outline is known, the details remain elusive. Sadly, part of this is because climate change is still viewed through a political lens and not a scientific glass. Consequently, the resources (students, scientists, equipment, laboratories, etc.) needed to address key questions are lacking. But scientists themselves also shoulder some responsibility. All too often, we revel in the technical and ignore the pragmatic.

While the details may be complex and nuanced, our goal in assembling this book is not. We want to draw attention to the 'big picture'; that global increases in CO_2 mean more than a warm summer day; that abrupt climatic change is likely to act synergistically with other ongoing changes, most notably invasive species biology, and that the subsequent degradation in natural and managed ecosystems should be an increased area of scientific and policy concern.

The challenge now is what to do with this knowledge – how to provide for the desires of a society of 7, soon to be 9, billion – while protecting the biodiversity and ecosystem processes that ensure the planet's capacity to continue to provide for us into the future. Minimizing the degradation of ecosystem services and biodiversity by invasive species is already a challenge; climate change is likely to heighten it. Continued global investment in societal awareness of the problem, the tools to combat it (both scientific and legal) and the active management of invasive species will be critical if we are to minimize irreversible environmental impacts and maintain the ecosystem services needed to satisfy our growing population's needs.

References

Arrhenius, S. (1896) On the influence of carbonic acid in the air upon temperature of the ground. *London, Edinburgh and Dublin Philosophical Magazine and Journal of Science* 5, 237–276.

Bradley, B.A., Blumenthal, D.M., Wilcove, D.S. and Ziska, L.H. (2010) Predicting plant invasions in an era of global change. *Trends in Ecology and Evolution* 25, 310–318.

Bradley, B.A., Blumenthal, D.M., Early, R., Grosholz, E.D., Lawler, J.J., Miller, L.P., Sorte, C.J.B., D'Antonio, C.M., Diez, J.M., Dukes, J.S., Ibanez, I. and Olden, J.D. (2012) Global change, global trade, and the next wave of plant invasions. *Frontiers in Ecology and the Environment* 10, 20–28.

Charles, H. and Dukes, J.S. (2007) Impacts of invasive species on ecosystem services. In: Nentwig, W. (ed.) *Biological Invasions*. Springer, Heidelberg, Germany, pp. 217–237.

Diez, J.M., *et al.* (2012) Will extreme climatic events facilitate biological invasions? *Frontiers in Ecology and the Environment* 10, 249–257.

Dukes, J.S. and Mooney, H.A. (1999) Does global change increase the success of biological invaders? *Trends in Ecology and Evolution* 14, 135–139.

Dukes, J.S. and Mooney, H.A. (2004) Disruption of ecosystem processes in western North America by invasive species. *Revista Chilena de Historia Natural* 77, 411–437.

Dukes, J.S., Chiariello, N.R., Loarie, S.R. and Field, C.B. (2011) Strong response of an invasive plant species (*Centaurea solstitialis* L.) to global environmental changes. *Ecological Applications* 21, 1887–1894.

Fourier, J. (1827) Memoire sur les temperatures du globe terrestre et des espaces planetaires. *Memoires de l'Academie Royale des Sciences* 7, 569–604.

Heimpel, G.E., Yang, Y., Hill, J.D. and Ragsdale, D.W. (2013) Environmental consequences of invasive species: greenhouse gas emissions of insecticide use and the role of biological control in reducing emissions. *PLoS ONE* 8 e72293, doi:10.1371/journal.pone.0072293.

Hellmann, J.J., Byers, J.E., Bierwagen, B.G. and Dukes, J.S. (2008) Five potential consequences of climate change for invasive species. *Conservation Biology* 22, 534–543.

Keenan, T.F., Hollinger, D.Y., Bohrer, G., Dragoni, D., Munger, J.W., Schmid, H.P. and Richardson, A.D. (2013) Increase in forest water-use efficiency as atmospheric carbon dioxide concentrations rise. *Nature* 499, 324–327.

Mohan, J.E., Ziska, L.H., Schlesinger, W.H., Thomas, R.B. and Sicher, R.C. (2006) Biomass and toxicity responses of poison ivy (*Toxicodendron radicans*) to elevated atmospheric CO_2. *Proceedings of the National Academy of Sciences* 103, 9086–9089.

Mooney, H.A. and Hobbs, R.J. (eds) (2000) *Invasive Species in a Changing World*. Island Press, Washington, DC.

Oerke, E.-C. (2006) Crop losses to pests. *Journal of Agricultural Science* 144, 31–43.

Pejchar, L. and Mooney, H.A. (2009) Invasive species, ecosystem services and human well-being. *Trends in Ecology and Evolution* 24, 497–504.

Pielke, R.A. Sr, Adegoke, J.O., Chase, T.N., Marshall, C.H., Matsui, T. and Niyogi, D. (2007) A new paradigm for assessing the role of agriculture in the climate system and in climate change. *Agricultural and Forest Meteorology* 142, 234–254.

Pimentel, D., Zuniga, R. and Morrison, D. (2005) Update on the environmental and economic costs associated with alien-invasive species in the United States. *Ecological Economics* 52, 273–288.

Powell, K.I., Chase, J.M. and Knight, T.M. (2013) Invasive plants have scale-dependent effects on diversity by altering species–area relationships. *Science* 339, 316–318.

Revelle, R. and Suess, H.S. (1957) Carbon dioxide exchanges between atmosphere and ocean and the question of an increase of atmospheric CO2 during the past decades. *Tellus* 9, 18–27.

Solomon, S., *et al.* (eds) (2007) *Climate Change 2007: The Physical Science Basis*. Cambridge University Press, Cambridge, UK.

Vilá, M., Espinar, J.L., Hejda, M., Hulme, P.E., Jarošik, V., Maron, J.L., Pergl, J., Schaffner, U., Sun, Y. and Pyšek, P. (2011) Ecological impacts of invasive alient plants: a meta-analysis of their effects on species, communities, and ecosystems. *Ecology Letters* 14, 702–708.

Walther, G.R., *et al.* (2009) Alien species in a warmer world: risks and opportunities. *Trends in Ecology and Evolution* 24, 686–693.

Webber, B.L. and Scott, J.K. (2012) Rapid global change: implications for defining natives and aliens. *Global Ecology and Biogeography* 21, 305–311.

Ziska, L.H. (2003) Evaluation of the growth response of six invasive species to past, present and future atmospheric carbon dioxide. *Journal of Experimental Botany* 54, 395–404.

Part I

The Dimensions of the Problem: Background and Science

2 Communicating the Dynamic Complexities of Climate and Ecology: Species Invasion and Resource Changes

John Peter Thompson[1] and Lewis H. Ziska[2]

[1]US National Invasive Species Advisory Committee, Washington, DC, USA, and Consultant – Bioeconomic Policy, Prince George's County, Maryland, USA; [2]Crop Systems and Global Change Laboratory, USDA-ARS, Beltsville, Maryland, USA

Les hommes ont oublié cette vérité, dit le renard. Mais tu ne dois pas l'oublier. Tu deviens responsable pour toujours de ce que tu as apprivoisé.[1]

The saddest aspect of life right now is that science gathers knowledge faster than society gathers wisdom.

(Asimov and Shulman, 1988)

Abstract

As the population reaches beyond 7 billion, the impact of human activities on the global environment will begin to alter substantially the complex biological systems necessary to support life. Of particular concern are anthropogenic changes in atmospheric composition that are altering the climatic processes associated with precipitation, temperature and weather disruptions. However, such direct disturbances can also distress ecosystem function indirectly by facilitating the spread and establishment of non-indigenous (invasive) species. Such species, in turn, can overwhelm biological stability by impacting native diversity negatively or, from a human perspective, by reducing resource availability (e.g. agriculture). As a consequence, system resources, from forests to streams to crops, can become increasingly transient, even as population pressure creates additional needs for such resources. Although these pressures are increasingly recognized, knowledge to address the basic and applied needs related to maintaining ecosystem resources is lacking, in part because of communication disparities between scientists and policy makers. Here, the science underlying climate change and invasive species is examined, in broad terms, and the difficulties in eliciting both the attention and means needed to sustain ecosystem services over time are outlined. Overall, societal awareness of the scientific issues will be necessary to provide the global solutions essential to address the dynamic challenges of a changing climate, invasive species and human resource needs.

1 'Men have forgotten this truth', said the fox. 'But you must not forget it. You become responsible, forever, for what you have tamed.'

Humans and Nature

The needs of humanity are linked inexorably to the resources of the earth. These resources, in turn, are provided by biological and physical systems that include food, feed, fibre and fuel, and the respective interactions with temperature, precipitation, soil, etc., that determine the temporal production of these resources. It is these relationships and interactions that provide the societal and functional underpinnings of the human population.

Immediate to human resource needs are the biological systems provided by plants. Plants are the only organisms that are autotrophs, i.e. capable of producing their own energy from physical inputs, and, consequently, will delineate the functional parameters of all biological systems. As such, plants are necessary for animal existence and ecosystem function.

Plants use sunlight, water, carbon dioxide and a suite of nutrients (e.g. nitrogen) to function and reproduce through time. With access of species-specific amounts of these four core ingredients, plant species and communities thrive in a physiological range determined by temperature thresholds. The biochemical and physical relationships of plant biology provide a crucial ecological keystone for the complex and interrelated life systems needed to support a human population of some 7 billion people.

However, given the current global population of 7 billion, human activities, particularly trade, are influencing species introductions and species selection greatly on a panoptic scale. In addition to species introduction, we recognize that humans can also alter the landscape in ways (e.g. deforestation, agriculture) that will, in turn, affect whether a given species introduction becomes established and successful at the regional level. By altering both biological habitat and physical chemistry, humans can reduce forests and grasslands to asphalt, and systematically manage biological landscapes through intentional species selection and accidental introduction as a feature of market externalization.

That human activity can alter species introductions and ecosystem services/ resources does not represent a new concept or insight. James Madison, a founding father and author of the US constitution, was convinced that:

> ... although no determinate limit presents itself to the increase of food, and to a population commensurate with it, other than the limited productiveness of the earth itself, we can scarcely be warranted in supposing that all the productive powers of its surface can be made subservient to the use of man, in exclusion of all the plants and animals not entering into his stock of subsistence; that all the elements and combinations of elements in the earth, the atmosphere, and the water, which now support such various and such numerous descriptions of created beings, animate and inanimate, could be withdrawn from that general destination, and appropriated to the exclusive support and increase of the human part of the creation; so that the whole habitable earth should be as full of people as the spots most crowded now are or might be made, and as destitute as those spots of the plants and animals not used by man.
>
> (Madison, 1818)

Humankind's tomorrows have been linked to plant biology and climate since *Homo sapiens* began to roam the world. Nature itself was arrayed asymmetrically against human endeavours until the coming of agricultural, and the ever-increasing speed of technological knowledge. Such knowledge, in turn, has permitted the economic growth and management of an increasing land area, often by large-scale altering of the existing environment (World Bank, 2012).

Madison knew that the 'faculty of cultivating the earth, and of rearing animals, by which food is increased beyond the spontaneous supplies of nature, belongs to man alone'. There was no doubt in 1818 that the:

> ... relation of the animal part and the vegetable part of the creation to each other, through the medium of the atmosphere, comes in aid of the reflection suggested by the general relation between the atmosphere

> and both. It seems to be now well understood, that the atmosphere, when respired by animals, becomes unfitted for their further use, and fitted for the absorption of vegetables; and that when evolved by the latter, it is refitted for the respiration of the former; an interchange being thus kept up, by which this breath of life is received by each, in a wholesome state, in return for it in an unwholesome one.
>
> (Madison, 1818)

As Madison's words fell mostly on the deaf ears of those unfamiliar with the scientific method, it was little surprise that consideration of the overwhelming complexities of human–environmental interactions were often misunderstood or avoided.

Human-induced Species Changes: A Global Perspective

Now we have arrived at a point when those complexities can no longer be ignored but must, in some capacity, be confronted and addressed as a means to ensure civilization's permanence. Why is this so? What is unique about the current set of circumstances that compels us to address recent anthropogenic impacts?

By an historically unprecedented global movement of plant and animal DNA across borders, and the adoption and proliferation of this DNA on every continent (save Antarctica), we have been able to provide the basic needs (food, clothing, shelter) of 7 billion individuals at the beginning of the 21st century. But these efforts may not be sufficient. In the next few decades, 2 billion more people will be added; and, as with the existing population, they will want their share of food, feed, fibre and fuel. But as global needs expand, environmental resources become constrained. There is not an unlimited supply of energy, of water, of land, of food.

And, as is becoming clear, the cultural and scientific efforts that have been made to meet the needs of the present 7 billion have come with an environmental price. The widespread introduction and distribution of economically desired plants and animals necessary to meet the needs of 7 billion has also resulted in the introduction of species that do great environmental and economic harm. Such species, in a very fundamental way, illustrate Madison's warning of not all plants and animals being 'subservient to the use of man'. Such species are referred to as invasive or, at times, 'exotic' or 'alien'. Officially, the USA defines 'invasive species' by Executive Order 13112 as 'an alien species whose introduction does or is likely to cause economic or environmental harm or harm to human health'. The term 'invasive species' is further clarified and defined as 'a species that is non-native to the ecosystem under consideration and whose introduction causes or is likely to cause economic or environmental harm or harm to human health' (ISAC, 2006; Beck *et al.*, 2008).

What is the extent of the harm or damage associated with such species? Does it really alter our ability to utilize natural resources to meet our needs? Consider two plant species, kudzu (*Pueraria lobata*) and cheatgrass (*Bromus tectorum*) in North America. Each occupies millions of acres of land in the USA, each becomes dominant in the landscape, with a subsequent loss of biodiversity, and each contributes directly to environmental damage: kudzu, as a precursor of tropospheric ozone; cheatgrass as a driver of fire frequency. The economic cost of kudzu is estimated at between US$100–500 million annually (Forseth and Innis, 2004). It has been estimated that cheatgrass infests up to 46 million acres of winter wheat, costing growers about US$300 million annually in lost crop yield, and another US$50–100 million annually in fire damage (Young and Evans, 1978).

Yet there are not just two but approximately 700 invasive plant species in the USA (Ziska and George, 2004). They currently infest around 100,000,000 acres of land. They are spreading at the rate of 3,000,000 acres year^{-1}. Invasive plants, in agriculture, are estimated to cost US$27 billion annually (Pimentel *et al.*, 2005). Invasive pathogens result in another US$20–30 billion in damage for crops, lawns and pastures in the USA; invasive insects another US$15 billion; invasive animals more US$ billions. This is for the USA alone. Estimates of global impacts can run as high as US$1.4 trillion annually, or almost 5% of global GDP (Pimentel *et al.*, 2005).

While the economic costs of invasive species are formidable, the greater expense may be our inability to match population needs with natural resources if, in turn, those resources are becoming increasingly limited by invasive species. How can you grow crops or livestock in South Africa once Parthenium weed (*Parthenium hysterophorus*) is introduced? The weed is toxic to domestic animals and, if eaten, results in tainted meat. It generates allelopathic effects in soils and outcompetes agronomic crops for available nutrients and moisture. *P. hysterophorus* can cover crops with its pollen, which prevents seed set, with productivity losses of up to 40% (Khosla and Sobti, 1979). How will the Middle East maintain wheat production if UG99, a new invasive wheat pathogen, becomes established in this region? What are the implications for global food security (or political stability)? In the USA, emerald ash borers were first detected in 2002. This insect has killed at least 50–100 million ash trees so far and threatens to kill most of the 7.5 billion ash trees throughout North America. The potential economic and environmental damage of this insect rivals that of Chestnut blight (*Cryphonectria parasitica* (Murrill) Barr) and Dutch elm disease (*Ophiostoma* spp.).

Perhaps a single invasive species is incapable of harm on a global scale, but their collective impact may have reached a threshold whereby the current and future resource needs of the global community, from shelter to food, are at risk.

Human-induced Environmental Changes: A Global Perspective

Exploitation and utilization of global resources to meet population needs has resulted in another constraint to natural systems. The need for energy and our primary strategy to meet that need, i.e. the combustion of fossil fuels, has resulted in human-generated increases in the concentration of certain atmospheric trace gases. These trace gases absorb energy in the infrared portion of the spectrum and, as such, are likely to contribute to increasing surface temperatures (IPCC, 2007). In addition to carbon dioxide (CO_2), methane (CH_4) and nitrous oxide (N_2O) are accumulating rapidly in the atmosphere as a result of human activities (IPCC, 2007).

The extent to which the accumulation of these gases will result in temperature increases and the potential consequences with respect to climatic change, from droughts to storm severity to sea level rise, have been noted extensively in the scientific and popular literature (IPCC, 2007). In addition to changes in the physical environment, there is also widespread agreement that the projected increases in atmospheric carbon dioxide can significantly stimulate the growth, development, reproduction and management of a wide range of plant species, including a number of invasive plant species (Ziska, 2003; Dukes and Mooney, 2004; Manea *et al.*, 2011). CO_2-induced changes in plant biology, in addition to impacting invasive plant species directly, will likely have an indirect effect on invasive insect and pathogen relationships with the plant hosts.

Globally, climate change and invasive species represent significant threats to our natural resource base, and the challenges of adapting to each are recognized in their own right. Less recognized, or appreciated, is that the twin threats to resource sustainability are not isolated from each other. That is, climate change and rising CO_2 levels will alter the biology of invasive species significantly, with subsequent changes regarding both resource availability and invasive species management.

Writing more than a decade ago, Mooney and Hobbs (2000) recognized the importance of these two anthropogenic factors and noted that human-induced rates of change for both climate and ecosystems were unprecedented in geological history. Each aspect of change is by itself, capable of significant biological change. But we are increasingly aware that it is their combined effects that are of key concern in the context of availability, development and allocation of the natural resources necessary for the growth and sustainability of human systems (Mooney and Hobbs, 2000).

Complexity and Uncertainty

As ecosystems are being transformed into new, non-historical configurations owing to a rapidly expanding variety of local and global anthropogenic changes (Hobbs *et al.*, 2009), the consequences regarding the introduction and spread of invasive species are putting enormous pressure on traditional ecosystem conservation values and policies (Collins and Crump, 2009; Minteer and Collins, 2010). Such unprecedented rapid changes are likely to alter ecosystem integrity and the subsequent resources and services accessible to human communities.

The synergistic linkage between climate change and invasive species biology is likely to be a new fundamental driver of ecosystem integrity and functionality. Indigenous species that have evolved together to form elaborate, complex patterns of strong and weak interactions that are temporally stable now compete with rapidly changing climatic conditions, as well as recently introduced species, some of which are more competitive, in part, because of those conditions. Should those conditions result in an overwhelming competitive advantage for one species, complex adaptive biological systems are likely to be reduced to simple systems where one or two species dominate (e.g. kudzu in the southern USA; Forseth and Innis, 2004). Loss of complexity, in turn, is likely to result in loss of diversity and ecosystem resilience to abrupt climatic change (Reich *et al.*, 2001).

The vulnerability of ecosystems to invasive introductions can vary with climate. Some introduced invasive species in certain ecosystems, such as the Argentine ant, (*Linepithema humile* Mayr), may be impacted negatively by changes in climatic patterns (Cooling *et al.*, 2012), while others, such as buffelgrass, *Pennisetum ciliare* L., may continue to increase their range, replacing current ecosystem communities (Franklin *et al.*, 2006). The sensitivity of species is not limited to temperature and precipitation; it is also affected by changes in atmospheric CO_2, such as those now altering the grasslands (savannahs) of Africa, which, within a century, may be transformed into forests (Higgins and Scheiter, 2012).

What factors underlie species transitions associated with abrupt changes in climate and the potential success of biological invaders for a given ecosystem? Specific outcomes will be difficult to predict since climate change and invasive species are likely to interact both spatially and temporally to alter functional ecosystem integrity. Characterizations of likely impacts would reflect the geospatial occurrence of the system (i.e. ecosystems over thousands of square kilometres versus those found in limited physical space), the degree of species diversity (i.e. rare species that occupy a small physical niche versus multi-species communities adapted to a wide range of abiotic conditions), as well as the rate of climatic change (i.e. gradual versus rapid extreme changes in temperature and precipitation) and, finally, human activity (i.e. the extent of global trade and the adoption and implementation of management to eliminate new pest threats). That all of these actions can occur singly and in combination illustrates the dynamic complexity and uncertainty associated with predicting climate and invasive impacts on system function. While specific outcomes may be difficult to forecast, there are some general empirical effects that are likely to occur.

Spatially, there is a range of biotic impacts related to specific native species, including physiological, phenological and distributional changes (Minteer and Collins, 2010). Climatic change not only would affect the composition of native species within ecosystems, but also would affect key aspects of invasive species biology such as introduction, establishment, demography and distribution (Bardsley and Edwards-Jones, 2007).

The temporal aspects of climatic change are also likely to alter biological systems by changing multiple and novel aspects that range from introduction to reproduction. For example, the ability of invasive species to become dispersed into new ecosystems is, in large part, determined by species movement between similar environmental parameters, as necessitated by the increasing global trade in goods and services (Olson, 2006). As the northern ice cap is reduced, and trade is expanded via the Northwest Passage, it is likely that species which previously had found this environment too extreme could become established. In addition, the speed of species change in a community is increased by the number of novel or non-indigenous species introductions (Olden *et al.*, 2010); for example, the rapid introduction of invasive species in the Arctic tundra with increased shipping. It is also possible that faster and more extreme changes in weather patterns could eliminate plant communities altogether, allowing greater opportunities for invasion (Hierro *et al.*, 2006); for example, increased tundra fires with warmer temperatures.

While such scenarios are plausible, the interconnectivity and interdependent relationships of climate, ecosystems and invasive species result in complexities that resist simplified absolute solutions. However, because ecosystems are defined by climate, anthropogenic behaviours (e.g. land-use change, anthropogenic climate change) will likely alter ecosystem function. As a consequence, the expected services and resources of a given ecosystem will be modified in response to a shifting biogeographical state, independently of any impact of invasives on ecosystem functioning per se. This points to a potential and rapid reshuffling of agricultural and environmental services and resources such as food, fuel, feed, fibre, flowers and forests that reflects a quickly changing environment and the resulting land-use decision pressures (Thompson, 2010). The risk of rapid system change with unpredictable short- and long-term outcomes related to a greater impact of invasive species is heightened in any case. This is likely to elevate the risk of economic or environmental harm, or harm to human health.

Emerging Consequences

Predictive, theoretical interactions aside, it is clear that there are, in fact, a number of ongoing regional examples where climatic change already appears to be impacting invasive species biology, specifically their establishment, spread and degree of ecosystem damage. In that regard, the following examples are meant to be illustrative and not inclusive.

Kudzu (*Pueraria montana* (Lour.) Merr. var. *lobata* (Willd.) Maesen and S. Almeida)

One of the most interesting forecasts regarding global warming and an invasive plant was made almost two decades ago in regard to the northward migration of kudzu (*P. lobata*), a ubiquitous invasive of the south-eastern USA. Tom Sasek and Boyd Strain at Duke University observed that the latitudinal distribution at that time was limited to regions south of the Mason–Dixon line by low winter temperatures of –15°C (Figure 7 in Sasek and Strain, 1990). A more recent evaluation of its current distribution suggests that kudzu has moved northward. There are now Midwestern US populations (and the recent discovery of kudzu in southern Canada), and the migration may be associated with an increase in minimum winter temperatures (Ziska *et al.*, 2011). Such migratory movements in response to reduced winter temperatures are likely to occur for other invasive plants. However, the extent and migratory range of invasive plant species in response to temperature is still uncertain. For example, Bradley *et al.* (2010) used bioclimatic envelope modelling to show that climate change could result in both range expansion and contraction for five widespread and dominant invasive plants in the western USA. Hence, while some invasive plant species could become an additional problem with climate change, it is also possible that a northern shift in their range could create restoration opportunities as well.

California grasslands

A study by Sandel and Dangremond (2012) used a trait climate relationship between invasive and native grasses to demonstrate that warmer areas contained a higher proportion of exotic species. The observable pattern was consistent with a simple model that was able to predict invasion severity. The study provided an *in situ* evaluation regarding rising temperatures and species invasions over a broad geographic and taxonomic scale.

Invasive insects in China

In recent decades, evidence of greater invasive species establishment has been observed for China (Lin *et al.*, 2007). Although increases in global trade are thought to be associated with the increased dispersion of such species, Chinese and USDA researchers have also examined the role of warming temperatures from 1900 through 2005 on invasive alien insect establishment (Huang *et al.*, 2011). Their findings indicated that for every 1°C increase in average annual surface temperature for mainland China, there was an increase in the establishment rate of invasive alien insects of approximately 0.5 species year^{-1}. This relationship was still significant even after accounting for increases in global trade from 1950 to 2005 (Huang *et al.*, 2011).

Pine bark beetle

Although the mountain pine bark beetle (MPB; *Dendroctonus ponderosae* Hopkins) is native to North America, human-induced changes in climate may also result in dispersion of the beetle into new geographical regions where it would be considered invasive. In recent decades, a widespread increase in tree mortality has been observed in western North America (Bentz *et al.*, 2010). This increase, in turn, has been associated with a large increase in MPB populations (Aukema *et al.*, 2008; Logan and Powell, 2009). The expansion of the MPB is a function of increased reproductive output, indicating that MPB is tracking climate change, specifically warming winters and a longer reproductive season, with subsequent consequences on tree mortality (Mitton and Ferrenberg, 2012). These consequences are not trivial. To date, more than 13 million ha of trees have been killed in British Columbia alone (Kurz *et al.*, 2008). If this dieback continues, conifer forests may change from regional carbon sinks to carbon sources, providing positive feedback for anthropogenic climate change.

Ready or Not

While additional information is greatly desired, the interactions of climate change and invasive species biology to date suggest that detrimental effects on ecosystem function and long-term health are likely to be exacerbated by invasive species. This is a particularly egregious occurrence, given that population pressures for additional resource needs from such systems are likely to increase dramatically during this century.

Overall, the scope and impact of these probable changes stress the tremendous need for a thorough scientific evaluation of probable links, potential outcomes and quantification of changes in ecosystem resources. There is an equally critical need to begin assessments of how land managers, consumers, policy makers, etc., can mitigate (long term) and adapt (short term) to the twin challenges of climate change and invasive species.

Given the continuous warnings issued by scientists, beginning with NASA's Jim Hansen in 1988 (Hansen *et al.*, 1988), and given the long-standing recognition of the environmental and economic damage inflicted by invasive species, human societies have had ample lead time to recognize and prepare appropriate cogent responses to adapt or lessen these impacts. Overall, rapid adaptability to geospatial and biotic circumstance has been a consistent feature of human societies (Diamond, 1997).

Lost in Translation

Yet, it appears that, even after decades, policy makers in the USA and elsewhere are ill-prepared to acknowledge the consequences of these global circumstances or to marshal the technical and scientific tools necessary to address them. While scientists, by large margins, recognize the severity of these global changes, they cannot motivate populations to take action unless they achieve the support of global leaders in industry, policy, etc. To achieve this type of motivation, issues raised by science must become internalized in economic and land-use policy.

James Madison noticed the tendency to lapse into habit when working with the land (Madison, 1818). The threat of change in the face of uncertain choice creates a state of political paralysis. 'The uncertainty that exists in evaluating strategies for adapting to climate change', according to Camacho (2010), 'is not merely scientific; there is considerable uncertainty regarding what should be the priorities and guiding principles for managing natural resources.' The problem is summarized succinctly by Mooney:

> In spite of this relatively precise language and three subsequent science assessments that have dramatically reduced the uncertainty that humans are driving these rapid changes, and in spite of the fact that CO_2 increase in the atmosphere is still accelerating, we still have not agreed upon an international action to meet the objectives of this convention, although there are the beginnings of dramatic changes in individual practices in many parts of the world; however, these are driven, initially at least, more by the rising price of energy rather than the threat of global warming.
>
> (Mooney, 2010)

How, then, can science translate and communicate the severity of the threat in a way that achieves consensus and support for the lay public? Science may offer an evaluation as to the severity of potential outcomes, but as is clear with climate change per se, policy will lag behind science. Is the response of policy makers simply caution in the face of uncertain predictions or a dangerous procrastination driven by political and economic constraints?

While scientific research may provide a perspective of the impacts of simultaneous changes in climate and invasive species on ecosystems, the meaning of these impacts with respect to altering human activities will be viewed with uncertainty. The desire for certainty can become a tool for opposition to scientific recommendations that might restrain the externalization of costs. When politically useful, the idea of 'good enough' is dismissed as not 'certain enough'. It is analogous to not knowing the exact value of pi, but knowing it well enough to calculate anything we might need to know, yet, in regard to climate change or invasion biology, needing much more rigorous and absolute answers before any action is supported. Unlike mathematical solutions, rapid changes in plant communities resulting from climatic patterns and invasive species will produce outcomes that are not entirely predictable; and are likely to reinforce uncertainty at the policy-maker level.

But as evidence of an anthropogenic signal associated with climatic change continues to grow, the rate of non-indigenous species movement into novel ecosystems is a significant concern. The persistent chronic disturbance of the land, the reshaping of landscapes, the externalization of economic development into the soil, air and water are altering the complex adaptive systems that support life as we know it. For example, some models of future plant distributions show that a temperature rise of 2–3°C over the next 100 years could result in half the world's plant species being threatened with extinction (Bramwell, 2007). Such a rate of extinction will be outside of any human experience, and the effect on higher animal species will be beyond calculation.

The fact of change, and the corresponding value of the resources that may be extracted following change, are often definitionally confounded. The measurable facts and the depth of belief obscure easy definitions and therefore colour *a priori* any policy or management debates. Often, for the lay public, definitions of the two systems, climatic and biological, will vary according to preconceived solutions. The result is a disconnected series of actions and inactions. The complexity of the two systems, taken individually or as a whole, create a critical and crucial challenge for the public as it works to identify key guidelines and metrics for decision making that minimizes risk under future climate scenarios and novel ecosystem complexes (Loss *et al.*, 2011).

Frustrating decision making at the political and sociological level is the relationship between climatic patterns and species distribution in ecosystems. These complex interconnections are neither continuously gradual nor continuously chaotic. Future valuations are most likely hyperbolic rather than exponential. Paraphrasing Bertrand Russell, the closer one looks for certainty, the less clear definitions and their boundary conditions become (Russell, 1918). Without certainty of a connection between climatic perturbation and species biology, policy makers and land managers will be hesitant to recognize and adapt accordingly.

There is no decision-making process, for example, to address the problem of climate change-induced species migration. Plant species on which the entirety of local ecosystems is based will respond by 'adapting their phenology or physiological responses to new climatic conditions via selection, migrate to a more suitable climate, or go extinct' (Davis and Shaw, 2001). Although some interdisciplinary research is moving towards a synergistic set of conservation, there is no policy support for sentinel sites in existing arboreta that could help address the concerns and issues of climate, migration and/or extinction (Coenen *et al.*, 2008; Donaldson, 2009; Primack and Miller-Rushing, 2009). Even specific examples of mounting economic loss from habitat change due to invasive species alone are obscured in the political sphere and the research left unconsidered (Pimentel *et al.*, 2005). Unfortunately, the rising level of landscape (scientific) illiteracy and the policy-makers' need for absolute assurance will continue to support public inaction on a broad array of issues important to climate, invasive species biology and resource utilization.

Bridging Uncertainty

Between the chasm of scientific consensus and public action, it is clear that there is empirical agreement on a fundamental environmental issue: humans are, in fact, altering the composition and integrity of the biosphere (IPCC, 2007). While the degree and rate of alteration is subject, rightfully, to debate, is quantification of every single variable and potential outcome necessary before society responds? How can scientific reality be translated into positive policy action?

Scientists are familiar with uncertainty; indeed, almost every published scientific paper contains data analysis and interpretation, providing 'confidence' regarding the probability of an outcome. However, the public and policy makers, not familiar with the nuances of statistics, often seek 'certainty' and prosaic solutions. It is the overlaying of these standards to those of science that can generate critical constraints that can prevent any coordinated marshalling of societal resources to meet a given threat (Table 2.1).

Scientists recognize that both climate change and invasive species involve large-scale, complex systems and possible outcomes are intrinsically more uncertain. Yet, it is also clear that this ambiguity, at least on the part of the scientific community, has diminished. For example, global scientific collaborations involving hundreds of scientists, as represented by the Intergovernmental Panel on Climate Change (IPCC), have reduced the degree of uncertainty regarding the impact of human activities on climatic trends with each successive report (IPCC, 2007, *inter alia*). Yes, these reports do not reflect the unanimous opinion of the entire scientific community. However, at present, there is a ~98% consensus among climatologists and agreement among the national academies of 32 countries as to a clear human influence

Table 2.1. Behavioural and attributional characteristics of science and government. Differences in problem perceptions can contribute to how environmental challenges are translated into policy decisions. (Adapted from Bradshaw and Borchers, 2000).

Science	Government
Acceptance of probability	Desire for certainty
Long-term interest	Interest may end with election
Replication necessary for belief	Beliefs may be situational
Oriented to solving problems	Service oriented
Acceptance of risk and failure	Risk and failure not tolerated
Innovation is esteemed	Innovation is suspect
Diffuse or non-existent clientele	Specific clientele, immediate and insistent

on global climate (Oreskes, 2004; Anderegg *et al.*, 2010). Such a concurrence among experts should provide, at least from a scientific viewpoint, the functional equivalent of 'the earth orbits the sun'.

Of course, there will be individuals, including scientists, whose strong personal beliefs will prevent the acceptance of any evidence not in conformity with those beliefs. The impact of individual beliefs constrains policy and the politicians that drive the process. John Locke wrote, '[n]ew opinions are always suspected, and usually opposed, without any other reason but because they are not already common' (Locke, 1639). Joseph Overton explained political resistance to new ideas in a given public policy area, such as agriculture, by noting that only a small, narrow range of potential policies could be politically acceptable at any given time (http://www.mackinac.org/OvertonWindow). The 'window' of politically acceptable options is described not by what politicians prefer, but rather by what they believe they can support and still win re-election. It follows, then, that science needs to infuse ideas and information within society before politicians can or will act. That is, society shifts to include different policy options not when ideas change among politicians, but when ideas change in the society that elects them.

There is a need, therefore, for a process that integrates scientific philosophy as well as data into societal thinking. The role of planetary gardeners and stewards of the earth is an uncomfortable one for many who see such adaptive solutions as little more than the 'familiar human arrogance toward nature that has carved deep gashes in the landscape and defined our modern environmental history' (Minteer and Collins, 2010). The non-linear nature of the problem and the fact that the two problems are intertwined with human-induced ecosystem disturbance as a means to maintain the societal status quo begs for a preliminary discussion of the philosophic differences between the precautionary principle and other context-setting systems such as the pro-actionary principle (More, 2004). The 'wicked' nature of both climate change and invasion biology provides a pathway for differing opinions and beliefs to work together towards a common objective. James Madison clearly saw what we did not: that what we take from an ecosystem must eventually be put back, or the system will take it back without us:

> The sum of these remarks on cultivating poor lands, and neglecting the means of keeping or making land rich, is, that if every thing grown on a soil is carried from it, it must become unproductive; that if every thing grown on it be directly or indirectly restored to it, it would not cease to be productive; and, consequently, that according to the degree in which the one or the other practice takes place, a farm must be impoverished, or be permanently productive and profitable.
>
> (Madison, 1818)

Fortunately, there are means to reconcile opposing points of view between science and policy. One of the most common is to increase public confidence by increasing communication. For example, since the 1960s, a number of environmental laws have been prompted by increasing scientific awareness of the issue by the public at large, particularly when the issue can be viewed with respect to public health (Wynne, 1995). When scientists and managers inform and involve public interest, policy outcomes are more likely to be consensus based and less likely to undergo legal challenges from disaffected groups (Johnson and Campbell, 1999). Bradshaw and Borchers (2000) have also proposed a realignment whereby the definition of scientific uncertainty could be regarded in a policy context as it is in science, i.e. as critical information for hypothesis building, setting priorities and making decisions.

The main challenge is that policy is incremental while science is episodic. Given the scope and scale of the environmental impacts that will alter the natural resource base on which humans depend, do we have sufficient time for our leaders to obtain scientific unanimity before a crucial threshold is irrevocably passed? Is it possible to decide in the face of uncertainty?

Moving Forward

Humankind, along with all the flora and fauna of the world, is living through a period of rapid ecological change brought about, in part, by their own actions. The quantity of non-indigenous species is increasing rapidly worldwide, causing both environmental and economic damage (Leung *et al.*, 2002; Hellmann *et al.*, 2008). If the numbers of invasive species are added to the predictions regarding climate change thresholds, cascades and tipping points (e.g. Lenton *et al.*, 2008), societies may have very little time to prepare for implementing, let alone understanding, needed adaptations and adjustments to agricultural practices, technologies and ecosystem service expectations.

Bridging the communication gap and conveying urgency to world leaders will be fundamental in addressing these environmental challenges. However, those researchers, environmentalists, land managers, ecologists and concerned citizens who are already familiar and working on adaptation and mitigation efforts related to climate change and invasive species can provide a suitable knowledge platform on which to begin. Their ongoing efforts, with a minimum of financial support and, at times, harassment from vested interests, must continue. In the spirit of bridging the disparities between science and policy, with the goal of making plain the science of climate change and invasive biology interactions, and the immediacy of their potential threats to humanity, such endeavours are incalculable to the world at large.

The information and discussion in this work begin the process of offering the productive strategies and tactics that will be needed to address the demands of tomorrow before they constrain the choices of today. Isaac Newton reportedly said that to 'explain all nature is too difficult a task for any one man or even for any one age', but, he continued, it is 'much better to do a little with certainty, and leave the rest for others that come after you, than to explain all things by conjecture without making sure of anything'.

References

Anderegg, W.R.L., Prall, J.W., Harold, J. and Schneider, S.H. (2010) Expert credibility in climate change. *Proceedings of the National Academy of Sciences* 107, 12107–12109.

Asimov, I. and Shulman, J.A. (1988) *Isaac Asimov's Book of Science and Nature Quotations.* Weidenfeld and Nicolson, New York.

Aukema, B.H., Carroll, A.L., Zheng, Y., Zhu, J., Raffa, K.F., Moore, R.D., Stahl, K. and Taylor, S.W. (2008) Movement of outbreak populations of mountain pine beetle: influences of spatiotemporal patterns an climate. *Ecography* 31, 348–358.

Bardsley, D.K. and Edwards-Jones, G. (2007) Invasive species policy and climate change: social perceptions of environmental change in the Mediterranean. *Environmental Science and Policy* 10(3), 230–242.

Beck, K.G., Zimmerman, K., Schardt, J.D., Stone, J., Lukens, R.R., Reichard, S., Randall, J., Cangelosi, A.A., Cooper, D. and Thompson, J.P. (2008) Invasive species defined in a policy context: recommendations from the Federal

Invasive Species Advisory Committee. *Invasive Plant Science and Management* 1, 414–421.

Bentz, B.J., Régnière, J., Fettig, C.J., Hansen, E.M., Hicke, J., Hayes, J.L., Kelsey, R., Negrón, J. and Seybold, S. (2010) Climate change and bark beetles of the western US and Canada: direct and indirect effects. *Bioscience* 60, 602–613.

Bradley, B.A., Blumenthal, D.M., Wilcove, D.S. and Ziska, L.H. (2010) Predicting plant invasions in an era of global change. *Trends in Ecology and Evolution* 25, 310–318.

Bradshaw, G.A. and Borchers, J.G. (2000) Uncertainty as information: narrowing the science–policy gap. *Conservation Ecology* 4, 7–14.

Bramwell, D. (2007) The response of botanic gardens to climate change. *Botanical Garden Journal* 4, 2.

Camacho, A.E. (2010) Assisted migration: redefining Nature and natural resource law under climate change. *Yale Journal on Regulation* 10, 171–253.

Coenen, D., Porzecanski, I. and Crisman, T.L. (2008) Future directions in conservation and development: incorporating the reality of climate change. *Biodiversity* 9, 106–113.

Collins, J.P. and Crump, M.L. (2009) *Extinction in Our Times: Global Amphibian Decline.* Oxford University Press, Oxford, UK.

Cooling, M., Hartley, S., Sim, D.A. and Lester, P.J. (2012) The widespread collapse of an invasive species: Argentine ants (*Linepithema humile*) in New Zealand. *Biological Letters* 8, 430–433.

Davis, M.B. and Shaw, R.G. (2001) Range shifts and adaptive responses to Quaternary climate change. *Advancement of Science* 292, 673–679.

Diamond, J.M. (1997) *Guns, Germs and Steel: The Fate of Human Societies.* W.W. Norton and Company, New York.

Donaldson, J.S. (2009) Botanic gardens science for conservation and global change. *Trends in Plant Science* 14, 608–613.

Dukes, J.S. and Mooney, H.A. (2004) Disruption of ecosystem processes in western North America by invasive species. *Revista Chilena de Historia Natural* 77, 411–437.

Forseth, I.N. and Innis, A.F. (2004) Kudzu (*Pueraria montana*): history, physiology, and ecology combine to make a major ecosystem threat. *Critical Reviews in Plant Science* 23, 401–413.

Franklin, K.A., Lyons, K., Nagler, P.L., Lampkin, D., Glenn, E.P., Molina-Freaner, F., Markow, T. and Huete, A.R. (2006) Buffelgrass (*Pennisetum ciliare*) land conversion and productivity in the plains of Sonora, Mexico. *Biological Conservation* 127, 62–71.

Hansen, J., Fung, I., Lacis, A., Rind, D., Lebedeff, S., Ruedy, R., Russell, G. and Stone, P. (1988) Global climate changes as forecast by Goddard Institute for Space Studies three-dimensional model. *Journal of Geophysical Research* 93, 9341–9364.

Hellmann, J.J., Byers, J.E., Bierwagen, B.G. and Dukes, J.S. (2008) Five potential consequences of climate change for invasive species. *Conservation Biology* 22, 534–543.

Hierro, J.L., Villarreal, D., Ozkan, E., Graham, J.M. and Callaway, R.M. (2006) Disturbance facilitates invasion: the effects are stronger abroad than at home. *American Naturalist* 168, 144–156.

Higgins, S.I. and Scheiter, S. (2012) Atmospheric CO_2 forces abrupt vegetation shifts locally, but not globally. *Nature* 27, 1–14.

Hobbs, R.J., Higgs, E. and Harris, J.A. (2009) Novel ecosystems: implications for conservation and restoration. *Trends in Ecology and Evolution* 24, 599–605.

Huang, D., Haack, R.A. and Zhang, R. (2011) Does global warming increase establishment rates of invasive alien species? A centurial time series analysis. *PLoS ONE* 6, e24733.

IPCC (2007) *Climate Change 2007: Impacts, Adaptation and Vulnerability.* IPCC Secretariat, Geneva, Switzerland.

ISAC (2006) Invasive Species Definition Clarification and Guidance White Paper (http://www.invasivespeciesinfo.gov/docs/council/isacdef.pdf, accessed 31 December 2013).

Johnson, B. and Campbell, R. (1999) Ecology and participation in landscape-based planning within the Pacific Northwest. *Journal of Policy Studies* 27, 502–509.

Khosla, S.N. and Sobti, S.N. (1979) Parthenium – a national health hazard, its control and utility – a review. *Pesticides* 13, 121–127.

Kurz, W.A., Stinson, G., Rampley, G.J., Dymond, C.C. and Neilson, E.T. (2008) Risk of natural disturbances makes future contribution of Canada's forests to the global carbon cycle highly uncertain. *Proceedings of the National Academy of Sciences* 105, 1551–1555.

Lenton, T.M., Held, H., Kriegler, E., Hall, J.W., Lucht, W., Rahmstorf, S. and Schellnhuber, J. (2008) Tipping elements in the Earth's climate system. *Proceedings of the National Academy of Sciences* 7, 1786–1793.

Leung, B., Lodge, D.M., Finnoff, D., Shogren, J.F., Lewis, M. and Lamberti, G. (2002) An ounce of prevention or a pound of cure: bioeconomic risk analysis of invasive species. *Proceedings of the Royal Society, Biology* 269, 2407–2413.

Lin, W., Zhou, G., Cheng, X. and Xu, R. (2007) Fast economic development accelerates biological invasions in China. *PLoS ONE* 2, e1208.

Locke, J. (1639) An Essay Concerning Human Understanding: With Thoughts on the Conduct of the Understanding (1639).

Logan, J.A. and Powell, J.A. (2009) Ecological consequences of climate change in altered forest insect disturbance regimes. In: Wagner, F.H. (ed.) *Climate Warming in Western North America: Evidence and Environmental Effects.* University of Utah Press, Salt Lake City, Utah.

Loss, S.R., Terwilliger, L.A. and Peterson, A.C. (2011) Assisted colonization: integrating conservation strategies in the face of climate change. *Biological Conservation* 144, 92–100.

Madison, J. (1818) Address to the Agricultural Society of Arberlmarle Virginia. In: *Letters and Other Writings of James Madison.* J.B. Lippincott & Co, by order of Congress 1865, Philadelphia, Pennsyvania, pp. 63–96.

Manea, A., Leishman, M.R. and Downey, P.O. (2011) Exotic C_4 grasses have increased tolerance to glyphosate under elevated carbon dioxide. *Weed Science* 59, 28–36.

Minteer, B.A. and Collins, J.P. (2010) Move it or lose it? The ecological ethics of relocating species under climate change. *Ecological Applications* 20, 1801–1804.

Mitton, J.B. and Ferrenberg, S.M. (2012) Mountain pine beetle develops an unprecedented summer generation in response to climate warming. *American Naturalist* 179, 163–171.

Mooney, H.A. (2010) The ecosystem-service chain and the biological diversity crisis. *Phiosophical Transactions of the Royal Society, Biology* 365, 31–39.

Mooney, H.A. and Hobbs, R.J. (2000) *Invasive Species in a Changing World.* Island Press, Washington, DC.

More, M. (2004) The Proactionary Principle, Version 1.0: Draft for public comment (http://www.extropy.org/proactionaryprinciple.htm, accessed 31 December 2013).

Olden, J.D., Kennard, M.J., Lawler, J.J. and Poff, N.L. (2010) Challenges and opportunities in implementing managed relocation for conservation of freshwater species. *Conservation Biology* 25, 10–17.

Olson, L.J. (2006) The economics of terrestrial invasive species: a review of the literature. *Agricultural and Resource Economics Review* 35, 178–194.

Oreskes, N. (2004) The scientific consensus on climate change. *Science* 306(5702), 1686, doi:10.1126/science.1103618.

Pimentel, D., Zuniga, R. and Morrison, D. (2005) Update on the environmental and economic costs associated with alien-invasive species in the United States. *Ecological Economics* 52, 273–288.

Primack, R.B. and Miller-Rushing, A.J. (2009) The role of botanical gardens in climate change research. *New Phytologist* 182, 303–313.

Reich, P.B., *et al.* (2001) Plant diversity enhances ecosystem responses to elevated CO_2 and nitrogen deposition. *Nature* 410, 809–812.

Russell, B. (1918) *The Philosophy of Logical Atomism (1918).* Routledge, 1985. 2009 edn, Taylor and Francis e-Library, New York and London.

Saint Exupéry, A. de (1943) Le Petit Prince: Chapitre XXI (http://microtop.ca/lepetitprince/chapitre21.html, accessed 31 December 2013).

Sandel, B. and Dangremond, E.M. (2012) Climate change and the invasion of California by grasses. *Global Change Biology* 18, 277–289.

Sasek, T.W. and Strain, B.R. (1990) Implications of atmospheric CO_2 enrichment and climatic change for the geographical distribution of two introduced vines in the USA. *Climatic Change* 16, 31–51.

Thompson, J.P. (2010) Climate Change and Invasive Species – a Dynamic Relationship (http://ipetrus.blogspot.com/2010/01/climate-change-invasive-species-dynamic.html, accessed 31 December 2013).

World Bank, (2012) *Inclusive Green Growth; The Pathway to Sustainable Development.* International Bank for Reconstruction and Development, Washington, DC.

Wynne, B. (1995) Public understanding of science. In: Jasanoff, S., Markle, G.E., Petersen, J.G. and Pinch, T. (eds) *Handbook of Science and Technology Studies.* Sage Publications, London, pp. 366–388.

Young, J.A. and Evans, R.A. (1978) Population dynamics after wildfires in sagebrush grasslands. *Journal of Range Management* 31, 283–289.

Ziska, L.H. (2003) Evaluation of the growth response of six invasive species to past, present and future carbon dioxide concentrations. *Journal of Experimental Botany* 54, 395–404.

Ziska, L.H. and George, K. (2004) Rising carbon dioxide and invasive, noxious plants: potential threats and consequences. *World Resource Review* 16, 427–447.

Ziska, L.H., Blumenthal, D.M., Runion, G.B., Hunt, E.R. Jr and Diaz-Soltero, H. (2011) Invasive species and climate change: an agronomic perspective. *Climatic Change* 105, 13–42.

3 Climate Change and Plant Pathogen Invasions

Karen A. Garrett,[1,3] Sara Thomas-Sharma,[1] Greg A. Forbes[2] and John Hernandez Nopsa[1,3]

[1]*Department of Plant Pathology, Kansas State University, Manhattan, Kansas, USA;* [2]*International Potato Center, Beijing, China;* [3]*Plant Biosecurity Cooperative Research Centre, Bruce, ACT, Australia*

Abstract

Climate has an important effect on plant disease and the probability of plant pathogen invasions, through effects on hosts and pathogen vectors as well as on the pathogens themselves. Aerially dispersed pathogens are an important group of plant pathogens, and their dispersal and invasion may be modified by changes in wind patterns. Pathogens vectored by arthropods may be affected by weather impacts on their vectors, often through the filter of vector behaviour. Soilborne pathogens have more challenges to rapid invasion, but human transport can introduce them quickly into novel settings. For pathogens, variability within a species may be of great importance, and many important pathogen invasions are invasions of new genotypes of ubiquitous pathogen species. The connectivity of a landscape for pathogen movement is determined by the spatial distribution of host, pathogen and environmental conditions, and connectivity may also be affected by climate change. Therefore, when there is a new invasion, it is often challenging to evaluate which of these factors was limiting prior to invasion. Most economically important pathogen invasions are influenced strongly by human decision making. Adaptation strategies are needed both to address increased disease risk and to manage pathogen species and important subpopulations before they are introduced and established.

Climate Change and Pathogen Invasions: An Overview

The relationship between weather and plant disease risk

Plant pathologists have long recognized that a susceptible host and virulent pathogen can cause plant disease epidemics only under favourable environmental conditions. Observations in this regard date back to 300 BC, when Theophrastus observed that wheat losses to rust varied with wind intensity (discussed by Dixon, 2012). Today, over 2000 years later, we continue to evaluate the influence of weather on disease epidemics, now under climate change scenarios. Studies of weather on plant disease vary in the scale of observation, i.e. microclimate changes versus macroclimate (local, regional, continental and global) changes. For example, dry microclimates are predicted to limit epidemics of bacteria that infect leaves by reducing leaf wetness, while rising global temperatures can promote epidemics of heat-loving bacteria like *Ralstonia solanacearum* (Jones and Barbetti, 2012). Similarly, wind speed and direction, greenhouse gas concentrations and altered rainfall patterns change the risk of plant disease epidemics by affecting the biology, physiology, population dynamics and ecological relationships among biota important for disease (hosts, pathogens, vectors, weed

  Invasive Species and Global Climate Change (eds L.H. Ziska and J.S. Dukes)

hosts, predators, symbionts, etc.) in these ecosystems (Coakley *et al.*, 1999; Garrett *et al.*, 2006).

The importance of plant disease

Global food insecurity, marked by a growth in the number of hungry people from 800 million in 1996 to over 1.02 billion in 2009 (Hoffmann, 2011), has many social, economic and political causes (Chakraborty and Newton, 2011). Agricultural uncertainties like poor market access, food price fluctuations, inferior food distribution systems, inadequate technology transfer and crop failure are some of the numerous problems; but plant diseases that can destroy 10–15% of the global harvest are a particular concern (Oerke, 2006; Chakraborty and Newton, 2011). The Irish potato famine of the 1840s, for example, was caused by potato late blight (causal agent: *Phytophthora infestans*) and resulted in the loss of millions of lives. Similarly, the great Bengal famine of 1943, caused by *Bipolaris oryzae* infecting rice, resulted in 2 million deaths. Now, under the pressures of climate change, plant diseases continue to threaten crops and food security (Gregory *et al.*, 2009; Chakraborty and Newton, 2011; Luck *et al.*, 2011; Dixon, 2012).

What makes disease different from other biological invasions?

Disease represents the interaction between at least two species, host and pathogen. SIR (susceptible-infected-resistant) models are often used to describe epidemics in terms of the processes that move host individuals from the susceptible-but-not-yet-infected class to the infected class and to the resistant (no longer susceptible because of conditioning) class (Gilligan and van den Bosch, 2008). Because of the key role of weather in determining the probability that host individuals will move between these classes, climate change may have important effects on these processes. For plant disease, there are many different forms of disease transmission, such as transmission by vectors or by passive movement of pathogen propagules. The likelihood of transmission may also be influenced by the environmental conditioning of the hosts, such as when temperature affects the function of disease resistance genes (Webb *et al.*, 2010).

Recent important invasions

Cassava, an important staple food crop of sub-Saharan Africa, has experienced two severe disease epidemics in recent years: cassava mosaic disease (CMD) and cassava brown streak disease (CBSD). CMD, endemic to cassava growing regions of Africa, was known to be caused predominantly by *African cassava mosaic virus* (ACMV) and *East African cassava mosaic virus* (EACMV), until recently (Legg, 1999). In the late 1980s, an unusually severe form of the disease, noticed first in Uganda, began to spread, and by 2005 had affected 3 million km^2 of cassava growing area across nine countries (Legg *et al.*, 2011). The epidemic was caused by a recombinant strain of EACMV-UG acting synergistically with the ACMV strain, radiating from the initial centre of origin at a rate of 100 km year^{-1} (Legg *et al.*, 2011). In contrast, CBSD is an example of a new encounter of a pathogen with its host. This epidemic began in 2004, but unlike CMD, which invaded in a contiguous zone, CBSD spread as multiple isolated foci of infection (Legg *et al.*, 2011). The disease was found to be caused by two species, *Cassava brown streak virus* (CBSV) and *Ugandan cassava brown streak virus* (UCBSV), both of which were found in the epidemic regions (Legg *et al.*, 2011). Thus, while the two pathogens originate in Uganda, and are both transmitted by the same whitefly vector (*Bemisia tabaci*), the epidemiological and biological characteristics of the diseases are quite different. Yet another virus, the *Banana bunchy top virus*, transmitted by the banana aphid (*Pentalonia nigronervosa*), is listed among the top 100 invasive alien species in the world (Lowe *et al.*, 2000). The virus causes an extremely destructive disease of banana that was first noticed in Fiji in 1889. Banana bunchy top disease is now found in 33 countries of Africa, Asia, Australia and the South Pacific islands.

Among emerging infectious plant diseases, 47% have been viral, 30% fungal and 16% bacterial (Anderson *et al.*, 2004). A recent example of the potential effects of climate change on disease invasions is seen in the Iberian Peninsula, where warming has been double the world average. Several viral diseases have emerged, starting with citrus tristeza, tomato leaf curl and tomato necrosis in the 1980s, and more recently (2000s), tomato mosaic and cucurbit vein yellowing (Canto *et al.*, 2009). New diseases continue to be a threat (Fisher *et al.*, 2012), and there are many other examples of recent invasions. *Phytophthora ramorum*, causal agent of sudden oak death in North American and European forests, also causes disease in more than 40 different plant genera worldwide (Rizzo *et al.*, 2005). *Phytophthora cinnamomi*, originally described in 1922 in Sumatra, has been spreading around the world, to the point of being considered one of the major problems for natural and managed ecosystems (Hardham, 2005); apparently limited by weather conditions (Bergot *et al.*, 2004). The list of important invasive pathogens in agricultural and unmanaged systems is large and growing. These organisms must be considered as 'major threats' for both the agricultural industry and unmanaged ecosystems, and climate change may lengthen this list. One example is the pathogen, *Candidatus liberibacter*, causing huanglongbing or 'citrus greening' (Bove, 2006; Gottwald, 2010), whose impact on the citrus industry is being felt throughout the southern USA (Bove, 2006).

For many emerging diseases, it is not obvious which previously limiting factor(s) has changed to allow emergence. For some cassava diseases, the limiting factor was apparently lack of range overlap for crop species and pathogen, such that when the crop species was introduced to a new continent, the encounter occurred and the disease could progress (Jones, 2009). In other cases, the abiotic environment may have been limiting, so that climate change could increase the likelihood of invasions. If pathogen propagules are produced more rapidly or more commonly in a region, this will increase the probability of unusual events such as long-range dispersal via wind (Brown and Hovmøller *et al.*, 2002) or accidental spread via human movement (Wellings *et al.*, 1987; Hovmøller *et al.*, 2008). For some pathogen invasions, the critical moment of introduction may be understood fairly well. For example, the soybean rust pathogen (*Phakopsora pachyrhizi*) was probably introduced to the continental USA by hurricane Ivan (Schneider *et al.*, 2005). If the frequency or severity of storm systems increases, this may also contribute to the likelihood of new pathogen invasions.

In addition to the effects of weather patterns, the impact of the increase of CO_2 in the atmosphere is conducive to a larger accumulation of plant biomass, and this can modify microclimatic conditions on plants, potentially promoting the development of fungal diseases (Eastburn *et al.*, 2011). A rise in CO_2 concentration could also lead to an increase in fecundity in some fungi (Dixon, 2012). With respect to control, rising CO_2 levels could enlarge epicuticular waxes, making it more difficult for fungicides and other chemical products to penetrate the plant and control pathogens. Other types of air pollution may also influence the relative success of pathogens, as suggested by changes in pathogen species associated with long-term wheat archives (Bearchell *et al.*, 2005).

Objectives of this Chapter

This chapter will provide an overview of the potential role of climate change in plant pathogen invasions, emphasizing the more recent literature. The challenge of interpreting the role of climate change in invasions lies in determining what had previously reduced the connectivity of pathosystems. Climate can limit connectivity, but so can other factors such as limited opportunities for encounters between different host, pathogen and vector populations. A summary of the factors influencing invasions is provided, including how different types of pathogens invade, the potential impacts on disease of host, pathogen and vector invasions, how the scale of consideration influences the perceptions of invasion and emergence, the effect of landscape structures on networks for invasion, the social–ecological context of invasions, and adaptation and risk management strategies for invasions.

Different Types of Invading Pathogens

Pathogens have a wide range of dispersal strategies. Plant pathogens differ from animal pathogens in the greater importance of fungal pathogens (Miller, 2012), many of which are dispersed aerially across great distances. Passive dispersal by wind is also important for many bacterial pathogens. Because they impact ice nucleation processes, airborne bacteria may also have feedback effects on weather and frost damage to plants. Many pathogens are dispersed by vectors, which add another element to the influence of climate change and potential behavioural responses to conditions. Soilborne pathogens generally invade more slowly, unless dispersed by humans.

Aerobiology of invasion

The movement of pathogens through aerobiological processes is influenced by a number of ecological and environmental processes (Fig. 3.1), and most of these are a function of weather and climate processes. The framework presented by Edmonds (1979) and Isard *et al.* (2005) emphasizes the different scales at which processes occur and the importance of phenology. The stages

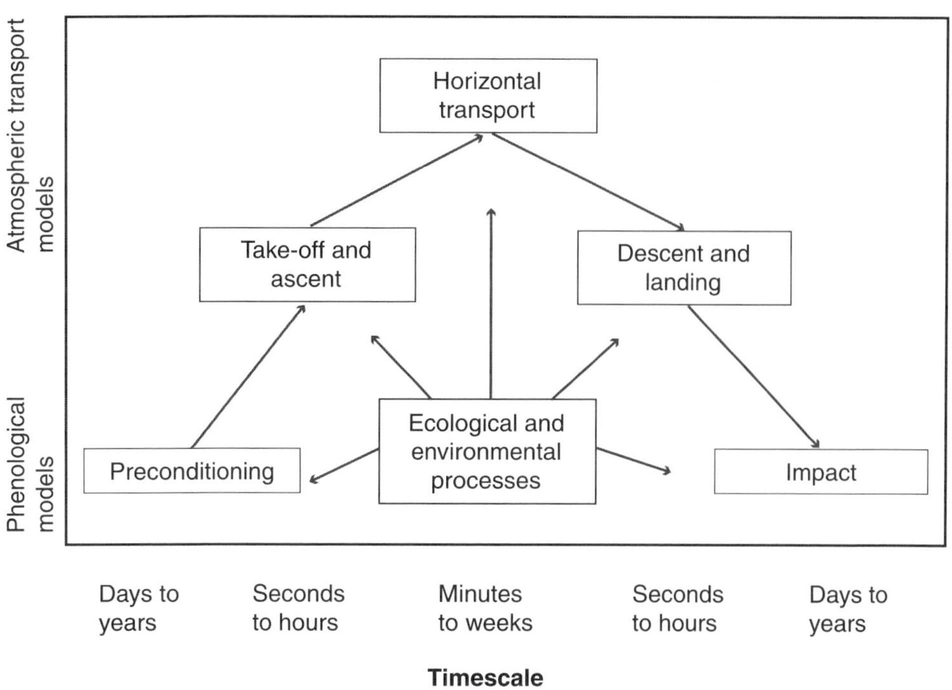

Fig. 3.1. Stages in movement of pathogen propagules through aerobiological processes (adapted from Edmonds, 1979, and Isard *et al.*, 2005). Each of these stages is influenced by weather patterns and thus potentially by climate change.

impacted by phenology – preconditioning and impact – are most readily studied in the context of controlled environment experiments such as the evaluation of reproduction or infection rates at different temperatures, relative humidity levels or atmospheric CO_2 concentrations. The other processes are influenced strongly by atmospheric transport, which is beyond the scope of most controlled environment experiments. Yet, these processes will be key in determining the risk of new aerobiological invasions. Consider in more detail the ecological and environmental processes of particular importance for foliar pathogens (Fig. 3.2). The processes important for preconditioning (propagule production) and impact (colonization of plants) are ones commonly studied in simulations of the climate change effects on pathogens: plant growth stage, disease stage, temperature and leaf wetness (Caubel *et al.*, 2012). Evaluating the processes important in atmospheric transport introduces a set of more problematic experimental possibilities. Pathogen propagule escape from the canopy, as well as deposition onto plants, will be influenced by factors such as canopy structure, winds and turbulence. Turbulent transport and dilution in the atmosphere will be influenced by turbulent diffusion and wind shear. Survival while airborne will be influenced by factors such as UV light, temperature and relative humidity.

The annual migrations of pathogenic fungi make up an important part of their epidemiology, as illustrated dramatically by the migrations of rust fungi between warmer and cooler regions on many continents (Aylor, 2003; Isard *et al.*, 2005, 2007, 2011). The mostly passive wind dispersal of pathogenic fungi is strongly dependent on weather conditions, though there may be active spore projection facilitating release from the canopy in some cases. There are

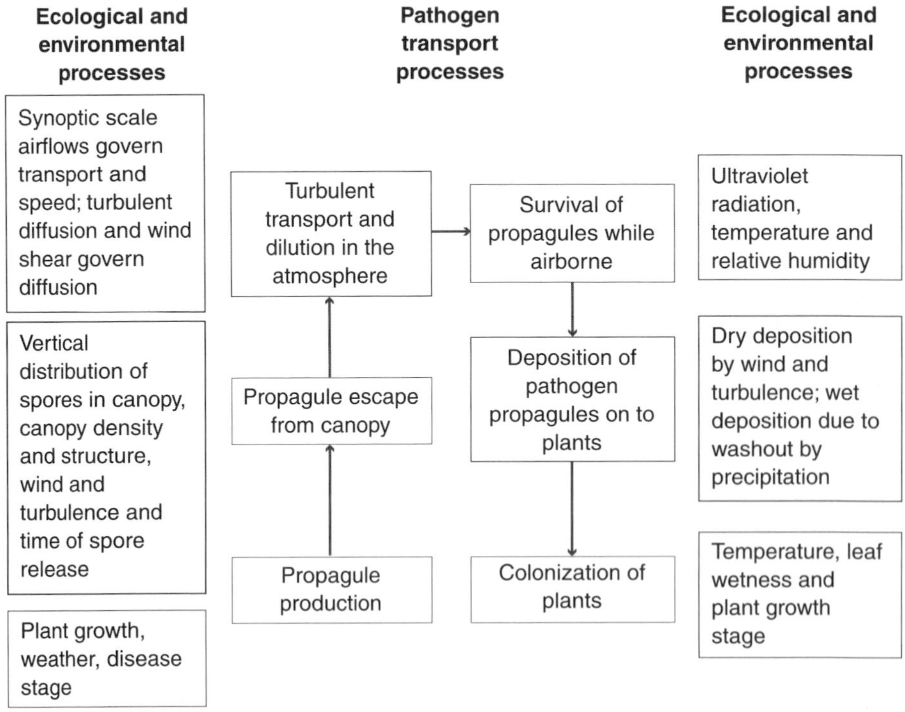

Fig. 3.2. Stages in the transport process for an aerially dispersed pathogen, generalized from the soybean rust example of Isard *et al.* (2005).

also examples in animal systems, such as the possibility that African or Asian dust is a source of Aspergillus infecting coral reefs (Garrison *et al.*, 2003; Weir-Brush *et al.*, 2004; Rypien, 2008).

There are now some efforts to evaluate near-surface wind speeds and gusts under climate change and their relationship to ENSO (El Niño-Southern Oscillation) cycles, often with an emphasis on evaluating the potential effects on wind power generation and/or damage to infrastructure (Haugen and Iversen, 2008; Abeysirigunawardena *et al.*, 2009; Najac *et al.*, 2009; McInnes *et al.*, 2011; Greene *et al.*, 2012; Pryor *et al.*, 2012). The occurrence of key wind events for pathogen dispersal may not be associated strongly with the same parameters that are determined for wind power generation. Bullock *et al.* (2012) evaluated projected wind speeds in southern England and found high variability among global circulation models, with a mean projection of decreased wind speeds. While such an outcome could make it more difficult for wind-dispersed native plant species to move to new locations (Bullock *et al.*, 2012), as environmental conditions that favour them move, it could also slow movement of aerially dispersed pathogens.

Much of the literature addressing aerobiology under climate change has focused on pollen and allergens, including fungal spores (Berman, 2011; Ziska and Beggs, 2012). How climate change will affect dispersal processes for aerially dispersed pathogens represents a research frontier for plant pathology. Studies of wind speed and turbulence for particular locations are likely to need downscaling attention to support realistic analyses. Analyses will need to advance beyond the emphasis on typical wind conditions under climate change, because it will be the extreme weather events (in terms of wind conditions and other weather variables) that may result in new introductions across long distances (Rosenzweig *et al.*, 2001; Brown and Hovmuller, 2002).

Another important consideration is the role of microbes in ice nucleation, and developing understanding of the dynamics of airborne microbes and their role in cloud formation, precipitation and radiative forcing (Mohler *et al.*, 2007; Morris *et al.*, 2011). Ice nucleation caused by bacteria distributed on plants also plays a role in frost injury and distribution of the bacteria on plants in nature (Lindow *et al.*, 1978; Lindow, 1983). It is intriguing to consider the possible feedbacks from changing plant-associated microbial communities on atmospheric processes, ice nucleation and rainfall patterns. Pratt *et al.* (2009) considered the role of aerosol particles in cloud formation to be a particularly important source of uncertainty in climate change projections.

Vectored pathogens

Diverse plant pathogens use fungi, nematodes and arthropods as vectors for transmission from plant to plant. Changes in weather and other climatic parameters can influence these multitrophic interactions greatly by altering the development, abundance, distribution and survival of vectors (Finlay and Luck, 2011). Aphids, an important group of virus vectors, are predicted to multiply by five additional generations each year in temperate regions under a global mean surface temperature increase of 2°C (Yamamura and Kiritani, 1998). Temperature-related range expansions of vectors, and consequently pathogens, into higher latitudes are predicted in many pathosystems such as Pierce's disease of grape, Stewart's wilt in maize and aster yellows in solanaceous crops (Jones and Barbetti, 2012). Increased precipitation in temperate regions would increase fungi-vectored virus epidemics by allowing better dissemination of the vector zoospores; while fewer such epidemics are predicted in tropical regions where a decline in rainfall is expected (Jones and Barbetti, 2012). Strong winds can similarly carry mites and certain insect vectors greater distances, along with the disease propagules they carry (Jones and Barbetti, 2012). Climate change can also expand the geographical range of vector reservoir hosts and alter vector predation.

In cotton, for example, elevated [CO_2] led to improved biological control of the cotton aphid (virus vector) by its coccinellid predator, *Leis axyridis* (Chen *et al.*, 2005). In addition, changes in pathogen biology such as increased pathogen accumulation in plants (for example, *Tobacco mosaic virus* (TMV) replication and movement rates increase at higher temperatures) and changes in transmission patterns can affect pathogen acquisition by vectors. Thus, in vectored pathogens, complex interactions between the host, vector, pathogen and surrounding environment make predictions of climate change effects more challenging (Garrett *et al.*, 2011).

Soilborne pathogens

Pathogens living in soil are considered soilborne (Shurtleff and Averre, 1997). The losses caused by these are estimated to be between 7% and 15% on average for the five main food staple crops (Raaijmakers *et al.*, 2009). Current understanding of the effects of climate change on soilborne pathogens lags behind others such as foliar pathogens (Eastburn *et al.*, 2011). Soil is a complex conglomerate of biological, physical and chemical networks, involving organisms from all levels in the food chain and maintaining processes such as carbon sequestration, nutrient recycling and regulation of pathogenic and parasitic populations (Pritchard, 2011; Chakraborty *et al.*, 2012). Given the number of species involved and the complexity of their relationships, studying and understanding these processes is challenging (Compant *et al.*, 2010). All trophic processes are vital for soil food webs, but only some specific combinations allow plants to tolerate particular stresses or result in disease-suppressive soils. The rhizosphere is the 'playgound' where pathogens and beneficial organisms interact and drive plants to either a healthy or a diseased state (Raaijmakers *et al.*, 2009). Changes in the soil network and in the relationships among all the elements involved – changes potentially driven by climate change, where increasing temperature and changes in humidity and rain periods could lead to increases in some pathogen populations and a break in soil homeostasis – may favour pathogen invasions and generate disease outbreaks (Sturrock *et al.*, 2011).

Soilborne pathogens have different reproductive strategies. For example, Fusarium wilt inoculum is frequently a cluster of spores and/or mycelia. *Pythium* and *Phytophthora* produce zoospores that are able to swim, and *Rhizoctonia* and *Gaeumannomyces* are present as mycelia and may grow through the soil. The movement of these pathogens is driven by water irrigation and rain runoff. Soil texture and water could be the most important factors in the dispersal of pathogen propagules in soil, implying that changes in these regimes have the potential to lead to pathogen invasions (Raaijmakers *et al.*, 2009; Roos *et al.*, 2011).

CO_2 enrichment stimulates plant growth, producing an increase in root biomass that may expand the amount of tissue available for new infections (Ghini *et al.*, 2008). It is expected that disease patterns change geographically as a consequence of climate change (Ghini *et al.*, 2011). Soil organisms are vectors of economically important plant viruses. Increases in soil temperature and humidity are expected to boost the activity of soil nematodes and fungi, promoting new invasive processes and diseases like *Potato mop-top virus* and *Beet necrotic yellow vein virus* in Sweden (Roos *et al.*, 2011). Changes in the weather pattern during the past 60 years are associated with an increase in outbreaks and higher severity of the disease Phytophthora root rot caused by *P. cinnamomi* (Sturrock *et al.*, 2011).

On the other hand, increased temperatures can be useful to certain management strategies like solarization, currently used to control soilborne pathogens (Juroszek and von Tiedemann, 2011). However, some new conservation agriculture practices (i.e. tillage reduction) can increase disease risk because organic matter and crop residue can serve as a pathogen reservoir (Garrett *et al.*, 2012). New technologies such as metabolomics, microarrays and next-generation sequencing methods are tools that will support better understanding of microbial interactions at the soil–plant level, as well as the analysis of microbial communities and their role in disease invasion and disease ecology. The characterization of microbial communities can be a baseline for understanding climate change and the dynamics of invasive species (Garrett *et al.*, 2012).

Host, Pathogen and Vector Invasions

Pathogen invasions can be influenced strongly by host and vector invasions. Climate change impacts on host and vector invasions can thus affect pathogen invasions.

Host invasions

When plants invade a new area (or a crop species is introduced to a new area), there are several potential outcomes. Under the enemy release hypothesis, a plant with a set of coevolved pathogens in its old range may experience fewer pathogens in its new range (Mitchell and Power, 2003). In some cases, however, exposure of a plant species to new pathogens may result in particularly destructive new diseases, e.g. cassava viruses (Jones, 2009). Alternatively, an invasive plant may serve as an important green bridge for pathogens of crop species, such as the role of the invasive weed, kudzu, in soybean rust epidemics (Li *et al.*, 2010) and the role of invasive grass species in barley yellow dwarf disease epidemics (Borer *et al.*, 2007; Malmstrom *et al.*, 2011). For vectored pathogens, an invasion may lead to an encounter with a more effective vector species, as again in the case of cassava viruses (Jones, 2009). There may be apparent competition among plant species through plant disease, where spillover of pathogens produced on one species affects the fitness of other species (Alexander and Holt, 1998; Power and Mitchell, 2004; Cobb *et al.*, 2010; Cox *et al.*, 2013).

Pathogen invasions

Pathogen invasions may similarly result in new configurations of host and vectors (Fitt *et al.*, 2006). Pathogen invasions may also result in new patterns of range overlap between different pathogen species subpopulations and between different pathogen species. When invasions lead to a new overlap of mating types within a species, potential new risks from sexual reproduction include both greater destructiveness to hosts and greater adaptation to environmental conditions. The potential for multiple types of genetic exchange among species is a risk somewhat specific to microbes. Interspecific hybridization in fungi can result in new types of destructiveness in invasive pathogens, as in the case of Dutch elm disease (Brasier, 2000, 2001). Co-occurrence of pathogens can cause important disease problems. For example, sweetpotato virus disease, caused by the synergistic interaction between *Sweetpotato chlorotic stunt virus* and *Sweetpotato feathery mottle virus*, can cause losses of up to 80–90% in many varieties (Clark *et al.*, 2012). When new forms of environmental adaptation or of adaptation to host genetic resistance develop, this can be the basis for new global epidemics. Wheat stripe rust is an example of new environmental adaptation, where populations with tolerance to higher temperatures have spread widely in recent years (Chen, 2005; Milus *et al.*, 2006; Hovmøller *et al.*, 2011). Wheat stripe rust also represents a cautionary tale for predicting disease scenarios under climate change. Prior to identification of the new heat-tolerant strains, a reasonable scenario analysis would have indicated that wheat stripe rust should gradually become less common under rising temperatures. A prime example of a 're-emerging' disease is a new race of wheat stem rust, Ug99, a more virulent and potentially devastating race to which the majority of wheat varieties were susceptible, making it a global agricultural threat. Since this disease was first reported in Uganda during 1999, it has spread to many different countries in Africa and South-east Asia (Singh *et al.*, 2011).

Vector invasions

Vector invasions result in expansions of the geographical and host range of pathogens, shifts in the relative severity of diseases and the emergence of new host–pathogen–vector combinations, as well as new insect assemblages (Worner and Gevrey, 2006). The invasion of the glassy-winged sharpshooter (*Homalodisca vitripennis*) into California in the late 1980s is one such example of range expansion. The vectored pathogen, *Xylella fastidiosa*, originally noticed on citrus trees in Argentina, now threatens the grape industry in southern California and is expected to move northwards with rising temperatures (Hopkins and Purcell, 2002; Jones and Barbetti, 2012). In China and the Andean region of South America, the sweet potato whitefly (*B. tabaci*) is expected to displace the less cold-tolerant glasshouse whitefly (*Trialeurodes vaporariorum*), causing shifts in associated viral diseases (Jones and Barbetti, 2012). Because a single plant can harbour multiple viruses, changes in the biotypes of vectors (such as aphids and whiteflies) are predicted to alter the transmission efficiencies of individual viruses, consequently affecting the composition of local viral populations (Canto *et al.*, 2009). Various factors predispose vectors to such invasions, including dietary flexibility, phenological plasticity and overwintering strategy (Ward and Masters, 2007). In a changing climate scenario, vectors with a broad host range, limited synchrony with host phenology and resistance to entering into diapause would have a greater advantage. Climate change would also shape the temporal and spatial spread of vectors by affecting generation time, extending the active season, promoting variability in vector fitness and behaviour, expanding the geographical range of reservoir hosts and altering natural predation. Especially in virus pathosystems, the role of vector invasions is likely to be extremely important in disease outbreaks (Anderson *et al.*, 2004).

Animal and plant diseases

For the most part, plant and animal diseases are studied by different groups of people, but interest in synthesizing these disciplines is growing because of the potential synergy (Mundt *et al.*, 2009; Wilkinson *et al.*, 2011; Borer *et al.*, 2012; Dunn *et al.*, 2012; Skelsey *et al.*, 2013). For example, there is the potential for sudden oak death to reduce Lyme disease risk by altering the quality of the environment for animal hosts (Swei *et al.*, 2012). Sudden aspen decline has been associated with increases in Sin Nombre virus in small mammals (Lehmer *et al.*, 2012).

Pathogen Invasions Versus Pathogen Emergence: Questions of Scale

For pathogens, the spatial resolution (or size of minimum units) being studied is an important consideration. A pathogen may be present in a particular area but relatively rare. When it infects a large number of units (individuals or small locations) in that area, it may be said to emerge. Taxonomic resolution is also an important consideration, because different subpopulations of the same pathogen species may have very important functional differences.

Disease emergence versus pathogen invasion

Diseases are generally said to emerge when they become of greater importance. Thus, emergence as a concept is sometimes criticized as being more journalistic than scientific. The unusual scales of interest for diseases, compared to other taxa addressed in this book, make it worth mentioning the context of emergence. Pathogen invasions and/or annual migrations (from overwintering or oversummering locations in milder areas) may produce epidemics with high inter-annual variability, often because of inter-annual variation in weather patterns

(Li *et al.*, 2010; Sutrave *et al.*, 2012). Emergence is generally associated with the ongoing importance of a pathogen. Invasion considered in a coarsely demarcated landscape, such as invasion into countries, may be the first step to emergence. Emergence might be thought of as the continued invasion of enough finely demarcated 'units' (locations or individuals) to constitute an important epidemic. In some cases, pathogens may need to overcome Allee effects, such that per capita reproduction is disproportionately smaller for low populations (Courchamp *et al.*, 2009). For pathogens, Allee effects may occur when pathogens themselves require high populations in order to find mates (Garrett and Bowden, 2002) or when their vectors require higher populations for greater reproductive success.

Invasions and intraspecific variation

Many plant pathogens that are already widely spread still pose a new threat to agricultural productivity through the emergence of variants with greater pathogenic fitness, or which otherwise pose new problems for disease management. A well-known recent example is Ug99 (Singh *et al.*, 2011). There have, however, been many others. Potato late blight, for example, is a serious disease of global importance on potato and tomato. There have been many long-standing efforts to develop useful levels of host plant resistance in potato but at present most cultivars grown worldwide are very susceptible (Cooke *et al.*, 2011). One factor which has contributed to the failure to control potato late blight with host resistance has been pathogen evolution toward specific virulence, which occurred very rapidly after the introduction of a resistant potato variety (Forbes, 2012). These changes in specific virulence within a pathogen population are often extreme, rendering a cultivar highly susceptible which had been virtually immune when it was released. However, pathogen evolution may lead to quantitative changes that are also significant for crop productivity. Recently, researchers in Europe described a new clonal lineage of *P. infestans* known as 13_A2, which rapidly has displaced other lineages, representing 75% of the pathogen population across Great Britain three years after its initial appearance (Cooke *et al.*, 2012).

Effects of climate on invasion and persistence risk factors

We can distinguish between invasion and persistence (Gilligan and van den Bosch, 2008), where invasion is the necessary first step (Table 3.1). If climate change at a location results in a shift so that a pathogen can overwinter or oversummer – for example, if minimum temperatures are higher so that they stay within the range a pathogen can tolerate, or if reduced precipitation allows spores in soil to overseason – this has the potential to influence greatly the likelihood of persistence. The increased probability of reproduction will have obvious effects on invasion and persistence risks. If the season extends so that more generations of pathogen reproduction are possible, this can have a large effect on the pathogen population because of the compound interest nature of population growth. Perhaps most importantly in the long run, climate change can shift an environment–pathogen interaction such that the evolutionary potential of the pathogen is altered (McDonald and Linde, 2002), with resulting impacts on both invasive and persistence risk.

Landscape Structures and Networks

Climate change can modify landscape structures and networks for potential invasion. These effects will be realized in combination with changes in transportation and trade networks.

Invasion networks

Network models include nodes (vertices) and the links (edges) that connect them

Table 3.1. Implications of potential shifts in pathogen life cycles and natural history for invasion and persistence when changing from one type of climate to another at a particular location.

Potential differences between climates with implications for pathogen life cycles and natural history		Implications of transition from Climate A to Climate B for invasion and persistence	
Climate A	Climate B	Invasion risk	Persistence risk
No overseasoning	Overseasoning	Site can act more readily as inoculum source for invasion to other areas	Persistence is much more likely; epidemics can begin earlier in the season
Lower probability of reproduction	Higher probability of reproduction	Higher numbers of propagules lead directly to higher invasion risk and indirectly to higher probability of new pathogen types adapted to new scenarios	Higher numbers of propagules lead directly to higher persistence risk and indirectly to higher probability of new pathogen types adapted to persistence
Conditions typically can support pathogen population only *below* Allee threshold	Conditions typically support pathogen population *above* Allee threshold	Gains in population are supported so that invasion is more likely	Gains in population make persistence feasible
Fewer generations annually	More reproductive cycles per year	Compound interest growth of pathogen population makes invasion more likely	Compound interest growth of pathogen population makes persistence more likely

(Barrat *et al.*, 2008; Newman, 2010). Nodes might be locations such as counties or fields, or entities such as individuals or species or genotypes in food webs. Links may represent the probability of movement between nodes, or other measures such as a measure of influence. Network models have been applied to plant disease relatively recently (Jeger *et al.*, 2007; Moslonka-Lefebvre *et al.*, 2011), to pathogens such as *Armillaria lutea* (Lamour *et al.*, 2007), *P. ramorum* (Harwood *et al.*, 2009) and *P. pachyrhizi* (Sutrave *et al.*, 2012). Landscape heterogeneity is an important factor for invasions by most plant pathogens (Gottwald *et al.*, 2002; Venette and Cohen, 2006; Parnell *et al.*, 2009; Pautasso *et al.*, 2010; Venette *et al.*, 2010; Mundt *et al.*, 2011; Filipe *et al.*, 2012), and network models offer a useful approach for describing landscape structure (Margosian *et al.*, 2009). For example, dynamic network models that incorporate changes in links over time can be used to evaluate the role of 'green bridges' or the maintenance of potential hosts through time to allow pathogens to persist and disperse. The phenology of node and link availability may be important. For example, Fusarium head blight is a pathosystem where changes in host (timing of anthesis) and pathogen phenology are particularly important for determining disease severity (Madgwick *et al.*, 2011). Understanding network structures also informs strategies for sampling and managing invasions (Baxter and Possingham, 2011; Chadès *et al.*, 2011; Sutrave *et al.*, 2012).

The topology of invasion

As discussed above, the disease triangle is often used to indicate the relationship between pathogen, host and environment (Fig. 3.3a). Typically, risk maps for pathogens have been developed on a point-by-point basis, where the risk of infection is determined for each point or pixel in a map (typically based primarily on weather variables) (Fig. 3.3b). A limitation of the point-by-point approach is that it does not take into account network structures that allow pathogens to be introduced or reintroduced more readily to a given point. For example, in Fig. 3.3c the black point that is not connected to any other points may have disease-conducive weather conditions, but disease risk will tend to be lower than for black points that are connected to other black points. Similarly, low-risk locations may end up having somewhat higher risk if they are linked to multiple high-risk locations. Under climate change, the topology of invasion may shift, both in terms of which points have 'locally' conducive weather for disease and in terms of how well connected the highly conducive points are (Fig. 3.3d). There are many possibilities for such dual shifts of point-wise changes in conduciveness and changes in the connectedness of highly conducive locations (Fig. 3.4). Some shifts may result in different 'hotspots' for disease and invasion risk, while the overall level of risk remains the same. In other scenarios, new alignments of conducive locations and links among locations can result in changes in invasion risk, even when the number of disease-conducive locations remains the same.

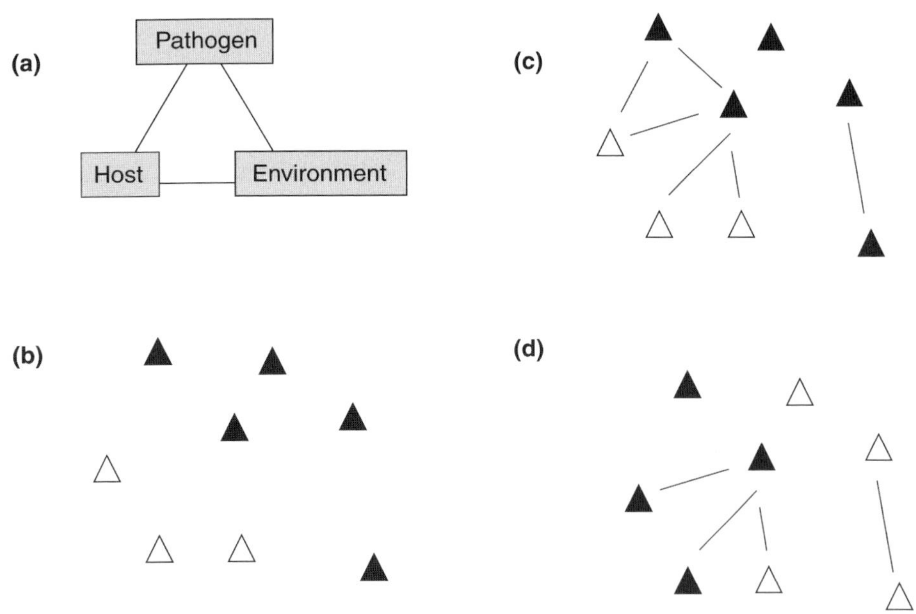

Fig. 3.3. (a) The disease triangle indicates how a susceptible host, a competent pathogen and a conducive environment interact to support the occurrence of disease. (b) Species distribution maps, or maps of disease occurrence, indicate where a pathogen species is likely to occur. In this schematic, each triangle represents a location, such as a field. The components of the disease triangle determine the likelihood of disease occurrence (or severity of disease) at each location, where occurrence is indicated here by black shading. (c) While pathogen distribution maps often are limited to point-by-point analyses of weather conditions, the links for likely movement between locations can also be included in maps as part of a network analysis. Here, links where movement is likely are indicated by lines connecting the triangles. (d) Under climate change, both the locations where disease is likely to occur and the links where movement is likely may change.

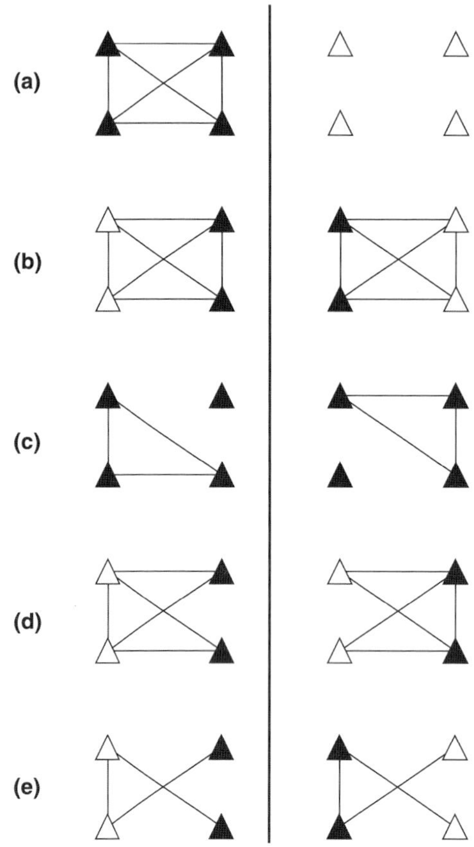

Fig. 3.4. There are many possible outcomes when potential invasion topology is altered by climate change. (a) In the extreme, the topology might change between a set of locations where disease is likely (indicated by black shading) that are all strongly connected for pathogen movement (indicated by lines joining pairs of locations) and a set of locations where disease is unlikely at all locations and movement is unlikely between all pairs. (b) Some types of changes in topology could result in the same overall degree of disease risk that is simply moved from one set of locations to another. (c) Likewise, patterns of likely movement could change, while linked locations have the same likelihood of disease occurrence. (d) In some cases, changes in which locations are linked could have important epidemic impacts if the change is between linked low-risk locations and linked high-risk locations. (e) Similarly, a change in which locations have a high likelihood of disease occurrence could have important epidemic impacts if this changes the likelihood of occurrence for the highly connected locations.

Social–Ecological Systems and Invasions

Disease across the spectrum of unmanaged to managed plant systems

Pathogens affect plants across the spectrum, from systems with little management to intensive agricultural systems, including new landscapes formed by bioenergy crops which may be derived from native grassland species (Tilman *et al.*, 2006; Crouch *et al.*, 2009; Schrotenboer *et al.*, 2011; Stewart and Cromey, 2011). Under climate change, unexpected roles for plant pathogens in unmanaged systems may be observed. For example, the carbon balance of Arctic tundra under increased snow cover may be mediated by a plant pathogen (Olofsson *et al.*, 2011). In some cases, the same generalist pathogens may span unmanaged and agricultural systems, such as *P. ramorum* in forests and the horticultural trade (Rizzo *et al.*, 2005) and *Macrophomina phaseolina* in tallgrass prairie and annual crop production (Saleh *et al.*, 2010). Models that take a systems approach will be important for linking social and biological systems to understand climate change, invasive potential and the full system implications of invasions (Cook *et al.*, 2007; Cheatham *et al.*, 2009; Epanchin-Niell *et al.*, 2010; Mills *et al.*, 2011; Newton *et al.*, 2011; Savary *et al.*, 2011a; Garrett, 2012). Invasive pathogens may disrupt natural systems, and invasive plant species may play a role. For example, *Barley yellow dwarf virus* in California grasslands reproduces in invasive grass species and also has important negative effects on native grass reproduction (Malmstrom *et al.*, 2005; Borer *et al.*, 2007).

Forest systems offer additional challenges for long-term management under climate change, whether for timber production or for historically natural conditions, because of the many sources of variability in scenario evaluation (Dukes *et al.*, 2009; Sturrock *et al.*, 2011; Meentemeyer *et al.*, 2012). The important pine diseases, pitch canker and Dothistroma needle blight, may experience overall reductions in suitable areas globally, with increases in some areas such as Europe and New Zealand (Watt *et al.*, 2011), while Dothistroma needle blight has increased in British Columbia to levels not previously recorded (Woods *et al.*, 2005).

Climate change and invasions through seed systems

The impact of climate change on seedborne diseases has received little attention. 'Seed' (sexual or asexually derived planting material) is a primary input in agriculture and can initiate large-scale disease epidemics if poor seed health is maintained. Overall, formal seed production and certification systems have accounted for only about 10% of seed use, with the remaining amount produced by farmers themselves (Almekinders *et al.*, 1994). Thus, when new cropping regions emerge due to climate change, infected planting material could serve as an important source for pathogen expansion into these areas. Seed trade (regional, national and international) would facilitate such expansions and allow for new disease emergence by reducing the gap between native hosts and exotic pathogens, or vice versa (Jones and Barbetti, 2012). Such species jumps have been documented in begomoviruses, and could be aggravated under climate change scenarios, especially for generalist pathogens (Jones and Barbetti, 2012). In contrast, prolonged heatwaves anticipated in the future could also diminish pathogen load in seeds. Various bacterial seed pathogens such as *Xanthomonas* and *Pseudomonas* species and viruses like the *Potato leaf roll virus* are susceptible to heat treatments and could have limited survival in seed material under heatwaves (Jones and Barbetti, 2012).

The seed systems for many crop plants are important components of disease management, but this is particularly true for vegetatively propagated crops, such as potato, sweet potato and cassava, which suffer yield reduction due to the accumulation of seedborne (propagule-borne) pathogens. Many of the most important of these degeneration pathogens are viruses, and their spread and invasiveness may be affected by climate change. The most obvious mechanism would appear to be via the direct effects of climate change on the spread of virus vectors (Roos *et al.*, 2011), but climate change may also affect the overall make-up of plant species in the area and host plant physiology, and thereby the host–virus interaction (Canto *et al.*, 2009).

Adaptation and Risk Management Strategies for Invasions

Forecasting plant disease invasions

Short-term forecasting models have commonly been developed to support within-season decision making by farmers. These models can be modified for application in large-scale scenario analysis (Sparks *et al.*, 2011). Variation in climate, including weather cycles across multiple years, affects diseases and the resulting losses (Coakley, 1979; Scherm and van Bruggen, 1994; Scherm and Yang, 1995; Kriss *et al.*, 2012; Garrett *et al.*, 2013). Forecasting diseases is challenging within the range of typical baseline weather conditions, and there is additional uncertainty in how pathogenic populations will react to changes in the environment (Fitzpatrick and Hargrove, 2009). It is important to create good models that help scientists (entomologists, plant pathologists, agronomists, biologists and ecologists, among others), producers, the general public, regulators and decision makers to take the best action possible to support adaptation. Knowing where and when an outbreak is likely to implement makes management decisions easier to apply and to adjust. Among the different adaptation strategies created to control pathogen invasions are models that predict the potential occurrence of new diseases (Gilbert *et al.*, 2012). Many models have been proposed and tested as tools in disease management programmes such as soybean rust (Pan *et al.*, 2006; Sutrave *et al.*, 2012), Fusarium head blight (Fusarium Head Blight Prediction Center, www.wheatscab.psu.edu) and wheat stem rust (Singh *et al.*, 2011).

Evaluating geographic risk

Addressing uncertainty and the effects of climate variability in scenarios for plant disease risk is an active research area (Garrett *et al.*, 2013; Gouache *et al.*, 2013). The field of 'decision making under uncertainty' offers insights into how people managing plant disease are likely to respond, where there are different perceptions of what risk is important (McRoberts *et al.*, 2011). There are a number of challenges for mapping plant disease risk under climate change because of the many interactions and sources of variability in these linked systems (Scherm and van Bruggen, 1994; Scherm, 2004; Garrett *et al.*, 2011; Shaw and Osborne, 2011; Pautasso *et al.*, 2012). As improved and high-resolution climate scenario data are developed, there will be the potential to apply more mechanistic models of plant pathogen dispersal and invasion (Aylor, 2003; Skelsey *et al.*, 2008). As global trade accelerates, there is increasing risk for the spread of invasive pathogens, while policies to contain invasions rarely keep pace (Brasier, 1996, 2008). Spatial aspects of epidemiology have been studied, but there is a lack of emphasis on studying the spatial structure of injuries on yield losses (Savary *et al.*, 2006). Barnes *et al.* (2010) demonstrated that there could be important benefits from adaptation processes for climate change where long-term and quantitative adaptation processes seem to be more efficient than short-term adaptation. Pest risk assessment (PRA) has been used to moderate the movement of commodities among countries. PRAs have been developed using lists of pathogens present in different regions or countries, along with extensive literature reviews, in order to evaluate the risk of disease. Knowing the susceptibility of a host to certain pests and pathogens is required to assess the phytosanitary risk in a determined area, and an analytical method has been proposed to predict new host–pathogen/pest interactions and to provide information about likely incidence in a particular host (Gilbert *et al.*, 2012).

Risk management

The options for management of plant disease invasions in major agricultural systems may be equal to the task of adaptation to climate change. There is potential for plant breeding to address the problem of pathogen emergence or re-emergence in major agricultural systems, although there may be continuing challenges, as in the example of new races of the wheat stem rust pathogen (Singh *et al.*, 2011). Some types of management are dependent on the regional inoculum load, so changes in season length, general conduciveness to disease, or regional management will influence individual farmers' abilities to manage disease (Garrett *et al.*, 2009; Newton *et al.*, 2011). Disease insurance is a novel option for managing the financial risks from invasions (Bekkerman *et al.*, 2008), which may become more common as subsidy programmes shift toward insurance in developed countries and new micro-insurance programmes are put in place in developing countries. Another approach that may support improved disease management is budgeting of greenhouse gas emissions. Crop disease can be thought of as contributing to greenhouse gas emissions when it reduces the efficiency of fertilizer use and farm operations (Carlton *et al.*, 2012). Stopping or slowing invasions is a form of climate change mitigation, assuming the management activities do not release more greenhouse gasses than are conserved (Mahmuti *et al.*, 2009; Barnes *et al.*, 2010; Hughes *et al.*, 2011).

Challenges for adaptation for resource-poor farmers

The risk of pathogen invasion or emergence will certainly be greater for smallholder farmers in developing countries who have less access to many adaptation strategies and less access to training. For example, potato late blight has recently caused severe damage at very high altitudes in the Andes in communities where farmers had not seen the disease before (Salazar, 2006; Perez *et al.*, 2010). The farmers grew susceptible cultivars and had no access to an agricultural extension service for support. Reducing risk in these smallholder systems will require a multifaceted approach including greater emphasis on integrated pest and pathogen management, development and use of varieties and species resistant to pests and diseases, and monitoring programmes (Howden *et al.*, 2007).

There are two particular challenges for adapting to climate change among resource-poor farmers. The first is limited appreciation of seed-based disease introductions. Anderson *et al.* (2004) indicated that the predominant driver for emerging infectious plant disease epidemics was pathogen introductions. Local and international trade of plant products, seed and live plants is an important route for such introductions. Although seed testing and regulatory systems are in place in some countries, these systems are often unstable in developing countries and in local communities. A second challenge is limitations in locally available technical inputs to mitigate plant disease epidemics. The poorest farmers, most affected by changing climates, live in disaster-prone areas with limited research inputs on handling climate change (Hoffmann, 2011). Providing these farmers with abiotic and biotic stress-tolerant cultivars, vector-monitoring techniques, quality seed and other improved agricultural inputs and technical information would improve their preparedness for new disease challenges. This in turn requires research activities to generate region- and crop-specific sustainable technology.

Global cooperation will be important in developing the databases needed to construct good models and to perform the key step of validating them (Scherm *et al.*, 2000; Jeger *et al.*, 2007; Jeger and Pautasso, 2008; Savary *et al.*, 2011b). Analysis of multiple diseases simultaneously, in combination with consideration of other global change factors such as availability of labour and water, can provide more complete analyses of likely impacts (Savary *et al.*, 2011a, 2012). Because of uncertainty in many components of scenario analyses, it may often be best to devise management systems that will be resilient to any of the potential types of climate change (Sutherst *et al.*, 2011).

Acknowledgements

We appreciate support from the CGIAR Consortium Research Program for Roots, Tubers and Bananas (RTB) and the CGIAR Research Program on Climate Change, Agriculture and Food Security (CCAFS), with the financial assistance of the European Union, Canadian International Development Agency, World Bank, New Zealand Ministry of Foreign Affairs and Trade and Danida, and with the technical support of IFAD. We also appreciate support by the Australian Government's Cooperative Research Centres Program, USDA APHIS Grant 11-8453-1483-CA, USDA NC RIPM Grant 2010-34103-20964, US NSF Grant EF-0525712, as part of the joint NSF-NIH Ecology of Infectious Disease programme, US NSF Grant DEB-0516046, USAID SANREM, Ceres Trust and the Kansas Agricultural Experiment Station (Contribution No 13-294-B). The views expressed herein can in no way be taken to reflect the official opinion of these agencies. Garrett dedicates this chapter to the memory of Doris Garfinkel.

References

Abeysirigunawardena, D.S., Gilleland, E., Bronaugh, D. and Wong, P. (2009) Extreme wind regime responses to climate variability and change in the Inner South Coast of British Columbia, Canada. *Atmosphere-Ocean* 47, 41–62.

Alexander, H.M. and Holt, R.D. (1998) The interaction between plant competition and disease. *Perspectives in Plant Ecology, Evolution and Systematics* 1, 206–220.

Almekinders, C.J.M., Louwaars, N.P. and Debruijn, G.H. (1994) Local seed systems and their importance for an improved seed supply in developing countries. *Euphytica* 78, 207–216.

Anderson, P.K., Cunningham, A.A., Patel, N.G., Morales, F.J., Epstein, P.R. and Daszak, P. (2004) Emerging infectious diseases of plants:

pathogen pollution, climate change and agrotechnology drivers. *Trends in Ecology and Evolution* 19, 535–544.

Aylor, D.E. (2003) Spread of plant disease on a continental scale: role of aerial dispersal of pathogens. *Ecology* 84, 1989–1997.

Barnes, A.P., Wreford, A., Butterworth, M.H., Semenov, M.A., Moran, D., Evans, N. and Fitt, B.D.L. (2010) Adaptation to increasing severity of phoma stem canker on winter oilseed rape in the UK under climate change. *Journal of Agricultural Science* 148, 683–694.

Barrat, A., Barthelemy, M. and Vespignani, A. (2008) *Dynamical Processes on Complex Networks.* Cambridge University Press, Cambridge, UK.

Baxter, P.W.J. and Possingham, H.P. (2011) Optimizing search strategies for invasive pests: learn before you leap. *Journal of Applied Ecology* 48, 86–95.

Bearchell, S.J., Fraaije, B.A., Shaw, M.W. and Fitt, B.D.L. (2005) Wheat archive links long-term fungal pathogen population dynamics to air pollution. *Proceedings of the National Academy of Sciences* 102, 5438–5442.

Bekkerman, A., Goodwin, B.K. and Piggott, N.E. (2008) Spatio-temporal risk and severity analysis of soybean rust in the United States. *Journal of Agricultural and Resource Economics* 33, 311–331.

Bergot, M., Cloppet, E., Perarnaud, V., Deque, M., Marcais, B. and Desprez-Loustau, M.L. (2004) Simulation of potential range expansion of oak disease caused by *Phytophthora cinnamomi* under climate change. *Global Change Biology* 10, 1539–1552.

Berman, D. (2011) Climate change and aeroallergens in South Africa. *Current Allergy and Clinical Immunology* 24, 65–71.

Borer, E.T., Hosseini, P.R., Seabloom, E.W. and Dobson, A.P. (2007) Pathogen-induced reversal of native dominance in a grassland community. *Proceedings of the National Academy of Sciences* 104, 5473–5478.

Borer, E.T., Antonovics, J., Kinkel, L.L., Hudson, P.J., Daszak, P., Ferrari, M.J., Garrett, K.A., Parrish, C.R., Read, A.F. and Rizzo, D.M. (2012) Bridging taxonomic and disciplinary divides in infectious disease. *Ecohealth* 8, 261–267.

Bove, J.M. (2006) Huanglongbing: a destructive, newly-emerging, century-old disease of citrus. *Journal of Plant Pathology* 88, 7–37.

Brasier, C.M. (1996) *Phytophthora cinnamomi* and oak decline in southern Europe. Environmental constraints including climate change. *Annales des Sciences Forestieres* 53, 347–358.

Brasier, C.M. (2000) Plant pathology – the rise of the hybrid fungi. *Nature* 405, 134–135.

Brasier, C.M. (2001) Rapid evolution of introduced plant pathogens via interspecific hybridization. *Bioscience* 51, 123–133.

Brasier, C.M. (2008) The biosecurity threat to the UK and global environment from international trade in plants. *Plant Pathology* 57, 792–808.

Brown, J.K.M. and Hovmøller, M.S. (2002) Aerial dispersal of pathogens on the global and continental scales and its impact on plant disease. *Science* 297, 537–541.

Bullock, J.M., White, S.M., Prudhomme, C., Tansey, C., Perea, R. and Hooftman, D.A.P. (2012) Modelling spread of British wind-dispersed plants under future wind speeds in a changing climate. *Journal of Ecology* 100, 104–115.

Canto, T., Aranda, M.A. and Fereres, A. (2009) Climate change effects on physiology and population processes of hosts and vectors that influence the spread of hemipteran-borne plant viruses. *Global Change Biology* 15, 1884–1894.

Carlton, R.R., West, J.S., Smith, P. and Fitt, B.D.L. (2012) A comparison of GHG emissions from UK field crop production under selected arable systems with reference to disease control. *European Journal of Plant Pathology* 133, 333–351.

Caubel, J., Launay, M., Lannou, C. and Brisson, N. (2012) Generic response functions to simulate climate-based processes in models for the development of airborne fungal crop pathogens. *Ecological Modelling* 242, 92–104.

Chadès, I., Martin, T.G., Nicol, S., Burgman, M.A., Possingham, H.P. and Buckley, Y.M. (2011) General rules for managing and surveying networks of pests, diseases, and endangered species. *Proceedings of the National Academy of Sciences* 108, 8323–8328.

Chakraborty, S. and Newton, A.C. (2011) Climate change, plant diseases and food security: an overview. *Plant Pathology* 60, 2–14.

Chakraborty, S., Pangga, I.B. and Roper, M.M. (2012) Climate change and multitrophic interactions in soil: the primacy of plants and functional domains. *Global Change Biology* 18, 2111–2125.

Cheatham, M.R., Rouse, M.N., Esker, P.D., Ignacio, S., Pradel, W., Raymundo, R., Sparks, A.H., Forbes, G.A., Gordon, T.R. and Garrett, K.A. (2009) Beyond yield: plant disease in the context of ecosystem services. *Phytopathology* 99, 1228–1236.

Chen, F.J., Ge, F. and Parajulee, M.N. (2005) Impact of elevated CO_2 on tri-trophic interaction of *Gossypium hirsutum, Aphis gossypii,* and *Leis axyridis. Environmental Entomology* 34, 37–46.

Chen, X.M. (2005) Epidemiology and control of stripe rust [*Puccinia striiformis* f. sp *tritici*] on wheat. *Canadian Journal of Plant Pathology-Revue Canadienne de Phytopathologie* 27, 314–337.

Clark, C.A., Davis, J.A., Mukasa, S.B., Abad, J.A., Tugume, A.K., Cuellar, W.J., Fuentes, S., Kreuze, J.F., Tairo, F.D., Gibson, R.W. and Valkonen, J.P.T. (2012) Sweetpotato viruses: 15 years of progress on understanding and managing complex diseases. *Plant Disease* 96, 168–185.

Coakley, S.M. (1979) Climate variability in the Pacific Northwest and its effect on stripe rust disease of winter wheat. *Climatic Change* 2, 33–51.

Coakley, S.M., Scherm, H. and Chakraborty, S. (1999) Climate change and plant disease management. *Annual Review of Phytopathology* 37, 399–426.

Cobb, R.C., Meentemeyer, R.K. and Rizzo, D.M. (2010) Apparent competition in canopy trees determined by pathogen transmission rather than susceptibility. *Ecology* 91, 327–333.

Compant, S., van der Heijden, M.G.A. and Sessitsch, A. (2010) Climate change effects on beneficial plant–microorganism interactions. *Fems Microbiology Ecology* 73, 197–214.

Cook, A.R., Otten, W., Marion, G., Gibson, G.J. and Gilligan, C.A. (2007) Estimation of multiple transmission rates for epidemics in heterogeneous populations. *Proceedings of the National Academy of Sciences* 104, 20392–20397.

Cooke, D.E.L., *et al.* (2012) Genome analyses of an aggressive and invasive lineage of the Irish Potato Famine pathogen. *Plos Pathogens* 8, e1002940.

Cooke, L.R., *et al.* (2011) Epidemiology and integrated control of potato late blight in Europe. *Potato Research* 54, 183–222.

Courchamp, F., Berex, L. and Gascoigne, J. (2009) *Allee Effects in Ecology and Conservation.* Oxford University Press, Oxford, UK.

Cox, C.M., Bockus, W.W., Holt, R.D., Fang, L. and Garrett, K.A. (2013) The spatial connectedness of plant species: potential links for apparent competition via plant diseases. *Plant Pathology* 62, 1195–1428.

Crouch, J.A., Beirn, L.A., Cortese, L.M., Bonos, S.A. and Clarke, B.B. (2009) Anthracnose disease of switchgrass caused by the novel fungal species *Colletotrichum navitas. Mycological Research* 113, 1411–1421.

Dixon, G.R. (2012) Climate change – impact on crop growth and food production, and plant pathogens. *Canadian Journal of Plant Pathology-Revue Canadienne de Phytopathologie* 34, 362–379.

Dukes, J.S., *et al.* (2009) Responses of insect pests, pathogens, and invasive plant species to climate change in the forests of northeastern North America: What can we predict? *Canadian Journal of Forest Research-Revue Canadienne de Recherche Forestière* 39, 231–248.

Dunn, A.M., *et al.* (2012) Indirect effects of parasites in invasions. *Functional Ecology* 26, 1262–1274.

Eastburn, D.M., McElrone, A.J. and Bilgin, D.D. (2011) Influence of atmospheric and climatic change on plant–pathogen interactions. *Plant Pathology* 60, 54–69.

Edmonds, R.L. (1979) *Aerobiology: The Ecological Systems Approach.* Dowden, Hutchinson, and Ross, Straudsburg, Pennsylvania.

Epanchin-Niell, R.S., Hufford, M.B., Aslan, C.E., Sexton, J.P., Port, J.D. and Waring, R.M. (2010) Controlling invasive species in complex social landscapes. *Frontiers in Ecology and the Environment* 8, 210–216.

Filipe, J.A.N., Cobb, R.C., Meentemeyer, R.K., Lee, C.A., Valachovic, Y.S., Cook, A.R., Rizzo, D.A. and Gilligan, C.A. (2012) Landscape epidemiology and control of pathogens with cryptic and long-distance dispersal: sudden oak death in Northern Californian Forests. *Plos Computational Biology* 8, e1002328, doi:10.1371/journal.pcbi.1002328.

Finlay, K.J. and Luck, J.E. (2011) Response of the bird cherry-oat aphid (*Rhopalosiphum padi*) to climate change in relation to its pest status, vectoring potential and function in a crop–vector–virus pathosystem. *Agriculture Ecosystems and Environment* 144, 405–421.

Fisher, M.C., Henk, D.A., Briggs, C.J., Brownstein, J.S., Madoff, L.C., McCraw, S.L. and Gurr, S.J. (2012) Emerging fungal threats to animal, plant and ecosystem health. *Nature* 484, 186–194.

Fitt, B.D.L., Huang, Y.J., van den Bosch, F. and West, J.S. (2006) Coexistence of related pathogen species on arable crops in space and time. *Annual Review of Phytopathology* 44, 163–182.

Fitzpatrick, M.C. and Hargrove, W.W. (2009) The projection of species distribution models and the problem of non-analog climate. *Biodiversity and Conservation* 18, 2255–2261.

Forbes, G.A. (2012) Using host resistance to manage potato late blight with particular reference to developing countries. *Potato Research* 55, 205–216.

Garrett, K.A. (2012) Information networks for plant disease: commonalities in human management networks and within-plant signaling networks. *European Journal of Plant Pathology* 133, 75–88.

Garrett, K.A. and Bowden, R.L. (2002) An Allee effect reduces the invasive potential of *Tilletia indica*. *Phytopathology* 92, 1152–1159.

Garrett, K.A., Dendy, S.P., Frank, E.E., Rouse, M.N. and Travers, S.E. (2006) Climate change effects on plant disease: genomes to ecosystems. *Annual Review of Phytopathology* 44, 489–509.

Garrett, K.A., Zúñiga, L.N., Roncal, E., Forbes, G.A., Mundt, C.C., Su, Z. and Nelson, R.J. (2009) Intraspecific functional diversity in hosts and its effect on disease risk across a climatic gradient. *Ecological Applications* 19, 1868–1883.

Garrett, K.A., *et al.* (2011) Complexity in climate-change impacts: an analytical framework for effects mediated by plant disease. *Plant Pathology* 60, 15–30.

Garrett, K.A., Jumpponen, A., Toomajian, C. and Gomez-Montano, L. (2012) Climate change and plant health: designing research spillover from plant genomics for understanding the role of microbial communities. *Canadian Journal of Plant Pathology* 34, 349–361.

Garrett, K.A., Dobson, A.D.M., Kroschel, J., Natarajan, B., Orlandini, S., Tonnang, H.E.Z. and Valdivia, C. (2013) The effects of climate variability and the color of weather time series on agricultural diseases and pests, and decision-making for their management. *Agricultural and Forest Meteorology* 170, 216–227.

Garrison, V.H., Shinn, E.A., Foreman, W.T., Griffin, D.W., Holmes, C.W., Kellogg, C.A., Majewski, M.S., Richardson, L.L., Ritchie, K.B. and Smith, G.W. (2003) African and Asian dust: from desert soils to coral reefs. *Bioscience* 53, 469–480.

Ghini, R., Hamada, E. and Bettiol, W. (2008) Climate change and plant diseases. *Scientia Agricola* 65, 98–107.

Ghini, R., Bettiol, W. and Hamada, E. (2011) Diseases in tropical and plantation crops as affected by climate changes: current knowledge and perspectives. *Plant Pathology* 60, 122–132.

Gilbert, G.S., Magarey, R., Suiter, K. and Webb, C.O. (2012) Evolutionary tools for phytosanitary risk analysis: phylogenetic signal as a predictor of host range of plant pests and pathogens. *Evolutionary Applications* 5, 869–878.

Gilligan, C.A. and van den Bosch, F. (2008) Epidemiological models for invasion and persistence of pathogens. *Annual Review of Phytopathology* 46, 385–418.

Gottwald, T.R. (2010) Current epidemiological understanding of citrus huanglongbing. *Annual Review of Phytopathology* 48, 119–139.

Gottwald, T.R., Sun, X., Riley, T., Graham, J.H., Ferrandino, F. and Taylor, E.L. (2002) Georeferenced spatiotemporal analysis of the urban citrus canker epidemic in Florida. *Phytopathology* 92, 361–377.

Gouache, D., Bensadoun, A., Brun, F., Pagé, C., Makowski, D. and Wallach, D. (2013) Modelling climate change impact on Septoria tritici blotch (STB) in France: accounting for climate model and disease model uncertainty. *Agricultural and Forest Meteorology* 170, 242–252.

Greene, J.S., Chatelain, M., Morrissey, M. and Stadler, S. (2012) Estimated changes in wind speed and wind power density over the western High Plains, 1971–2000. *Theoretical and Applied Climatology* 109, 507–518.

Gregory, P.J., Johnson, S.N., Newton, A.C. and Ingram, J.S.I. (2009) Integrating pests and pathogens into the climate change/food security debate. *Journal of Experimental Botany* 60, 2827–2838.

Hardham, A.R. (2005) *Phytophthora cinnamomi*. *Molecular Plant Pathology* 6, 589–604.

Harwood, T.D., Xu, X., Pautasso, M., Jeger, M.J. and Shaw, M.W. (2009) Epidemiological risk assessment using linked network and grid based modelling: *Phytophthora ramorum* and *Phytophthora kernoviae* in the UK. *Ecological Modelling* 220, 3353–3361.

Haugen, J.E. and Iversen, T. (2008) Response in extremes of daily precipitation and wind from a downscaled multi-model ensemble of anthropogenic global climate change scenarios. *Tellus Series A: Dynamic Meteorology and Oceanography* 60, 411–426.

Hoffmann, U. (2011) Assuring food security in developing countries under the challenges of climate change: Key trade and development issues of a fundamental transformation of agriculture. United Nations Conference on Trade and Development (http://unctad.org/en/docs/osgdp20111_en.pdf, accessed 10 February 2014).

Hopkins, D.L. and Purcell, A.H. (2002) *Xylella fastidiosa*: cause of Pierce's disease of grapevine and other emergent diseases. *Plant Disease* 86, 1056–1066.

Hovmøller, M.S., Yahyaoui, A.H., Milus, E.A. and Justesen, A.F. (2008) Rapid global spread of two aggressive strains of a wheat rust fungus. *Molecular Ecology* 17, 3818–3826.

Hovmøller, M.S., Sorensen, C.K., Walter, S. and Justesen, A.F. (2011) Diversity of *Puccinia striiformis* on cereals and grasses. *Annual Review of Phytopathology* 49, 197–217.

Howden, S.M., Soussana, J.-F., Tubiello, F.N., Chhetri, N., Dunlop, M. and Meinke, H. (2007) Adapting agriculture to climate change. *Proceedings of the National Academy of Sciences* 104, 19691–19696.

Hughes, D.J., West, J.S., Atkins, S.D., Gladders, P., Jeger, M.J. and Fitt, B.D.L. (2011) Effects of disease control by fungicides on greenhouse gas emissions by UK arable crop production. *Pest Management Science* 67, 1082–1092.

Isard, S.A., Gage, S.H., Comtois, P. and Russo, J.M. (2005) Principles of the atmospheric pathway for invasive species applied to soybean rust. *Bioscience* 55, 851–861.

Isard, S.A., Russo, J.M. and Ariatti, A. (2007) The Integrated Aerobiology Modeling System applied to the spread of soybean rust into the Ohio River valley during September 2006. *Aerobiologia* 23, 271–282.

Isard, S.A., Barnes, C.W., Hambleton, S., Ariatti, A., Russo, J.M., Tenuta, A., Gay, D.A. and Szabo, L.J. (2011) Predicting soybean rust incursions into the North American continental interior in 2007 and 2008 using crop monitoring, spore trapping, and aerobiological modeling. *Plant Disease* 95, 1346–1357.

Jeger, M.J. and Pautasso, M. (2008) Plant disease and global change – the importance of long-term data sets. *New Phytologist* 177, 8–11.

Jeger, M.J., Pautasso, M., Holdenrieder, O. and Shaw, M.W. (2007) Modelling disease spread and control in networks: implications for plant sciences. *New Phytologist* 174, 279–297.

Jones, R.A.C. (2009) Plant virus emergence and evolution: origins, new encounter scenarios, factors driving emergence, effects of changing world conditions, and prospects for control. *Virus Research* 141, 113–130.

Jones, R.A.C. and Barbetti, M.J. (2012) Influence of climate change on plant disease infections and epidemics caused by viruses and bacteria. *CAB Reviews* 7, 1–33.

Juroszek, P. and von Tiedemann, A. (2011) Potential strategies and future requirements for plant disease management under a changing climate. *Plant Pathology* 60, 100–112.

Kriss, A.B., Paul, P.A. and Madden, L.V. (2012) Variability in Fusarium head blight epidemics in relation to global climate fluctuations as represented by the El Niño–Southern Oscillation and other atmospheric patterns. *Phytopathology* 102, 55–64.

Lamour, A., Termorshuizen, A.J., Volker, D. and Jeger, M.J. (2007) Network formation by rhizomorphs of *Armillaria lutea* in natural soil: their description and ecological significance. *Fems Microbiology Ecology* 62, 222–232.

Legg, J.P. (1999) Emergence, spread and strategies for controlling the pandemic of cassava mosaic virus disease in east and central Africa. *Crop Protection* 18, 627–637.

Legg, J.P., *et al.* (2011) Comparing the regional epidemiology of the cassava mosaic and cassava brown streak virus pandemics in Africa. *Virus Research* 159, 161–170.

Lehmer, E.M., Korb, J., Bombaci, S., Ghachu, J., Hart, L., Kelly, A., Jara-Molinar, E., O'Brien, C. and Wright, K. (2012) The interplay of plant and animal disease in a changing landscape: the role of sudden aspen decline in moderating Sin Nombre virus prevalence in natural deer mouse populations. *Ecohealth* 9, 205–216.

Li, X., Esker, P.D., Pan, Z., Dias, A.P., Xue, L. and Yang, X.B. (2010) The uniqueness of the soybean rust pathosystem: an improved understanding of the risk in different regions of the world. *Plant Disease* 94, 796–806.

Lindow, S.E. (1983) The role of bacterial ice nucleation in frost injury to plants. *Annual Review of Phytopathology* 21, 363–384.

Lindow, S.E., Arny, D.C. and Upper, C.D. (1978) Distribution of ice nucleation-active bacteria on plants in nature. *Applied and Environmental Microbiology* 36, 831–838.

Lowe, S., Browne, M., Boudjelas, S. and De Poorter, M. (2000) 100 of the world's worst invasive alien species: a selection from the global invasive species database. Invasive Species Specialist Group, Auckland, New Zealand.

Luck, J., Spackman, M., Freeman, A., Trebicki, P., Griffiths, W., Finlay, K. and Chakraborty, S. (2011) Climate change and diseases of food crops. *Plant Pathology* 60, 113–121.

McDonald, B.A. and Linde, C. (2002) Pathogen population genetics, evolutionary potential, and durable resistance. *Annual Review of Phytopathology* 40, 349–379.

McInnes, K.L., Erwin, T.A. and Bathols, J.M. (2011) Global Climate Model projected changes in 10 m wind speed and direction due to anthropogenic climate change. *Atmospheric Science Letters* 12, 325–333.

McRoberts, N., Hall, C., Madden, L.V. and Hughes, G. (2011) Perceptions of disease risk: from social construction of subjective judgments to rational decision making. *Phytopathology* 101, 654–665.

Madgwick, J.W., West, J.S., White, R.P., Semenov, M.A., Townsend, J.A., Turner, J.A. and Fitt, B.D.L. (2011) Impacts of climate change on wheat anthesis and fusarium ear blight in the UK. *European Journal of Plant Pathology* 130, 117–131.

Mahmuti, M., West, J.S., Watts, J., Gladders, P. and Fitt, B.D.L. (2009) Controlling crop disease contributes to both food security and climate change mitigation. *International Journal of Agricultural Sustainability* 7, 189–202.

Malmstrom, C.M., McCullough, A.J., Johnson, H.A., Newton, L.A. and Borer, E.T. (2005) Invasive annual grasses indirectly increase virus incidence in California native perennial bunchgrasses. *Oecologia* 145, 153–164.

Malmstrom, C.M., Melcher, U. and Bosque-Perez, N.A. (2011) The expanding field of plant virus ecology: historical foundations, knowledge gaps, and research directions. *Virus Research* 159, 84–94.

Margosian, M.L., Garrett, K.A., Hutchinson, J.M.S. and Wirth, K.A. (2009) Connectivity of the American agricultural landscape: assessing the national risk of crop pest and disease spread. *Bioscience* 59, 141–151.

Meentemeyer, R.K., Haas, S.E. and Vaclavik, T. (2012) Landscape epidemiology of emerging infectious diseases in natural and human-altered ecosystems. *Annual Review of Phytopathology* 50, 379–402.

Miller, Z.J. (2012) Fungal pathogen species richness: Why do some plant species have more pathogens than others? *American Naturalist* 179, 282–292.

Mills, P., Dehnen-Schmutz, K., Ilbery, B., Jeger, M., Jones, G., Little, R., MacLeod, A., Parker, S., Pautasso, M., Pietravalle, S. and Maye, S. (2011) Integrating natural and social science perspectives on plant disease risk, management and policy formulation. *Philosophical Transactions of the Royal Society B-Biological Sciences* 366, 2035–2044.

Milus, E.A., Seyran, E. and McNew, R. (2006) Aggressiveness of *Puccinia striiformis* f. sp *tritici* isolates in the South-Central United States. *Plant Disease* 90, 847–852.

Mitchell, C.E. and Power, A.G. (2003) Release of invasive plants from fungal and viral pathogens. *Nature* 421, 625–627.

Mohler, O., DeMott, P.J., Vali, G. and Levin, Z. (2007) Microbiology and atmospheric processes: the role of biological particles in cloud physics. *Biogeosciences* 4, 1059–1071.

Morris, C.E., Sands, D.C., Bardin, M., Jaenicke, R., Vogel, B., Leyronas, C., Ariya, P.A. and Psenner, R. (2011) Microbiology and atmospheric processes: research challenges concerning the impact of airborne micro-organisms on the atmosphere and climate. *Biogeosciences* 8, 17–25.

Moslonka-Lefebvre, M., Finley, A., Dorigatti, I., Dehnen-Schmutz, K., Harwood, T., Jeger, M.J., Xiangming, X., Holdenrieder, O. and Pautasso, M. (2011) Networks in plant epidemiology: from genes to landscapes, countries, and continents. *Phytopathology* 101, 392–403.

Mundt, C.C., Sackett, K.E., Wallace, L.D., Cowger, C. and Dudley, J.P. (2009) Long-distance dispersal and accelerating waves of disease: empirical relationships. *American Naturalist* 173, 456–466.

Mundt, C.C., Sackett, K.E. and Wallace, L.D. (2011) Landscape heterogeneity and disease spread: experimental approaches with a plant pathogen. *Ecological Applications* 21, 321–328.

Najac, J., Boe, J. and Terray, L. (2009) A multi-model ensemble approach for assessment of climate change impact on surface winds in France. *Climate Dynamics* 32, 615–634.

Newman, M. (2010) *Networks: An Introduction.* Oxford University Press, New York.

Newton, A.C., Johnson, S.N. and Gregory, P.J. (2011) Implications of climate change for diseases, crop yields and food security. *Euphytica* 179, 3–18.

Oerke, E.C. (2006) Crop losses to pests. *Journal of Agricultural Science* 144, 31–43.

Olofsson, J., Ericson, L., Torp, M., Stark, S. and Baxter, R. (2011) Carbon balance of Arctic tundra under increased snow cover mediated by a plant pathogen. *Nature Climate Change* 1, 220–223.

Pan, Z., Yang, X.B., Pivonia, S., Xue, L., Pasken, R. and Roads, J. (2006) Long-term prediction of soybean rust entry into the Continental United States. *Plant Disease* 90, 840–846.

Parnell, S., Gottwald, T.R., van den Bosch, F. and Gilligan, C.A. (2009) Optimal strategies for the eradication of Asiatic citrus canker in heterogeneous host landscapes. *Phytopathology* 99, 1370–1376.

Pautasso, M., Dehnen-Schmutz, K., Holdenrieder, O., Pietravalle, S., Salama, N., Jeger, M.J., Lange, E. and Hehl-Lange, S. (2010) Plant health and global change – some implications for landscape management. *Biological Reviews* 85, 729–755.

Pautasso, M., Doring, T.F., Garbelotto, M., Pellis, L. and Jeger, M.J. (2012) Impacts of climate change on plant diseases – opinions and trends. *European Journal of Plant Pathology* 133, 295–313.

Perez, C., Nicklin, C., Dangles, O., Vanek, S., Sherwood, S., Halloy, S., Garrett, K. and Forbes, G. (2010) Climate change in the High Andes: implications and adaptation strategies for small-scale farmers. *International Journal of Environmental, Cultural, Economic and Social Sustainability* 6, 71–88.

Power, A.G. and Mitchell, C.E. (2004) Pathogen spillover in disease epidemics. *American Naturalist* 164, S79–S89.

Pratt, K.A., DeMott, P.J., French, J.R., Wang, Z., Westphal, D.L., Heymsfield, A.J., Twohy, C.H.,

Prenni, A.J. and Prather, K.A. (2009) *In situ* detection of biological particles in cloud ice-crystals. *Nature Geoscience* 2, 397–400.

Pritchard, S.G. (2011) Soil organisms and global climate change. *Plant Pathology* 60, 82–99.

Pryor, S.C., Barthelmie, R.J., Clausen, N.E., Drews, M., MacKellar, N. and Kjellstrom, E. (2012) Analyses of possible changes in intense and extreme wind speeds over northern Europe under climate change scenarios. *Climate Dynamics* 38, 189–208.

Raaijmakers, J.M., Paulitz, T.C., Steinberg, C., Alabouvette, C. and Moenne-Loccoz, Y. (2009) The rhizosphere: a playground and battlefield for soilborne pathogens and beneficial microorganisms. *Plant and Soil* 321, 341–361.

Rizzo, D.M., Garbelotto, M. and Hansen, E.A. (2005) *Phytophthora ramorum*: integrative research and management of an emerging pathogen in California and Oregon forests. *Annual Review of Phytopathology* 43, 309–335.

Roos, J., Hopkins, R., Kvarnheden, A. and Dixelius, C. (2011) The impact of global warming on plant diseases and insect vectors in Sweden. *European Journal of Plant Pathology* 129, 9–19.

Rosenzweig, C., Iglesias, A., Yang, X.B., Epstein, P.R. and Chivian, E. (2001) Climate change and extreme weather events; implications for food production, plant diseases, and pests. *Global Change and Human Health* 2, 90–104.

Rypien, K.L. (2008) African dust is an unlikely source of *Aspergillus sydowii*, the causative agent of sea fan disease. *Marine Ecology Progress Series* 367, 125–131.

Salazar, L.F. (2006) Emerging and re-emerging potato diseases in the Andes. *Potato Research* 49, 43–47.

Saleh, A.A., Ahmed, H.U., Todd, T.C., Travers, S.E., Zeller, K.A., Leslie, J.F. and Garrett, K.A. (2010) Relatedness of *Macrophomina phaseolina* isolates from tallgrass prairie, maize, soybean and sorghum. *Molecular Ecology* 19, 79–91.

Savary, S., Teng, P.S., Willocquet, L. and Nutter, F.W. (2006) Quantification and modeling of crop losses: a review of purposes. *Annual Review of Phytopathology* 44, 89–112.

Savary, S., Mila, A., Willocquet, L., Esker, P.D., Carisse, O. and McRoberts, N. (2011a) Risk factors for crop health under global change and agricultural shifts: a framework of analyses using rice in tropical and subtropical Asia as a model. *Phytopathology* 101, 696–709.

Savary, S., *et al.* (2011b) International agricultural research tackling the effects of global and climate changes on plant diseases in the developing world. *Plant Disease* 95, 1204–1216.

Savary, S., Nelson, A., Willocquet, L., Pangga, I. and Aunario, J. (2012) Modeling and mapping potential epidemics of rice diseases globally. *Crop Protection* 34, 6–17.

Scherm, H. (2004) Climate change: can we predict the impacts on plant pathology and pest management? *Canadian Journal of Plant Pathology-Revue Canadienne de Phytopathologie* 26, 267–273.

Scherm, H. and van Bruggen, A.H.C. (1994) Global warming and non-linear growth: how important are changes in average temperature? *Phytopathology* 84, 1380–1384.

Scherm, H. and Yang, X.B. (1995) Interannual variations in wheat rust development in China and the United States in relation to the El Niño/Southern Oscillation. *Phytopathology* 85, 970–976.

Scherm, H., Sutherst, R.W., Harrington, R. and Ingram, J.S.I. (2000) Global networking for assessment of impacts of global change on plant pests. *Environmental Pollution* 108, 333–341.

Schneider, R.W., Hollier, C.A., Whitam, H.K., Palm, M.E., McKemy, J.M., Hernandez, J.R., Levy, L. and De Vries-Paterson, R. (2005) First report of soybean rust caused by *Phakopsora pachyrhizi* in the continental United States. *Plant Disease* 89, 774.

Schrotenboer, A.C., Allen, M.S. and Malmstrom, C.M. (2011) Modification of native grasses for biofuel production may increase virus susceptibility. *Global Change Biology Bioenergy* 3, 360–374.

Shaw, M.W. and Osborne, T.M. (2011) Geographic distribution of plant pathogens in response to climate change. *Plant Pathology* 60, 31–43.

Shurtleff, M.C. and Averre, C.W. III (1997) *Glossary of Plant Pathological Terms.* APS Press, St Paul, Minnesota.

Singh, R.P., Hodson, D.P., Huerta-Espino, J., Jin, Y., Bhavani, S., Njau, P., Herrera-Foessel, S., Singh, P.K, Singh, S. and Govindan, V. (2011) The emergence of Ug99 races of the stem rust fungus is a threat to world wheat production. *Annual Review of Phytopathology* 49, 465–481.

Skelsey, P., Holtslag, A.A.M. and van der Werf, W. (2008) Development and validation of a quasi-Gaussian plume model for the transport of botanical spores. *Agricultural and Forest Meteorology* 148, 1383–1394.

Skelsey, P., With, K.A. and Garrett, K.A. (2013) Why dispersal should be maximized at intermediate scales of heterogeneity. *Theoretical Ecology* 6, 203–211.

Sparks, A.H., Forbes, G.A., Hijmans, R.J. and Garrett, K.A. (2011) A metamodeling framework for extending the application domain of process-based ecological models. *Ecosphere* 2, art90.

Stewart, A. and Cromey, M. (2011) Identifying disease threats and management practices for bio-energy crops. *Current Opinion in Environmental Sustainability* 3, 75–80.

Sturrock, R.N., Frankel, S.J., Brown, A.V., Hennon, P.E., Kliejunas, J.T., Lewis, K.J., Worrall, J.J. and Woods, A.J. (2011) Climate change and forest diseases. *Plant Pathology* 60, 133–149.

Sutherst, R.W., Constable, F., Finlay, K.J., Harrington, R., Luck, J. and Zalucki, M.P. (2011) Adapting to crop pest and pathogen risks under a changing climate. *Wiley Interdisciplinary Reviews-Climate Change* 2, 220–237.

Sutrave, S., Scoglio, C., Isard, S.A., Hutchinson, J.M.S. and Garrett, K.A. (2012) Identifying highly connected counties compensates for resource limitations when sampling national spread of an invasive pathogen. *PLoS ONE* 7, e37793.

Swei, A., Briggs, C.J., Lane, R.S. and Ostfeld, R.S. (2012) Impacts of an introduced forest pathogen on the risk of Lyme Disease in California. *Vector-Borne and Zoonotic Diseases* 12, 623–632.

Tilman, D., Hill, J. and Lehman, C. (2006) Carbon-negative biofuels from low-input high-diversity grassland biomass. *Science* 314, 1598–1600.

Venette, R.C. and Cohen, S.D. (2006) Potential climatic suitability for establishment of *Phytophthora ramorum* within the contiguous United States. *Forest Ecology and Management* 231, 18–26.

Venette, R.C., *et al.* (2010) Pest risk maps for invasive alient species: a roadmap for improvement. *Bioscience* 60, 349–362.

Ward, N.L. and Masters, G.J. (2007) Linking climate change and species invasion: an illustration using insect herbivores. *Global Change Biology* 13, 1605–1615.

Watt, M.S., Ganley, R.J., Kriticos, D.J. and Manning, L.K. (2011) Dothistroma needle blight and pitch canker: the current and future potential distribution of two important diseases of Pinus species. *Canadian Journal of Forest Research-Revue Canadienne de Recherche Forestière* 41, 412–424.

Webb, K.M., Ona, I., Bai, J., Garrett, K.A., Mew, T., Vera Cruz, C.M. and Leach, J.E. (2010) A benefit of high temperature: increased effectiveness of a rice bacterial blight disease resistance gene. *New Phytologist* 185, 568–576.

Weir-Brush, J.R., Garrison, V.H., Smith, G.W. and Shinn, E.A. (2004) The relationship between gorgonian coral (Cnidaria: Gorgonacea) diseases and African dust storms. *Aerobiologia* 20, 119–126.

Wellings, C.R., McIntosh, R.A. and Walker, J. (1987) *Puccinia striiformis* f.sp. *tritici* in Eastern Australia – possible means of entry and implications for plant quarantine. *Plant Pathology* 36, 239–241.

Wilkinson, K., Grant, W.P., Green, L.E., Hunter, S., Jeger, M.J., Lowe, P., Medley, G.F., Mills, P., Phillipson, J., Poppy, G.M. and Waage, J. (2011) Infectious diseases of animals and plants: an interdisciplinary approach Introduction. *Philosophical Transactions of the Royal Society B-Biological Sciences* 366, 1933–1942.

Woods, A., Coates, K.D. and Hamann, A. (2005) Is an unprecedented Dothistroma needle blight epidemic related to climate change? *Bioscience* 55, 761–769.

Worner, S.P. and Gevrey, M. (2006) Modelling global insect pest species assemblages to determine risk of invasion. *Journal of Applied Ecology* 43, 858–867.

Yamamura, K. and Kiritani, K. (1998) A simple method to estimate the potential increase in the number of generations under global warming in temperate zones. *Applied Entomology and Zoology* 33, 289–298.

Ziska, L.H. and Beggs, P.J. (2012) Anthropogenic climate change and allergen exposure: the role of plant biology. *Journal of Allergy and Clinical Immunology* 129, 27–32.

4 Analysis of Invasive Insects: Links to Climate Change

Andrew Paul Gutierrez[1,2] and Luigi Ponti[1,3]

[1]*Center for the Analysis of Sustainable Agricultural Systems (CASAS Global NGO), Kensington, California, USA;* [2]*Division of Ecosystem Science, College of Natural Resources, University of California, Berkeley, California, USA;* [3]*Laboratorio Gestione Sostenibile degli Agro-Ecosistemi (UTAGRI-ECO), Agenzia Nazionale per le Nuove Tecnologie, l'Energia e lo Sviluppo Economico Sostenibile (ENEA), Centro Ricerche Casaccia, Rome, Italy*

Abstract

Climate change is expected to alter the geographic distribution and abundance of many species, to increase the invasion of new areas by exotic species and, in some cases, to lead to species extinction. This chapter reviews some of the links between invasive insects and climate change. The effects of climate change on insect pest populations can be direct, through impacts on their physiology and behaviour, or indirect, through biotic interactions (i.e. bottom-up and top-down effects). Anthropogenic climate and global change is expected to be a major driver in the introduction, establishment, distribution, impact and changes in the efficacy of mitigation strategies for invasive species. To address these problems, we must be able to predict climate change impacts on species distribution and abundance. Commonly used ecological niche modelling approaches have implicit assumptions about the biology of the target species and attempt to characterize the ecological niche using aggregate weather and other factors in the area of recorded distribution. More holistic physiologically based demographic modelling approaches explicitly describe the biological and physiological responses of species to weather and the species they interact with on fine temporal and spatial scales. The geographic distribution and relative abundance of four invasive insect pests are reviewed under observed and +2°C weather scenarios across the USA and Mexico: the tropical New World screwworm, the pink bollworm, the Mediterranean fruit fly (i.e. medfly) and the olive fly. The distribution of the olive fly is examined across the Mediterranean basin to illustrate the transferability of the model to analyses of new regions and climate change scenarios.

Introduction

Climate change is expected to alter the distribution and abundance of many species and increase the invasion of new areas by exotic species, with consequences for ecosystem function. Invasive species collectively cause in excess of US$140 billion in losses annually in the USA (Pimentel *et al.*, 2000) and US$1 trillion worldwide (Oerke and Dehne, 2004). This is nearly an order of magnitude higher than losses caused by natural disasters (see Ricciardi *et al.*, 2011). Among invasive alien species, most plant introductions have been intentional (e.g. crops, ornamentals) or as contaminants in seed and soil (e.g. yellow starthistle, *Centaurea solstitialis* L.). In contrast, most arthropod/insect introductions have been accidental (Pimentel *et al.*, 2005). Insects as a major taxonomic group have a long history of invasion and establishment, and have garnered considerable research focus due to their often high economic impact and visibility (Pyšek *et al.*, 2008). Overall, a substantial share, 30% or more, of insect pests are alien species, but they make up a small fraction of the total insect fauna (Ward and Masters, 2007). A strong geographical bias exists in invasion biology, with most efforts addressing invasive species in the USA and other developed countries, with Africa and Asia being understudied and likely reflecting differences in regional wealth (Pyšek *et al.*, 2008). This bias has serious ecological and economic consequences for rapidly globalizing countries such as India that face high risk of biological invasions, with the potential consequences being felt globally as these understudied areas are not only the recipients but also major global sources of invasive species (Khuroo *et al.*, 2011).

A recent review of the advances in the analysis of the responses of invasive species to changing climate can be found in Dukes (2011). Currently, the ability to deal with the new challenges posed by the effects of climate change on invasive species relies heavily on experimental results that tend to be short term, of limited availability and based on factors (e.g. warming, [CO_2]) applied separately to single trophic levels (Ward and Masters, 2007). It is obvious that new approaches are needed to bridge the gap between field experiments used to study invasive insect biology and climate envelope approaches commonly used in macroecology to assess the distribution of species (Kerr *et al.*, 2007). This chapter reviews some of the links between species invasions and climate change, with a focus on specific biological characteristics common to arthropods and insects. While the chapter emphasizes invasive insects, many of the arguments and methods of analysis also apply to the native and invasive species of any taxa.

Examining Links between Species Invasions and Climate Change

The effects of climate change on insect pest populations can be direct, through impacts on their physiology and behaviour, or indirect, through biotic interactions with other species (i.e. bottom-up and top-down effects; Bale *et al.*, 2002). Hellmann *et al.* (2008) identified five potential areas affecting insect invasions with respect to anthropogenic climate and global change: (i) mechanisms of transport and introduction; (ii) enhanced establishment; (iii) changes in the distribution of existing invasive species; (iv) the impact of established invasive species; and (v) changes in the efficacy of management strategies.

Transport and introduction

Global trade is generally recognized as the primary driver for the introduction and spread of exotic insect species (e.g. Tatem and Hay, 2007; Tatem, 2009). On broad spatial scales and for a range of taxonomic groups including insects, Essl *et al.* (2011) posit that the consequences of the current high levels of international trade on the extent of biological invasions will probably be completely realized only decades into the future, pointing out that the numbers of alien species currently established correlate more closely to past indicators of socio-economic activity (e.g. trade). Current invasions may, to some extent, be seen as a socio-economic legacy of past economic activities.

Climate change may also alter the transport and introduction of species directly. For example, wind is a significant factor in the distribution and colonization of many species, and projected increases in storm frequency and severity are likely to augment the number of invasive pest (plant seeds, pathogens and insects) introductions and the distance they travel. The longest weather-associated migration of an insect occurred in 1988, when desert locusts from Africa were transported 4500 km by a subtropical low pressure wave and a resulting hurricane to the Caribbean islands and the east coast of South America (Rosenberg and Burt, 1999), but less spectacular windborne movements of pests and pathogens are a common occurrence.

Establishment

Surface warming may be a key factor in the establishment of new invasive species. Using the first record of occurrence for 394 invasive and non-invasive alien insect species introduced in China, the USA and the UK during the period 1900–2005, Huang *et al.* (2011) found a significant positive association between the establishment rates of invasive insects and the increase in mean annual temperature. This association was significant even when adjusted for changes in international trade, suggesting that climate warming could boost the establishment rates of invasive insects even without an increased probability of introduction. Climate warming is also projected to expand the range of invasive species (see below). For example, the globally invasive tiger mosquito (*Aedes albopictus*; see Benedict *et al.*, 2007) has expanded its range into new alpine areas in Europe from its current range in northern Italy (Roiz *et al.*, 2011). Warmer nights are increasing the flight activity of the native winter pine processionary moth (*Thaumetopoea pityocampa* Denis & Schiffermüller) in Europe, enabling the adults to disperse greater distances (Battisti *et al.*, 2006). There are no doubt countless other developing examples.

Distribution

A common problem in biological control is that invasive pathogens, insects and other pests in new geographical ranges are likely to have few natural enemies, and as a result, their distribution tends to be limited more by climate than by biotic interactions (Scherm and Coakley, 2003). Warming climate may alter the distribution of existing invasive insects by helping their successful spread into new areas (Robinet and Roques, 2010) or by restricting the spread in areas close to the upper range of favourability (e.g. Gutierrez and Ponti, 2013). The rate of climate change (i.e. geographic shifts of isotherms over time; see Loarie *et al.*, 2009) is of concern for biodiversity conservation, as the dispersal rates of many species, especially perennial plants, are up to an order of magnitude slower than for many annual plant species (Burrows *et al.*, 2011), increasing the risk of extinction of perennials. The rate of climate change is less of a problem for invasive insects (Hulme, 2012) because their rates of spread are often an additional order of magnitude higher (Liebhold and Tobin, 2008) than for annual plants. For this reason, Ohlemüller (2011) posited that invasive insect species were less likely to run out of suitable climatic conditions (i.e. areas within the species' climatic niche space) than native species in the face of climate change. Climate change affects the physiology and behaviour of insects (and plants), and even modest warming can affect dramatically the energy budget of insects in diapause, and affect the species' geographic distributions and success of invasion (Hahn and Denlinger, 2011). Warming could affect all stages via increased respiration and decreases in growth, reproduction and survival. These changes would also appear to complicate biological control efforts. Methods for predicting the potential geographic distribution and relative abundance must include the above factors.

Impact of established invaders

Established insect invaders may impact ecosystems both directly, via trophic interactions (herbivory, predation or parasitism), and indirectly, via more complex mechanisms such as competition for resources, disease transmission, apparent competition, or pollination disruption, and also in other ways (Kenis *et al.*, 2009; Schweiger *et al.*, 2010). Pollination is a key ecosystem service in natural and agricultural systems, and adapted local pollinators may be prone to disruption by the interaction of climate change and invasive species (Schweiger *et al.*, 2008, 2010). The impact of invasive insect pests may also be mediated by the effects of climate change on their natural enemies (Thomson *et al.*, 2010) and via species interactions (Yang and Rudolf, 2010). The effects of climate change on the majority of invasive species and their biotic interactions have not been considered and conceal significant unforeseen vulnerability (Burkett *et al.*, 2005).

Management efficacy

A review of 136 case studies of eradication efforts against 75 invasive species indicated that local campaigns were more successful than regional or national ones, whereas reaction time, level of knowledge and insularity were unrelated to eradication success (Pluess *et al.*, 2012). The effectiveness of management strategies for invasive species is mostly a function of the capacity to assess the dynamics of the invaded biological systems, with difficulty likely increasing in the face of climate change (see Gilman *et al.*, 2010) and rapid change in abiotic factors, insect migrations and demographic pressures. Complicating this is the fact that the eradication/containment of invasive species is often attempted without the benefit of sound analysis of the potential range and invasiveness of the invasive species (Myers *et al.*, 1998, 2000; Gutierrez and Ponti, 2013).

Biological control is a viable option in many cases, but requires sound analysis before natural enemies are introduced. At a more pragmatic level, climate uncertainty related to temperature, rainfall, wind speed and other factors is also likely to change current pest control practices, by altering the timing of pesticide applications and their effectiveness (e.g. persistence, drift and deposition). Although experimental data are lacking, statistical evaluations suggest increased pesticide usage in many crops in response to temperature and/or precipitation (Chen and McCarl, 2001). Rising CO_2 per se may increase competition between plants (Manea and Leishman, 2011) and reduce the efficacy of certain herbicides (Ziska *et al.*, 1999). The indirect effects of CO_2 on insecticide and fungicide efficacy have not been characterized.

Assessing Climate Change Effects: Geographic Distribution and Abundance of Invasive Insects

Plants and arthropods (e.g. Phylum Arthropoda: Class Insecta) are heterothermic, indicating that their developmental rates vary with the temperature of their immediate environment, but other factors such as moisture and nutrition may also influence development. In the short run, the phenology, dynamics and abundance of invasive heterotherms in an area is determined largely by the weather, interactions with other species and the resources in their food webs and chains. Climate (i.e. long-run weather) and interactions with other species and resources determine the average abundance and geographic range. Climate change, specifically the effects of changes in surface temperatures, precipitation and rising $[CO_2]$ on plant hosts are likely to alter the distribution and abundance of many insects directly and via impacts on trophic interactions.

CO_2 levels (George *et al.*, 2007) and temperatures (Meineke *et al.*, 2013) are known to be higher in urban areas, with temperatures being of a similar magnitude to the global warming predicted for the next 50 years. These conditions provide microcosm climate warming experiments. For example, Meineke *et al.* (2013) found that the scale insect, *Parthenolecanium quercifex*, was 13 times more abundant on willow oak trees in the hottest parts of Raleigh, North Carolina, in the south-eastern USA, than in cooler areas within the city, despite similar parasitism rates. Furthermore, scale collected from trees in hot areas became more abundant in hot greenhouses than in cool greenhouses, whereas the abundance of scale collected from cooler areas remained low in hot and cool greenhouses. This suggested that some scale demes had adapted to high temperatures. The synergism of climate with biotic and human-mediated components, including habitat modification, is illustrated by the wide distribution of the ubiquitous invasive Argentine ant (*Linepithema humile* (Mayr); Roura-Pascual *et al.*, 2011). The direct effect of global climate and environmental change is thought to add considerable uncertainty to the structure, function and stability of native and invaded ecosystems (Tylianakis *et al.*, 2008).

However, although the biological implications of climate change are often posited, how climate change will alter the risk, spread and impact of invasive species is poorly understood (see Yemshanov *et al.*, 2009). Ziska *et al.* (2011) posed the following questions in that regard: how vulnerable is agriculture to invasive species; are current pest management strategies sufficient to control invasive outbreaks in the future; what are the knowledge gaps; can we provide initial recommendations for scientists, land managers and policy makers in regard to available resources? We note that the same questions and methods of analysis should apply to species in natural ecosystems that will also be affected by climate change.

It is generally acknowledged that the impact of climate change on native and invasive species biology can be understood best via holistic analysis (e.g. Sutherst and Bourne, 2009; Gilman *et al.*, 2010), but rarely is this done, and further, the question of what methods should be used in the analyses remains largely unresolved (Gallien *et al.*, 2010). A critical first step to assess the invasiveness of a species is to predict accurately the potential changes in its populations in time and space. Commonly used methods to predict the geographic distribution of species are classed as ecological niche models (ENMs), but holistic realistic physiologically based demographic modelling (PBDMs) offering a larger suite of insights is also available. The conceptual underpinnings of ENMs and PBDMs are introduced below, with an exposé on their ecological and bioeconomic applications given in greater detail in Chapter 17.

Ecological Niche Models (ENMs)

ENMs are developed to characterize the ecological niche space of a species based on the aggregate weather and other factors in their recorded geographic range, and are often used to predict the geographic/climatic range in new areas. ENMs assume the current geographic distribution of a species is the best indicator of its climatic requirements, that the distribution is in equilibrium with the current climate and climate niche conservatism is maintained in both space and time (Beaumont *et al.*, 2009; see Elith and Leathwick, 2009).

But, ENMs have important deficiencies: difficulty incorporating trophic interactions (Davis *et al.*, 1998), the use of aggregate weather data that miss important short-term weather effects (see Lozier *et al.*, 2009), implicit ecological and mathematical assumptions having no mechanistic basis (Soberón and Nakamura, 2009); the results and transferability (validation) are conjectural, with a predictive power potentially lower than simple spatial interpolation (Bahn and McGill, 2007). Further, some caution that ENM predictions of species' responses to climate change based on projections in altered environmental conditions may be unsound, as climate change may influence the abundance of species and the direction of selection, which may also disrupt trophic interactions (van der Putten *et al.*, 2010; Wardle *et al.*, 2011). Despite shortcomings, ENMs have been used widely to assess the potential distribution of invasive species under climate change (Jeschke and Strayer, 2008). Thuiller *et al.* (2005) suggest that ENMs provide unbiased first steps for screening the potential range of invasive plants in new areas and, we add, may also be useful for the initial evaluation of arthropods and insects as well (e.g. Lozier and Mills, 2011).

Physiologically Based Demographic Models (PBDMs)

Instead of using distribution records from the native range of an invasive species as the basis for estimating the ecological niche (see above), PBDMs explicitly model the mechanistic, weather-driven biology of the species, and of the relevant species in its food chain or web, and use the model to predict prospectively the phenology, dynamics and distribution of the species across diverse climatic and geographic areas. The PBDM approach is holistic and assumes that while weather may change (e.g. climate change), the response of species to weather factors does not, except on a longer timescale, but there may be exceptions (see Meineke *et al.*, 2013). This physiological approach has early roots in the work of Fitzpatrick and Nix (1968) and de Wit and Goudriaan (1978): the underpinning modelling concepts are summarized by Gutierrez and Baumgärtner (1984, 2007), Gutierrez (1992, 1996) and Gutierrez *et al.* (1994). In their most complete form, PBDMs capture the weather-driven daily age-structured number and dry matter allocation dynamics. Although the allocation quantities and proportions may differ radically among species, allocation is first corrected for egestion, then respiration and conversion costs, and then to reproduction, if the organism is mature, or growth, if the organism is immature, and to reserves if there is a surplus. This allows the same model to be used to model plant and animal species, despite very different apparent biology and the widely differing allocation rates. PBDMs may vary in complexity, but must include the developmental rate of the organism (Brière *et al.*, 1999), resource acquisition rates (e.g. photosynthesis, water and nitrogen acquisition by plants, and resource acquisition by higher trophic levels; Gutierrez and Baumgärtner, 1984, 2007; Gutierrez, 1992, 1996), temperature-dependent rates of mortality, age-specific fecundity, survivorship and dormancy as appropriate (see Gutierrez and Ponti, 2013). PBDM parameters may be estimated from dry matter allocation studies and from age-specific life-table studies conducted across gradients of ecological factors, with temperature being a dominant one. PBDMs are time-varying life tables (*sensu* Gilbert *et al.*, 1976), the parameters of which have physiological nuance and may vary over time with weather and other factors. Because PBDMs are weather driven, applications to climate change scenarios are a natural application.

Examples of PBDMs: a plant and an insect

PBDMs may be used to model individual growth and development (e.g. Fig. 4.1a,b), at the population level, in a tri-trophic context, in a meta-population context if the rules of movement are known (e.g. say between plants or locations; Gutierrez *et al.*, 1999) and at a regional or larger level using a geographic information system (GIS) kriging to smooth data between locations (see below). More complete details for developing PBDMs are presented in Chapter 17.

PBDMs for a plant may consist of linked age-mass structured subpopulation dynamics models for leaves, stems and shoots, and roots, and healthy and infested fruit. The plant model predicts the phenology and dormancy if appropriate. PBDMs for herbivores or their natural enemies may be age-structured models of the different life stages and may also have attributes of mass, stage, sex, dormancy and other factors as required. PBDMs for the per capita mass and number dynamics of cotton (*Gossypium hirsutum* L.) and a ladybird beetle (*Hippodamia convergens* Guérin-Méneville) are illustrated in Fig. 4.1a,b, respectively (see other examples in Gutierrez *et al.*, 1987). Each figure illustrates the daily dry matter allocation to respiration, egestion, growth and reproduction, and together show the level of biological detail provided by PBDMs, the dynamics of which change with weather. Note that the figure legends for the different components of dry matter allocation are the same for both species, despite radically different allocations and scales.

Regional Analysis of Four Invasive Insect Species with Global Warming

Seven invasive pests have been reviewed by Gutierrez and Ponti (2013, 2014) with the goal of evaluating ongoing eradication/

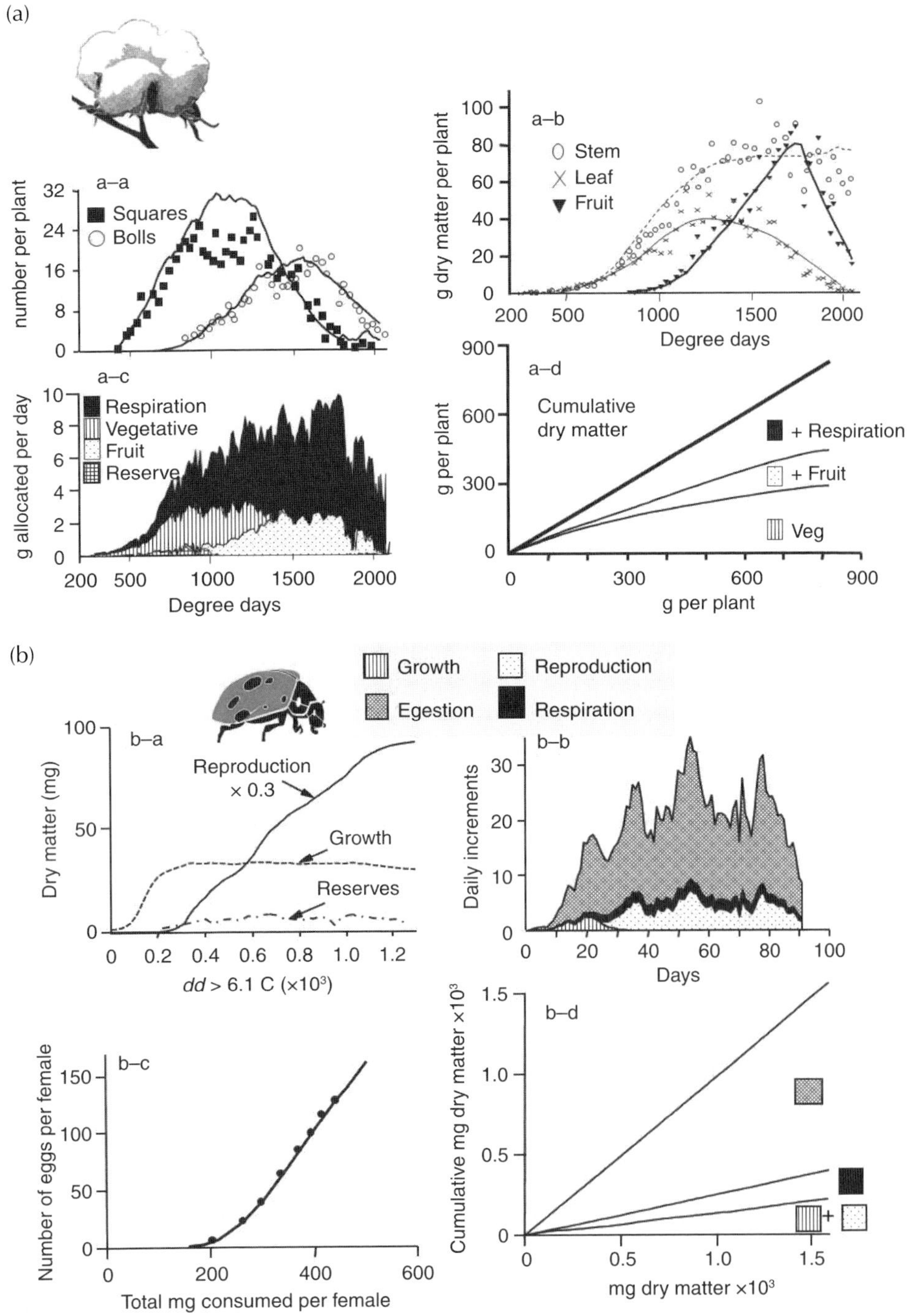

Fig. 4.1. PBDMs: (a) cotton ((a–a) number of fruit by stage, (a–b) dry matter, (a–c) daily allocation of dry matter and (a–d) cummulative allocation of dry matter); (b) a ladybeetle ((b–a) dry matter, (b–b) daily dry matter allocation, (b–c) eggs per day and (b–d) cummulative allocation of dry matter). (See Gutierrez *et al.*, 1987, for details and other examples.) Symbols are data and lines are simulation results.

containment efforts against them. Here, we examine the role of weather/climate and biotic interactions on the distribution and relative abundance of four of the insect species, employing a GRASS GIS-based regional analyses (GRASS Development Team, 2010) using observed weather and a +2°C warming scenario: the tropical New World screwworm (*Cochliomyia hominivorax* (Coquerel)) and the pink bollworm (*Pectinophora gossypiella* (Saunders)) on cotton (*G. hirsutum* L.); the Mediterranean fruit fly (*Ceratitis capitata* (Wiedemann) (i.e. medfly)) on fruit tree hosts; and the olive fly (*Bactrocera oleae* (Rossi)) on olive (*Olea europaea* L.). A +2°C threshold (European Commission, 2007; Zickfeld *et al.*, 2009) is widely used to assess the risks associated with dangerous levels of global warming and as a policy reference (Lowe *et al.*, 2009; Rogelj *et al.*, 2010).

Screwworm

The native tropical–subtropical New World screwworm causes myiasis in many vertebrate species, including humans. Data to develop a PBDM for screwworm are sparse in the literature, despite more than US$750 million having been spent during a 45 year effort to eradicate the fly in the southern USA and Mexico using the sterile insect technique (SIT). The current eradication programme is designed to limit the fly's distribution to the south of the eradication barrier at the Darian Gap in southern Panama (Krafsur *et al.*, 1986).

The fly has high lower and upper thermal thresholds, making it susceptible to moderately cold temperatures but robust to high temperatures (Gutierrez and Ponti, 2014). The field data suggest that cold temperatures are limiting in the northern reaches of its recorded distribution, with moisture being a further limiting factor in arid regions (see Parman, 1945; Rahn and Barger, 1973; Krafsur, 1987). The 450 mm isocline for annual rainfall separates arid regions from more humid subtropical and tropical areas, and this threshold was used to develop an index of moisture favourability for the pest (see Gutierrez and Ponti, 2013, 2014).

Including only the daily effects of temperature on screwworm development, reproduction and survival, the PBDM predicts a wide distribution (not shown). However, if we assume favourability for screwworm decreases with decreasing annual rainfall ($0 \leq lx_{rain}$ = rainfall mm/450 mm ≤ 1), the model yields the area of permanent winter survival based on field observations posited by Bushland (1985) (Fig. 4.2a). The highest average densities of pupae per year are predicted in the tropical areas of Mexico. Areas with densities above the mid-range include south Texas and south Florida, suggesting that these areas may be suitable for the year-round survival of screwworm. Densities in the desert areas of Arizona, California and the northern highlands of Mexico are in the lower quartile of the range, and are likely unsuitable for permanent screwworm persistence. The results support field observations that current cold temperatures limit winter survival northward and at higher elevations (e.g. central Mexico), while low rainfall and cold temperatures greatly limit the fly's distribution in the desert areas of the south-west USA and large areas of Mexico. Invasions of the fly into areas of Texas and elsewhere in the western USA during favourable periods of weather are aided by the fly's high vagility and by prevailing winds (see Hightower *et al.*, 1965; Mayer and Atzeni, 1993). The effects of other sources of moisture were not included in the model. Hence, the model failed to capture the favourable microclimate created by available river water and irrigation in the arid region of the lower Colorado River (Fig. 4.2a), but this factor can be easily included. Climate warming of +2°C is predicted to increase the area of permanent favourability for the fly northward, especially into the wetter areas of the south-east USA (Fig. 4.2b).

Pink bollworm

The highly destructive, stenophagous pink bollworm (PBW) was first discovered in

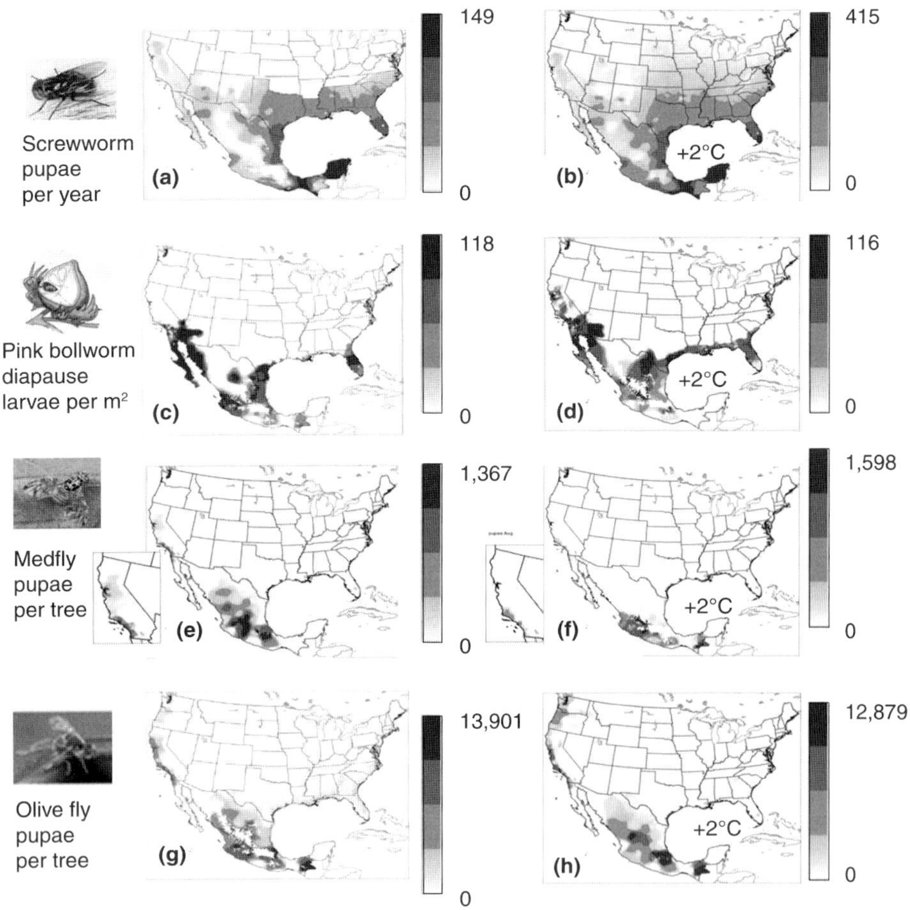

Fig. 4.2. The prospective distribution and relative abundance (cumulative per year) of four invasive species below 3000 m across the continental USA and Mexico using observed 1983–2003 weather data (a, c, e, g) and +2°C climate warming scenarios (b, d, f, h): (a, b) screwworm pupae per year, (c, d) diapause pink bollworm larvae per plant per year, (e, f) medfly pupae per year and (g, h) olive fly pupae per year (see Gutierrez and Ponti, 2013, 2014, for greater detail). Note the insets for California in e, f.

Florida in 1932 and spread to cotton in other areas of the USA and Mexico, especially into desert cotton in Arizona and southern California. Dispersal of the pest is aided by summer monsoon winds that annually transport adults long distances from the desert valleys of southern California northward into the southern reaches of the Great Central Valley and elsewhere (Stern and Sevacherian, 1978). The USDA and the California Department of Agriculture (CDFA) began an eradication programme against PBW in 1968 in Arizona and California (Staten *et al.*, 1992) using SIT, but eradication proved elusive (Chu *et al.*, 1996) and the programme was redirected with the new goal of preventing the establishment of PBW in the Central Valley. In 1997, genetically modified cotton expressing genes from the bacterium *Bacillus thuringensis* to produce an endotoxin (*Cry1Ac δ-endotoxin*) was introduced across much of the cotton belt. Bt cotton is highly effective against PBW and greatly reduces its density in desert cotton (e.g. Tabashnik *et al.*, 2010), and this reduction in density was thought favourable for renewal of PBW eradication efforts. More than US$450 million have been spent on the ongoing PBW eradication programme that is now piggybacked on the Bt cotton technology (see Gutierrez and Ponti, 2013).

A PBDM for cotton–pink bollworm was developed by Gutierrez *et al.* (1977), Stone and Gutierrez (1986) and Gutierrez *et al.* (2006). This system model was used to examine the potential distribution of pink bollworm in the USA and Mexico. Abundant biological data were available to parameterize nearly all aspects of the system model. Key elements of PBW biology are that diapause is initiated in late summer in response to decreasing photoperiod and temperature (Gutierrez *et al.*, 1981), and diapause larvae are cold intolerant (Gutierrez *et al.*, 1977, 2006; see data in Venette *et al.*, 2000). Further, we note that not all areas with temperatures favourable for pink bollworm have sufficient rainfall or irrigation for cotton production (e.g. lower Baja California).

Using total diapause larvae per plant in the absence of control measures as a metric of favourability, the model predicts that the highest populations (and survival) occur in the southern desert regions of Arizona, California and north-western Mexico and Baja California, where, prior to the advent of Bt cotton, high summer populations and large numbers of autumn diapause larvae were common, with winter survival in the Central Valley of California being exceedingly low (Gutierrez *et al.*, 2006; Gutierrez and Ponti, 2013; Fig. 4.2c). Although PBW has been recorded from southern New Mexico and west Texas, average winter survival is predicted to be low and annually quite variable due to freezing winter temperatures common in the area. Pink bollworm in much of the south-eastern USA is also limited by winter temperatures, and only southern Texas and central and south Florida are predicted to be favourable during the critical winter period. Central Mexico is moderately favourable, while the Yucatan is unfavourable due to the interaction of high temperatures and short photoperiod that adversely affects diapause development.

Of special interest is that climate warming is projected to increase the distribution of the pest into the Central Valley of California (see Gutierrez *et al.*, 2006), along the US gulf coast and in northern Florida and eastern Central Mexico (Fig. 4.2d).

Medfly

The polyphagous Mediterranean fruit fly (medfly) is a tropical species of East African origins that is intolerant of moderately low and high temperatures (see Gutierrez and Ponti, 2011). The fly is established in sub-Saharan Africa, the Mediterranean Basin, Argentina, Western Australia, Hawaii, Mexico and Central America, Italy and possibly southern California (Carey, 1991; Sutherst *et al.*, 2007; Gutierrez and Ponti, 2011). Low numbers of adult medfly have been detected periodically in the Los Angeles Basin and elsewhere in California since 1975 (Carey, 1991, 1996), but no permanent populations have been found. An intensive area-wide insecticide spray programme to eradicate the fly was begun after it was first detected in southern California in 1975. The fly was not detected again until 1980 (Myers *et al.*, 2000), when an intensive detection and eradication programme based on protein bait and insecticide sprays was initiated. An ongoing SIT programme was begun in 1994, costing more than US$270 million.

The medfly system model developed by Gutierrez and Ponti (2011) was used to examine the potential distribution of the fly in Arizona–California and Italy, and later for North America, to assess where the fly could establish and spread (Gutierrez and Ponti, 2013). As observed, tropical south-western Mexico and areas bordering Guatemala are highly favourable (Fig. 4.2e), while vast areas of northern Mexico are largely unfavourable. Compared to tropical Mexico, predicted fly populations in near coastal southern California (see insets) are about one-third, with Florida predicted still less favourable, while the remaining areas of the USA are unfavourable.

Climate warming is predicted to decrease the currently limited range of the fly in California (Gutierrez and Ponti, 2011), and in Mexico (Fig. 4.2f).

Olive fly

Drought-tolerant olive is of African origins and has been planted worldwide. Olive is widely planted in coastal California, the Central Valley and southern desert areas of California, with some cultivation occurring in central Arizona and other areas of the USA (e.g. Texas, Florida). The olive fly was first detected in the Los Angeles Basin of California in 1998 and quickly spread and established in the state. The olive fly has not established in other olive growing areas of the USA. Eradication using SIT has been attempted but failed in the Mediterranean Basin (Estes *et al.*, 2012), but has not been attempted in California.

A PBDM model for olive and olive fly, developed by Gutierrez *et al.* (2009), was used to predict the distribution of olive fly successfully in California (see field survey data, Wang *et al.*, 2009) and in Italy, including its distribution in the microclimates along the northern lakes. The model was also used in a fine-scale analysis of the fly's distribution and abundance in Sardinia (see Ponti *et al.*, 2009a) and across much of the Mediterranean Basin (Ponti *et al.*, 2009b). Of special note is that olive has wider thermal limits than the fly.

At the scale of the USA and Mexico, the PBDM predicts a wide potential distribution for olive (see Gutierrez and Ponti, 2013), while the potential distribution of olive fly is considerably smaller (Fig. 4.2g). In the USA, the highest fly densities are predicted in coastal south and central California, with penetration into the northern and central reaches of the Central Valley of California. The fly is limited by high temperatures in the southern Central Valley and the desert areas of the western USA and Mexico. Despite olive production in central Florida, olive fly densities are predicted prospectively to be half those of coastal southern California, while areas of coastal Texas and Louisiana are predicted to be marginal. The areas of highest favourability for the fly in North America are south and central Mexico.

Climate warming of +2°C in California is predicted to restrict the distribution and abundance of the fly in the Central Valley, but increase northward along the coast in California (see Gutierrez *et al.*, 2009) into Oregon and into the higher elevations of Central Mexico (Fig. 4.2h).

Olive fly and climate change in the Mediterranean Basin

Daily weather data for climate warming scenarios are increasingly available, but each has its own set of assumptions. Here, weather data for the Mediterranean Basin from the ERA40 re-analysis of meteorological observations for years 1958–2000 (http://www.ecmwf.int/research/era/) was used as the observed weather data to drive the PBDM (see Ponti *et al.*, 2009b). For heuristic purposes, a climate-warming scenario that increases daily temperatures +2°C with all other weather variables remaining unchanged is used to assess the effects on olive and olive fly dynamics. No reliable trend is predicted for rainfall in the region, and hence the effects of rainfall are ignored. Climate model projections for future weather could also be used (see Ponti *et al.*, 2009a), but we use this simpler approach to illustrate the process and relative changes that may occur.

Under present climate, mild coastal areas across the Mediterranean Basin are predicted as highly favourable for olive fly, whereas low favourability is predicted at higher inland elevations with low winter temperatures (e.g. areas of Europe), and areas with high summer temperatures close to the upper thermal limits of the fly that decrease reproduction, induce reproductive dormancy and increase mortality (e.g. areas of Morocco and the Middle East, Fig. 4.3a).

With +2°C climate warming, olive fly abundance decreases in many of the hot areas that previously were favourable under

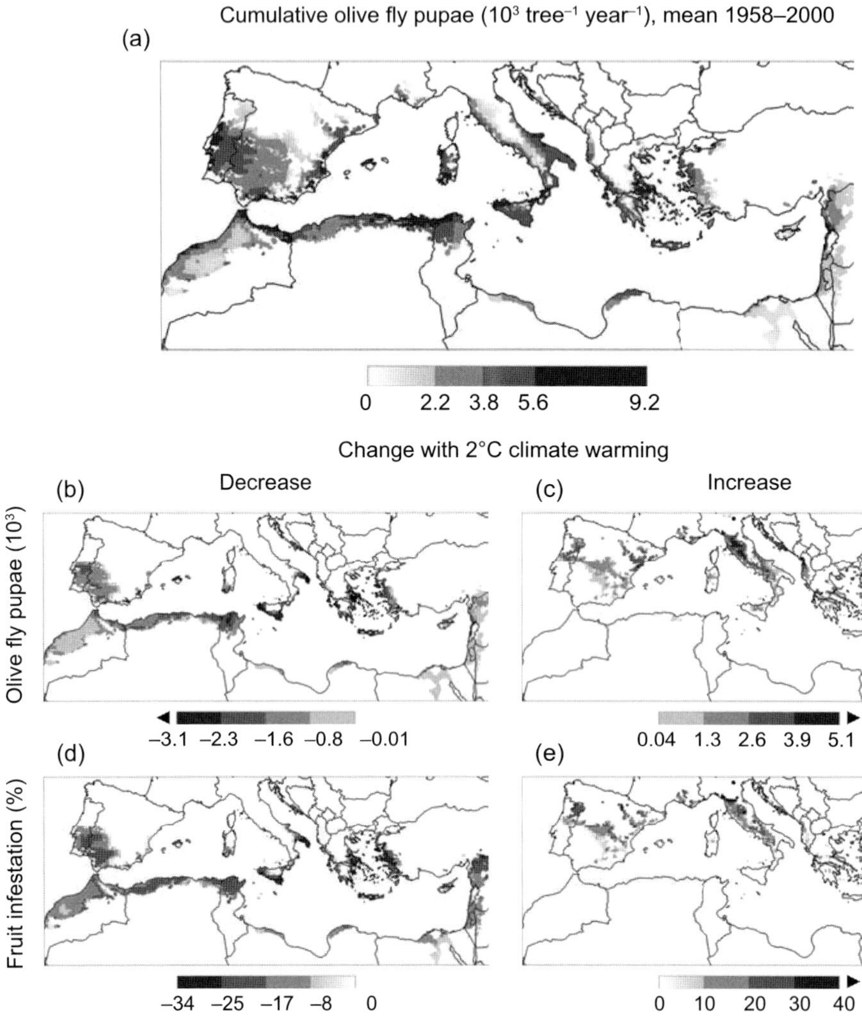

Fig. 4.3. Distribution and abundance (*Bactrocera oleae*) as influenced by olive growth and development as driven by the present climate (1958–2000) and a +2°C warming scenario: (a) cumulative number of pupae (10^3 tree^{-1} year^{-1}) under the present climate, and compared to the present climate the negative (b) and positive (c) changes in pupal number with +2°C climate warming, and the effects on the per cent of olive fruit infestation by the olive fly (d, e). Breakpoints in the greyscale bars at the bottom of subfigure (a) are equally spaced breaks estimated using the Fisher–Jenks classIntervals function in the classInt package of the R statistical software (R Development Core Team, 2010). The triangles at the ends of some greyscale bars indicate that a few outliers were found in the data using the boxplot function in R and are the darkest areas in the maps.

the present climate (e.g. areas of Spain, coastal North Africa and Palestine/Israel) (Fig. 4.3b). In contrast, fly populations increase in higher altitude inland areas as temperatures warm (e.g. central Spain and Italy) (Fig. 4.3c). Fruit infestation levels decline in hot areas (Fig. 4.3d), whereas infestation levels increase in milder inland areas, especially at high altitudes that become more favourable for both olive (not shown) and the fly (Fig. 4.3e). Changes in olive yield and fly abundance can increase or decrease, affecting fruit attack rates and fruit quality, and control costs. Combining all of these factors in a bioeconomic analysis enables estimations of the economic impact of climate change on olive.

Summary

This chapter reviews some of the links between invasive insect biology and climate change. Invasive species collectively cause in excess of US$140 billion in losses annually in the USA (Pimentel *et al.*, 2000) and US$1 trillion worldwide (Oerke and Dehne, 2004). The effects of climate change on insect pest populations can be direct, through impacts on their physiology/ behaviour, or indirect, through biotic interactions (i.e. bottom-up and top-down effects; Bale *et al.*, 2002). Hellmann *et al.* (2008) identified five potential areas affecting insect invasions with respect to anthropogenic climate and global change: (i) mechanisms of transport and introduction; (ii) enhanced establishment; (iii) changes in the distribution of existing invasive species; (iv) the impact of established invasive species; and (v) changes in the efficacy of management strategies.

A critical first step to assess the degree of species invasiveness is to predict accurately the changes in its distribution and abundance in space and time, but how to do this has been a difficult recurring problem (Gallien *et al.*, 2010). Commonly used methods to predict the geographic distribution of species are classed as ecological niche models (ENMs) that use occurrence records and aggregate weather data to characterize the ecological niche space of species. ENMs make implicit mathematical and biological assumptions not linked to biological data, and have limitations when used to predict the range of species and the effects of climate change. More holistic physiologically based demographic models (PBDMs) may be developed that model explicitly the biological and physiological responses of species to weather and the species with which they interact. PBDMs are independent of distribution records and offer a larger suite of insights concerning the potential distribution of the species under current and future climate change scenarios.

Four invasive insect pests were reviewed under observed and +2°C weather scenarios across the USA and Mexico: the tropical New World screwworm, the pink bollworm on cotton, the Mediterranean fruit fly (i.e. medfly) on fruit tree hosts and the olive fly on olive. The PBDMs predict fine-scale distributions and relative abundance for the pests under current and +2°C climate warming. Climate change is predicted to increase the distribution of the screwworm and the pink bollworm, reduce the geographic distribution of the medfly, and alter the distribution of the olive fly. Without modification, the PBDM for the olive fly was used to examine the distribution of the pest in the Mediterranean Basin, illustrating the transferability of the PBDMs across space and time and weather scenarios.

References

Bahn, V. and McGill, B.J. (2007) Can niche-based distribution models outperform spatial interpolation? *Global Ecology and Biogeography* 16, 733–742.

Bale, J.S., *et al.* (2002) Herbivory in global climate change research: direct effects of rising temperature on insect herbivores. *Global Change Biology* 8, 1–16.

Battisti, A., Stastny, M., Buffo, E. and Larsson, S. (2006) A rapid altitudinal range expansion in the pine processionary moth produced by the 2003 climatic anomaly. *Global Change Biology* 12, 662–671.

Beaumont, L.J., Gallagher, R.V., Thuiller, W., Downey, P.O., Leishman, M.R. and Hughes, L. (2009) Different climatic envelopes among invasive populations may lead to under estimations of current and future biological invasions. *Diversity and Distributions* 15, 409–420.

Benedict, M.Q., Levine, R.S., Hawley, W.A. and Lounibos, L.P. (2007) Spread of the tiger: global risk of invasion by the mosquito *Aedes albopictus*. *Vector-Borne and Zoonotic Diseases* 7, 76–85.

Brière, J.F., Pracros, P., Le Roux, A.Y. and Pierre, J.S. (1999) A novel rate model of temperature-dependent development for arthropods. *Environmental Entomology* 28, 22–29.

Burkett, V.R., Wilcox, D.A., Stottlemyer, R., Barrow, W., Fagre, D., Baron, J., Price, J., Nielsen, J.L., Allen, C.D., Peterson, D.L., Ruggerone, G. and Doyle, T. (2005) Nonlinear dynamics in ecosystem response to climatic change: case studies and policy implications. *Ecological Complexity* 2, 357–394.

Burrows, M.T., *et al.* (2011) The pace of shifting climate in marine and terrestrial ecosystems. *Science* 334, 652–655.

Bushland, R.C. (1985) Eradication program in the southwestern United States. *Miscellaneous Publications of the Entomological Society of America* 62, 12–15.

Carey, J.R. (1991) Establishment of the Mediterranean fruit fly in California. *Science* 253, 1369–1373.

Carey, J.R. (1996) Incipient Mediterranean fruit fly population in California: implications for invasion biology. *Ecology* 77, 1690–1697.

Chen, C.C. and McCarl, B.A. (2001) An investigation of the relationship between pesticide usage and climate change. *Climatic Change* 50(4), 475–487, doi:10.1023/A:1010655503471.

Chu, C.C., Henneberry, T.J., Weddle, R.C., Natwick, E.T., Carson, J.R., Valenzuela, C., Birdsall, S.L. and Staten, R.T. (1996) Reduction of pink bollworm (Lepidoptera: Gelechiidae) populations in the Imperial Valley, California, following mandatory short-season cotton management systems. *Journal of Economic Entomology* 89, 175–182.

Davis, A.J., Jenkinson, L.S., Lawton, J.H., Shorrocks, B. and Wood, S. (1998) Making mistakes when predicting shifts in species range in response to global warming. *Nature* 391, 783–786.

De Wit, C.T. and Goudriaan, J. (1978) *Simulation of Ecological Processes*, 2nd edn. Pudoc Publishers, Wageningen, Netherlands.

Dukes, J.S. (2011) Responses of invasive species to a changing climate and atmosphere. In: Richardson, D.M. (ed.) *Fifty Years of Invasion Ecology: The Legacy of Charles Elton*. Wiley-Blackwell, Chichester, UK, pp. 345–357.

Elith, J. and Leathwick, J.R. (2009) Species distribution models: ecological explanation and prediction across space and time. *Annual Review of Ecology and Evolutionary Systematics* 40, 677–697.

Essl, F., *et al.* (2011) Socioeconomic legacy yields an invasion debt. *Proceedings of the National Academy of Sciences* 108, 203–207.

Estes, A.M., Nestel, D., Belcari, A., Jessup, A., Rempoulakis, P. and Economopoulos, A.P. (2012) A basis for the renewal of sterile insect technique for the olive fly, *Bactrocera oleae* (Rossi). *Jourrnal of Applied Entomology* 136, 1–16.

European Commission (2007) Limiting global climate change to 2°C: the way ahead for 2020 and beyond. COM(2007) 2 final (http://eur-lex.europa.eu/LexUriServ/LexUriServ.do?uri=COM:2007:0002:FIN:EN:PDF, accessed 13 January 2014).

Fitzpatrick, E.A. and Nix, H.A. (1968) The climatic factor in Australian grasslands ecology. In: Moore, R.M. (ed.) *Australian Grasslands*. Australian National University Press, Canberra, pp. 3–26.

Gallien, L., Münkemüller, T., Albert, C.H., Boulangeat, I. and Thuiller, W. (2010) Predicting potential distributions of invasive species: where to go from here? *Diversity and Distributions* 16, 331–342.

George, K., Ziska, L.H., Bunce, J.A. and Quebedeaux, B. (2007) Elevated atmospheric CO_2 concentration and temperature across an urban–rural transect. *Atmospheric Environment* 41, 7654–7665.

Gilbert, N., Gutierrez, A.P., Frazer, B.D. and Jones, R.E. (1976) *Ecological Relationships*. Freeman and Co, New York.

Gilman, S.E., Urban, M.C., Tewksbury, J., Gilchrist, G.W. and Holt, R.D. (2010) A framework for community interactions under climate change. *Trends in Ecology and Evolution* 25, 325–331.

GRASS Development Team (2010) Geographic Resources Analysis Support System (GRASS) Software, Version 6.4.0. Open Source Geospatial Foundation (http://grass.osgeo.org, accessed 3 January 2014).

Gutierrez, A.P. (1992) The physiological basis of ratio-dependent predator–prey theory: the metabolic pool model as a paradigm. *Ecology* 73, 1552–1563.

Gutierrez, A.P. (1996) *Applied Population Ecology: A Supply–Demand Approach*. John Wiley and Sons, New York.

Gutierrez, A.P. and Baumgärtner, J.U. (1984) Multitrophic level models of predator–prey energetics: II. A realistic model of plant–herbivore–parasitoid–predator interactions. *Canadian Entomology* 116, 933–949.

Gutierrez, A.P. and Baumgärtner, J.U. (2007) Modeling the dynamics of tritrophic population interactions. In: Kogan, M. and Jepson, P. (eds) *Perspectives in Ecological Theory and Integrated Pest Management*. Cambridge University Press, Cambridge, UK, pp. 301–360.

Gutierrez, A.P. and Ponti, L. (2011) Assessing the invasive potential of the Mediterranean fruit fly in California and Italy. *Biological Invasions* 13, 2661–2676.

Gutierrez, A.P. and Ponti, L. (2013) Eradication of invasive species: why the biology matters. *Environmental Entomology* 42, 395–411.

Gutierrez, A.P. and Ponti, L. (2014) The new world screwworm: exploring effects of temperature and rainfall on the prospective distribution in North America and the Mediterranean Basin. *Agricultural and Forest Entomology* 16, 158–173.

Gutierrez, A.P., Butler, G.D. Jr, Wang, Y.H. and Westphal, D. (1977) The interaction of pink bollworm (Lepidoptera: Gelichiidae), cotton, and weather: a detailed model. *Canadian Entomology* 109, 1457–1468.

Gutierrez, A.P., Butler, G.D. Jr and Ellis, C.K. (1981) Pink bollworm: diapause induction and termination in relation to fluctuating temperatures and decreasing photophases. *Environmental Entomology* 10, 936–942.

Gutierrez, A.P., Schulthess, F., Wilson, L.T., Villacorta, A.M., Ellis, C.K. and Baumgärtner, J.U. (1987) Energy acquisition and allocation in plants and insects: a hypothesis for the possible role of hormones in insect feeding patterns. *Canadian Entomology* 119, 109–129.

Gutierrez, A.P., Mills, N.J., Schreiber, S.J. and Ellis, C.K. (1994) A physiologically based tritrophic perspective on bottom-up-top-down regulation of populations. *Ecology* 75, 2227–2242.

Gutierrez, A.P., Yaninek, J.S., Neuenschwander, P. and Ellis, C.K. (1999) A physiologically based metapopulation dynamics: the tritrophic cassava system as a case study. *Ecological Modelling* 123, 225–242.

Gutierrez, A.P., d'Oultremont, T., Ellis, C.K. and Ponti, L. (2006) Climatic limits of pink bollworm in Arizona and California: effects of climate warming. *Acta Oecologia* 30, 353–364.

Gutierrez, A.P., Ponti, L. and Cossu, Q.A. (2009) Effects of climate warming on olive and olive fly (*Bactrocera oleae* (Gmelin)) in California and Italy. *Climatic Change* 95, 195–217.

Hahn, D.A. and Denlinger, D.L. (2011) Energetics of insect diapause. *Annual Review Entomology* 56, 103–121.

Hellmann, J.J., Byers, J.E., Bierwagen, B.G. and Dukes, J.S. (2008) Five potential consequences of climate change for invasive species. *Conservation Biology* 22, 534–543.

Hightower, B.G., Adams, A.L. and Alley, D.A. (1965) Dispersal of released irradiated laboratory-reared screw-worm flies. *Journal of Economic Entomology* 58, 373–374.

Huang, D., Haack, R.A. and Zhang, R. (2011) Does global warming increase establishment rates of invasive alien species? A centurial time series analysis. *PLoS ONE* 6, e24733.

Hulme, P.E. (2012) Invasive species unchecked by climate. *Science* 335, 537–538.

Jeschke, J.M. and Strayer, D.L. (2008) Usefulness of bioclimatic models for studying climate change and invasive species. *Annals of New York Acadamy of Science* 1134, 1–24.

Kenis, M., Auger-Rozenberg, M.A., Roques, A., Timms, L., Péré, C., Cock, M.J.W., Settele, J., Augustin, S. and Lopez-Vaamonde, C. (2009) Ecological effects of invasive alien insects. *Biological Invasions* 11, 21–45.

Kerr, J.T., Kharouba, H.M. and Currie, D.J. (2007) The macroecological contribution to global change solutions. *Science* 316, 1581–1584.

Khuroo, A.A., Reshi, Z.A., Rashid, I. and Dar, G.H. (2011) Towards an integrated research framework and policy agenda on biological invasions in the developing world: a case-study of India. *Environmental Research* 111(7), 999–1006, doi:10.1016/j.envres.2011.02.011.

Krafsur, E.S. (1987) Climatological correlates of screwworm (*Cochliomyia hominivorax*) abundance in Texas, USA. *Medical and Veterinary Entomology* 1, 71–80.

Krafsur, E.S., Townson, H., Davidson, G. and Curtis, C.F. (1986) Screwworm eradication is what it seems. *Nature* 323, 495–496.

Liebhold, A.M. and Tobin, P.C. (2008) Population ecology of insect invasions and their management. *Annual Review of Entomology* 53, 387–408.

Loarie, S.R., Duffy, P.B., Hamilton, H., Asner, G.P., Field, C.B. and Ackerly, D.D. (2009) The velocity of climate change. *Nature* 462, 1052–1055.

Lowe, J.A., Huntingford, C., Raper, S.C.B., Jones, C.D., Liddicoat, S.K. and Gohar, L.K. (2009) How difficult is it to recover from dangerous levels of global warming? *Environmental Research Letters* 4, 014012, doi:10.1088/1748-9326/4/1/014012.

Lozier, J.D. and Mills, N.J. (2011) Predicting the potential invasive range of the light brown apple moth (*Epiphyas postvittana*) using biologically informed and correlative species models. *Biological Invasions* 13, 2409–2421.

Lozier, J.D., Aniello, P. and Hickerson, M.J. (2009) Predicting the distribution of Sasquatch in western North America: anything goes with ecological niche modeling. *Journal of Biogeography* 36, 1623–1627.

Manea, A. and Leishman, M.R. (2011) Competitive interactions between native and exotic plant species are altered under elevated carbon-dioxide. *Oecologia* 164, 735–744.

Mayer, D.G. and Atzeni, M.G. (1993) Estimation of dispersal distances for *Cochliomyia hominivorax* (Diptera: Calliphoridae). *Environmental Entomology* 22, 368–374.

Meineke, E.K., Dunn R.R., Sexton J.O. and Frank, S.D. (2013) Urban warming drives insect pest abundance on street trees. *PLoS ONE* 8(3), e59687, doi:10.1371/ journal.pone.0059687.

Myers, J.H., Savoie, A. and van Randen, E. (1998) Eradication and pest management. *Annual Review of Entomology* 43, 471–491.

Myers, J.H., Simberloff, D., Kuris, A.M. and Carey, J.R. (2000) Eradication revisited: dealing with exotic species. *Trends in Ecology and Evolution* 15, 316–320.

Oerke, E.C. and Dehne, H.W. (2004) Safeguarding production losses in major crops and the role of crop protection. *Crop Protection* 23, 275–285.

Ohlemüller, R. (2011) Running out of climate space. *Science* 334, 613–614.

Parman, D.C. (1945) Effect of weather on *Cochliomyia americana* and a review of methods and economic applications of the study. *Journal of Economic Entomology* 38, 66–76.

Pimentel, D., Lach, L., Zuniga, R. and Morrison, D. (2000) Environmental and economic costs of non-indigenous species in the United States. *BioScience* 50, 53–65.

Pimentel, D., Zuniga, R. and Morrison, D. (2005) Update on the environmental and economic costs associated with alien-invasive species in the United States. *Ecological Economics* 52, 273–288.

Pluess, T., Cannon, R., Jarošík, V., Pergl, J., Pyšek, P. and Bacher, S. (2012) When are eradication campaigns successful? A test of common assumptions. *Biological Invasions* 14(7), 1365–1378, doi:10.1007/s10530-011-0160-2.

Ponti, L., Cossu, Q.A. and Gutierrez, A.P. (2009a) Climate warming effects on the *Olea europaea–Bactrocera oleae* system in Mediterranean islands: Sardinia as an example. *Global Change Biology* 15, 2874–2884.

Ponti, L., Gutierrez, A.P. and Ruti, P.M. (2009b) The olive–*Bactrocera oleae* (Diptera Tephritidae) system in the Mediterranean Basin: a physiologically based analysis driven by the ERA-40 climate data. *Notiziario sulla Protezione delle Piante – III Serie* 1, 113–128.

Pyšek, P., Richardson, D.M., Pergl, J., Jarošík, V., Sixtová, Z. and Weber, E. (2008) Geographical and taxonomic biases in invasion ecology. *Trends in Ecology and Evolution* 23, 237–244.

R Development Core Team (2010) R: a language and environment for statistical computing. R Foundation for Statistical Computing, Vienna (http://www.R-project.org, accessed 3 January 2014).

Rahn, J.J. and Barger, G.L. (1973) Weather conditions and screwworm activity. *Agricultural Meteorology* 11, 197–211.

Ricciardi, A., Palmer, M.E. and Yan, N.D. (2011) Should biological invasions be managed as natural disasters? *Bioscience* 61, 312–317.

Robinet, C. and Roques, A. (2010) Direct impacts of recent climate warming on insect populations. *Integrative Zoology* 5, 132–142.

Rogelj, J., Nabel, J., Chen, C., Hare, W., Markmann, K., Meinshausen, M., Schaeffer, M., Macey, K. and Hohne, N. (2010) Copenhagen accord pledges are paltry. *Nature* 464, 1126–1128.

Roiz, D., Neteler, M., Castellani, C., Arnoldi, D. and Rizzoli, A. (2011) Climatic factors driving invasion of the tiger mosquito (*Aedes albopictus*) into new areas of Trentino, northern Italy. *PLoS ONE* 6, e14800.

Rosenberg, J. and Burt, P.J.A. (1999) Windborne displacements of desert locusts from Africa to the Caribbean and South America. *Aerobiologia* 15, 167–175.

Roura-Pascual, N., *et al.* (2011) Relative roles of climatic suitability and anthropogenic influence in determining the pattern of spread in a global invader. *Proceedings of the National Academy of Sciences* 108, 220–225.

Scherm, H. and Coakley, S.M. (2003) Plant pathogens in a changing world. *Australasian Plant Pathology* 32(2), 157–165.

Schweiger, O., Settele, J., Kudrna, O., Klotz, S. and Kühn, I. (2008) Climate change can cause spatial mismatch of trophically interacting species. *Ecology* 89, 3472–3479.

Schweiger, O., Biesmeijer, J.C., Bommarco, R., Hickler, T., Hulme, P.E., Klotz, S., Kühn, I., Moora, M., Nielsen, A. and Ohlemüller, R. (2010) Multiple stressors on biotic interactions: how climate change and alien species interact to affect pollination. *Biology Reviews* 85, 777–795.

Soberòn, J. and Nakamura, M. (2009) Niches and distibutional areas: concepts, methods and assumptions. *Proceedings of the National Academy of Science* 106, 19644–19650.

Staten, R.T., Rosander, R.W. and Keaveny, D.F. (1992) Genetic control of cotton insects, 1992: the PBW as a working programme. In: *Proceedings of An International Symposium On Management of Insect Pests*, Vienna, 19–23 October 1992. STI/PUB/909, IAEA, Vienna, pp. 269–283.

Stern, V. and Sevacherian, V. (1978) Long-range dispersal of pink bollworm into the San Joaquin Valley. *California Agriculture* 32, 4–5.

Stone, N.D. and Gutierrez, A.P. (1986) Pink bollworm control in southwestern desert cotton. I. A field-oriented simulation model. *Hilgardia* 54, 1–24.

Sutherst, R.W. and Bourne, A.S. (2009) Modelling non-equilibrium distributions of invasive species: a tale of two modelling paradigms. *Biological Invasions* 11, 1231–1237.

Sutherst, R.W., Maywald, G.F. and Bourne, A.S. (2007) Including species interactions in risk assessments for global change. *Global Change Biology* 13, 1843–1859.

Tabashnik, B.E., Sisterson, M.S., Ellsworth, P.C., Dennehy, T.J., Antilla, L., Liesner, L., Whitlow, M., Staten, R.T., Fabrick, J.A. and Unnithan, G.C. (2010) Suppressing resistance to Bt cotton with sterile insect releases. *Nature Biotechnology* 28, 1304–1307.

Tatem, A.J. (2009) The worldwide airline network and the dispersal of exotic species: 2007–2010. *Ecography* 32, 94–102.

Tatem, A.J. and Hay, S.I. (2007) Climatic similarity and biological exchange in the worldwide airline transportation network. *Proceedings of the Royal Society, B* 274, 1489–1496.

Thomson, L.J., Macfadyen, S. and Hoffmann, A.A. (2010) Predicting the effects of climate change on natural enemies of agricultural pests. *Biological Control* 52, 296–306.

Thuiller, W., Richardson, D.M., Pysek, P., Midgley, G.F., Hughes, G.O. and Rouget, M. (2005) Niche-based modelling as a tool for predicting the risk of alien plant invasions at a global scale. *Global Change Biology* 11, 2234–2250.

Tylianakis, J.M., Didham, R.K., Bascompte, J. and Wardle, D.A. (2008) Global change and species interactions in terrestrial ecosystems. *Ecology Letters* 11, 1351–1363.

van der Putten, W.H., Macel, M. and Visser, M.E. (2010) Predicting species distribution and abundance responses to climate change: why it is essential to include biotic interactions across trophic levels. *Philosophical Transactions of the Royal Society (B)* 365, 2025–2034.

Venette, R.C., Naranjo, S.E. and Hutchison, W.D. (2000) Implications of larval mortality at low temperatures and high soil moistures for establishment of pink bollworm (Lepidoptera: Gelechiidae) in southeastern United States cotton. *Environmental Entomology* 29, 1018–1026.

Wang, X.G., Johnson, M.W., Daane, K.M. and Nadel, H. (2009) High summer temperatures affect the survival and reproduction of olive fruit fly (Diptera: Tephritidae). *Environmental Entomology* 38, 1496–1504.

Ward, N.L. and Masters, G.J. (2007) Linking climate change and species invasion: an illustration using insect herbivores. *Global Change Biology* 13, 1605–1615.

Wardle, D.A., Bardgett, R.D., Callaway, R.M. and Van der Putten, W.H. (2011) Terrestrial ecosystem responses to species gains and losses. *Science* 332, 1273–1277.

Yang, L.H. and Rudolf, V.H.W. (2010) Phenology, ontogeny and the effects of climate change on the timing of species interactions. *Ecological Letters* 13, 1–10.

Yemshanov, D., McKenney, D.W., Pedlar, J.H., Koch, F.H. and Cook, D. (2009) Toward an integrated approach to modeling the risks and impact of invasive forest species. *Environmental Reviews* 17, 163–178.

Zickfeld, K., Eby, M., Matthews, H.D. and Weaver, A.J. (2009) Setting cumulative emissions targets to reduce the risk of dangerous climate change. *Proceedings of the National Academy of Sciences* 106, 16129–16134.

Ziska, L.H., Teasdale, J.R. and Bunce, J.A. (1999) Future atmospheric carbon dioxide may increase tolerance to glyphosate. *Weed Science* 47, 608–615.

Ziska, L.H., Blumenthal, D.M., Runion, G.B., Hunt, E.R. Jr and Diaz-Soltero, H. (2011) Invasive species and climate change: an agronomic perspective. *Climatic Change* 105, 13–42.

5 Climate Change, Plant Traits and Invasion in Natural and Agricultural Ecosystems

Dana M. Blumenthal and Julie A. Kray

Rangeland Resources Research Unit, USDA-ARS, Fort Collins, Colorado, USA

Abstract

Invasive plant species often thrive in new and resource-rich environments and may therefore benefit from global changes that create such environments. Global change effects on invasion risk will depend both on the environment in which competition between invasive and resident plant species occurs and on the physiological and life history traits of the competing species. In agricultural environments, risks include the northward movement of many problematic invasive species, decreased biotic resistance to invasion with extreme climatic events and strong global change responses of particular invasive species. However, due to their similar rapid resource acquisition and growth strategies, invasive species and crops may respond similarly to changes that increase resources. Furthermore, management adaptations, such as planting crops or varieties suited to new climate conditions, will help to maintain biotic resistance to invasion in crops. In natural ecosystems, global changes that increase resources rapidly may favour fast-growing invasive plant species and inhibit the slower-growing native species adapted to current climate conditions. In addition, management options to help native species compete with invasive species in novel environments, such as assisted migration, are relatively limited. Thus, with the predicted increases in CO_2, precipitation (in some regions) and extreme events, natural ecosystems may be particularly vulnerable to invasion.

Introduction

Invasive plants include a tremendous array of species, from spiny, spreading forbs (e.g. Canada thistle, *Cirsium arvense*) to grasses, large (giant reed, *Arundo donax*) and small (cheatgrass, *Bromus tectorum*), nitrogen-fixing species (blue-leafed wattle, *Acacia saligna*), fire-promoting species (buffelgrass, *Pennisetum ciliare*) and prolific vines and trees (kudzu, *Pueraria lobata*; tree of heaven, *Ailanthus altissima*). The reasons they are so successful vary widely, both among species and ecosystems (Catford *et al.*, 2009; Gurevitch *et al.*, 2011). It is not surprising, therefore, that invasive plant species' responses to the variety of global changes can be similarly variable (Sorte *et al.*, 2013). For example, elevated carbon dioxide (CO_2) increased the growth of yellow starthistle (*Centaurea solstitialis*) sixfold in an annual grassland in California (Dukes *et al.*, 2011), but decreased the success of redroot pigweed (*Amaranthus retroflexus*) grown in competition with soybean (*Glycine max*) (Ziska, 2010). What accounts for this difference? These responses to elevated CO_2 likely depend on the relative responses of the invasive species and their competitors, which in turn depend on the context in which competition takes place. Yellow starthistle's deep root system may have allowed greater access to limiting soil resources, enabling it to respond more strongly to elevated CO_2 than the annual grasses against which it was competing. In contrast, soil resources are less limiting in cropping systems, and redroot pigweed may be less responsive than soybean to elevated CO_2 because its C_4 photosynthesis is saturated at lower CO_2 concentrations.

This chapter explores the characteristics of both invasive plants and invaded ecosystems to search for generalizations that may allow us to predict when and where climate change will influence invasion strongly. It builds on reviews that have considered global change–plant invasion interactions broadly (Dukes and Mooney, 1999; Vila *et al.*, 2007; Thuiller *et al.*, 2007; Bradley *et al.*, 2010a; Ziska and Dukes, 2011; Sorte *et al.*, 2013) or examined particular changes, types of species or regions in more detail (D'Antonio and Vitousek, 1992; Richardson *et al.*, 2000; Walther *et al.*, 2002, 2009; Weltzin *et al.*, 2003; Diez *et al.*, 2012). The focus here is on elevated CO_2 and the resulting changes in temperature and precipitation (henceforth 'global change'). Invasive species are defined broadly to include species that invade cropland as well as natural or semi-natural ecosystems. The primary objective is to delineate the ways in which the effects of climate change depend on the traits of both invasive plants and the crops or native species against which they compete.

Invasive Plants and Change

Despite considerable variability in invasive species' responses to global change, there is concern that elevated CO_2 and climate change may favour invasive species more often than not (Dukes and Mooney, 1999; Bradley *et al.*, 2010a). The simplest reason for this is that global change involves change, altering the environmental conditions to which resident species are accustomed. Native species are suited to their environment due to a combination of community assembly and natural selection. Crops are suited to their environment due to the choice of crop and artificial selection. Changes to those environments are therefore likely to make resident species less well suited to their environments, reducing their competitive ability and therefore biotic resistance to invasion (Byers, 2002; D'Antonio and Thomsen, 2004; Levine *et al.*, 2004).

Invasive species, in contrast, are well suited to change. To become invasive, a plant has to get to, grow in and reproduce in new environments. Successful invasive plants tend to have traits that help them cope with new and, by extension, changing environments, including strong dispersal abilities, broad environmental tolerances, rapid growth rates, low reliance on mutualists and rapid evolution due to short generation times (Fig. 5.1) (Dukes and Mooney, 1999; Pyšek and Richardson, 2007; van Kleunen *et al.*, 2008, 2010; Clements and DiTommaso, 2011; Davidson *et al.*, 2011). Furthermore, while invasive species are also suited to their current environments, they have had less time to develop local adaptations (Parker *et al.*, 2003; Lambrinos, 2004; Friedman *et al.*, 2008) and may suffer less than other residents when conditions change.

The net effect of environmental change on invasion is likely to depend on both the degree to which resident species are inhibited by the change and the availability of invasive species that can thrive in the new environment. In the case of cropland, intensive management is likely to help maintain biotic resistance. As environmental conditions change, farmers can plant new crops or varieties that are suited to new environments. They can also adjust production practices to help crops thrive and compete effectively against invasive species. In fact, changing environmental conditions, usually by rotating crops, is often used to manage cropland invaders. Invasive species typically respond to crop rotation with shifts in community composition as species suited to the new environment become more abundant. Similarly, with long-term global change, the diversity of invasive species already within croplands is likely to enable

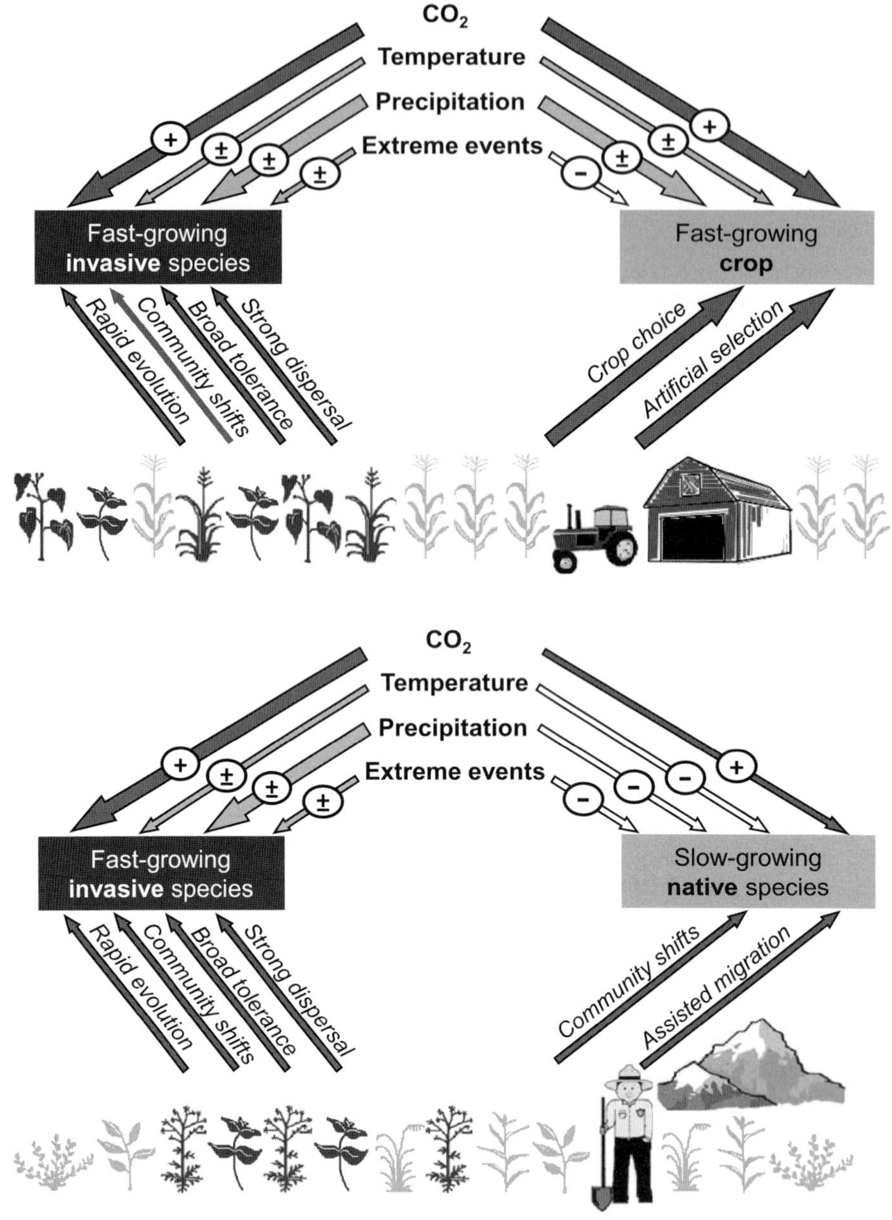

Fig. 5.1. Effects of climatic and atmospheric changes on resident species (lighter shades), and invasive species (darker shades). Arrows represent positive (black), negative (white) or variable (gray) effects of changes (above), or responses to change (below). Thicker arrows represent stronger effects. In cropland (upper panel), similarities between fast-growing invasive species and crops lead to similar positive effects of changes that directly increase resources, and intensive management helps match crops to novel environments. In natural ecosystems (lower panel), slow-growing native species differ from invasive species, and are more often inhibited than favored by changes. Furthermore, management options for matching native species to novel environments are relatively limited. Climatic and atmospheric changes may therefore decrease biotic resistance, and favor invasive species, more in natural ecosystems than in cropland.

invasive species as a group to adjust to new environments. Given the counteracting forces of human management and invader community shifts, therefore, the net effect of change on plant invasion in cropping systems may be relatively small (Fig. 5.1a).

Biotic resistance in natural systems, in contrast, may decrease with rapid global change (Fig. 5.1b). Unlike crops, native species must rely on natural dispersal or evolution to keep up with changing climates. Such processes are thought to be too slow relative to the pace of climate change for many species, particularly sessile species such as plants (Sorte, 2013). As a result, species turnover, the proportion of species replaced by other species with climate change, is expected to be substantial (e.g. 27–63% of European plant species) (Thuiller *et al.*, 2005), creating colonization opportunities for both native and invasive species suited to the new environments. In natural ecosystems, diversity may help native species maintain biotic resistance. However, management options for maintaining biotic resistance are more limited in natural systems than in cropland, as most species are not planted, and assisted migration, or moving species to help them keep up with climate change, involves risks as well as benefits (Richardson *et al.*, 2009). Decreased biotic resistance, coupled with invasive species that can cope with changing environments, may lead to an overall increase in invasion risk in natural systems.

Invasive Plants and Changing Resources

Global change affects plants more directly than it affects other organisms. In addition to simply changing the environment within which plants live, several of the atmospheric and climatic factors that are changing are key plant resources, notably CO_2 and water. This is not coincidental. Global change is caused primarily by burning fossil fuels derived from plant material (Dukes, 2003). In 2010, carbon (C) emissions from fossil fuels and cement were 9.1 Pg C (Peters *et al.*, 2012), which equated to approximately 16% of terrestrial C fixation (Field *et al.*, 1998). The resulting increase in the atmospheric concentration of CO_2 (from approximately 278 ppm in 1750 to 390 ppm in 2010) (Peters *et al.*, 2012) increases photosynthetic efficiency and acts as a fertilizer, increasing the growth of most plants (Ainsworth and Long, 2005). Although it is outside the scope of this chapter, nitrogen released from fossil fuels also enters the atmosphere, primarily as nitric oxide, contributing to atmospheric N deposition and again fertilizing plants (Vitousek *et al.*, 1997; for discussion of N and invasion see Perry *et al.*, 2004; Bradley *et al.*, 2010a; Gonzalez *et al.*, 2010). It is coincidental, however, that as the atmospheric concentration of CO_2 increases, it leads to changes in climate that influence the availability of water, another resource that often limits plant growth. Precipitation is expected to increase at high latitudes and decrease at lower latitudes (Meehl *et al.*, 2007), and warming is expected to increase evapotranspiration and reduce the water available for plant growth. Thus, for plants, global change directly changes three key resources, recycling previously immobilized C and N and redistributing them globally and regionally, respectively, and changing the availability of water.

Changes in these resources can have dramatic effects on plant communities (Tilman, 1990; Stevens *et al.*, 2004). For example, both water and N addition can transform semi-arid shortgrass steppe from an ecosystem dominated by perennial grasses into one dominated by invasive forbs (Lauenroth *et al.*, 1978; Milchunas and Lauenroth, 1995). Plant community responses to changing resource availability are determined in part by plant traits, and may therefore be related to invasion (Davis *et al.*, 2000). Plant species are arrayed consistently along a spectrum from slow-growing species, adapted to the efficient use of limited resources, to fast-growing species, adapted to the rapid use of abundant resources (Reich *et al.*, 1997; Diaz *et al.*, 2004; Wright *et al.*, 2004). Many species from relatively undisturbed ecosystems lie at the efficient resource use end of this spectrum, while many crops and invasive species are adapted to rapid resource use and acquisition (Chapin, 1980).

As with change in general, the net effect of global changes that alter resources depends on the relative effects on resident and invasive species. Because crops and cropland invaders often have similar resource-use strategies, their relative success may change little with changing resource availability (Fig. 5.1a). Important exceptions to this occur when crops and invaders differ, such as in systems with cool-season (C_3) crops and warm-season (C_4) invaders (discussed below) (Ainsworth and Long, 2005). In contrast, in natural ecosystems, invasive species differ more in resource-use strategy from co-occurring native species (Leishman *et al.*, 2007; Penuelas *et al.*, 2010). As a result, increases in CO_2 and water can favour invasive over native species (Fig. 5.1b) (Smith *et al.*, 2000; Daehler, 2003; Bradley *et al.*, 2010a). Increases in resource availability may also increase enemy release indirectly, and therefore the success of non-native species, by favouring poorly defended fast-growing species that can be released strongly from natural enemies in their introduced range (Blumenthal *et al.*, 2009). Global changes do not always increase resource availability, however. Reductions in water availability with warming, decreased precipitation and reductions in N availability with elevated CO_2 may inhibit invasion in natural systems (Bradley *et al.*, 2010a). A meta-analysis examining native versus exotic global change responses found that resource availability predicted the performance of both native and exotic species, and weak evidence that invasive species responded more strongly than natives. Elevated CO_2 significantly increased the performance of exotic but not native plants, while warming significantly increased the performance of native but not exotic plants (Sorte *et al.*, 2013). Thus, understanding how global changes influence resource availability may be helpful, but not always sufficient, in predicting how invasion will respond to global change.

In sum, increased CO_2 and climate change seem likely to alter invasion most in natural ecosystems, where native species have limited abilities to move and adapt, and often differ in resource-use strategies from invasive species. In crop ecosystems, crop choice, artificial selection and similarities between crops and invaders may keep crops more competitive with invasive species. There are many exceptions to these broad predictions, however, which necessitate more detailed examination of particular global change drivers, ecosystems and species. The following sections assess how plant invasion is likely to be influenced by elevated CO_2, warming and altered precipitation, individually and in combination.

Elevated Atmospheric CO_2

Unlike other environmental variables, all plant species have historically experienced very similar atmospheric CO_2 concentrations. Consequently, species do not have an array of adaptations to different levels of CO_2, as they do, for example, to water. Nevertheless, some traits incidentally influence plant species' ability to take advantage of high CO_2 concentrations, and therefore their success relative to other species with elevated CO_2. The clearest case of this is the photosynthetic pathway. Warm-season plants with C_4 photosynthesis concentrate CO_2 in bundle-sheath cells and rely on different enzymes for photosynthesis than C_3 plants, decreasing water loss and reducing photorespiration at high temperatures. Consequently, their photosynthetic and growth rates tend to be less sensitive to CO_2 than those of cool-season species with C_3 photosynthesis. Across studies, C_3 species respond more strongly to elevated CO_2 than do C_4 species (Wand *et al.*, 1999; Ainsworth and Long, 2005).

Such differences can influence the outcome of competition between invasive species and residents. In cropping systems, an array of studies have found that elevated CO_2 provides an advantage to C_3 invaders competing with C_4 crops or, conversely, C_3 crops competing with C_4 invaders (Patterson *et al.*, 1984; Alberto *et al.*, 1996; Ziska, 2000, 2001, 2003b; Davis and Ainsworth, 2012). Elevated CO_2 has also been observed to favour invasive species over crops with the same photosynthetic pathway (Ziska, 2001, 2003b, 2010), or vice versa (Davis and Ainsworth, 2012), suggesting that other traits also influence the relative responses of crops and invasive species to elevated CO_2. While the photosynthetic pathway can also influence CO_2 effects on natural plant communities (Polley, 1997; Morgan *et al.*, 2011), in the one relevant experimental study, elevated CO_2 did not favour C_3 invasive forbs relative to a C_4 native grass (Williams *et al.*, 2007).

Growth rates can also affect CO_2 responses. While most plant species initially increase photosynthesis in response to elevated CO_2, limited sinks for the resulting photosynthate can limit longer-term photosynthetic responses (Körner, 2011). Rapid growth can increase sink strength, and fast-growing species tend to have relatively strong responses to elevated CO_2 (Hunt *et al.*, 1993; Poorter and Navas, 2003). This may help to explain strong CO_2 responses in both crops and invasive species (Ziska, 2003a; Ainsworth and Long, 2005). In contrast, slower-growing species from relatively undisturbed ecosystems can have lower responses to elevated CO_2 (Hunt *et al.*, 1993), potentially leading to strong CO_2 effects on invasion in such systems. Facilitation of invasion by elevated CO_2 has been observed in field studies in a number of ecosystems (Hattenschwiler and Korner, 2003; Belote *et al.*, 2004; Dukes *et al.*, 2011). For example, elevated CO_2 stimulated seed production of the invasive winter-annual grass, red brome (*Bromus madritensis* subsp. *rubens*), more strongly than that of a native winter-annual grass in the Nevada desert (Smith *et al.*, 2000). The strong growth response of red brome may have been due to low construction costs, which allowed red brome to make use of additional C without greatly increasing N acquisition (Nagel *et al.*, 2004). Elevated CO_2 does not always facilitate invasion (Williams *et al.*, 2007), but has not been observed to decrease invasion in relatively undisturbed ecosystems.

Water use may also influence CO_2 effects on invasion (Patterson, 1995b; Dukes and Mooney, 1999; Davis and Ainsworth, 2012). Stronger CO_2 gradients between atmosphere and leaf with elevated CO_2 allow plants to reduce stomatal conductance and increase water-use efficiency (Patterson, 1995b; Morgan *et al.*, 2011). As a result, elevated CO_2 can increase two resources, C and water, which could favour invasive species that have relatively high water requirements relative to native species in natural ecosystems (Dukes, 2002; Cavaleri and Sack, 2010). This indirect, water-mediated effect of elevated CO_2 may also influence invasion in cropland (Patterson, 1995b). For example, it has been suggested as an explanation for both the increased and decreased success of redroot pigweed relative to crops with elevated CO_2 (tomato, *Solanum lycopersicum*, and soybean, respectively) (Valerio *et al.*, 2011; Davis and Ainsworth, 2012).

Warming

Temperature is expected to increase globally, with the largest increases over land and near the poles (Meehl *et al.*, 2007). The influence of warming on plant invasion is also likely to vary regionally, with both latitude and precipitation. The most consistent prediction to date is that invasive species, like native species, will move poleward and upward, following the temperature conditions to which they are adapted. This is a common result from bioclimate envelope models, which use information about where species exist currently, in conjunction with predicted future environmental conditions, to map likely future species distributions (Bradley *et al.*, 2010a; see also Part III, Chapter 15). Given the poleward and upward movement predicted for many species, it is likely that very high latitude and high-elevation regions, which have to date been less invaded than warmer regions, may become more susceptible to invasion (Pauchard *et al.*, 2009; Alexander *et al.*, 2011; Marini *et al.*, 2012). Specific predictions for high-altitude regions include greater risk of invasion by invasive hawkweeds (*Hieracium* spp.) in Australia (Beaumont *et al.*, 2009), 29 exotic annual grass species in South Africa (Parker-Allie *et al.*, 2009) and spotted knapweed (*Centaurea biebersteinii*) in North America (Broennimann and Guisan, 2008; Bradley *et al.*, 2009).

Specific predictions have also been made regarding latitude. In the USA, many of the most problematic invasive species are currently either limited to or more problematic in southern regions (Patterson, 1995b; Ziska and Dukes, 2011). As temperatures increase, these species are predicted to move northwards, increasing control costs and/or yield loss (Patterson, 1995b; Fuhrer, 2003; Clements and DiTommaso, 2011). In maize (*Zea mays*) production, for example, the historically damaging cold-tolerant weed, velvetleaf (*Abutilon theophrasti*), will likely become less problematic than the warm-season weed, Johnsongrass (*Sorghum halepense*), which is projected to expand 200–600 km poleward into the north-east and Great Lakes states (McDonald *et al.*, 2009). Kudzu, privet (*Ligustrum sinense*; *Ligustrum vulgare*) and cogongrass (*Imperata cylindrica*) are also likely to spread 200–300 km northward, while continuing to occupy their currently invaded ranges in the south-eastern USA (Jarnevich and Stohlgren, 2009; Bradley *et al.*, 2010b). In the UK, black grass (*Alopecurus myosuroides*) may spread northward with higher temperatures, but could also become less competitive with deeper-rooted cereal crops as drought events become more frequent (Stratonovitch *et al.*, 2012). Within the California grass flora, invasive species richness increases with temperature, and invasives are more likely than natives to have high values for traits that correlate positively with warm temperatures, such as height, leaf size, seed size, specific leaf area and N content (Sandel and Dangremond, 2012). Similar trends have been identified experimentally, with warming increasing the growth of invasives more than the native Belgian grassland species, perhaps because many of the more common invasive species are native to areas further south (Verlinden and Nijs, 2010).

Conversely, at the other temperature extreme, tropical areas have been relatively uninvaded, perhaps due to strong biotic resistance from abundant plant life (Lonsdale, 1999; Sax, 2001; Fine, 2002), suggesting that invasion resistance could increase with warming in subtropical regions. Simultaneously, some invasive species may retreat from their lowest latitudinal extents as temperatures rise (Richardson *et al.*, 2000; Bradley, 2009; Bradley *et al.*, 2009).

The influence of warming on invasion may also vary with precipitation, favouring invaders in mesic areas, where it increases growth through direct effects on physiological processes, and inhibiting invaders in water-limited areas, where it decreases growth by increasing evapotranspiration (Bradley *et al.*, 2010a; Hoeppner and Dukes, 2012). Among the few experimental studies of warming and invasion in natural systems, warming facilitated invasion at a mesic site in Belgium, had no effect on invasion at a somewhat drier site in California and inhibited invasion at the driest site, in New Zealand (Williams *et al.*, 2007; Verlinden and Nijs, 2010; Dukes *et al.*, 2011).

Finally, warming, like elevated CO_2, can influence the relative success of C_3 and C_4 plants, as warming increases photorespiration and therefore decreases photosynthesis more in C_3 than in C_4 species. The New Zealand site in which warming inhibited invasion is not only relatively dry but also is dominated by a mixture of C_3 and C_4 grasses (Williams *et al.*, 2007). Although the photosynthetic pathway did not explain CO_2 responses, it could explain warming responses: positive responses of the native C_4 grass to warming may have increased biotic resistance to invasion by the C_3 forbs (Williams *et al.*, 2007). Similarly, experimental warming can favour C_4 invaders growing in competition with C_3 crops. For example, in a greenhouse competition experiment with the C_3 crop, soybean, warming by 3°C increased the relative yield of the C_4 weed, redroot pigweed, by 54%, but had no effect on the relative yield of the C_3 weed, common cocklebur (*Xanthium strumarium*) (Flint and Patterson, 1983).

Changing Precipitation

Unlike elevated CO_2 and warming, precipitation is predicted to increase in some regions, particularly at higher latitudes, and decrease in others, particularly at lower latitudes (Meehl *et al.*, 2007). The seasonality, frequency and intensity of precipitation are also likely to change and, again, changes are likely to vary by region. Finally, there is more uncertainty in predictions of precipitation than in predictions of other aspects of climate change (Meehl *et al.*, 2007). Together, these sources of variability make it more difficult to make general predictions about how precipitation change will influence plant invasion.

Despite the uncertainty surrounding changing precipitation and invasion, the traits of native and invasive species may provide some predictive ability. Relative to co-occurring native species, invasive species tend to use more water (Cavaleri and Sack, 2010), suggesting that they may benefit more from increases in precipitation and, conversely, suffer more from decreases in precipitation (Bradley *et al.*, 2010a). Field experiments provide some evidence of such patterns. For example, increases in precipitation favoured, and decreases in precipitation inhibited, the invasive grass, ripgut brome (*Bromus diandrus*), more than native species in the California Channel Islands (Levine *et al.*, 2010). Increases in precipitation also increased the relative success of invasive species in grasslands in Wyoming, Utah and California (Miller *et al.*, 2006; Thomsen and D'Antonio, 2007; Blumenthal *et al.*, 2008; Levine *et al.*, 2010) (Fig. 5.2).

Such patterns are not universal, however. Increases in precipitation had little effect on the relative success of the native and invasive species in the Mojave Desert, Mediterranean shrubland and a planted Montana grassland (Garcia-Serrano *et al.*, 2004; Maron and Marler, 2007; Rao and Allen, 2010). And in some cases, drier conditions may inhibit native more than invasive species. In riparian ecosystems of the south-western USA, where dams have reduced flooding, native species adapted for rapid growth following flooding appear to be less drought tolerant than invasive species such as Russian olive (*Elaeagnus angustifolia*) and tamarisk

Fig. 5.2. Dalmatian toadflax in mixed-grass prairie near Cheyenne, Wyoming, USA. Experimental snow addition promotes Dalmatian toadflax invasion in this semi-arid grassland, an effect that is compounded by increases in nitrogen deposition (Blumenthal *et al.*, 2008). (Photo: Stephen Asmus.)

(*Tamarix ramosissima*, *Tamarix chinensis* and hybrids) (Glenn and Nagler, 2005). Consequently, predicted decreases in stream flow with decreased precipitation and snowpack may favour invasive rather than native species in this system (Perry *et al.*, 2012). Similarly, decreases in summer precipitation could increase the competitive ability of cheatgrass, which grows in the autumn and spring, relative to native species that grow in the summer (Fig. 5.3) (Bradley, 2009).

By comparison, cropland invaders, being relatively similar to crops, might be expected to have more similar responses to changing precipitation. In a survey of water-manipulation experiments involving competition between invasive weeds and maize or soybean, decreased water availability had widely variable effects and was at least as likely to favour the crop as the invader

Fig. 5.3. Cheatgrass invading sagebrush steppe near Lander, Wyoming, USA (Blumenthal *et al.*, 2006). One of the most invasive plants in North America, cheatgrass may become even more invasive with elevated CO_2 (Ziska *et al.*, 2005), and could become more or less invasive with changing temperature and precipitation (Bradley, 2009). Warmer temperatures and more frequent droughts are also expected to increase fire frequency and intensity, exacerbating positive feedbacks between cheatgrass invasion and fire (Abatzoglou and Kolden, 2011).

(Patterson, 1995a). For particular combinations of crops and invaders, however, the effects of altered precipitation can be quite strong. In an irrigated competition experiment, Johnsongrass reduced the biomass of maize by approximately 15% at field capacity, but by 65% under drought conditions (Leguizamon *et al.*, 2011). The opposite pattern was observed in competition between green foxtail (*Setaria viridis*) and maize, where foxtail greatly reduced the yield of irrigated maize but had no effect on the yield of dryland maize in a relatively dry year (McGiffen *et al.*, 1997). Such incidences of reduced invasive species impact under drought conditions may be common where invasive plants compete with crops for light and drought reduces the intensity of light competition (Patterson, 1995a).

Extreme Events

Droughts, floods, severe storms and heatwaves are predicted to become more frequent and intense with climate change, and are likely to have increasingly strong effects on plant communities (Smith, 2011). These extreme climatic events may influence invasive species in multiple ways. Extreme events can act as disturbances, killing or damaging resident species, increasing resource availability and providing opportunities for invasive species (Hobbs and Huenneke, 1992; Davis *et al.*, 2000; Diez *et al.*, 2012). For example, in the south-eastern USA, invasive vines increased in abundance following Hurricanes Andrew and Katrina (Horvitz *et al.*, 1998; Brown *et al.*, 2011). Similarly, in the south-western USA, severe drought in 2004 and 2005 killed many native species, providing an opportunity for the invasive grass, Lehmann lovegrass (*Eragrostis lehmanniana*) to spread and, in many areas, dominate plant communities (Scott *et al.*, 2010).

The effects of extreme events on crops are well known, and likely to intensify with climate change (Tubiello *et al.*, 2007). For example, using the CERES-maize model, Rosenzweig *et al.* (2002) estimated that reductions in maize growth in the USA due to heavy precipitation were likely to double under climate change. Such changes are likely to reduce biotic resistance from crops and provide opportunities for invasive plants, particularly those that escape damage due to phenological or trait differences from those of the crop. For example, flooded sections of maize and soybean fields are often invaded by flood-tolerant species such as barnyardgrass (*Echinochloa crus-galli*). Less severe stress imposed by drought, heavy precipitation or heat may also influence crop–invader competition, but overall is as likely to favour the crop as the invader (Patterson, 1995a).

Strong winds or floods can also influence invasive species dispersal (Ziska and Dukes, 2011; Diez *et al.*, 2012). For example, seeds of Santa Maria feverfew (*Parthenium hysterophorus*), a problematic invader of both cropland and rangeland, were widely dispersed across Swaziland by Cyclone Demonia (Burgiel and Muir, 2010). Similarly, an extreme flood in 1997 allowed the fast-growing poisonous shrub tree, tobacco (*Nicotania glauca*), to invade semi-arid rangeland in Australia (Florentine and Westbrooke, 2005). Furthermore, flood-driven invasion was particularly severe in experimental plots where grazing was allowed, an example of how extreme events can interact with land-use change to influence invasion. Finally, extreme events may even influence the impact of invasive species on resident species. For example, droughts and heatwaves may compound the effects of invasive grasses on fire frequency (discussed below) (D'Antonio and Vitousek, 1992; Abatzoglou and Kolden, 2011).

Combined Effects of Climatic and Atmospheric Changes

CO_2 and temperature

Only a handful of studies have examined the combined effects of elevated CO_2 and warming on plant invasion. In New Zealand grassland, warming strongly reduced the success of two invasive composites (*Hypochaeris radicata* and *Leontodon taraxacoides*), while elevated CO_2 had only minor effects and did not interact with warming (Williams *et al.*, 2007). In contrast, elevated CO_2 strongly increased invasion by another composite, yellow starthistle, irrespective of warming, which had only minor effects on starthistle invasion (Dukes *et al.*, 2011). In the only similar study in cropping systems, warming increased and elevated CO_2 decreased the success of the C_4 invader, barnyardgrass, relative to the C_3 crop, rice (*Oryza sativa*) (Alberto *et al.*, 1996). The combination of warming and elevated CO_2 led to no net change in the relative growth of the two species, but decreased the seed yield of rice relative to barnyardgrass.

CO_2 and precipitation

Given its indirect effect on water-use efficiency, elevated CO_2 might also interact with changes in precipitation to affect invasion (Patterson, 1995b). For example, elevated CO_2 decreased the competitive ability of the C_4 invader redroot pigweed, with the C_3 crop tomato under well-watered conditions, but increased the competitive ability of pigweed under drought conditions, perhaps due to the strong increases in pigweed leaf water potential (Valerio *et al.*, 2011). In contrast, altered precipitation did not influence the positive effects of elevated CO_2 on invasive species in California annual grassland, or a pasture community within a controlled environment (Newton *et al.*, 1996; Dukes *et al.*, 2011).

Temperature and precipitation

Given that water availability can influence the effects of warming, and warming influences water availability, simultaneous changes in temperature and precipitation might be expected to interact to affect invasion. The combined effects of temperature and precipitation changes have been studied primarily through bioclimate envelope modelling. Models have revealed that precipitation can be very important in determining where invasive species will be able to live in the future. For example, modelled future distributions of five western US invasive plant species were influenced strongly by precipitation, suggesting that projections based on temperature alone were insufficient, especially for water-limited regions (Bradley, 2009; Bradley *et al.*, 2009). The distributions of yellow starthistle and tamarisk are likely to expand with climate change, while those of spotted knapweed and cheatgrass may expand in some regions and contract in others. Leafy spurge (*Euphorbia esula*), on the other hand, is likely to retreat from areas of its current range as they become climatically unsuitable.

Interactions Between Global Change and Local Processes Influencing Invasion

Bioclimate envelope models have also shown that the effects of temperature and precipitation on invasive species' ranges depend on the availability of suitable habitat within a species' future climatic niche. While climate exerts primary controls on plant distributions at a broad scale, local landscape-scale factors such as land use, cover and soil type can strongly influence climate effects on invasion. In South Africa, soil type, through its influence on water and nutrient availability, constrained the projected distributions of five invaders greatly under future climate conditions (Richardson *et al.*, 2000). In the north-eastern USA, land use and forest cover strongly influenced the predicted future ranges of three invasive plant species, restricting predicted ranges in climatically suitable locations without optimal landscape attributes (Ibáñez *et al.*, 2009). Combinations of temperature, precipitation and soil type have also been predicted to influence invasion in crops. While the distribution of black grass could increase with climate change in cereal croplands in the UK, its impact on crops may decrease, due to relative drought intolerance (Stratonovitch *et al.*, 2012).

Another important interaction is likely to be that between global change, fire and invasion. While fire effects on invasion vary widely, in many parts of the world fire is more likely to facilitate than to inhibit invasive species (D'Antonio, 2000). Invasive species, particularly grasses, can also increase fire frequency and severity, leading to positive feedbacks between fire and invasion (D'Antonio and Vitousek, 1992). As global changes are expected to influence the likelihood of both fire and invasion, they have the potential to accelerate or decelerate fire-invasion feedbacks (Smith *et al.*, 2000; Abatzoglou and Kolden, 2011; Ziska and Dukes, 2011). For example, in western North America, many problematic invasive grasses, such as cheatgrass, buffelgrass (*P. ciliare*), Lehmann lovegrass, medusahead (*Taeniatherum caput-medusa*) and red brome (*B. madritensis* subsp. *rubens*), both promote and benefit from fire (D'Antonio and Vitousek, 1992; Brooks and Chambers, 2011). Elevated CO_2 and severe droughts have both been found to favour one or more of these species (Smith *et al.*, 2000; Ziska *et al.*, 2005; Scott *et al.*, 2010), while the predicted effects of precipitation and temperature changes vary by region and climate change scenario (Bradley, 2009). At the same time, warmer temperatures, decreases in precipitation (at southern latitudes) and increases in the frequency or duration of droughts may increase the frequency and intensity of fires, while elevated CO_2 may partially offset these effects in some areas (Westerling *et al.*, 2006; Seager and Vecchi, 2010; Abatzoglou and Kolden, 2011; Morgan *et al.*, 2011). In combination, strong positive responses of invasive grasses to climate change and environmental conditions more suitable for fire may increase invasion dramatically in regions invaded by fire-tolerant grasses.

While global change may often increase fire, it is important to note that the opposite can also occur. For example, elevated CO_2 is thought to decrease fire frequency in South African savannahs by allowing tree seedlings to grow more quickly, escape periodic fires and outcompete flammable grasses (Bond *et al.*, 2003; Buitenwerf *et al.*, 2012). Although decreases in fire frequency relative to historical levels have also been linked to invasion, there is less evidence to suggest that this might be a common pattern (D'Antonio, 2000).

Conclusions

Past success in novel environments and strong responses to increases in resources suggest that many invasive species will thrive as atmospheric CO_2 increases and climates change. In cropland, increases in atmospheric CO_2 and the frequency/intensity of extreme events may increase invasion risk more often than not, warming may lead to problematic new invaders in temperate regions and altered precipitation will likely have varied effects on invasion. At the same time, intensive management is likely to provide crops with advantages that will help them adjust to climate change and provide biotic resistance to invasion. Maintaining such advantages may require specific management adaptations to global change, however, such as the use of crops or varieties suited to new and variable climates (see Part IV). In natural ecosystems, changes that increase resource availability, including increases in CO_2, extreme events and (in some regions) precipitation, may often favour invasive over native species. Furthermore, managers have relatively few options for helping native species keep up with invaders in natural ecosystems. Assisted migration may be important not only for saving species from extinction but also for maintaining biotic resistance in novel ecosystems (Seastedt *et al.*, 2008; Richardson *et al.*, 2009). Managers may have the greatest opportunity to influence biotic resistance in natural systems during ecosystem restoration. Restoration strategies need to be developed that allow managers to maximize biotic resistance under future as well as current climatic conditions (Bradley and Wilcove, 2009). In sum, global change may often increase the relative success of invasive species, particularly in natural ecosystems. Adapting management to limit invasion will require increasing biotic resistance, which will in turn require careful matching of species and genotypes for both crops and native species to future environments.

References

Abatzoglou, J.T. and Kolden, C.A. (2011) Climate change in western US deserts: potential for increased wildfire and invasive annual grasses. *Rangeland Ecology and Management* 64, 471–478.

Ainsworth, E.A. and Long, S.P. (2005) What have we learned from 15 years of free-air CO_2 enrichment (FACE)? A meta-analytic review of the responses of photosynthesis, canopy properties and plant production to rising CO_2. *New Phytologist* 165, 351–371.

Alberto, A.M.P., Ziska, L.H., Cervancia, C.R. and Manalo, P.A. (1996) The influence of increasing carbon dioxide and temperature on competitive interactions between a C_3 crop, rice (*Oryza sativa*), and a C_4 weed (*Echinochloa glabrescens*). *Australian Journal of Plant Physiology* 23, 795–802.

Alexander, J.M., Kueffer, C., Daehler, C.C., Edwards, P.J., Pauchard, A. and Seipel, T. (2011) Assembly of non-native floras along elevational gradients explained by directional ecological filtering. *Proceedings of the National Academy of Sciences* 108, 656–661.

Beaumont, L.J., Gallagher, R.V., Downey, P.O., Thuiller, W., Leishman, M.R. and Hughes, L. (2009) Modelling the impact of *Hieracium* spp. on protected areas in Australia under future climates. *Ecography* 32, 757–764.

Belote, R.T., Weltzin, J.F. and Norby, R.J. (2004) Response of an understory plant community to elevated [CO_2] depends on differential responses of dominant invasive species and is mediated by soil water availability. *New Phytologist* 161, 827–835.

Blumenthal, D., Chimner, R.A., Welker, J.M. and Morgan, J.A. (2008) Increased snow facilitates plant invasion in mixedgrass prairie. *New Phytologist* 179, 440–448.

Blumenthal, D., Mitchell, C.E., Pysek, P. and Jarosik, V. (2009) Synergy between pathogen release and resource availability in plant invasion. *Proceedings of the National Academy of Sciences* 106, 7899–7904.

Bond, W.J., Midgley, G.F. and Woodward, F.I. (2003) The importance of low atmospheric CO_2 and fire in promoting the spread of grasslands and savannas. *Global Change Biology* 9, 973–982.

Bradley, B.A. (2009) Regional analysis of the impacts of climate change on cheatgrass invasion shows potential risk and opportunity. *Global Change Biology* 15, 196–208.

Bradley, B.A. and Wilcove, D.S. (2009) When invasive plants disappear: transformative restoration possibilities in the Western United States resulting from climate change. *Restoration Ecology* 17, 715–721.

Bradley, B.A., Oppenheimer, M. and Wilcove, D.S. (2009) Climate change and plant invasions: restoration opportunities ahead? *Global Change Biology* 15, 1511–1521.

Bradley, B.A., Blumenthal, D.M., Wilcove, D.S. and Ziska, L.H. (2010a) Predicting plant invasions in an era of global change. *Trends in Ecology and Evolution* 25, 310–318.

Bradley, B.A., Wilcove, D.S. and Oppenheimer, M. (2010b) Climate change increases risk of plant invasion in the Eastern United States. *Biological Invasions* 12, 1855–1872.

Broennimann, O. and Guisan, A. (2008) Predicting current and future biological invasions: both native and invaded ranges matter. *Biology Letters* 4, 585–589.

Brooks, M.L. and Chambers, J.C. (2011) Resistance to invasion and resilience to fire in desert shrublands of North America. *Rangeland Ecology and Management* 64, 431–438.

Brown, D.R., Sherry, T.W. and Harris, J. (2011) Hurricane Katrina impacts the breeding bird community in a bottomland hardwood forest of the Pearl River basin, Louisiana. *Forest Ecology and Management* 261, 111–119.

Buitenwerf, R., Bond, W.J., Stevens, N. and Trollope, W.S.W. (2012) Increased tree densities in South African savannas: >50 years of data suggests CO_2 as a driver. *Global Change Biology* 18, 675–684.

Burgiel, S.W. and Muir, A.A. (2010) *Invasive Species, Climate Change and Ecosystem-based Adaptation: Addressing Multiple Drivers of Global Change.* Global Invasive Species Programme, Washington, DC, and Nairobi, Kenya.

Byers, J.E. (2002) Impact of non-indigenous species on natives enhanced by anthropogenic alteration of selection regimes. *Oikos* 97, 449–458.

Catford, J.A., Jansson, R. and Nilsson, C. (2009) Reducing redundancy in invasion ecology by integrating hypotheses into a single theoretical framework. *Diversity and Distributions* 15, 22–40.

Cavaleri, M.A. and Sack, L. (2010) Comparative water use of native and invasive plants at multiple scales: a global meta-analysis. *Ecology* 91, 2705–2715.

Chapin, F.S.I. (1980) The mineral nutrition of wild plants. *Annual Review of Ecology and Systematics* 11, 233–260.

Clements, D.R. and DiTommaso, A. (2011) Climate change and weed adaptation: can evolution of invasive plants lead to greater range expansion than forecasted? *Weed Research* 51, 227–240.

Daehler, C.C. (2003) Performance comparisons of co-occurring native and alien invasive plants: implications for conservation and restoration. *Annual Review of Ecology Evolution and Systematics* 34, 183–211.

D'Antonio, C.M. (2000) Fire, plant invasions, and global changes. In: Mooney, H.A. and Hobbs, R.J. (eds) *Invasive Species in a Changing World.* Island Press, Washington, DC, pp. 65–93.

D'Antonio, C.M. and Thomsen, M. (2004) Ecological resistance in theory and practice. *Weed Technology* 18, 1572–1577.

D'Antonio, C.M. and Vitousek, P.M. (1992) Biological invasions by exotic grasses, the grass fire cycle, and global change. *Annual Review of Ecology and Systematics* 23, 63–87.

Davidson, A.M., Jennions, M. and Nicotra, A.B. (2011) Do invasive species show higher phenotypic plasticity than native species and, if so, is it adaptive? A meta-analysis. *Ecology Letters* 14, 419–431.

Davis, A.S. and Ainsworth, E.A. (2012) Weed interference with field-grown soybean decreases under elevated [CO_2] in a FACE experiment. *Weed Research* 52, 277–285.

Davis, M.A., Grime, J.P. and Thompson, K. (2000) Fluctuating resources in plant communities: a general theory of invasibility. *Journal of Ecology* 88, 528–534.

Diaz, S., *et al.* (2004) The plant traits that drive ecosystems: evidence from three continents. *Journal of Vegetation Science* 15, 295–304.

Diez, J.M., *et al.* (2012) Will extreme climatic events facilitate biological invasions? *Frontiers in Ecology and the Environment* 10, 249–257.

Dukes, J.S. (2002) Comparison of the effect of elevated CO_2 on an invasive species (*Centaurea solstitialis*) in monoculture and community settings. *Plant Ecology* 160, 225–234.

Dukes, J.S. (2003) Burning buried sunshine: human consumption of ancient solar energy. *Climatic Change* 61, 31–44.

Dukes, J.S. and Mooney, H.A. (1999) Does global change increase the success of biological invaders? *Trends in Ecology and Evolution* 14, 135–139.

Dukes, J.S., Chiariello, N.R., Loarie, S.R. and Field, C.B. (2011) Strong response of an invasive plant species (*Centaurea solstitialis* L.) to global environmental changes. *Ecological Applications* 21, 1887–1894.

Field, C.B., Behrenfeld, M.J., Randerson, J.T. and Falkowski, P. (1998) Primary production of the biosphere: integrating terrestrial and oceanic components. *Science* 281, 237–240.

Fine, P.V.A. (2002) The invasibility of tropical forests by exotic plants. *Journal of Tropical Ecology* 18, 687–705.

Flint, E.P. and Patterson, D.T. (1983) Interference and temperature effects on growth in soybean (*Glycine max*) and associated C-3 and C-4 weeds. *Weed Science* 31, 193–199.

Florentine, S.K. and Westbrooke, M.E. (2005) Invasion of the noxious weed *Nicotiana glauca* R. Graham after an episodic flooding event in the arid zone of Australia. *Journal of Arid Environments* 60, 531–545.

Friedman, J.M., Roelle, J.E., Gaskin, J.F., Pepper, A.E. and Manhart, J.R. (2008) Latitudinal variation in cold hardiness in introduced *Tamarix* and native *Populus*. *Evolutionary Applications* 1, 598–607.

Fuhrer, J. (2003) Agroecosystem responses to combinations of elevated CO_2, ozone, and global climate change. *Agriculture Ecosystems and Environment* 97, 1–20.

Garcia-Serrano, H., Escarre, J. and Sans, F.X. (2004) Factors that limit the emergence and establishment of the related aliens *Senecio inaequidens* and *Senecio pterophorus* and the native *Senecio malacitanus* in Mediterranean climate. *Canadian Journal of Botany-Revue Canadienne de Botanique* 82, 1346–1355.

Glenn, E.P. and Nagler, P.L. (2005) Comparative ecophysiology of *Tamarix ramosissima* and native trees in western US riparian zones. *Journal of Arid Environments* 61, 419–446.

Gonzalez, A.L., Kominoski, J.S., Danger, M., Ishida, S., Iwai, N. and Rubach, A. (2010) Can ecological stoichiometry help explain patterns of biological invasions? *Oikos* 119, 779–790.

Gurevitch, J., Fox, G.A., Wardle, G.M., Inderjit and Taub, D. (2011) Emergent insights from the synthesis of conceptual frameworks for biological invasions. *Ecology Letters* 14, 407–418.

Hattenschwiler, S. and Korner, C. (2003) Does elevated CO_2 facilitate naturalization of the non-indigenous *Prunus laurocerasus* in Swiss temperate forests? *Functional Ecology* 17, 778–785.

Hobbs, R.J. and Huenneke, L.F. (1992) Disturbance, diversity, and invasion: implications for conservation. *Conservation Biology* 6, 324–337.

Hoeppner, S.S. and Dukes, J.S. (2012) Interactive responses of old-field plant growth and composition to warming and precipitation. *Global Change Biology* 18, 1754–1768.

Horvitz, C.C., Pascarella, J.B., McMann, S., Freedman, A. and Hofstetter, R.H. (1998) Functional roles of invasive non-indigenous plants in hurricane-affected subtropical hardwood forests. *Ecological Applications* 8, 947–974.

Hunt, R., Hand, D.W., Hannah, M.A. and Neal, A.M. (1993) Further responses to CO_2 enrichment in British herbaceous species. *Functional Ecology* 7, 661–668.

Ibáñez, I., Silander, J.A., Wilson, A.M., Lafleur, N., Tanaka, N. and Tsuyama, I. (2009) Multivariate forecasts of potential distributions of invasive plant species. *Ecological Applications* 19, 359–375.

Jarnevich, C.S. and Stohlgren, T.J. (2009) Near term climate projections for invasive species distributions. *Biological Invasions* 11, 1373–1379.

Körner, C. (2011) The grand challenges in functional plant ecology. *Frontiers in Plant Science* 2, 1–3.

Lambrinos, J.G. (2004) How interactions between ecology and evolution influence contemporary invasion dynamics. *Ecology* 85, 2061–2070.

Lauenroth, W.K., Dodd, J.L. and Sims, P.L. (1978) The effects of water- and nitrogen-induced stresses on plant community structure in a semiarid grassland. *Oecologia* 36, 211–222.

Leguizamon, E.S., Yanniccari, M.E., Guiamet, J.J. and Acciaresi, H.A. (2011) Growth, gas exchange and competitive ability of *Sorghum halepense* populations under different soil water availability. *Canadian Journal of Plant Science* 91, 1011–1025.

Leishman, M.R., Haslehurst, T., Ares, A. and Baruch, Z. (2007) Leaf trait relationships of native and invasive plants: community- and global-scale comparisons. *New Phytologist* 176, 635–643.

Levine, J.M., Adler, P.B. and Yelenik, S.G. (2004) A meta-analysis of biotic resistance to exotic plant invasions. *Ecology Letters* 7, 975–989.

Levine, J.M., McEachern, A.K. and Cowan, C. (2010) Do competitors modulate rare plant response to precipitation change? *Ecology* 91, 130–140.

Lonsdale, W.M. (1999) Global patterns of plant invasions and the concept of invasibility. *Ecology* 80, 1522–1536.

McDonald, A., Riha, S., DiTommaso, A. and DeGaetano, A. (2009) Climate change and the geography of weed damage: analysis of US maize systems suggests the potential for significant range transformations. *Agriculture Ecosystems and Environment* 130, 131–140.

McGiffen, M.E., Forcella, F., Lindstrom, M.J. and Reicosky, D.C. (1997) Covariance of cropping systems and foxtail density as predictors of weed interference. *Weed Science* 45, 388–396.

Marini, L., Battisti, A., Bona, E., Federici, G., Martini, F., Pautasso, M. and Hulme, P.E. (2012) Alien and native plant life-forms respond differently to human and climate pressures. *Global Ecology and Biogeography* 21, 534–544.

Maron, J. and Marler, M. (2007) Native plant diversity resists invasion at both low and high resource levels. *Ecology* 88, 2651–2661.

Meehl, G.A., *et al.* (2007) Global climate projections. In: Solomon, S., Qin, D., Manning, M., Chen, Z., Marquis, M., Averyt, K.B., Tignor, M. and Miller, H.L. (eds) *Climate Change 2007: The Physical Science Basis. Contribution of Working Group I to the Fourth Assessment Report of the Intergovernmental Panel on Climate Change.*

Cambridge University Press, Cambridge, UK, and New York, USA, pp. 747–846.

Milchunas, D.G. and Lauenroth, W.K. (1995) Inertia in plant community structure: state changes after cessation of nutrient-enrichment stress. *Ecological Applications* 5, 452–458.

Miller, M.E., Belnap, J., Beatty, S.W. and Reynolds, R.L. (2006) Performance of *Bromus tectorum* L. in relation to soil properties, water additions, and chemical amendments in calcareous soils of southeastern Utah, USA. *Plant and Soil* 288, 1–18.

Morgan, J.A., Lecain, D.R., Pendall, E., Blumenthal, D.M., Kimball, B.A., Carrillo, Y., Williams, D.G., Heisler-White, J., Dijkstra, F.A. and West, M. (2011) C-4 grasses prosper as carbon dioxide eliminates desiccation in warmed semi-arid grassland. *Nature* 476, 202–205.

Nagel, J.M., Huxman, T.E., Griffin, K.L. and Smith, S.D. (2004) CO_2 enrichment reduces the energetic cost of biomass construction in an invasive desert grass. *Ecology* 85, 100–106.

Newton, P.C.D., Clark, H., Bell, C.C. and Glasgow, E.M. (1996) Interaction of soil moisture and elevated CO_2 on the above-ground growth rate, root length density and gas exchange of turves from temperate pasture. *Journal of Experimental Botany* 47, 771–779.

Parker, I.M., Rodriguez, J. and Loik, M.E. (2003) An evolutionary approach to understanding the biology of invasions: local adaptation and general-purpose genotypes in the weed *Verbascum thapsus*. *Conservation Biology* 17, 59–72.

Parker-Allie, F., Musil, C.F. and Thuiller, W. (2009) Effects of climate warming on the distributions of invasive Eurasian annual grasses: a South African perspective. *Climatic Change* 94, 87–103.

Patterson, D.T. (1995a) Effects of environmental stress on weed/crop interactions. *Weed Science* 43, 483–490.

Patterson, D.T. (1995b) Weeds in a changing climate. *Weed Science* 43, 685–700.

Patterson, D.T., Flint, E.P. and Beyers, J.L. (1984) Effects of CO_2 enrichment on competition between a C-4 weed and a C-3 crop. *Weed Science* 32, 101–105.

Pauchard, A., *et al.* (2009) Ain't no mountain high enough: plant invasions reaching new elevations. *Frontiers in Ecology and the Environment* 7, 479–486.

Penuelas, J., Sardans, J., Llusia, J., Owen, S.M., Carnicer, J., Giambelluca, T.W., Rezende, E.L., Waite, M. and Niinemets, U. (2010) Faster returns on 'leaf economics' and different biogeochemical niche in invasive compared with native plant species. *Global Change Biology* 16, 2171–2185.

Perry, L.G., Galatowitsch, S.M. and Rosen, C.J. (2004) Competitive control of invasive vegetation: a native wetland sedge suppresses *Phalaris arundinaceae* in carbon-enriched soil. *Journal of Applied Ecology* 41, 151–162.

Perry, L.G., Andersen, D.C., Reynolds, L.V., Nelson, S.M. and Shafroth, P.B. (2012) Vulnerability of riparian ecosystems to elevated CO_2 and climate change in arid and semiarid western North America. *Global Change Biology* 18, 821–842.

Peters, G.P., Marland, G., Le Quere, C., Boden, T., Canadell, J.G. and Raupach, M.R. (2012) CORRESPONDENCE: rapid growth in CO_2 emissions after the 2008–2009 global financial crisis. *Nature Climate Change* 2, 2–4.

Polley, H.W. (1997) Implications of rising atmospheric carbon dioxide concentration for rangelands. *Journal of Range Management* 50, 562–577.

Poorter, H. and Navas, M.-L. (2003) Plant growth and competition at elevated CO_2: on winners, losers and functional groups. *New Phytologist* 157, 175–198.

Pyšek, P. and Richardson, D.M. (2007) Traits associated with invasiveness in alien plants: Where do we stand? In: Nentwig, W. (ed.) *Biological Invasion, Ecological Studies*, Vol 193. Springer-Verlag, Berlin and Heidelberg, Germany, pp. 97–126.

Rao, L.E. and Allen, E.B. (2010) Combined effects of precipitation and nitrogen deposition on native and invasive winter annual production in California deserts. *Oecologia* 162, 1035–1046.

Reich, P.B., Walters, M.B. and Ellsworth, D.S. (1997) From tropics to tundra: global convergence in plant functioning. *Proceedings of the National Academy of Sciences* 94, 13730–13734.

Richardson, D.M., Bond, W.J., Dean, W.R.J., Higgins, S.I., Midgley, G.F., Milton, S.J., Powrie, L.W., Rutherford, M.C., Samways, M.J. and Schulze, R.E. (2000) Invasive alien species and global change: a South African perspective. In: Mooney, H.A. and Hobbs, R.J. (eds) *Invasive Species in a Changing World*. Island Press, Washington, DC, pp. 303–349.

Richardson, D.M., *et al.* (2009) Multidimensional evaluation of managed relocation. *Proceedings of the National Academy of Sciences* 106, 9721–9724.

Rosenzweig, C., Tubiello, F.N., Goldberg, R., Mills, E. and Bloomfield, J. (2002) Increased crop damage in the US from excess precipitation under climate change. *Global Environmental*

Change: Human and Policy Dimensions 12, 197–202.

Sandel, B. and Dangremond, E.M. (2012) Climate change and the invasion of California by grasses. *Global Change Biology* 18, 277–289.

Sax, D.F. (2001) Latitudinal gradients and geographic ranges of exotic species: implications for biogeography. *Journal of Biogeography* 28, 139–150.

Scott, R.L., Hamerlynck, E.P., Jenerette, G., Moran, M. and Barron-Gafford, G.A. (2010) Carbon dioxide exchange in a semidesert grassland through drought-induced vegetation change. *Journal of Geophysical Research: Biogeosciences* 115, G03026, doi:10.1029/2010JG001348.

Seager, R. and Vecchi, G.A. (2010) Greenhouse warming and the 21st century hydroclimate of southwestern North America. *Proceedings of the National Academy of Sciences* 107, 21277–21282.

Seastedt, T.R., Hobbs, R.J. and Suding, K.N. (2008) Management of novel ecosystems: are novel approaches required? *Frontiers in Ecology and the Environment* 6, 547–553.

Smith, M.D. (2011) An ecological perspective on extreme climatic events: a synthetic definition and framework to guide future research. *Journal of Ecology* 99, 656–663.

Smith, S.D., Huxman, T.E., Zitzer, S.F., Charlet, T.N., Housman, D.C., Coleman, J.S., Fenstermaker, L.K., Seemann, J.R. and Nowak, R.S. (2000) Elevated CO_2 increases productivity and invasive species success in an arid ecosystem. *Nature* 408, 79–81.

Sorte, C.J.B. (2013) Predicting persistence in a changing climate: flow direction and limitations to redistribution. *Oikos* doi:10.1111/j.1600-0706.2012.00066.x.

Sorte, C.J.B., Ibáñez, I., Blumenthal, D.M., Molinari, N., Miller, L.P., Grosholz, E.D., Diez, J.M., D'Antonio, C.M., Olden, J.D., Jones, S.J. and Dukes, J.S. (2013) Poised to prosper? A cross-system comparison of climate change impacts on native and non-native species performance. *Ecology Letters* doi:10.1111/ele.12017.

Stevens, C.J., Dise, N.B., Mountford, J.O. and Gowing, D.J. (2004) Impact of nitrogen deposition on the species richness of grasslands. *Science* 303, 1876–1879.

Stratonovitch, P., Storkey, J. and Semenov, A.A. (2012) A process-based approach to modelling impacts of climate change on the damage niche of an agricultural weed. *Global Change Biology* 18, 2071–2080.

Thomsen, M.A. and D'Antonio, C.M. (2007) Mechanisms of resistance to invasion in a California grassland: the roles of competitor identity, resource availability, and environmental gradients. *Oikos* 116, 17–30.

Thuiller, W., Lavorel, S., Araujo, M.B., Sykes, M.T. and Prentice, I.C. (2005) Climate change threats to plant diversity in Europe. *Proceedings of the National Academy of Sciences* 102, 8245–8250.

Thuiller, W., Richardson, D.M. and Midgley, G. (2007) Will climate change promote alien plant invasions? In: Nentwig, W. (ed.) *Biological Invasion, Ecological Studies*, Vol 193. Springer-Verlag, Berlin and Heidelberg, Germany, pp. 197–211.

Tilman, D. (1990) Constraints and tradeoffs: toward a predictive theory of competition and succession. *Oikos* 58, 3–15.

Tubiello, F.N., Soussana, J.F. and Howden, S.M. (2007) Crop and pasture response to climate change. *Proceedings of the National Academy of Sciences* 104, 19686–19690.

Valerio, M., Tomecek, M.B., Lovelli, S. and Ziska, L.H. (2011) Quantifying the effect of drought on carbon dioxide-induced changes in competition between a C-3 crop (tomato) and a C-4 weed (*Amaranthus retroflexus*). *Weed Research* 51, 591–600.

van Kleunen, M., Manning, J.C., Pasqualetto, V. and Johnson, S.D. (2008) Phylogenetically independent associations between autonomous self-fertilization and plant invasiveness. *American Naturalist* 171, 195–201.

van Kleunen, M., Weber, E. and Fischer, M. (2010) A meta-analysis of trait differences between invasive and non-invasive plant species. *Ecology Letters* 13, 235–245.

Verlinden, M. and Nijs, I. (2010) Alien plant species favoured over congeneric natives under experimental climate warming in temperate Belgian climate. *Biological Invasions* 12, 2777–2787.

Vila, M., Corbin, J.D., Dukes, J.S., Pino, J. and Smith, S.D. (2007) Linking plant invasions to global environmental change. In: Canadell, J., Pataki, D. and Pitelka, L. (eds) *Terrestrial Ecosystems in a Changing World.* Springer, New York, pp. 93–102.

Vitousek, P.M., Aber, J.D., Howarth, R.W., Likens, G.E., Matson, P.A., Schindler, D.W., Schlesinger, W.H. and Tilman, G.D. (1997) Human alteration of the global nitrogen cycle: sources and consequences. *Ecological Applications* 7, 737–750.

Walther, G.R., Post, E., Convey, P., Menzel, A., Parmesan, C., Beebee, T.J.C., Fromentin, J.M., Hoegh Guldberg, O. and Bairlein, F. (2002) Ecological responses to recent climate change. *Nature* 416, 389–395.

Walther, G.R., *et al.* (2009) Alien species in a warmer world: risks and opportunities. *Trends in Ecology and Evolution* 24, 686–693.

Wand, S.J.E., Midgley, G.F., Jones, M.H. and Curtis, P.S. (1999) Responses of wild C4 and C3 grass (Poaceae) species to elevated atmospheric CO_2 concentration: a meta-analytic test of current theories and perceptions. *Global Change Biology* 5, 723–741.

Weltzin, J.F., Belote, R.T. and Sanders, N.J. (2003) Biological invaders in a greenhouse world: will elevated CO_2 fuel plant invasions? *Frontiers in Ecology and the Environment* 1, 146–153.

Westerling, A.L., Hidalgo, H.G., Cayan, D.R. and Swetnam, T.W. (2006) Warming and earlier spring increase western US forest wildfire activity. *Science* 313, 940–943.

Williams, A.L., Wills, K.E., Janes, J.K., Schoor, J.K.V., Newton, P.C.D. and Hovenden, M.J. (2007) Warming and free-air CO_2 enrichment alter demographics in four co-occurring grassland species. *New Phytologist* 176, 365–374.

Wright, I.J., *et al.* (2004) The worldwide leaf economics spectrum. *Nature* 428, 821–827.

Ziska, L.H. (2000) The impact of elevated CO_2 on yield loss from a C-3 and C-4 weed in field-grown soybean. *Global Change Biology* 6, 899–905.

Ziska, L.H. (2001) Changes in competitive ability between a C-4 crop and a C-3 weed with elevated carbon dioxide. *Weed Science* 49, 622–627.

Ziska, L.H. (2003a) Evaluation of the growth response of six invasive species to past, present and future atmospheric carbon dioxide. *Journal of Experimental Botany* 54, 395–404.

Ziska, L.H. (2003b) Evaluation of yield loss in field sorghum from a C-3 and C-4 weed with increasing CO_2. *Weed Science* 51, 914–918.

Ziska, L.H. (2010) Elevated carbon dioxide alters chemical management of Canada thistle in no-till soybean. *Field Crops Research* 119, 299–303.

Ziska, L.H. and Dukes, J.S. (2011) *Weed Biology and Climate Change.* Wiley-Blackwell, Ames, Iowa.

Ziska, L.H., Reeves, J.B. and Blank, B. (2005) The impact of recent increases in atmospheric CO_2 on biomass production and vegetative retention of cheatgrass (*Bromus tectorum*): implications for fire disturbance. *Global Change Biology* 11, 1325–1332.

Part II

Case Studies

6 Non-native Species in Antarctic Terrestrial Environments: The Impacts of Climate Change and Human Activity

Kevin A. Hughes and Peter Convey

British Antarctic Survey, Natural Environment Research Council, Cambridge, UK

Abstract

Antarctic terrestrial biodiversity is simple compared to other regions of the Earth, with many higher taxonomic groups not represented, due to the continent's isolation, the severe climatic conditions and the relative scarcity of suitable habitats. So far, Antarctic biodiversity has been little affected by non-native species introductions, due to (i) the late arrival of humans on the continent (*c.*1820), (ii) the overall low intensity of human activity and (iii) the concentration of most of that activity around a limited number of research stations and tourist sites, such as exist onthe Antarctic Peninsula. However, human activity is increasing, and Antarctica is increasingly vulnerable to the human-mediated importation of non-native species and the redistribution of indigenous Antarctic species. While the Antarctic Peninsula is one of the most rapidly warming regions on the planet, the Antarctic continent has, so far, experienced relatively little climatic change, but this is expected to change over the next century. Consequently, terrestrial communities are increasingly vulnerable, as climate change increases the risk of non-native species establishment and dispersal. This chapter describes non-native species in Antarctica that have already become established. Also described are the eradications that have been attempted and the practicality of minimizing microbial introductions. Finally, the chapter discusses recent policy developments relating to non-native species and suggests that more needs to be done by the Antarctic Treaty Parties to implement biosecurity practices and eradicate existing non-native colonists, before fragile Antarctic communities are changed irreversibly.

Introduction

Antarctica was the last continent to be touched by the direct influence of humans, with the first people landing in 1819 and overwintering in 1897/98 (Headland, 2009). The region's isolation and extreme climate have limited the level of human activity, in contrast to the other continents, where humans have been present for tens of thousands of years at least and large population migrations have occurred (Hughes *et al.*, 2006, 2011a; Mellars, 2006; Headland, 2009). As a result of the brief and limited influence of humans on the region, relatively few non-native species have been introduced to Antarctica deliberately or accidentally, with the majority of those that have become established occurring on the sub-Antarctic islands (Frenot *et al.*, 2005; Convey, 2008; Greenslade and Convey, 2012; Hughes and Convey, 2012; UK, 2012a). Two major factors currently threaten to change this situation. More people are visiting the continent, assisted by increasingly rapid means of access via ships and aircraft; and climate change, particularly in the Antarctic Peninsula region, is occurring at one of the fastest rates globally. Both these factors, individually and in synergy, make the establishment of species originating from more temperate areas more likely (Convey *et al.*, 2009, 2012; Turner *et al.*, 2009a; Convey, 2011; Chown *et al.*, 2012a,b).

This chapter describes the vulnerability of native Antarctic species and communities to competition from, and invasion by, non-native species, particularly in light of rapid regional climate change, and also discusses whether the actions available to the Antarctic Treaty Consultative Parties (ATCP) are sufficient to reduce the risk of non-native species introductions and the resulting impacts on native biota and ecosystems.

Legislative Protection

Legislation that protects Antarctica from human impact is centred around the Protocol on Environmental Protection to the Antarctic Treaty (also known as the Madrid Protocol or the Environmental Protocol), which was signed in 1991 and came into force in 1998 (ATCP, 1991). The Protocol designates Antarctica as a 'natural reserve, devoted to peace and science' (Article 2). Legislation relating to non-native species is found in Annex II Conservation of Fauna and Flora, which specifically prohibits the intentional introduction of non-native plants and animals to land, ice shelves or into water within the Antarctic Treaty area, unless for a defined scientific purpose and in accordance with a permit (Annex II, Article 4). Non-native species can be introduced in accordance with a permit for laboratory studies, but they must be removed or destroyed at the end of the permitted period, to prevent their release into the local Antarctic environment. Sledge dogs, which were the mainstay of Antarctic field party travel for several decades, are no longer permitted within the Antarctic Treaty area, the last being removed from the continent before 1998. Parties are also required to take precautions to prevent the introduction of microorganisms. The importation of fresh foodstuffs (meats – including poultry products – and vegetables) is permitted, although efforts should be made to ensure that they are free of disease, and the importation of associated non-sterile soil is minimized as far as possible.

One notable omission from Annex II concerns the issue of the unintentional introduction of non-native species. Neither does Annex II mention explicitly the transfer of native Antarctic organisms between biogeographic regions within the continent, although it could be argued that this is implicit within the requirement that activities be limited to ensure that 'the diversity of species, as well as the habitats essential to their existence, and the balance of the ecological systems existing within the Antarctic Treaty areas be maintained' (Annex II, Article 3(3c)).

Antarctic Terrestrial Environments

Antarctica is isolated from the rest of the Earth's continents both by the atmospheric circulation and the Southern Ocean's Antarctic Circumpolar Current that formed after the Antarctic Peninsula separated from South America *c.*30 million years ago in the final stages of the break-up of the super-continent, Gondwana (Fig. 6.1) (Clarke *et al.*, 2005; Barnes *et al.*, 2006; Bergstrom *et al.*, 2006). These isolated Antarctica from lower latitudes, accelerating the cooling of the continent and resulting in the development of continental-scale ice sheets, which today reach a maximum thickness of over 4000 m and an average of over 2000 m across the entire continent (see Convey *et al.*, 2009, and references therein). Today, only *c.*0.34% (or 44,000 km^2) of Antarctica is ice free, with less than 6000 km^2 in the climatically less extreme areas within 5 km of the coast, where the bulk of terrestrial flora and fauna are found (Convey *et al.*, 2009). As a result, sites where terrestrial communities can develop are rare compared to the size of the continent as a whole, and are vulnerable to impacts from humans, who,

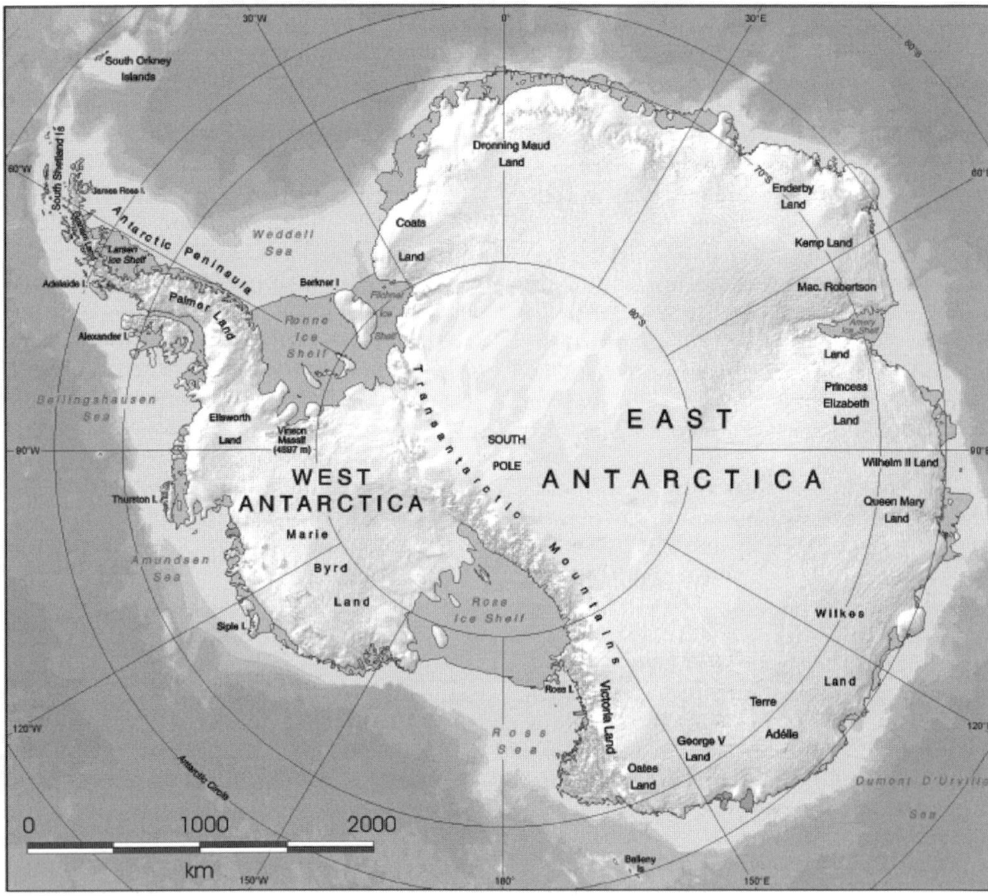

Fig. 6.1. Map of Antarctica.

in general, have favoured these locations for research stations and other infrastructure. Terrestrial communities are generally small and island-like, with distances between communities with distinct biodiversity varying on a scale of hundreds of metres to hundreds of kilometres, separated by ice, ice shelves, sea ice and/or open water (Bergstrom and Chown, 1999). Building on recent realization of the biogeographic complexity within Antarctic terrestrial diversity and biogeography (Convey and Stevens, 2007; Convey *et al.* 2008), Terauds *et al.* (2012) used available biodiversity data to propose the recognition of 15 distinct biogeographic regions within Antarctica, with each containing broadly similar macrobiotic diversity. Given that some national operators and the tourism industry operate over several of these regions, this chapter highlights the need for biosecurity precautions to prevent the inter-regional transfer of species within the Antarctic Treaty area (Chown and Convey, 2007; Lee and Chown, 2011).

Antarctic terrestrial biodiversity is restricted both by the extreme climatic conditions and by the isolation of the continent, which has limited the level of natural colonization of the region from other landmass since the Last Glacial Maximum *c.*20,000 years ago (Anderson *et al.*, 2002). However, substantial evidence is now emerging that natural refugia have existed for at least several million years (and in some cases, tens of millions of years, spanning several glacial cycles) that have allowed the persistence of ancient lineages within the continent (Chown and Convey, 2007; Convey and Stevens, 2007; Convey *et al.*, 2008; Vyverman *et al.*, 2010; Fraser *et al.*, 2012).

Nevertheless, contemporary Antarctic terrestrial biodiversity is restricted. There are only two native higher plants (the Antarctic hair grass, *Deschampsia antarctica*, and the pearlwort, *Colobanthus quitensis*), and the flora is dominated by cryptogams (mosses, liverworts, lichens) (Smith, 1984; Convey, 2007; Ochyra *et al.*, 2008). Lichens are the dominant macroscopic biological group in many parts of Antarctica, and some are present even in areas in the continent's interior that experience some of the planet's most severe climatic conditions (Øvstedal and Smith, 2001). Other microscopic microbial species abound, with some studies suggesting high levels of endemism and biodiversity, at least in certain groups (de Wever *et al.*, 2009; Vyverman *et al.*, 2010; Peeters *et al.*, 2011). Notably, bacterial, fungal and cyanobacterial mat communities may represent the extent of life in some locations (Hodgson *et al.*, 2010). Where environmental conditions reach their most extreme, such as in the ice-free Dry Valleys of the Ross Sea region, some species exist in the rock (Friedmann, 1982; Hughes and Lawley, 2003; Thomas *et al.*, 2008). These endolithic communities, consisting of lichens, cyanobacteria, fungi and bacteria, are protected by the rock from the external conditions, yet there is sufficient penetration of light and retention of water to allow photosynthesis, which sustains the community. Like the plants, animals have a restricted biodiversity, with only two native higher insects on the continent (the chironomid midges, *Belgica antarctica* and *Parochlus steinenii*), both limited to the Antarctic Peninsula region (Convey and Block, 1996). Other common invertebrate groups include collembola (springtails), arthropods (mites), tardigrades (water-bears), rotifers and nematodes, (Block, 1984; Convey and McInnes, 2005; Adams *et al.*, 2006; Convey, 2007). In general, competition for habitat and food sources is thought to be low, and there are few true predators, with most species being microbivores or detritivores (Convey, 1996a; Hogg *et al.*, 2006).

Humans in Antarctica

Human interest in the southern polar regions commenced in the late 18th century when reports of the seal populations attracted sealers to the sub-Antarctic islands (Convey and Lebouvier, 2009; Headland, 2009; Tin *et al.*, 2009; Trathan and Reid, 2009). As seal populations were decimated rapidly in these areas, sealers moved further south and commenced operations in islands north-west of the Antarctic Peninsula in the 1820s (Headland, 2009). Subsequently, in the early 20th century, whales started to be harvested on a commercial scale, with several whaling stations established in the sub-Antarctic islands and on Deception Island (South Shetland Islands) and Signy Island (South Orkney Islands) (Trathan and Reid, 2009; Dibbern, 2010).

The heroic age of Antarctic exploration (*c.*1895–1922) brought small numbers of private adventurers and national government-funded expeditions that started to explore the Antarctic coastline and interior and established some of the earliest huts and other structures on the continent. Following this period, with expeditions by Mawson, Shirase and the drama associated with Scott and Amundsen's attempts to reach the South Pole, interest in Antarctica was reignited with the International Geophysical Year (IGY; 1957–1958), which focused international research on Antarctica and led to unprecedented levels of international cooperation, even at the height of the Cold War (Walton and Clarkson, 2011). Research stations were established throughout Antarctica, including at the South Pole (by the USA) and the Pole of Inaccessibility (by the USSR). The success of the IGY gave impetus to negotiations to transcend potential disputes over national territorial claims, leading to the Antarctic Treaty, which was signed in 1959 and came into force on 23 June 1961. The 12 initial signatory nations agreed that Antarctica should be demilitarized, disputes over sovereignty should be set aside and that international scientific cooperation should be promoted in the region. Other countries have since acceded to the Treaty. Many of the research stations established for the IGY are still occupied, and today national governmental operators, representing the current 29 Consultative Parties and 21 Non-consultative Parties to the Antarctic Treaty, run *c.*75 year-round and summer-only stations in the Antarctic Treaty area, which includes all land and sea south of latitude 60°S (COMNAP, 2012; Convey *et al.*, 2012). Currently, around 4000 science and support staff can operate out of these research stations at any one time during the austral summer, with the numbers dropping to *c.*1000 personnel during winter (Jabour, 2009; COMNAP, 2012). The other main source of human activity within terrestrial Antarctica comes from tourism, predominantly in the form of cruises around the Antarctic Peninsula and Scotia Arc archipelagos, with daily landings at sites of historical, scenic and/or wildlife interest. Tourists have been visiting Antarctica since the late 1950s, but visitation started to increase substantially between 2000/01 and 2007/08, when individual tourist landing on Antarctica rose to over 33,000 (Aronson *et al.*, 2011; Convey *et al.*, 2012; IAATO, 2012a). Numbers have declined slightly since then, predominantly due to the global economic slowdown.

Nevertheless, since 1992, tourists have visited over 400 locations, with more than 2 million individual tourist landings, and the expectation is that the industry will grow again in line with economic recovery. Most tourists visit a few popular sites, such as Whalers Bay (Deception Island) or Port Lockroy (western Antarctic Peninsula), where up to 16,000 landings have been recorded annually (Lynch *et al.*, 2010; IAATO, 2012a). At some inland locations, land-based tourism is still expensive and exclusive, but there is already relatively inexpensive land-based infrastructure provided by some National Antarctic Programmes on King George Island, South Shetland Islands, to support tourists using commercial airlinks to Punta Arenas (Chile). In the future, more permanent infrastructure may be established formally to accommodate this growing market (Chown *et al.*, 2012b; UK, 2012b).

Human-mediated Introduction of Non-native Species

While examination of existing Antarctic biodiversity might suggest that natural rates of colonization for some biological groups is low, we have little direct information on the levels of propagule input for terrestrial biota by birds, wind and marine mammals, or via marine currents or debris (Marshall, 1996; Pearce *et al.*, 2009; Hughes and Convey, 2010, 2012; Parnikoza *et al.*, 2012). However, it is estimated that the frequency of human-mediated introductions of non-native species to Antarctica and the sub-Antarctic islands may exceed natural colonization rates by at least two orders of magnitude (Pugh, 1997; Frenot *et al.*, 2005). In contrast, the risk of plant propagule, invertebrate and non-sterile soil introductions by anthropogenic means has been better characterized. Human-mediated transport of propagules and/or soil has been recorded associated with cargo (Lee and Chown, 2009a,b; Osyczka, 2010; Tsujimoto *et al.*, 2010), vehicles, including aircraft and ships (Hughes *et al.*, 2010a,b), wood, sand and aggregate (Osyczka *et al.*, 2012), fresh foods (Hughes *et al.*, 2005, 2011a) and with visitors' clothing and personal possessions (Whinam *et al.*, 2005; Chown *et al.*, 2012a), and risk assessments have been attempted for certain groups such as springtails (Greenslade and Convey, 2011) and flowering plants (Chown *et al.*, 2012a).

The sub-Antarctic islands, which form a ring around Antarctica in the Southern Ocean north of latitude 60°S, have been visited by humans for over two centuries, and in that time over 200 non-native plants, vertebrates (including fish, rodents, cats, reindeer, sheep and mouflon) and invertebrates have been introduced, with some islands now having more non-native plant species than indigenous plant species (Frenot *et al.*, 2005). The contrast with Antarctica is marked, with only around 11 known non-native species, some at multiple locations, within the terrestrial environment (Hughes and Convey, 2012; Greenslade *et al.*, 2012; Molina-Montenegro *et al.*, 2012; UK, 2012a). Further south, the barriers to invasion by non-native species include the severe climatic conditions and the continent's extreme isolation. Although the introduction of non-native microorganisms has received relatively little scientific investigation (Hughes and Nobbs, 2004; Convey, 2008; Cowan *et al.*, 2011; Hughes *et al.*, 2011b), the presence of humans in Antarctica will result in human skin commensal bacteria being shed (Cowan *et al.*, 2011). Humans may also act as unintentional vectors of microorganisms through the soil and dirt attached to cargo, clothing and fresh foods distributed throughout the region (Hughes *et al.*, 2011b).

Many non-native species introductions occurred with the development of land-based infrastructure to service the needs of the early sealing and later whaling activities that occurred on an industrial scale in the early 20th century (Convey and Lebouvier, 2009). Following on from this period, there has been a series of accidental and deliberate introduction events since the 1930s. The grass, *Poa annua*, was found close to the whaling station at Whalers Bay (Deception Island, South Shetland Islands) possibly as early as January 1936, presumably introduced on/in cargo (Smith, 1996). This population was destroyed during a series of volcanic eruptions on the island in the late 1960s and early 1970s. In the 1950s and 1960s, long before the Environmental Protocol was developed and came into force, a series of transplantation experiments were performed by scientists from several nations to see if plant species from lower latitudes and also the Arctic had the ability to survive and establish in Antarctica. Although many of these species did not survive or thrive (and were mostly eventually removed), a number of species introduced unintentionally alongside them did. For example, southern beech trees transplanted by Argentine scientists from Tierra del Fuego to Cierva Point on the northern Antarctic Peninsula in the mid-1950s did not survive, yet the grass, *Poa pratensis*, introduced with the trees is still found at the transplantation site (Fig. 6.2) (Corte, 1961; Smith, 1996; Spain *et al.*, 2012). Similarly, plant species transplanted by British scientists from South Georgia, the Falkland Islands and the Arctic to Signy Island (South Orkney Islands) did not thrive and were later destroyed. However, the flightless chironomid midge, *Eretmoptera murphyi*, which is endemic to South Georgia (Fig. 6.3), and the enchytraeid worm, *Christensenidrilus blocki*, became established in the local soil (Block *et al.*, 1984; Dózsa-Farkas and Convey, 1997). The worm has remained restricted to the vicinity of the introduction site, but the distribution of the midge has expanded up to 220 m away from the original plot (Hughes and Worland, 2010).

Rats and mice would almost certainly have been introduced to Antarctica and lived synanthropically in whaling stations, as happened in the sub-Antarctic. However, these vertebrate species are not thought to be capable of overwintering in Antarctica in isolation from human infrastructure, probably due to the low temperatures, lack of winter food sources and lack of vegetation cover required for rodent nest building and insulation. Invertebrates are also known to live in association with station buildings in Antarctica, with two non-native dipteran populations from the genus *Lycoriella* (black fungus midges) found at Rothera and Casey Stations, which are on opposite sides of the continent (Hughes *et al.*, 2005). The flies at Rothera were found in the alcohol bond and were readily eradicated; however, the Casey flies were residing in the sewage system and sub-floor areas of the station and, despite considerable eradication efforts, persist at the station. The boreal dipteran *Trichocera maculipennis* was found initially in high numbers associated with the Uruguayan Artigas Base on King George Island, South Shetland Islands, in 2005/6, but have also been observed in the natural environment outside the base (Volonterio *et al.*, 2013).

Fig. 6.2. The non-native grass, *Poa pratensis*, was introduced unintentionally to ground close to the Argentine research station at Cierva Point, Graham Land, Antarctic Peninsula, during the 1954/55 austral summer. (Photograph: L. Pertierra.)

Fig. 6.3. The flightless chironomid midge, *Eretmoptera murphyi*, was introduced to Signy Island, South Orkney Islands, from South Georgia, probably during the late 1960s, and has now spread up to 220 m away from the introduction site. (Photograph: P. Bucktrout/British Antarctic Survey.)

Given the life cycle, biogeographic distribution and physiological capacity of this species, it is feasible that it may have colonized the area outside the station, and could spread further to other ice-free locations with equivalent climates.

Most of the other known non-native species, including several species of introduced collembola (Wise, 1971; Greenslade and Wise, 1984; Greenslade, 1995; Greenslade and Convey, 2012; Greenslade *et al.*, 2012), are not dependent on human infrastructure for survival, although may have been introduced initially to sites disturbed by human activity or partially sheltered by buildings. For example, *P. annua* was first introduced to Arctowski Station, Admiralty Bay, King George Island, South Shetland Islands, in *c.*1985/86, where it was found in metal grating at the entrance to the main building. By 1990, it had spread to a greenhouse area and above subterranean hot water pipes within a restricted area of *c.*0.4 km^2, and 2 years later it was found in a number of locations with disturbed ground. The grass was found growing among indigenous plant communities for the first time in 2005/6, and in 2008/9 it had spread to a glacier forefield 1.5 km from the station (Olech, 1996, 2003; Smith, 1996; Frenot *et al.*, 2005; Chwedorzewska, 2008; Olech and Chwedorzewska, 2011).

Eradication

As discussed earlier, the Environmental Protocol prohibits the introduction of non-native species to Antarctica, and, given the potential threat to indigenous communities posed by invasive species, the Antarctic Treaty Consultative Meeting (ATCM) Committee for Environmental Protection (CEP) encourages their rapid eradication. However, if eradications are not performed at the earliest opportunity, they may not be effective once a species has become more widely distributed. The *P. annua* distribution around Arctowski Station is now extensive and eradication may be difficult. *P. annua* has also been found near other research stations on the northern Antarctic Peninsula (General Bernardo O'Higgins Station, Trinity Peninsula, Gabriel González Videla Station, Paradise Bay and Almirante Brown Station, Paradise Bay) and, given what happened at Arctowski Station, eradication should be undertaken as soon as possible (Molina-Montenegro *et al.*, 2012). A recent study has reported a second potential non-native vascular plant species, *Juncus bufonius*, at Admiralty Bay, at least in the form of viable propagules – the species was cultivated in the laboratory from soil taken from near Arctowski Station within the Antarctic Specially Protected Area (ASPA) No. 128 Western Shore of Admiralty Bay, King George Island, South Shetland Islands (Cuba-Diaz *et al.*, 2012).

Eradications of non-native species have been undertaken within the Antarctic Treaty area, but only on a small scale. As discussed previously, *Lycoriella* sp. were eradicated from Rothera Research Station, and several pest species have been eradicated from within hydroponic facilities at stations where these systems have been established (e.g. McMurdo (USA), Davis and Mawson Stations (Australia)) (Hughes and Convey, 2010). To date, eradications of non-synanthropic species include several plant species near Progress II Station, Larsemann Hills, East Antarctica (Russia, 1999), and grasses, other graminoids and *Cerastium* sp. from near Great Wall Station, Fildes Peninsula, King George Island, Antarctic Peninsula (H.-U. Peter, personal communication, 2006). A single plant, thought to be *Puccinellia* sp., was removed from near a field hut close to Syowa Station, Enderby Land, East Antarctica, in 2007 (Japan, 1996; S. Imura, personal communication, 2007; Tsujimoto *et al.*, 2010), and a single specimen of *Nassauvia magellanica* (a flowering plant in the aster family) was eradicated by the authors in January 2010 from Whalers Bay, Deception Island (UK and Spain, 2010; Smith and Richardson, 2011; Fig. 6.4). A second non-native plant, *Gamochaeta nivalis*, was found initially alongside the *N. magellanica*, but was apparently washed away between discovery and eradication. Due to the cost and logistical difficulties associated with non-native species eradi-

Fig. 6.4. The suspected non-native vascular plant, *Nassauvia magellanica*, was eradicated from Deception Island in January 2010. (Photograph: K. Hughes.)

cations within Antarctica, the removal of such species when numbers are low and distribution is restricted is better, both financially and practically. However, Treaty Parties have not to date been proactive in the eradication of non-native species, and it has been left to individual scientists, working largely independently of environmental managers, to take the initiative and remove these plants. Nevertheless, at the ATCM CEP in Hobart in July 2012, Spain, Argentina and the UK reported on the colonization status of *P. pratensis* at Cierva Point, Palmer Archipelago, Antarctic Peninsula (Spain *et al.*, 2012). Following the inadvertent introduction of the grass in 1954/55 during transplantation experiments, they noted that the distribution of the grass had expanded from a single patch *c.*40 cm across in 1995 to a patch *c.*1 m across in 2012, and suggested that eradication might be attempted once an appropriate methodology was identified. Currently, the distribution of *P. annua* near Arctowski Station, as well as that of most non-native invertebrates known from Antarctica, is such that eradication may be difficult or impossible without substantial disruption and damage to the terrestrial habitat colonized. The CEP has yet to establish guidelines to help Parties judge whether or not to eradicate species and how to recognize that eradication attempts are unlikely to succeed or where the 'cost', in terms of damage to local communities, exceeds the benefits of removing the non-native species. However, this does not prevent individual Parties from acting on advice from their own scientists and environmental managers (Hughes and Convey, 2012).

Microorganisms

One issue that, to date, has received little attention is the introduction of non-native microorganisms into Antarctica and their impacts on the indigenous microbial communities which constitute the great majority of the biomass on the continent and which are a major subject of scientific study. Non-native microorganisms can be introduced to Antarctica using the same routes as other macrobiota, but the fact that they exist as symbionts of humans means that, wherever humans travel, non-native microorganisms are introduced (Cowan *et al.*, 2011, and references therein). Over the past century, humans have visited most of the major coastal ice-free areas and nunatak groups, but little or no consideration has been given to preventing non-native microorganism dispersal (Cowan *et al*, 2011; Hughes *et al.*, 2011b). Undertaking Antarctic fieldwork while simultaneously preventing microbial release is, in most cases, unfeasible, although it is attempted at a small number of particularly sensitive and exceptional locations, such as the areas of geothermal activity within ASPA. 130 Tramway Ridge on Mount Erebus. However, with the rapid increases in the sophistication of molecular and biochemical technology, science in the future will need reference sites free of human microbial contamination. Few areas of Antarctica have been protected with this in mind. Two restricted zones have been designated within ASPA No. 126 Byers Peninsula (Livingstone Island, South Shetland Islands), specifically to preserve the area for future molecular analysis of the impacts of climate change on native microbial communities (Pertierra *et al.*, 2013). In contrast, considerable efforts have been made to prevent chemical and biological contamination during the penetration of Antarctica's subglacial lakes, although it may be difficult to assess the success of these measures (Cowan *et al.*, 2011; Schiermeier, 2011; UK, 2011).

The wide daily and annual fluctuations in Antarctic soil temperatures mean that soils may be substantially above freezing for much of the austral summer (Walton, 1984; Davey *et al.*, 1992; Bokhorst *et al.*, 2007). In line with this, many native soil microorganisms are psychrotolerant rather than obligate psychrophiles, and a proportion are mesophiles, although comprehensive studies of microbial ecophysiology are lacking. It is therefore likely that some microbial strains imported to Antarctica may be able to function and interact with existing microbial communities, thereby increasing species richness and ecosystem functions such as biogeochemical cycling (Cowan *et al.*, 2011). The Environmental Protocol does not permit the importation of non-sterile soil within the Antarctic Treaty area, which could be a major source of non-native fungi, bacteria and other microbial groups (ATCP, 1991). However, fresh foods are imported routinely to Antarctica by most national programmes, and root vegetables can often be associated with soil and plant material infected by food spoilage microorganisms (Hughes *et al.*, 2011a). Should such species be released accidentally into the Antarctic terrestrial environment, they could cause disease in indigenous plants and/or alter existing microbial communities (Klopper and Smith, 1998; Hughes *et al.*, 2011a).

Climate Change

The Antarctic Peninsula is one of the three most rapidly warming regions on the planet, and the fastest in the southern hemisphere, while the bulk of the Antarctic continent has so far experienced relatively small, inconsistent or non-significant change (Convey *et al.*, 2009; Turner *et al.*, 2009a). This situation is expected to change over the next century, as the continent appears to have been largely protected from the full impacts of global climatic change by the parallel formation of the Antarctic ozone hole (Turner *et al.*, 2009b). As this repairs over the next century, the climate of the continent will change more rapidly. This may be of limited biological significance inland, where temperatures will still remain far below zero, but is likely to have far greater impacts around the continent's coast, where temperatures are close to or just above freezing during the austral summer.

Climate change in the Antarctic Peninsula region has generally reduced the physical stresses affecting terrestrial species. Two important changes are warming – which increases the energy available to organisms – and the increased availability of liquid water, either through longer periods of melt or changes in precipitation from snow to rain. Warming and the associated widespread reductions in ice and snow will also expose new ground for colonization. However, in some circumstances, stress levels can also increase, for instance through changes in radiation levels (Newsham *et al*., 2002), local cooling (Doran *et al*., 2002), the frequency of freeze–thaw events, changes in wind direction or strength (Hodgson *et al*., 2006), precipitation (Chown and Smith, 1993; Tweedie and Bergstrom, 2000; Smith, 2002; Chapuis *et al*., 2004) and/or extreme events (Nielsen *et al*., 2011). It is clear that better information on local physical environments, environmental drivers and ecosystem response is required.

As a broad generalization, the relaxation of current environmental limits, particularly through warming and greater water availability, is predicted to lead to greater biological productivity in Antarctic terrestrial and freshwater communities, resulting in population growth, distributions expanding at the local scale and increased development of more complex communities, even among the native biota (reviewed by Convey, 2006, 2011; Nielsen *et al*., 2011). This will be accompanied by changes in the distributions of native species and, importantly, by increasing rates of colonization and subsequent invasion by non-native species. Globally, the establishment of non-native species is often accompanied by fundamental changes in native ecosystems. The sensitivity of Antarctic terrestrial ecosystems is demonstrated by past introductions to sub-Antarctic islands (Frenot *et al*., 2005; Convey and Lebouvier, 2009).

There are few studies of biological responses in natural terrestrial ecosystems to climate change in Antarctica (Convey, 2006; Nielsen *et al*., 2011). Rapid expansions in local populations of the two native flowering plants on the Argentine Islands are widely quoted (Fowbert and Smith, 1994; Smith, 1994; Parnikoza *et al*., 2009). Similar changes are assumed, and supported by anecdotal observations, to be taking place elsewere in the northern Antarctic Peninsula/Scotia Arc region, although such conclusions are hampered by limited historical data. Observations on the Argentine Islands, and field environmental manipulation studies, suggest that warmer temperatures and the increasing availability of liquid water encourage the growth of established plants, more frequent seed maturation, improved germination and seedling establishment (Fowbert and Smith, 1994; Smith, 1994; Convey, 1996b; McGraw and Day, 1997; Ruhland and Day, 2001; Gerighausen *et al*., 2003; Convey and Smith, 2006). Nielsen *et al*. (2011) similarly conclude that studies of the important soil faunal group, nematode worms, indicate responses to longer-term climate trends and rare 'extreme' events. However, basic survey and monitoring of the dominant cryptogamic and soil microbial communities of Antarctica and their associated fauna is lacking (Convey, 2006, 2010; Barrett *et al*., 2008; Wall *et al*., 2011).

With this background, the Antarctic Peninsula is both the region of Antarctica most highly visited by tourists and national Antarctic programmes and is also currently experiencing the most rapid rates of climate and associated environmental changes. This combination of factors is likely to have serious implications for indigenous communities as the barriers to invasion by non-native species are eroded (Hughes and Convey, 2010; Chown *et al*., 2012a,b). As part of the International Polar Year (IPY) 'Aliens in Antarctica' project, Chown *et al*. (2012a) undertook a continent-wide evaluation of the risks to Antarctica posed by non-native vascular plants. They quantified the number of seeds transported inadvertently to Antarctica by different types of visitors and found that, on average, each visitor carried 9.5 seeds, although individual scientists and tourist support staff carried more seeds than tourists or ships' crew. As tourist numbers are greater (33,054) than those of national operator personnel (7085 in the year of study), the two types transported a similar number of seeds overall, with tourist operations estimated to have transported *c*.31,732 seeds to Antarctica compared with *c*.38,897 by national operator staff during the 2007/8 summer season. Other important dispersal mechanisms for seeds include on/in cargo transported to Antarctica by national operators (Hughes *et al*., 2010a; Lee and Chown 2009a,b; Tsujimoto and Imura, 2012). Thus, current data suggest that the cumulative risk from the import of non-native biota is greater for national operator than tourism-related activities, a pattern supported by the observation that there are no known instances of a non-native species becoming established in Antarctica linked directly with the tourism industry (Frenot *et al*., 2005; Tin *et al*., 2009).

Seeds identified in these studies included many species known to be invasive in the sub-Antarctic islands and the Arctic (Chown *et al*., 2012a; Tsujimoto and Imura, 2012). By calculating annual cumulative degree days for plant growth throughout Antarctica and plotting where tourists and national governmental operators visited, Chown *et al*. (2012a) created a map showing the areas most vulnerable to non-native plant species establishment and invasion. Unsurprisingly, the coast of the western Antarctic Peninsula and the islands off the coast of the Peninsula had the highest current risk, although the western Ross Sea region and some sites around the coast of east Antarctica were also at risk, due to their popularity with visitors. By estimating cumulative degree days across ice-free Antarctica under climate forecasts for 2100, it was also possible to predict which areas would become more vulnerable with climate change. The western Antarctic Peninsula region was most at risk, but coastal areas to the west of the Amery Ice Shelf and in the Ross Sea region will become increasingly vulnerable. The prediction that currently the western Antarctic Peninsula is at greatest risk of non-native species introductions is supported by several recent reports (Olech and Chwedorzewska, 2011; Smith and Richardson, 2011; Cuba-Diaz *et al*., 2012; Greenslade *et al*., 2012; Molina-Montenegro *et al*., 2012; Volonterio *et al*., 2013). In particular, Deception Island in the South Shetland Islands is the location most vulnerable to invasion in Antarctica, due to high visitation, its relatively benign climate compared to other regions of Antarctica and the presence of geothermally influenced soils. Deception Island has six recorded non-native collembolan species, compared to just eight native species, with many non-natives found at sites visited regularly by tourists and national operators (Greenslade *et al*., 2012). Furthermore, the recently discovered and eradicated plant, *N. magellanica*, was found in a protected area close to Whalers Bay, which receives up to 16,000 tourist visitors annually, plus regular visits from national operator personnel. This species may have been introduced by humans (Dibbern, 2010; Smith and Richardson, 2011), but as it is native to Terra del Fuego, it also could have colonized by natural means. With the precautionary principle in mind, the plant was removed on this occasion, but this highlights the difficulty in establishing the origin of a newly discovered species in the Antarctic environment in the absence of either observation of the initial colonization event or objective data assessing the relative probabilities of different introduction routes (see Hughes and Convey, 2012, for discussion). Local climatic changes may also be affecting the distribution of *P. annua* near Arctowski Station on King George Island, which has recently colonized new ground created with the climate change-related retreat of the Ecology Glacier (Olech and Chwedorzewska, 2011).

Climate change may also affect the potential colonization and establishment of non-native invertebrates. Climatic conditions in coastal areas of Antarctica, and the northern Antarctic Peninsula in particular, are well within the physiological limits of some Arctic and alpine species (Sømme, 1989; Denlinger and Lee, 2010). The wingless parthenogenetic chironomid midge, *E. murphyi*, that was introduced to Signy Island, South Orkney Islands from sub-Antarctic South Georgia, has been there for *c*.43 years, and in that time its distribution has expanded relatively slowly and is currently restricted to the moss slope within 220 m of its introduction site

(Hughes and Worland, 2010). However, recent work has examined the potential impact of climate change on the midge life cycle (Hughes *et al.*, 2013). *E. murphyi* remains in the larval stage for most of its life cycle, emerging briefly as a non-feeding adult at the height of summer to lay eggs. It is thought that the midge has a 2-year life cycle on Signy Island, as there are different larval size classes in the soil (Convey, 1992; Convey and Block, 1996; Hughes *et al.*, 2013). Growth rate and microhabitat climatic modelling confirmed that temperature constraints on larval development at Signy Island meant there was currently insufficient time for the midge to complete its life cycle in 1 year. However, if warming continues in the region, the midge could, in theory, complete its life cycle in 1 year within the next *c.*170 years, which might increase its rate of reproduction and invasive potential.

Further, Hughes *et al.* (2013) also found that temperature constraints on larval development on Anchorage Island, Marguerite Bay, off Adelaide Island (68°S), were theoretically similar to those on Signy Island (61°S), even though this location was ~750 km further south. Therefore, the establishment of this non-native midge at climatically similar intervening locations along the western Antarctic Peninsula is also plausible. A lack of effective natural dispersal mechanisms is likely to limit the natural spread of the midge. However, the risk of inadvertent transfer of such organisms was highlighted by a breach of operational procedures within the British Antarctic Survey's logistic system. In 2005, unwashed construction vehicles were transported to Rothera Research Station from Bird Island, South Georgia, where *E. murphyi* was resident and endemic (Hughes *et al.*, 2010b). The failure to follow correct procedure, including pressure washing the vehicles before transportation, resulted in the discovery of over 132 kg of South Georgia soil still attached to the vehicles 3 weeks after their arrival and offloading, with soil containing living and viable non-native vascular plants, bryophytes, microinvertebrates, nematodes, fungi, bacteria and *c.*40,000 seeds and numerous moss propagules. The invertebrates found included *E. murphyi* larvae, although there is currently no evidence that the species has colonized the area successfully, probably due to the lack of appropriate habitat in the immediate vicinity of the research station and its jetty.

The Antarctic Treaty System's Committee for Environmental Protection (CEP) *Non-native Species Manual* (2011) defines invasive species to be 'non-native species that are extending their range in the colonised Antarctic region, displacing native species and causing significant harm to biological diversity or ecosystem functioning'. Given the limited biodiversity found within Antarctic terrestrial ecosystems and the slow rate of natural transfer and colonization by species from outside the region since the gradual improvement of conditions after the Last Glacial Maximum, it is thought there are low levels of functional redundancy in native communities, and not all niches are filled. This may provide opportunities for new colonists, irrespective of whether they are introduced by natural or anthropogenic means (Convey, 2006, 2010, 2011). In addition to this, native species, due to their long isolation and typically adversity selected life history characteristics, are generally poor competitors and vulnerable to competition from more aggressive colonizers (Kennedy, 1993; Convey, 1996a). There is little evidence confirming that any non-native species currently in Antarctica have become invasive, although this is clearly the case on various sub-Antarctic islands (Frenot *et al.*, 2005). However, evidence suggests that it is possible. Hughes *et al.* (2013) showed that at Signy Island the mean biomass of *E. murphyi* larvae in the soil might exceed that of the entire indigenous microarthropod and micro-invertebrate fauna by between around two and five times. Furthermore, the mean population of 21,000 *E. murphyi* larvae m^{-2} could increase annual litter turnover by circa nine times, compared with estimates for the indigenous microarthropods (Collembola and Acari) (Davis, 1981). No indigenous higher insects are found in the South

Orkney Islands, and this marked increase in nutrient turnover rate has unknown impacts on the native ecosystem. Invasive plants may also have negative impacts on indigenous species. Molina-Montenegro *et al.* (2012) examined the competitive effects of *P. annua* on the native vascular plants, *D. antarctica* and *C. quitensis*. Growth chamber manipulation experiments showed that the presence of *P. annua* reduced the biomass and photochemical efficiency of both native species, while the native species had only a weak competitive effect on *P. annua*. With climate change, even native species may outcompete others, with Hill *et al.* (2011) demonstrating that the native *D. antarctica* is a highly effective scavenger for nitrogen uptake, and therefore is expected to outcompete local bryophytes under climate change scenarios affecting nitrogen availability.

Policy and the Future

The risks to Antarctic biodiversity from non-native species have only become widely recognized within the Antarctic community in the past decade. The benchmark paper by Frenot *et al.* (2005) raised the profile of the issue and catalysed a workshop on non-native species in Antarctica, hosted by New Zealand in Christchurch in 2006 (Rogan-Finnemore, 2008) and reported at the Antarctic Treaty Consultative meeting in Edinburgh, UK, later that year. This propelled the issue of non-native species into the policy-makers' spotlight (New Zealand, 2006). At that meeting, the ATCM's Committee for Environmental Protection made non-native species the highest priority for work over the following 5-year period. Following this, work by several Parties as well as the Scientific Committee on Antarctic Research (SCAR), Council of Managers of National Antarctic Programs (COMNAP), the International Association of Antarctica Tour Operators (IAATO) and the International Union for Conservation of Nature (IUCN) was amalgamated into the *CEP Non-native Species Manual* (Edition 2011), which will be revised regularly to accommodate new information and developments (Committee for Environmental Protection, 2011). One particularly accessible output was the COMNAP/SCAR guidelines for supply chain managers of National Antarctic Programmes (COMNAP/SCAR, 2012), which presented simple guidelines aimed at reducing propagule transfer into Antarctica from cargo, ships, aircraft, food and expeditioners' clothing and personal possessions. However, despite all this activity, no mandatory changes in practices have been agreed, and Parties are not obliged to undertake the biosecurity practices recommended within the CEP manual, although some Parties and organizations have produced their own guidelines (Potter and Maggs, 2008; SCAR, 2009; UK, 2009; IAATO, 2012b). Furthermore, in the absence of sustained monitoring, it is difficult to demonstrate whether the biosecurity practices of National Antarctic Programmes have improved. If operators have not reduced substantially the risk of introducing non-native species, then the value of the preceding policy work and the ability of the CEP to act in a unified manner may be questioned (Chown *et al.*, 2012b; Convey *et al.*, 2012). The increase in new reports of non-native species within Antarctica certainly suggests that biosecurity practices are still far from fully effective (Chown *et al*, 2012a; UK, 2012a). Without rapid and coordinated action from the Antarctic Treaty Parties and other interested groups, including the tourism industry, indigenous Antarctic biodiversity is likely to be at significant risk.

Conclusions

Local human activities and climate change are increasing the risk of non-native species introductions and establishment in Antarctica, with likely negative impacts on native communities. National governmental operators and the tourism industry are increasing the footprint of their activities, including accessing ever more remote locations, and are increasing, therefore, the risk of non-native species introductions in a wider range of terrestrial habitats. In addition, climate change in the Antarctic Peninsula region, which is a popular tourist destination and contains around half of all research stations, will increase the chance that introduced species can become established. Predicted climate warming and increased visitation over larger parts of the continent will also increase the risk of non-native species introductions. To date, none of the non-native species in Antarctica have been classified formally as invasive. However, the grass, *P. annua*, is a threat to indigenous species, due to its rapid rate of expansion around the Arctowski Station and its presence at other research stations. The Antarctic Treaty Consultative Meeting has gone some way to highlight the risks from non-native species to the Treaty Parties in recent years, but, as yet, little practical effort has been committed to the eradication of known non-native species. Unless biosecurity measures are implemented more widely, and the Treaty Parties take determined steps to eradicate non-native species when discovered, then the future of Antarctic terrestrial communities, some of which have developed in isolation for millions of years, is in doubt.

Acknowledgements

This chapter contributes to the British Antarctic Survey's Polar Science for Planet Earth programme Ecosystems, and its Environment Office Long Term Monitoring and Survey project (EO-LTMS), and the international SCAR EBA (Evolution and Biodiversity in Antarctica) research programme.

References

Adams, B., *et al.* (2006) Diversity and distribution of Victoria Land biota. *Soil Biology and Biochemistry* 38, 3003–3018.

Anderson, J.B., Shipp, S.S., Lowe, A.L., Wellner, J.S. and Mosola, A.B. (2002) The Antarctic Ice Sheet during the Last Glacial Maximum and its subsequent retreat history: a review. *Quaternary Science Reviews* 21, 49–70.

Aronson, R.B., Thatje, S., McClintock, J.B. and Hughes, K.A. (2011) Anthropogenic impacts on marine ecosystems in Antarctica. *Annals of the New York Academy of Sciences* 1223, 82–107.

ATCP (Antarctic Treaty Consultative Parties) (1991) *Protocol on Environmental Protection to the Antarctic Treaty.* CM 1960. Her Majesty's Stationery Office, London.

Barnes, D.K.A., Hodgson, D.A., Convey, P., Allen, C.S. and Clarke, A. (2006) Incursion and excursion of Antarctic biota: past, present and future. *Global Ecology and Biogeography* 15, 121–142.

Barrett, J.E., Virginia, R.A., Wall, D.H., Doran, P.T., Fountain, A.G., Welch, K.A. and Lyons, W.B. (2008) Persistent effects of a discrete warming event on a polar desert ecosystem. *Global Change Biology* 14, 2249–2261.

Bergstrom, D.M. and Chown, S.L. (1999) Life at the front: history, ecology and change on southern ocean islands. *Trends in Ecology and Evolution* 14, 472–477.

Bergstrom, D.M., Hodgson, D.A. and Convey, P. (2006) The physical setting of the Antarctic. In: Bergstrom, D.M., Convey, P. and Huiskes, A.H.L. (eds) *Trends in Antarctic Terrestrial and Limnetic Ecosystems: Antarctica as a Global Indicator.* Springer, Dordrecht, Netherlands, pp. 15–33.

Block, W. (1984) Terrestrial microbiology, invertebrates and ecosystems. In: Laws, R.M. (ed.) *Antarctic Ecology.* Academic Press, London, pp. 163–236.

Block, W., Burn, A.J. and Richard, K.J. (1984) An insect introduction to the maritime Antarctic. *Biological Journal of the Linnean Society* 23, 33–39.

Bokhorst, S., Huiskes, A., Convey, P. and Aerts, R. (2007) Climate change effects on organic matter decomposition rates in ecosystems from the Maritime Antarctic and Falkland Islands. *Global Change Biology* 13, 2642–2653.

Chapuis, J.L., Frenot, Y. and Lebouvier, M. (2004) Recovery of native plant communities after eradication of rabbits from the subantarctic Kerguelen Islands, and influence of climate change. *Biological Conservation* 117, 167–179.

Chown, S.L. and Convey, P. (2007) Spatial and temporal variability across life's hierarchies in the terrestrial Antarctic. *Philosophical Transactions of the Royal Society of London, Series B* 362, 2307–2331.

Chown, S.L. and Smith, V.R. (1993) Climate change and the short-term impact of feral house mice at the sub-Antarctic Prince Edward Islands. *Oecologia* 96, 508–518.

Chown, S.L., *et al.* (2012a) Continent-wide risk assessment for the establishment of non-

indigenous species in Antarctica. *Proceedings of the National Academy of Sciences* 109, 4938–4943.
Chown, S.L., *et al.* (2012b) Challenges to the future conservation of the Antarctic. *Science* 337, 158–159.
Chwedorzewska, K.J. (2008) *Poa annua* L. in Antarctic: searching for the source of introduction. *Polar Biology* 31, 263–268.
Clarke, A., Barnes, D.K.A. and Hodgson, D.A. (2005) How isolated is Antarctica? *Trends in Ecology and Evolution* 20, 1–3.
Committee for Environmental Protection (2011) *Committee for Environmental Protection (CEP) Non-native Species Manual* (edn 2011) (http://www.ats.aq/e/cep_handbook.htm, accessed 20 January 2014).
COMNAP (Council of Managers of National Antarctic Programs) and the Scientific Committee on Antarctic Research (SCAR) (2012) Checklist for supply chain managers of National Antarctic Programmes for the reduction in risk of transfer of non-native species (https://www.comnap.aq/Publications/Comnap%20Publications/COMNAP_SCAR_Checklists_for_Supply_Chain_Managers.pdf, accessed 20 January 2014).
Convey, P. (1992) Aspects of the biology of the midge, *Eretmoptera murphyi* Schaeffer (Diptera: Chironomidae), introduced to Signy Island, maritime Antarctica. *Polar Biology* 12, 653–657.
Convey, P. (1996a) The influence of environmental characteristics on the life history attributes of Antarctic terrestrial biota. *Biological Reviews* 71, 191–225.
Convey, P. (1996b) Reproduction of Antarctic flowering plants. *Antarctic Science* 8, 127–134.
Convey, P. (2006) Antarctic climate change and its influences on terrestrial ecosystems. In: Bergstrom, D.M., Convey, P. and Huiskes, A.H.L. (eds) *Trends in Antarctic Terrestrial and Limnetic Ecosystems: Antarctica as a Global Indicator.* Springer, Dordrecht, Netherlands, pp. 253–272.
Convey, P. (2007) Antarctic ecosystems. In: Levin, S.A., (ed.) *Encyclopedia of Biodiversity.* Elsevier, San Diego, California, 12 pp.
Convey, P. (2008) Non-native species in Antarctic terrestrial and freshwater environments: presence, sources, impacts and predictions. In: Rogan-Finnemore, M. (ed.) *Non-native Species in the Antarctic – Proceedings.* Gateway Antarctica, Christchurch, New Zealand, pp. 97–130.
Convey, P. (2010) Terrestrial biodiversity in Antarctica – recent advances and future challenges. *Polar Science* 4, 135–147.
Convey, P. (2011) Antarctic terrestrial biodiversity in a changing world. *Polar Biology* 34, 1629–1641.
Convey, P. and Block, W. (1996) Antarctic Diptera: ecology, physiology and distribution. *European Journal of Entomology* 93, 1–13.
Convey, P. and Lebouvier, M. (2009) Environmental change and human impacts on terrestrial ecosystems of the sub-Antarctic islands between their discovery and the mid-twentieth century. *Papers and Proceedings of the Royal Society of Tasmania* 143, 33–44.
Convey, P. and McInnes, S.J. (2005) Exceptional tardigrade-dominated ecosystems in Ellsworth Land, Antarctica. *Ecology* 86, 519–527.
Convey, P. and Smith, R.I.L. (2006) Responses of terrestrial Antarctic ecosystems to climate change. *Plant Ecology* 182, 1–10.
Convey, P. and Stevens, M.I. (2007) Antarctic biodiversity. *Science* 317, 1877–1878.
Convey, P., Gibson, J.A.E., Hillenbrand, C.-D., Hodgson, D.A., Pugh, P.J.A., Smellie, J.L. and Stevens, M.I. (2008) Antarctic terrestrial life – challenging the history of the frozen continent? *Biological Reviews* 83, 103–117.
Convey, P., Stevens, M.I., Hodgson, D.A., Smellie, J.L., Hillenbrand, C.-D., Barnes, D.K.A, Clarke, A., Pugh, P.J.A., Linse, K. and Cary, S.C. (2009) Exploring biological constraints on the glacial history of Antarctica. *Quaternary Science Reviews* 28, 3035–3048.
Convey, P., Hughes, K.A. and Tin, T. (2012) Continental governance and environmental management mechanisms under the Antarctic Treaty System: sufficient for the biodiversity challenges of this century? *Biodiversity* 13, 1–15.
Corte, A. (1961) La primera fanerogama adventicia hallada en el continente Antartico. *Contribución del Instituto Antártico Argentino* 62, 1–14.
Cowan, D.A., Chown, S.L., Convey, P., Tuffin, M., Hughes, K.A., Pointing, S. and Vincent, W.F. (2011) Non-indigenous microorganisms in the Antarctic – assessing the risks. *Trends in Microbiology* 19, 540–548.
Cuba-Díaz, M., Troncoso, J.M., Cordero, C., Finot, V.L. and Rondanelli-Reyes, M. (2012) *Juncus bufonius* L., a new non-native vascular plant in King George Island, South Shetland Islands. *Antarctic Science* doi:10.1017/S0954102012000958.
Davey, M.C., Pickup, J. and Block, W. (1992) Temperature-variation and its biological significance in fell field habitats on a maritime Antarctic island. *Antarctic Science* 4, 383–388.
Davis, R.C. (1981) Structure and function of two Antarctic terrestrial moss communities. *Ecological Monographs* 51, 125–143.

Denlinger, D.L. and Lee, R.E. (2010) *Low Temperature Biology of Insects*. Cambridge University Press, Cambridge, UK.

Dibbern, S. (2010) Fur seals, whales and tourists: a commercial history of Deception Island, Antarctica. *Polar Record* 46, 210–221.

Doran, P.T., *et al.* (2002) Antarctic climate cooling and terrestrial ecosystem response. *Nature* 415, 517–520.

Dózsa-Farkas, K. and Convey, P. (1997) *Christensenia*, a new enchytraeid genus from Antarctica. *Polar Biology* 17, 482–486. (Paper subsequently modified – see Erratum (1998) *Polar Biology* 20, 292.)

Fowbert, J.A. and Smith, R.I.L. (1994) Rapid population increase in native vascular plants in the Argentine Islands, Antarctic Peninsula. *Arctic and Alpine Research* 26, 290–296.

Fraser, C.I., Nikula, R., Ruzzante, D.E. and Waters, J.M. (2012) Poleward bound: biological impacts of southern hemisphere glaciations. *Trends in Ecology and Evolution* 26, 462–471.

Frenot, Y., Chown, S.L., Whinam, J., Selkirk, P.M., Convey, P., Skotnicki, M. and Bergstrom, D.M. (2005) Biological invasions in the Antarctic: extent, impacts and implications. *Biological Reviews* 80, 45–72.

Friedmann, E.I. (1982) Endolithic microorganisms in the Antarctic cold desert. *Science* 215, 1045–1053.

Gerighausen, U., Bräutigam, K., Mustafa, O. and Peter, H.-U. (2003) Expansion of vascular plants on an Antarctic island – a consequence of climate change? In: Huiskes, A.H.L., Gieskes, W.W.C., Rozema, J., Schorno, R.M.C., van der Vies, S.M. and Wolff, W.S. (eds) *Antarctic Biology in a Global Context*. Backhuys, Leiden, Netherlands, pp. 79–83.

Greenslade, P. (1995) Collembola from the Scotia Arc and Antarctic Peninsula including descriptions of two new species and notes on biogeography. *Polskie Pismo Entomologiczne* 64, 305–319.

Greenslade, P. and Convey, P. (2012) Exotic Collembola on subantarctic islands: pathways, origins and biology. *Biological Invasions* 14, 405–417.

Greenslade, P. and Wise, K.A.J. (1984) Additions to the collembolan fauna of the Antarctic. *Transactions of the Royal Society of South Australia* 108, 203–205.

Greenslade, P., Potapov, M., Russell, D. and Convey, P. (2012) Global collembola on Deception Island. *Journal of Insect Science* 12, Article 111 (http://www.insectscience.org/12.111, accessed 20 January 2014).

Headland, R. (2009) *A Chronology of Antarctic Exploration*. Quaritch, London.

Hill, D., Farrar, J., Roberts, P., Farrell, M., Grant, H., Newsham, K.K., Hopkins, D.W., Bardgett, R.D.and Jones, D.J. (2011) Vascular plant success in a warming Antarctic may be due to efficient nitrogen acquisition. *Nature Climate Change* 1, 50–53.

Hodgson, D.A., Roberts, D., McMinn, A., Verleyen, E., Terry, B., Corbett, C. and Vyverman, W. (2006) Recent rapid salinity rise in three East Antarctic lakes. *Journal of Paleolimnology* 36, 385–406.

Hodgson, D.A., *et al.* (2010) The limnology and biology of the Dufek Massif, Transantarctic Mountains 82° south. *Polar Science* 4, 197–214.

Hogg, I.D., Cary, S.C., Convey, P., Newsham, K.K., O'Donnell, A.G., Adams, B.J., Aislabie, J., Frati, F., Stevens, M.I. and Wall, D.H. (2006) Biotic interactions in Antarctic terrestrial ecosystems: are they a factor? *Soil Biology and Biochemistry* 38, 3035–3040.

Hughes, K.A. and Convey, P. (2010) The protection of Antarctic terrestrial ecosystems from inter- and intra-continental transfer of non-indigenous species by human activities: a review of current systems and practices. *Global Environmental Change* 20, 96–112.

Hughes, K.A. and Convey, P. (2012) Determining the native/non-native status of newly discovered terrestrial and freshwater species in Antarctica – current knowledge, methodology and management action. *Journal of Environmental Management* 93, 52–66.

Hughes, K.A. and Lawley, B. (2003) A novel Antarctic microbial endolithic community within gypsum crusts. *Environmental Microbiology* 5, 555–565.

Hughes, K.A. and Nobbs, S. (2004) Long-term survival of human faecal microorganisms on the Antarctic Peninsula. *Antarctic Science* 16, 293–297.

Hughes, K.A. and Worland, M.R. (2010) Spatial distribution, habitat preference and colonisation status of two alien terrestrial invertebrate species in Antarctica. *Antarctic Science* 22, 221–231.

Hughes, K.A., Walsh, S., Convey, P., Richards, S. and Bergstrom, D.M. (2005) Alien fly populations established at two Antarctic research stations. *Polar Biology* 28, 568–570.

Hughes, K.A., Ott, S., Bolter, M. and Convey, P. (2006) Colonisation processes. In: Bergstrom, D.M., Convey, P. and Huiskes, A.H.L. (eds) *Trends in Antarctic Terrestrial and Limnetic Ecosystems: Antarctica as a Global Indicator*. Springer, Dordrecht, Netherlands, pp. 35–54.

Hughes, K.A., Lee, J.E., Ware, C., Kiefer, K. and Bergstrom, D.M. (2010a) Impact of anthropogenic transportation to Antarctica on alien seed viability. *Polar Biology* 33, 1123–1130.

Hughes, K.A., Convey, P., Maslen, N.R. and Smith, R.I.L. (2010b) Accidental transfer of non-native soil organisms into Antarctica on construction vehicles. *Biological Invasions* 12, 875–891.

Hughes, K.A., Lee, J.E., Tsujimoto, M., Imura, S., Bergstrom, D.M., Ware, C., Lebouvier, M., Huiskes, A.H.L., Gremmen, N.J.M., Frenot, Y., Bridge, P.D. and Chown, S.L. (2011a) Food for thought: risks of non-native species transfer to the Antarctic region with fresh produce. *Biological Conservation* 144, 1682–1689.

Hughes, K.A., Fretwell, P., Rae, J., Holmes, K. and Fleming, A. (2011b) Untouched Antarctica: mapping a finite and diminishing environmental resource. *Antarctic Science* 23, 537–548.

Hughes, K.A., Worland, M.R., Thorne, M. and Convey, P. (2013) The non-native chironomid *Eretmoptera murphyi* in Antarctica: erosion of the barriers to invasion. *Biological Invasions* 15, 269–281.

IAATO (International Association of Antarctica Tour Operators) (2012a) Tourism statistics (http://iaato.org/tourism-statistics, accessed 20 January 2014).

IAATO (International Association of Antarctica Tour Operators) (2012b) Boot, clothing and equipment decontamination guidelines for small boat operations (http://iaato.org/decontamination-guidelines, accessed 20 January 2014).

Jabour, J. (2009) National Antarctic programs and their impact on the environment. In: Kerry, K.R. and Riddle, M.J. (eds) *Health of Antarctic Wildlife – A Challenge for Science and Policy.* Springer, Heidelberg, Germany, pp. 211–229.

Japan (1996) *A Grass (Seed Plant) Found in Syowa Station Area, East Antarctica.* Information Paper 66, Antarctic Treaty Consultative Meeting XX, 29 April–10 May 1996, Utrecht, Netherlands (http://www.ats.aq/index_e.htm).

Kennedy, A. (1993) Water as a limiting factor in the Antarctic terrestrial environment: a biogeographical synthesis. *Arctic and Alpine Research* 25, 308–315.

Klopper, F.J. and Smith, V.R. (1998) First report of *Botryotinia fuckeliana* on Kerguelen cabbage on the sub-Antarctic Marion Island. *Plant Disease* 82, 710.

Lee, J.E. and Chown, S.L. (2009a) Breaching the dispersal barrier to invasion: quantification and management. *Ecological Applications* 19, 1944–1959.

Lee, J.E. and Chown, S.L. (2009b) Quantifying the propagule load associated with the construction of an Antarctic research station. *Antarctic Science* 21, 471–475.

Lee, J.E. and Chown, S.L. (2011) Quantification of intra-regional propagule movements in the Antarctic. *Antarctic Science* 23, 337–342.

Lynch, H.J., Crosbie, K., Fagan, W.F. and Naveen, R. (2010) Spatial patterns of tour ship traffic in the Antarctic Peninsula region. *Antarctic Science* 22, 123–130.

McGraw, J.B. and Day, T.A. (1997) Size and characteristics of a natural seed bank in Antarctica. *Arctic and Alpine Research* 29, 213–216.

Marshall, W.A. (1996) Biological particles over Antarctica. *Nature* 383, 680.

Mellars, P. (2006) Why did modern human populations disperse from Africa *ca.* 60 000 years ago? A new model. *Proceedings of the National Academy of Sciences* 103, 9381–9386.

Molina-Montenegro, M., Carrasco-Urra, F., Rodrigo, C., Convey, P., Valladares, F. and Gianoli, E. (2012) Occurrence of the non-native annual bluegrass (*Poa annua*) on the Antarctic mainland and its negative effects on native plants. *Conservation Biology* 26, 717–723.

New Zealand (2006) *Non-native Species in the Antarctic. Report of a Workshop.* Working Paper 13, Antarctic Treaty Consultative Meeting XXIX, 12–23 June 2006, Edinburgh, UK (http://www.ats.aq/index_e.htm).

Newsham, K.K., Hodgson, D.A., Murray, A.W.A., Peat, H.J. and Smith, R.I.L. (2002) Response of two Antarctic bryophytes to stratospheric ozone depletion. *Global Change Biology* 8, 972–983.

Nielsen, U.N., Wall, D.H., Adams, B.J. and Virginia, R.A. (2011) Antarctic nematode communities: observed and predicted responses to climate change. *Polar Biology* 34, 1701–1711.

Ochyra, R., Smith, R.I.L. and Bednarek-Ochyra, H. (2008) *The Illustrated Moss Flora of Antarctica.* Cambridge University Press, Cambridge, UK.

Olech, M. (1996) Human impact on terrestrial ecosystems in west Antarctica. *Proceedings of the NIPR Symposium on Polar Biology* 9, 299–306.

Olech, M. (2003) Expansion of alien vascular plant *Poa annua* L. in the vicinity of the Henryk Arctowski Station – a consequence of climate change? In: Olech, M. (ed.) *The Functioning of Polar Ecosystems as Viewed Against Global Environmental Changes.* XXIX International Polar Symposium, Kraków, Poland, 19–21 September 2003. Jagiellonian University, Krakow, pp. 89–90.

Olech, M. and Chwedorzewska, K.J. (2011) The first appearance and establishment of an alien vascular plant in natural habitats on the forefield of a retreating glacier in Antarctica. *Antarctic Science* 23, 153–154.

Osyczka, P. (2010) Alien lichens unintentionally transported to the 'Arctowski' station (South Shetlands, Antarctica). *Polar Biology* 33, 1067–1073.

Osyczka, P., Mleczko, P., Karasiński, D. and Chlebicki, A. (2012) Timber transported to Antarctica: a potential and undesirable carrier for alien fungi and insects. *Biological Invasions* 14, 15–20.

Øvstedal, D.O. and Smith, R.I.L. (2001) *Lichens of Antarctica and South Georgia. A Guide to Their Identification and Ecology.* Cambridge University Press, Cambridge, UK.

Parnikoza, I., Convey, P., Dykyy, I., Trakhimets, V., Milinevsky, G., Tyschenko, O., Inozemtseva, D. and Kozeretska, I. (2009) Current status of the Antarctic herb tundra formation in the central Argentine Islands. *Global Change Biology* 15, 1685–1693.

Parnikoza, I., Dykyy I., Ivanets V., Kozeretska, I., Kunakh, V., Rozhok, A., Ochyra, R. and Convey, P. (2012) Use of *Deschampsia antarctica* for nest building by the kelp gull in the Argentine Island area (Maritime Antarctica). *Polar Biology* 35, 1753–1758.

Pearce, D.A., Bridge, P.D., Hughes, K., Sattler, B., Psenner, R. and Russell, N.J. (2009) Microorganisms in the atmosphere over Antarctica. *FEMS Microbiology Ecology* 69, 143–157.

Peeters, K., Hodgson, D.A., Convey, P. and Willems, A. (2011) Culturable diversity of heterotrophic bacteria in Forlidas Pond (Pensacola Mountains) and Lundstrom Lake (Shackleton Range), Antarctica. *Microbial Ecology* 62, 399–413.

Pertierra, L.R., Hughes, K.A., Benayas, J., Justel, A. and Quesada, A. (2013) Environmental management of a scientific field camp in Maritime Antarctica: reconciling research impacts with conservation goals in remote ice-free areas. *Antarctic Science* doi:10.1017/S0954102012001083.

Potter, S. and Maggs, T. (2008) Antarctic quarantine management: Australia's framework and practice. In: Rogan-Finnemore, M. (ed.) *Non-native Species in the Antarctic – Proceedings.* Gateway Antarctica, Christchurch, New Zealand, pp. 165–182.

Pugh, P.J.A. (1997) Acarine colonization of Antarctica and the islands of the Southern Ocean: the role of zoohoria. *Polar Record* 33, 113–122.

Rogan-Finnemore, M. (ed.) (2008) *Non-native Species in the Antarctic, Proceedings.* Gateway Antarctica, Christchurch, New Zealand.

Ruhland, C.T. and Day, T.A. (2001) Size and longevity of seed banks in Antarctica and the influence of ultraviolet-B radiation on survivorship, growth and pigment concentrations of *Colobanthus quitensis* seedlings. *Environmental and Experimental Botany* 45, 143–154.

Russia (1999) *Initial Environmental Evaluation: Compacted Snow Runway at the Larseman Hills.* Information Paper 79, Antarctic Treaty Consultative Meeting XXIII, 24 May–4 June 1999, Lima, Peru (http://www.ats.aq/index_e.htm).

SCAR (Scientific Committee on Antarctic Research) (2009) *SCAR's Environmental Code of Conduct for Terrestrial Scientific Field Research in Antarctica.* Information Paper 4, Antarctic Treaty Consultative Meeting XXXII, 6–17 April 2009, Baltimore, MD, USA.

Schiermeier, Q. (2011) Race against time for raiders of the lost lake. *Nature* 469, 275.

Smith, R.I.L. (1984) Terrestrial biology of the Antarctic and sub-Antarctic. In: Laws, R.M. (ed.) *Antarctic Ecology.* Academic Press, London, pp. 61–162.

Smith, R.I.L. (1994) Vascular plants as indicators of regional warming in Antarctica. *Oecologia* 99, 322–328.

Smith, R.I.L. (1996) Introduced plants in Antarctica: potential impacts and conservation issues. *Biological Conservation* 76, 135–146.

Smith, R.I.L. and Richardson, M. (2011) Fuegian plants in Antarctica: natural or anthropogenically assisted immigrants. *Biological Invasions* 13, 1–5.

Smith, V.R. (2002) Climate change in the subantarctic: an illustration from Marion Island. *Climate Change* 52, 345–357.

Sømme, L. (1989) Adaptations of terrestrial arthropods to the alpine environment. *Biological Reviews* 64, 367–407.

Spain, Argentina and United Kingdom (2012) *Colonisation Status of the Non-native Grass* Poa pratensis *at Cierva Point, Danco Coast, Antarctic Peninsula.* Information Paper 13, Antarctic Treaty Consultative Meeting XXXV, 11–20 June 2012, Hobart, Australia (http://www.ats.aq/index_e.htm).

Terauds, A., Chown, S.L., Morgan, F., Peat, H.J., Watts, D.J., Keys, H., Convey, P. and Bergstrom, D.M. (2012) Conservation biogeography of the Antarctic. *Diversity and Distribution* 18, 726–741.

Thomas, D., Fogg, G.E., Convey, P., Fritsen, C.H., Gili, J.-M., Gradinger, R., Laybourn-Parry, J., Reid, K. and Walton, D.W.H. (2008) *The Biology of Polar Regions.* Oxford University Press, Oxford, UK.

Tin, T., Fleming, Z.L., Hughes, K.A., Ainley, D.G., Convey, P., Moreno, C.A., Pfeiffer, S., Scott, J. and Snape, I. (2009) Impacts of local human activities on the Antarctic environment. *Antarctic Science* 21, 3–33.

Trathan, P.N. and Reid, K. (2009) Exploitation of the marine ecosystem in the sub-Antarctic: historical impacts and current consequences. *Proceedings of the Royal Society of Tasmania* 143, 9–14.

Tsujimoto, M. and Imura, S. (2012) Does a new transportation system increase the risk of importing non-native species to Antarctica? *Antarctic Science* 24, 441–449.

Tsujimoto, M., Imura, S. and Kanda, H. (2010) Molecular systematics of a non-native vascular plant found near the Syowa Station, Antarctica. Poster at the International Polar Year Oslo Science Conference, Oslo, Norway (unpublished).

Turner, J., Bindschadler, R., Convey, P., di Prisco, G., Fahrbach, E., Gutt, J., Hodgson, D., Mayewski, P. and Summerhayes, C. (eds) (2009a) *Antarctic Climate Change and the Environment.* Scientific Committee on Antarctic Research (SCAR), Cambridge, UK.

Turner, J., Comiso, J.C., Marshall, G.J., Lachlan-Cope, T.A., Bracegirdle, T., Maksym, T., Meredith, M.P., Wang, Z. and Orr, A. (2009b) Non-annular atmospheric circulation change induced by stratospheric ozone depletion and its role in the recent increase of Antarctic sea ice extent. *Geophysical Research Letters* 36, L08502.

Tweedie, C.E. and Bergstrom, D.M. (2000) A climate change scenario for surface air temperature at subantarctic Macquarie Island. In: Davison, W., Howard-Williams, C. and Broady, P.A. (eds) *Antarctic Ecosystems: Models for Wider Ecological Understanding.* New Zealand Natural Sciences, Christchurch, New Zealand, pp. 272–281.

UK (2009) *Procedures for Vehicle Cleaning to Prevent Transfer of Non-native Species Into and Around Antarctica.* Working Paper 32, Antarctic Treaty Consultative Meeting XXXII, 6–17 April 2009, Baltimore, MD, USA (http://www.ats.aq/index_e.htm).

UK (2011) *Draft Comprehensive Environmental Evaluation (CEE) for the Proposed Exploration of Subglacial Lake Ellsworth, Antarctica.* Working Paper 16, Antarctic Treaty Consultative Meeting XXXIV, 20 June–1 July 2011, Buenos Aires, Argentina (http://www.ats.aq/index_e.htm).

UK (2012a) *The Assessment of Land-Based Activities in Antarctica.* Working Paper 4, Antarctic Treaty Consultative Meeting XXXV, 11–20 June 2012, Hobart, Australia (http://www.ats.aq/index_e.htm).

UK (2012b) *Colonisation Status of Known Non-native Species in the Antarctic Terrestrial Environment (Updated 2012).* Information Paper 29, Antarctic Treaty Consultative Meeting XXXV, 11–20 June 2012, Hobart, Australia (http://www.ats.aq/index_e.htm).

UK and Spain (2010) *Eradication of a Vascular Plant Species Recently Introduced to Whaler's Bay, Deception Island.* Information Paper 43, Antarctic Treaty Consultative Meeting XXXIII, 3–14 May 2010, Punta de Este, Uruguay (http://www.ats.aq/index_e.htm).

Volonterio, O., Ponce de León, R., Convey, P. and Krzeminska, E. (2013) First record of Trichoceridae (Diptera) in the maritime Antarctic. *Polar Biology* 36, 1125–1131.

Vyverman, W., Verleyen, E., Wilmotte, A., Hodgson, D.A., Willem, A., Peeters, K., Van de Vijver, B., De Wever, A., Leliaert, F. and Sabbe, K. (2010) Evidence for widespread endemism among Antarctic micro-organisms. *Polar Science* 4, 103–113.

Wall, D.H., Lyons, W.B., Convey, P., Howard-Williams, C., Quesada, A. and Vincent, W.F. (2011) Long-term ecosystem networks to record change: an international imperative. *Antarctic Science* 23, 209.

Walton, D.W.H. (1984) The terrestrial environment. In: Laws, R.M. (ed.) *Antarctic Ecology.* Academic Press, London, pp. 1–60.

Walton, D.H.W. and Clarkson, P.D. (2011) *Science in the Snow: Fifty Years of International Collaboration through the Scientific Committee on Antarctic Research.* Scientific Committee on Antarctic Research, Cambridge, UK.

Wever, A. de, Leliaert, F., Verleyen, E., Vanormelingen, P., Van der Gucht, K., Hodgson, D.A., Sabbe, K. and Vyverman, W. (2009) Hidden levels of phylodiversity in Antarctic green algae: further evidence for the existence of glacial refugia. *Proceedings of the Royal Society of London Series B* 276, 3591–3599.

Whinam, J., Chilcott, N. and Bergstrom, D.M. (2005) Subantarctic hitchhikers: expeditioners as vectors for the introduction of alien organisms. *Biological Conservation* 121, 207–219.

Wise, K.A.J. (1971) The Collembola of Antarctica. *Pacific Insects Monograph* 25, 57–74.

7 Synergies between Climate Change and Species Invasions: Evidence from Marine Systems

Cascade J.B. Sorte

Department of Ecology and Evolutionary Biology, University of California, Irvine, California, USA

Abstract

The hypothesis that climate change will facilitate species invasions has recently received increasing focus in studies of marine systems. Over the past decade, approaches to testing this hypothesis have shifted from time-series observations of concomitant increases in both processes to experimental tests that are beginning to reveal the mechanisms underlying the synergies between these two aspects of global change. The results of many studies conform to expectations that under climate change, invasive species' abundances, ranges and per capita effects – collectively indicative of invader impacts – will increase. However, there remain significant gaps in our understanding of responses to non-thermal factors (such as changes in ocean pH, dissolved oxygen and storm events) and how species-specific idiosyncrasies will manifest in changes at the community level. Continued exploration of the synergies between climate change and species invasions will help us to anticipate potential indirect effects (mediated through the responses of interacting invaders) on native species, which may themselves also be directly affected by climate change.

Introduction

Both climate change and species invasions have been recognized independently as major drivers of biodiversity loss (Sala *et al.*, 2000), and their total impact could be compounded by synergistic effects (Brook *et al.*, 2008). For example, climate change could drive an increase in invasive species' abundances and, consequently, density-dependent effects on native prey species (Fig. 7.1). On the other hand, invasive predators could drive population declines of native prey and, therefore, decrease the potential of these prey species to cope with climatic changes. Despite the likelihood that such reciprocal interactions will increasingly threaten the Earth's systems, a decade ago, few experimental studies had 'investigated the responses of different plant species to global warming in an ecosystem context, and none [had] examined the response of biological invaders' (Dukes and Mooney, 1999). Furthermore, although the terrestrial literature now contains in excess of 1000 papers discussing the topic, study of the interaction between climate change and species invasions in marine systems lags about a decade behind that in terrestrial systems (Fig. 7.2; cf. Rahel and Olden, 2008, for a review of freshwater systems). This discrepancy in publishing rate is despite evidence that marine systems appear, in some cases, to be threatened more greatly by climate change, invasions and their interaction than their terrestrial counterparts (Cohen and Carlton, 1998; Sorte *et al.*, 2010a, 2013; Burrows *et al.*, 2011; Sunday *et al.*, 2012).

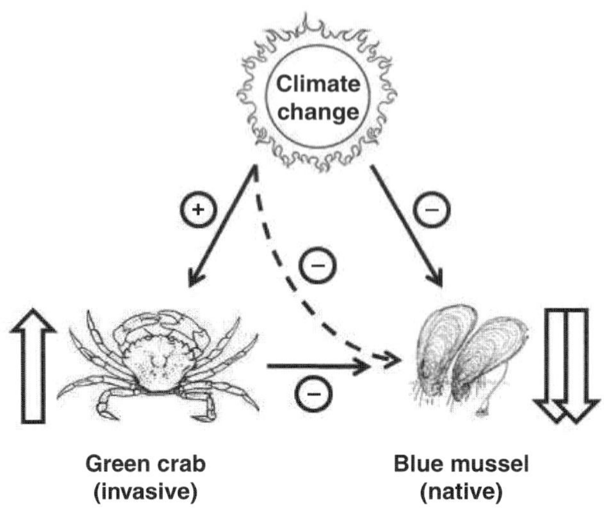

Fig. 7.1. Climate change can have both direct (solid lines) and indirect (dashed line) effects on species. For example, if climate change has a positive effect (e.g. increases the growth and survival) on the invasive green crab, then this will lead to an indirect negative effect on the crab's prey, the native blue mussel. Furthermore, if climate change also has a negative direct effect on the blue mussel, then this will lead to a doubly negative impact (via both direct and indirect effects) on the native species.

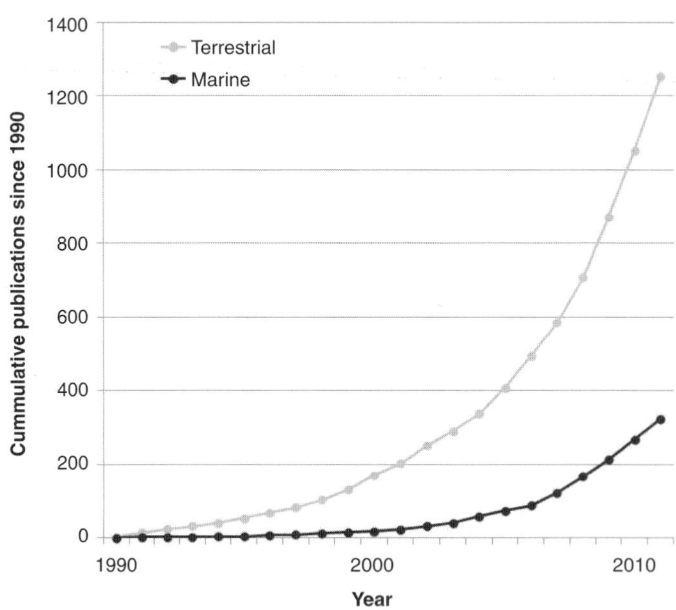

Fig. 7.2. Terrestrial (grey) and marine (black) studies on climate change and species invasions. Values are cumulative publications since 1990 and are based on a topics search using ISI Web of Knowledge.

This chapter first summarizes recent and predicted alterations in the ocean associated with climate change, as well as the main hypotheses for why these environmental alterations are predicted to impact invasive species less negatively than co-occurring native species. Second, the marine literature is reviewed to assess whether our current understanding of climate change effects on invasive species conforms to the expectations of these hypotheses. Finally, the chapter synthesizes the literature to explore how the responses of invasive species to climate change will, in turn, determine their future impacts on native species.

Climatic Changes in the Marine Realm

The oceans, which cover about 70% of the Earth's surface, have absorbed an estimated 26% of the excess CO_2 (Sabine *et al.*, 2004) and 84% of the increased heat held within the Earth system via greenhouse warming (between 1955 and 1998; Levitus *et al.*, 2005). Recent and future predicted increases in CO_2 levels and temperature feedback to influence other physical properties of marine systems (Table 7.1; Meehl *et al.*, 2007). While an increase in dissolved CO_2 might be expected to fertilize the growth of oceanic primary producers, it also drives decreases in pH (i.e. ocean acidification) and shifts in carbonate saturation states that can impede calcification (Doney *et al.*, 2012). Increased temperature leads to a decrease in dissolved oxygen and can increase stratification, further preventing oxygen diffusion (Keeling *et al.*, 2010). Global warming also causes the expansion of ocean water and melting of terrestrial ice, leading to sea level rise, while the melting of sea ice and changes in evaporation rates are associated with changes in ocean salinity. At the same time as average conditions are shifting, extreme climatic events are and will be increasing in frequency and intensity. Although global warming has been best studied, each of these environmental alterations has the potential to affect invasive species and their impacts.

Why Might Climate Change Favour Invaders over Native Species?

Changes in marine climatic conditions might be expected to favour invaders over native species because some of the same species' characteristics that could facilitate persistence under future conditions are those associated with invasion success. Specifically, both climate change 'winners' (species likely to benefit most or suffer least) and successful invaders are expected to (i) have environmental tolerance ranges that are either relatively broad or encompass future conditions, and/or (ii) have life-history characteristics favouring fast growth, high reproductive rates and rapid utilization of available resources.

Species differ in their physiological tolerances, with functioning being optimal over a small range of conditions, declining as conditions become increasingly suboptimal and ceasing completely (eventually leading to mortality) outside of tolerance limits (Monaco and Helmuth, 2011). Within this framework, then, climatic changes will or will not benefit a particular species based on whether conditions become more physiologically optimal or more stressful, respectively. I would note, however, that whereas considering the species level is appropriate for this discussion, tolerance

Table 7.1. Observed and predicted changes in the marine environment associated with climate change. Unless noted, values are from the Fourth Assessment Report of the Intergovernmental Panel on Climate Change and are based on the A1B scenario, when multiple projections were given (Meehl *et al.*, 2007).

Physical factor	Past changes (observed)	Changes predicted by 2100
pH	↓ 0.1 unit since pre-industrial time	↓ 0.25 units
Sea surface temperature	↑ 0.4°C between 1955 and 1998[b]	↑ 1–4°C
Dissolved oxygen	↓ in several regions[c]	↓ 1–7%[c]
Salinity		↓↑ depending on location
Sea level	↑ 0.08 m between 1961 and 2003	↑ 0.35 m
Extreme climatic events[a]	↑ frequency and intensity	↑ frequency and intensity

Notes: [a]Extreme heat events and storms creating high waves; [b]Levitus *et al.*, 2005; [c]Keeling *et al.*, 2010.

also varies intra-specifically, based on individuals' differential histories of adaptation and acclimation. Several marine studies, although not focusing directly on species invasions, have shown predictable responses to climate change associated with geographic affinity (see Helmuth *et al.*, 2006), and geographic range can be considered a proxy for environmental tolerance range. For example, when researchers from a field station in California, USA, compared abundances of coastal marine species between the early and late 20th century, they found that local ocean warming was associated with an increase in more warm-adapted species, a decrease in more cool-adapted species and no significant change in cosmopolitan (widely distributed and therefore likely to be broadly tolerant) species (Barry *et al.*, 1995; Sagarin *et al.*, 1999). However, patterns related to tolerance characteristics may sometimes be obscured or overshadowed by other factors, such as indirect effects via increases or decreases in key (e.g. foundation) species (Schiel *et al.*, 2004).

There is some evidence that species with broad environmental tolerances have been more successful as invaders (see Dukes and Mooney, 1999; Kolar and Lodge, 2001). Tolerance range could be related to invasion success because broad tolerances allow species to inhabit a wide native range, causing them to be transported more often, as well as increasing the probability that species will survive and establish in an introduced habitat (Theoharides and Dukes, 2007). Furthermore, invaders may enjoy greater success under shifting environmental conditions when they have been transported from a source location more similar to the future climate in their introduced habitat. Thus, the colder an introduced habitat, the more likely that invasive species have come from relatively warmer locations and are more temperature tolerant than the local native species, suggesting that study location should be acknowledged as a possible factor modifying the strength of this pattern. For example, in a recent meta-analysis, which indicated that invasive species would be favoured under warmer and more acidic conditions in marine systems, all of the studies included were conducted in relatively cool, temperate locations, whereas most of the invasive species considered had originated in warmer locations (Sorte *et al.*, 2013).

In addition to the relative breadth of physiological tolerances, 'pioneering' life-history characteristics (see Sakai *et al.*, 2001) could underlie both invasion success and the ability to capitalize on changing climates. Invasive species that can use available resources most efficiently, grow and mature quickly, and produce large numbers of offspring may be best able to colonize, establish and spread in their introduced habitat (Theoharides and Dukes, 2007). At the same time, climate change could increase resources either directly (e.g. via increased CO_2) or indirectly, when it acts as a disturbance, freeing up food or space for resistant or newly colonizing species, favouring marine species with high resource-use efficiencies. Together, this reasoning leads to the expectations that physiological tolerance and resource-use characteristics will partly determine the success of species, both as invaders and under climate change.

Responses of Marine Invaders to Changing Climatic Conditions

Whether the relative responses of invasive and native species to changing climatic conditions conform to these expectations has been examined primarily by comparisons focused on a subset of the most accessible intertidal and shallow subtidal communities and easily manipulated species. Foundation species from temperate communities are best represented, which, in contrast to terrestrial systems where foundation species are predominantly plants, include filter-feeding animals (e.g. bivalves) in addition to primary producers (seagrasses and seaweeds). There have also been a number of studies conducted in the epibenthic 'fouling' community, a highly tractable system of primarily sessile species including tunicates (sea squirts) and bryozoans (moss animals) that colonize submerged substrata such as docks, boat hulls and natural rocky reefs. Well-studied consumer species include several snails and crustaceans such as the green crab, *Carcinus maenas*. Here, the term 'invasive' is used synonymous with 'non-native', although the majority of non-native species used as examples have had demonstrated impacts on native species and communities. Native species undergoing range expansions are not considered invasive, although there may be cases where expanders and invaders are difficult to differentiate (such as for pelagic species) or where a 'new' invader was previously overlooked (see Nehring, 1998).

Observed changes in invasive species prevalence

The earliest and most numerous studies to invoke climate change as a cause of increases in invasive species have been descriptions and time series of observations. Such studies report on shifts in abundance coincident with a change in climatic conditions along a temporal or spatial axis, including an increase in invasive species as a particular location warms over time or across a geographic gradient of increasing temperature. Invasive species' responses observed concurrent with ocean warming are best represented. These have included increases in a suite of invasive epibenthic species in shallow subtidal communities on the north-east coast of the USA (Harris and Tyrrell, 2001), several species of non-native phytoplankton, zooplankton and fishes in the North Sea (Nehring, 1998; Beare *et al.*, 2004; Johns *et al.*, 2005), the proportion of alien species in the zooplankton and benthos of estuaries in Finland (Ojaveer *et al.*, 2011), the abundance and range of an introduced cordgrass in northern Europe (Loebl *et al.*, 2006; Nehring and Hesse, 2008) and a green alga invading the Canary Islands from across the Atlantic (Sangil *et al.*, 2012). Observing abundance changes across space and time, Carlton (2000) identified nine invasive species across five invertebrate phyla whose distributions had shifted northward concurrent with an increase in water temperature along the North American west coast.

However, Carlton (2000) also pointed out the difficulty of ascribing causation from observational studies, given that multiple factors may vary concurrently, limiting conclusions about the increase in introduction frequency or intensity that has occurred at the same time as accelerating climate change (Ruiz *et al.*, 1997; Wonham and Carlton, 2005). Thus, for example, although it is tempting to attribute the tenfold increase of invasive species cover in a California epibenthic community to local warming of ~1°C over the past 40 years, several of the most dominant invaders have only been introduced in the past 20 years (Sorte and Stachowicz, 2011). In order to resolve the relationship between climate change and invasions, researchers have partly circumvented these limitations of observational data in two ways.

First, an interaction between climate change and invasions is best supported when changes in the two are tightly linked in time, such as when abrupt shifts in invasive species abundances are associated with pronounced changes in climate. Diez *et al.* (2012) reviewed the effects of extreme climatic events on invasive species across systems, and invasives were favoured over natives in all marine studies included. For example, heatwave-related mortality was up to ten times higher for the invasive Mediterranean mussel, *Mytilus galloprovincialis*, than a native New Zealand mussel (Petes *et al.*, 2007). In addition, disturbance events associated with water flow have caused disproportionate mortality of native species, favouring *M. galloprovincialis* (Erlandsson *et al.*, 2006), the Asian clam (Nichols *et al.*, 1990) and invasive seagrass species (Steiner *et al.*, 2010). Temperature regime shifts have, in general, also been related to a 150% increase in invasions within the eastern Mediterranean (Raitsos *et al.*, 2010; Pancucci-Papadopoulou *et al.*, 2012; also see Galil, 2008).

Second, researchers have endeavoured to demonstrate the causal relationships between invasive species' abundances and environmental conditions across non-linear time series, such as when warmer and cooler years are interspersed over time. Using interspersed time-series data, Stachowicz *et al.* (2002) showed that invasive epibenthic species recruited earlier and in higher numbers during years of warmer winter temperatures in Connecticut, USA. Positive relationships between temperature and invasive species recruitment were subsequently demonstrated among epibenthic species in California, USA (Sorte and Stachowicz, 2011), and Nova Scotia, Canada (Saunders and Metaxas, 2007), as well as for the cordgrass, *Spartina alterniflora*, in Washington, USA (Buhle *et al.*, 2012). Similarly, Minchinton (2002a) found that the growth and fecundity of the common reed, *Phragmites australis*, increased during a year of high precipitation (leading to decreased salinity stress) relative to the low precipitation years before and after. Finally, capitalizing on variation in pH across a portion of the Mediterranean containing subtidal CO_2 vents, Hall-Spencer *et al.* (2008) found that invasive seaweeds were among the species resistant to (i.e. for which abundances did not decline in the face of) high CO_2 and low pH.

Incorporating variation across both space and time, Braby and Somero (2006a) related abundance of the invasive Mediterranean mussel and a native congener to temperature and salinity conditions across years and sites in California, USA. They found that abundance of the invasive mussel was related positively to salinity and related negatively to temperature, whereas the opposite patterns held for the native species. Their results, however, are surprising in light of surveys finding the invader to be more prevalent in warmer microhabitats (Schneider and Helmuth, 2007), as well as the evidence for high thermal tolerance discussed below. Given that temperature and salinity covary negatively across sites, it seems that the abundance of this invasive mussel is tied more strongly to salinity than to temperature. This example highlights the possibility that warming – although the best-studied environmental change – is not always the most important driver of biological alterations.

Overall, results from observational studies conform to expectations and largely support a facilitative effect of climate change on species invasions. However, there are exceptions to this general pattern, and observations including multiple species also highlight the idiosyncrasy of individual invaders' responses, including across climate drivers within a single species. In a survey of epibenthic species in estuaries of Australia, abundance patterns of only one of six invasive species were related to environmental variables in a manner suggesting an increase with climate change, whereas three species declined in warmer or acidic conditions (Dafforn *et al.*, 2009). There are several other reports of invasive species that have not increased (e.g. Ojaveer *et al.*, 2011) and native species that have increased as much as, or more than, non-natives (e.g. Sorte and Stachowicz, 2011; Sangil *et al.*, 2012) under climate change. To better understand the drivers behind the observed responses and to inform attempts to forecast future ecological changes, researchers have turned increasingly to manipulative studies to elucidate the mechanisms driving the relationship between changing climatic conditions and invasive species' abundances.

Single species studies in the laboratory and in the field

The results of controlled laboratory experiments provide more rigorous support for hypotheses linking invasive species' responses to changing environmental conditions. Importantly, for informing attempts to forecast future ecological changes, they have also allowed researchers to test the effects of likely future conditions across a broader range of conditions than are observed – at least predictably – in present-day climates. Using short-term laboratory experiments, several studies have demonstrated the higher performance of recent invaders than co-occurring native species under altered conditions, including survival and growth under osmotic stress (i.e. reduced salinity) for the tunicate, *Didemnum vexillum*, in the UK (Gröner *et al.*, 2011), survival under hypoxia and increased temperature for a gammarid amphipod in Finland (Sareyka *et al.*, 2011) and prey capture under increased temperature for the green crab, *C. maenas*, in Newfoundland, Canada (Matheson and Gagnon, 2012). Some of these studies have included a number and range of treatments sufficient to assess invasive species response curves. Such studies indicate the range of conditions that are optimal (where survival is 100%), stressful (survival <100%) and lethal (0% survival), as well as the rate of decline in survival across a change in conditions. For example, lethal temperature response curves indicate that in a California epibenthic community, declines in survival will be relatively abrupt for Didemnid tunicates relative to, for example, two co-occurring bryozoans (Zerebecki and Sorte, 2011), and populations of these species on the east and west coasts differ not only in their thermal limits but also in their likely rate of response to climate warming (Sorte *et al.*, 2011).

Beyond lethal tolerance limits and demographic rates of survival, growth and fecundity, responses of interest have also included subcellular modifications that, themselves, underlie patterns of organismal physiology (see Somero, 2012). To illustrate this reductionist approach, I work 'backwards' through processes by which climate change could impact the invasive tunicate, *Diplosoma listerianum*, a species which has increased in abundance in the epibenthic community of Bodega Harbor, California, concurrent with ocean warming (Sorte and Stachowicz, 2011). A link between local temperature increase and this species' population growth is supported by the results from field observations, indicating a positive correlation between recruitment rates and temperature (Sorte and Stachowicz, 2011). Furthermore, laboratory experiments showed increased growth and no change in mortality at increased temperature (Sorte *et al.*, 2010b), and the species has a 24-h lethal temperature of 27.9°C (Zerebecki and Sorte, 2011), which is almost 10°C higher than the highest 24-h temperature recorded locally over the previous 5-year period (Sorte, 2010). At the molecular level, Hsp70, a molecular chaperone responsible for rescuing cellular proteins that have started to denature under stress, was present at higher levels in the cells of this species than in a related native species that was less thermotolerant, potentially indicating a subcellular mechanism of enhanced tolerance in this invasive species (Zerebecki and Sorte, 2011).

The physiological mechanisms of responses to climatic change have been particularly well studied in the invasive and native species of *Mytilus* mussels, to the point that they have already been the subject of their own review (Lockwood and Somero, 2011; also see Somero, 2012). The two species are very similar in habitat, function and morphology, the latter so much so that they must be identified via genetic methods. The invasion of the Mediterranean mussel appears to have been facilitated by increased temperatures (Hilbish *et al.*, 2012), a relationship that is also supported by an observed range contraction when the Pacific Decadal Oscillation shifted to a cold-phase period (Hilbish *et al.*, 2010). This invasive species is more temperature tolerant than the related native species in all comparisons (Lockwood and Somero, 2011), including of survival (Schneider, 2008) and cardiac function (Braby and Somero, 2006b), as well as stress protein production and gene expression (Lockwood *et al.*, 2010; Fields *et al.*, 2012). Given their consistency across experiments, these findings of high physiological tolerance to environmental variation clearly help to explain the increasing success of the Mediterranean mussel as global temperatures increase. Across studies, there is support for the tolerance hypothesis, although whether invasive species have broader tolerances or tolerances shifted in the direction of future climates is more rarely known.

Studies incorporating community and ecosystem context

Compared to the number of correlative observations and single-species experiments describing climate change impacts on invasive species, manipulative studies at the community level are exceedingly rare. This follows the overall trend of climate-related publications focused on marine systems, in which only 20% of studies have considered biological responses at the community level or higher (Harley *et al.*, 2006). The explanation for this trend may be largely logistical, as species vary in how amenable they are to experimental manipulation. However, the results available to date indicate that realistic predictions for species abundances and community composition are very sensitive to the effects of species interactions. When growth rates were compared for invasive epibenthic species maintained in the laboratory as either species monocultures or as part of whole communities, the direction of warming effects switched from positive to negative for some species, and the outcomes of simulation models differed greatly when interactions were included in the parameterization (Sorte and White, 2013).

Although it is challenging to replicate realistic climatic changes in the field, researchers have had some success increasing temperatures with passive warming chambers in intertidal habitats (Charles and Dukes, 2009; Gedan and Bertness, 2009) and heated substrata in shallow subtidal systems (Smale *et al.*, 2011; Smale and Wernberg, 2012), although invasive species have only been included in studies using the latter technique. The most common species on heated settlement panels was the white crust tunicate, a recent invader in Australian waters, which tended to be over twice as abundant on heated than on unheated panels (Smale *et al.*, 2011). A subsequent experiment found a significant increase in this invasive colonial tunicate on heated panels, whereas an invasive bryozoan was more abundant on unheated tiles, and an invasive solitary tunicate showed contrasting responses between two sites (Smale and Wernberg, 2012). Finally, Incera *et al.* (2010) tested the effects of disturbance intensity and variability (as would be associated with changes in storm events) on the invasive seaweed, *Caulerpa racemosa*, by manipulating the presence of cobbles in tide pools, and they found that under high disturbance, *C. racemosa* fronds were smaller and less dense, whereas the number and cover of native species was increased. At present, manipulative experiments are so rare as to prevent specific conclusions, but the results to date indicate that community interactions and ecosystem context are likely to significantly alter predictions of invader responses to climate change.

Integrative Forecasts of Invader Impacts Under Climate Change

Forecasting the impacts of invasive species under climate change requires integrating knowledge gained from both field observations and manipulative experiments designed to test the mechanistic underpinnings of invaders' responses. Parker *et al.* (1999) defined impact as a function of local abundance or population size, geographic range and per capita effect on native species. Thus, any climatic change that drives invasive species to increase in abundance, expand their ranges or impact their native compatriots more strongly would be predicted to increase their overall impacts. Although just a few years ago there were 'few good predictions of which invasive species will have greater effects under climate change' in any system (Hellmann *et al.*, 2008), several recent studies have attempted to integrate emerging results into specific predictions for a number of invasive species.

Will abundances increase?

Increases in future population sizes at a single site are likely when climatic changes are related to increases in invasive species' demographic rates, such as fecundity or recruitment (e.g. Stachowicz *et al.*, 2002; Saunders and Metaxas, 2007; Sorte and Stachowicz, 2011; Buhle *et al.*, 2012), survival (e.g. Gröner *et al.*, 2011; Sareyka *et al.*, 2011) and growth (e.g. Minchinton, 2002a; Stachowicz *et al.*, 2002; Sorte *et al.*, 2010b). Fully parameterized population models are even more reliable because results from unmeasured responses could be contradictory, with, for example, negative effects on survival balancing positive effects on growth (e.g. Sorte *et al.*, 2010b). Saunders *et al.* (2010) parameterized an individual-based population model for an invasive bryozoan using empirical estimates of temperature effects on recruitment and growth rates. Their model results, which were relatively consistent with historical, sampled values at ambient temperatures, projected 9- and 62-fold increases in per cent cover upon warming of 1°C and 2°C, respectively. Similarly, Cockrell and Sorte (2013) developed stage-based population models for three invasive species at ambient and increased temperatures, which involved field surveys and laboratory mesocosm experiments to quantify the temperature dependence of seven transitions between three life stages. Likely increases in population growth rate were detected for the tunicate, *Botryllus schlosseri*, but not an invasive bryozoan or the tunicate, *Botrylloides violaceus*. These findings for *B. violaceus* were consistent with those of Grey (2011), who projected the population growth rate of this species based on demographic rates measured across four field sites and did not find a direct effect of temperature. Thus, species likely to increase in abundance with future climate change are those for which environmental changes have been found consistently (i.e. across multiple responses) to drive increases in historical population sizes and/or in demographic rates.

Furthermore, until now, I have focused primarily on the direct effects of changing climate on invasive species; however, climate-driven decreases in native species can also lead to the indirect facilitation of invasive species. Given that resource availability is related directly to invasion success (Davis *et al.*, 2000), climatic conditions that increase the mortality of native competitors could also lead to an increase in invasive species. For example, declines in particularly a native seagrass species have been implicated in allowing the spread of the invasive seaweed, *Caulerpa taxifolia*, in the Mediterranean (Ceccherelli and Cinelli, 1999; Occhipinti-Ambrogi, 2003). Similarly, the mortality of a native cordgrass initially increases wrack (i.e. dead plant material), which facilitates invasion spread by smothering marsh turf and increasing colonization by the invasive *P. australis* (Minchinton, 2002b). There are, thus, both direct and indirect pathways by which climate change should increase the abundance and, by extension, impacts of many invasive species.

Will ranges expand?

Whether or not climate change will drive the continued spread of an invasive species depends on whether that species has already expanded to fill its climate space and, if not, whether there is another, non-climate factor limiting its distribution. For example, based on lethal temperature and salinity tolerances, an invasive caprellid amphipod does not inhabit its full climate space (Ashton *et al.*, 2007); thus, it should continue to spread unless it is currently limited by non-climate factors. Similarly, the ability of the Japanese eelgrass to sustain growth at temperatures beyond the limits of its current habitat has been invoked to infer the potential for range expansions (Shafer *et al.*, 2008). Range projections have been made based on a single, or a few, threshold value(s), such as limits to survival, germination, photosynthesis, growth or reproduction (e.g. Carlton and Cohen, 2003; Loebl *et al.*, 2006), and overwhelmingly have predicted future range expansions. In addition, although species distribution models have been underutilized in marine studies (Robinson *et al.*, 2011), increases in future ranges have also been predicted for four invasive marine invertebrates using a climate envelope technique based on the species' current ranges (de Rivera *et al.*, 2011). However, the ranges of many of these invaders are not currently in equilibrium with (i.e. extending to the limits of) habitat climatic conditions, limiting the applicability of species distribution models (Jones *et al.*, 2010), and forecasted range expansions would require that the species have either the ability or the opportunity to spread under their own power or via transport vectors.

There are a few species for which current conditions have been demonstrated as setting contemporary range boundaries, suggesting that changing conditions could facilitate range shifts. For example, whereas cold-water temperatures appear to limit the population growth – and thus range expansion – at the poleward-range boundary of the invasive slipper limpet in northern Europe (Thieltges *et al.*, 2004), ocean warming is linked to increased reproductive success of this species and could facilitate its expansion (Valdizan *et al.*, 2011). Initially, as global mean temperatures increase, a concurrent increase in the frequency of extreme cold events could continue to limit invasive species' abundances. Severe population declines and range contractions associated with episodic or periodic cooling have been observed for the Asian green mussel (Firth *et al.*, 2011; also see Urian *et al.*, 2011) and an invasive porcelain crab (Canning-Clode *et al.*, 2011) in the south-eastern USA and for the Mediterranean mussel (Hilbish *et al.*, 2010) and green crab (Behrens Yamada and Kosro, 2010) along the US west coast. However, it is also clear that there are invasive species essentially 'lying in wait' for climatic conditions to change and, from the physiological perspective, improve (e.g. Diederich *et al.*, 2005; Witte *et al.*, 2010) in order to allow population growth and continued spread.

Will per capita effects on native species be strengthened?

Per capita effects on native species have been assessed for a small minority of marine invaders and, for these, the effects on native species tend to be negative (Sorte *et al.*, 2010a). A global review of introduced seaweeds, for example, found that impacts had been studied for only 17 of 277 species, and negative effects (found in almost 50% of studies) were three times as common as positive effects (Williams and Smith, 2007; also see Thomsen *et al.*, 2009). Even fewer studies have detected alterations in per capita effects under changing climatic conditions. Temperature-dependent predation has been demonstrated for an invasive gammarid amphipod in Poland (Van der Velde, 2009) and the Indo-Pacific lionfish in the Caribbean (Côté and Green, 2012), although for the latter, this per capita effect was predicted to be minor in comparison to the density-dependent effects of continuing local population growth. Höffle *et al.* (2012) conducted a factorial manipulation of temperature and the presence of the invasive mud snail, and found that the impacts on a native seagrass were additive, not synergistic. Thus, the evidence for climate-related changes in per capita effects remains equivocal, and whether the lack of a general pattern is driven by low data availability will be unclear until more impacts studies are undertaken.

Conclusions and Route Forward

The literature review above provides several lines of evidence that climate change has increased and will continue to increase the impacts of many invasive species in marine systems. However, only a handful of the total marine invaders have been studied, and this handful likely contains primarily species that are the most visible, easiest to observe and manipulate, and potentially also most impactful. Thus, there is still much research needed to evaluate whether – and when – responses to climate change will be more often and more strongly positive for invasive than for native species.

The clearest gaps in our current knowledge are related to (i) the mechanisms underlying species' seemingly idiosyncratic responses to climate change and (ii) the complexity inherent in forecasting biological changes. Mechanistically, we have made some progress in comparing physiological tolerances between invasive and native species, although a better understanding would require ecophysiological studies spanning – and reporting – species' complete tolerance ranges. Hypotheses regarding resource use and related life-history characteristics should be increasingly testable as marine biologists shift focus from the impacts of solely climate warming to other changes associated directly with resource availability such as ocean acidification. Furthermore, as invasive–native comparisons accrue in the literature, there will be more adequate studies for comparing life-history traits as van Kleunen *et al.* (2010) have done for plants.

An ultimate goal of understanding the processes underlying the responses to climate change is to inform attempts to predict future species abundances and community structure. This is a daunting prospect, particularly given that responses will integrate across multiple climatic factors, demographic responses and interacting species. To identify climatic drivers eliciting the strongest responses, as well as interactive effects between these drivers, marine biologists, again, need to increase efforts focused on non-temperature factors such as ocean pH, hypoxia and salinity. Researchers should continue to embrace complexity by conducting studies incorporating multiple demographic responses, life stages and species – both in seclusion and in a community context. Targeted studies should also test the degree to which climate change would be expected to cause population growth and the spread of already problematic, noxious species versus species that are currently relatively restricted (and perhaps have not advanced beyond a lag phase; Witte *et al.*, 2010). There may also be cases (e.g. see Sorte *et al.*, 2010b) in which climate change will be disproportionately detrimental to invasive species, which presents intriguing opportunities for restoration (Bradley and Wilcove, 2009). Overall, there remains a paucity of work documenting the interaction between climate change and invasions in marine systems, and an increase in studies across species and locations will allow us to draw broader conclusions about the cases in which invaders are likely to be favoured in future climates.

In conclusion, although invasive species have long been known as threats to intact marine systems (e.g. Thompson *et al.*, 2002), concern about their potential impacts has recently been overshadowed by climate change (Williams and Grosholz, 2008). Invasions deserve continued and increased consideration, given building evidence that invader abundances, ranges, and thus impacts, are likely to increase under changing climate conditions. The indirect effects of climate change on invasive species therefore have the potential to deliver a 'double whammy' to native species already experiencing direct negative effects of climate change.

Acknowledgements

I thank M. Bracken and B. Bradley for comments and suggestions and L.H. Ziska and J.S. Dukes for the opportunity to contribute to this project.

References

Ashton, G.V., Willis, K.J., Burrows, M.T. and Cook, E.J. (2007) Environmental tolerance of *Caprella mutica*: implications for its distribution as a marine non-native species. *Marine Environmental Research* 64, 305–312.

Barry, J.P., Baxter, C.H., Sagarin, R.D. and Gilman, S.E. (1995) Climate-related, long-term faunal changes in a California rocky intertidal community. *Science* 267, 672–675.

Beare, D.J., Burns, F., Greig, A., Jones, E.G., Peach, K., Kienzle, M., McKenzie, E. and Reid, D.G. (2004) Long-term increases in prevalence of North Sea fishes having southern biogeographic affinities. *Marine Ecology Progress Series* 284, 269–278.

Behrens Yamada, S. and Kosro, P.M. (2010) Linking ocean conditions to year class strength of the invasive European green crab, *Carcinus maenas*. *Biological Invasions* 12, 1791–1804.

Braby, C.E. and Somero, G.N. (2006a) Ecological gradients and relative abundance of native (*Mytilus trossulus*) and invasive (*Mytilus galloprovincialis*) blue mussels in the California hybrid zone. *Marine Biology* 148, 1249–1262.

Braby, C.E. and Somero, G.N. (2006b) Following the heart: temperature and salinity effects on heart rate in native and invasive species of blue mussels (genus *Mytilus*). *Journal of Experimental Biology* 209, 2554–2566.

Bradley, B.A. and Wilcove, D.S. (2009) When invasive plants disappear: transformative restoration possibilities in the Western United States resulting from climate change. *Restoration Ecology* 17, 715–721.

Brook, B.W., Sodhi, N.S. and Bradshaw, C.J.A. (2008) Synergies among extinction drivers under global change. *Trends in Ecology and Evolution* 23, 453–560.

Buhle, E.R., Feist, B.E. and Hilborn, R. (2012) Population dynamics and control of invasive *Spartina alterniflora*: inference and forecasting under uncertainty. *Ecological Applications* 22, 880–893.

Burrows, M.T., *et al.* (2011) The pace of shifting climate in marine and terrestrial ecosystems. *Science* 334, 652–655.

Canning-Clode, J., Fowler, A.E., Byers, J.E., Carlton, J.T. and Ruiz, G.M. (2011) '*Caribbean Creep*' chills out: climate change and marine invasive species. *PLoS ONE* 6, e29657.

Carlton, J.T. (2000) Global change and biological invasions in the oceans. In: Mooney, H.A. and Hobbs, R.J. (eds) *Invasive Species in a Changing World.* Island Press, Washington, DC, and Covelo, California, pp. 31–53.

Carlton, J.T. and Cohen, A.N. (2003) Episodic global dispersal in shallow water marine organisms: the case history of the European shore crabs *Carcinus maenas* and *C. aestuarii*. *Journal of Biogeography* 30, 1809–1820.

Ceccherelli, G. and Cinelli, F. (1999) Effects of *Posidonia oceanica* canopy on *Caulerpa taxifolia* size in a north-western Mediterranean Bay. *Journal of Experimental Marine Biology and Ecology* 240, 19–36.

Charles, H. and Dukes, J.S. (2009) Effects of warming and altered precipitation on plant and nutrient dynamics of a New England salt marsh. *Ecology* 19, 1758–1773.

Cockrell, M.L. and Sorte, C.J.B. (2013) Predicting climate-induced changes in population dynamics of invasive species in a marine epibenthic community. *Journal of Experimental Marine Biology and Ecology* 440, 42–48.

Cohen, A.N. and Carlton, J.T. (1998) Accelerating invasion rate in a highly invaded estuary. *Science* 279, 555–558.

Côté, I.M. and Green, S.J. (2012) Potential effects of climate change on a marine invasion: the importance of current context. *Current Zoology* 58, 1–8.

Dafforn, K.A., Glasby, T.M. and Johnston, E.L. (2009) Links between estuarine condition and spatial distributions of marine invaders. *Diversity and Distributions* 15, 807–821.

Davis, M.A., Grime, J.P. and Thompson, K. (2000) Fluctuating resources in plant communities: a general theory of invasibility. *Journal of Ecology* 88, 528–534.

Diederich, S., Nehls, G., van Beusekom, J.E.E. and Reise, K. (2005) Introduced Pacific oysters (*Crassostrea gigas*) in the northern Wadden Sea: invasion accelerated by warm summers? *Helgoland Marine Research* 59, 97–106.

Diez, J.M., *et al.* (2012) Will extreme climatic events facilitate biological invasions? *Frontiers in Ecology and the Environment* 10, 249–257.

Doney, S.C., *et al.* (2012) Climate change impacts on marine ecosystems. *Annual Review of Marine Science* 4, 4.1–4.27.

Dukes, J.S. and Mooney, H.A. (1999) Does global change increase the success of biological invaders? *Trends in Ecology and Evolution* 14, 135–139.

Erlandsson, J., Pal, P. and McQuaid, C.D. (2006) Re-colonisation rate differs between co-existing indigenous and invasive intertidal mussels following major disturbance. *Marine Ecology Progress Series* 320, 169–176.

Fields, P.A., Zuzow, M.J. and Tomanek, L. (2012) Proteomic responses of blue mussel (*Mytilus*) congeners to temperature acclimation. *Journal of Experimental Biology* 215, 1106–1116.

Firth, L.B., Knights, A.M. and Bell, S.S. (2011) Air temperature and winter mortality: implications for the persistence of the invasive mussel, *Perna viridis* in the intertidal zone of the south-eastern United States. *Journal of Experimental Marine Biology and Ecology* 400, 250–256.

Galil, B.S. (2008) Alien species in the Mediterranean Sea – which, when, where, why? *Hydrobiologia* 606, 105–116.

Gedan, K.B. and Bertness, M.D. (2009) Experimental warming causes rapid loss of plant diversity in New England salt marshes. *Ecology Letters* 12, 842–848.

Grey, E.K. (2011) Relative effects of environment and direct species interactions on the population growth rate of an exotic ascidian. *Oecologia* 166, 935–947.

Gröner, F., Lenz, M., Wahl, M. and Jenkins, S.R. (2011) Stress resistance is two colonial ascidians from the Irish Sea: the recent invader *Didemnum vexillum* is more tolerant to low salinity than the cosmopolitan *Diplosoma listerianum*. *Journal of Experimental Marine Biology and Ecology* 409, 48–52.

Hall-Spencer, J.M., Rodolfo-Metalpa, R., Martin, S., Ransome, E., Fine, M., Turner, S.M., Rowley, S.J., Tedesco, D. and Buia, M.-C. (2008) Volcanic carbon dioxide vents show ecosystem effects of ocean acidification. *Nature* 454, 96–99.

Harley, C.D.G., Hughes, A.R., Hultgren, K., Miner, B.G., Sorte, C.J.B., Thornber, C.S., Rodriguez, L.F., Tomanek, L. and Williams, S.L. (2006) The impacts of climate change in coastal marine systems. *Ecology Letters* 9, 228–241.

Harris, L.G. and Tyrrell, M.C. (2001) Changing community states in the Gulf of Maine: synergism between invaders, overfishing and climate change. *Biological Invasions* 3, 9–21.

Hellmann, J.J., Byers, J.E., Bierwagen, B.G. and Dukes, J.S. (2008) Five potential consequences of climate change for invasive species. *Conservation Biology* 22, 534–543.

Helmuth, B., Mieszkowska, N., Moore, P. and Hawkins, S.J. (2006) Living on the edge of two changing worlds: forecasting the responses of rocky intertidal ecosystems to climate change. *Annual Review of Ecology, Evolution, and Systematics* 37, 373–404.

Hilbish, T.J., Brannock, P.M., Jones, K.R., Smith, A.B., Bullock, B.N. and Wethey, D.S. (2010) Historical changes in the distributions of invasive and endemic marine invertebrates are contrary to global warming predictions: the effects of decadal climate oscillations. *Journal of Biogeography* 37, 423–431.

Hilbish, T.J., Lima, F.P., Brannock, P.M., Fly, E.K., Rognstad, R.L. and Wethey, D.S. (2012) Change and stasis in marine hybrid zones in response to climate warming. *Journal of Biogeography* 39, 676–687.

Höffle, H., Wernberg, T., Thomsen, M.S. and Holmer, M. (2012) Drift algae, an invasive snail and elevated temperature reduce ecological performance of a warm-temperate seagrass, through additive effects. *Marine Ecology Progress Series* 450, 67–80.

Incera, M., Bertocci, I. and Benedetti-Cecchi, L. (2010) Effects of mean intensity and temporal variability of disturbance on the invasion of *Caulerpa racemosa* var. *cylindracea* (Caulerpales) in rock pools. *Biological Invasions* 12, 501–514.

Johns, D.G., Edwards, M., Greve, W. and John, A.W.G. (2005) Increasing prevalence of the marine cladoceran *Penilia avirostris* (Dana, 1852) in the North Sea. *Helgoland Marine Research* 59, 214–218.

Jones, C.C., Acker, S.A. and Halpern, C.B. (2010) Combining local- and large-scale models to predict the distributions of invasive plant species. *Ecological Applications* 20, 311–326.

Keeling, R.F., Körtzinger, A. and Gruber, N. (2010) Ocean deoxygenation in a warming world. *Annual Review of Marine Science* 2, 199–229.

Kolar, C.S. and Lodge, D.M. (2001) Progress in invasion biology: predicting invaders. *Trends in Ecology and Evolution* 16, 199–204.

Levitus, S., Antonov, J. and Boyer, T. (2005) Warming of the world ocean, 1955–2003. *Geophysical Research Letters* 32, L02604, doi:10.1029/2004GL021592.

Lockwood, B.L. and Somero, G.N. (2011) Invasive and native blue mussels (genus *Mytilus*) on the California coast: the role of physiology in a biological invasion. *Journal of Experimental Marine Biology and Ecology* 400, 167–174.

Lockwood, B.L., Sanders, J.G. and Somero, G.N. (2010) Transcriptomic responses to heat stress in invasive and native blue mussels (genus *Mytilus*): molecular correlates of invasive success. *Journal of Experimental Biology* 213, 3548–3558.

Loebl, M., van Beusekom, J.E.E. and Reise, K. (2006) Is spread of the neophyte *Spartina anglica* recently enhanced by increasing temperatures? *Aquatic Ecology* 40, 315–324.

Matheson, K. and Gagnon, P. (2012) Effects of temperature, body size, and chela loss on competition for a limited food resource between indigenous rock crab (*Cancer irroratus* Say) and recently introduced green crab (*Carcinus maenas* L.). *Journal of Marine Biology and Ecology* 428, 49–56.

Meehl, G.A., *et al.* (eds) (2007) *Climate Change 2007: The Physical Science Basis. Contribution of Working Group I to the Fourth Assessment Report of the Intergovernmental Panel on Climate Change* Cambridge University Press, Cambridge, UK, and New York, pp. 747–845.

Minchinton, T.E. (2002a) Precipitation during El Niño correlates with increasing spread of *Phragmites australis* in New England, USA, coastal marshes. *Marine Ecology Progress Series* 242, 305–309.

Minchinton, T.E. (2002b) Disturbance by wrack facilitates spread of *Phragmites australis* in a coastal marsh. *Journal of Experimental Marine Biology and Ecology* 281, 89–107.

Monaco, C.J. and Helmuth, B. (2011) Tipping points, thresholds and the keystone role of physiology in marine climate change research. *Advances in Marine Biology* 60, 123–160.

Nehring, S. (1998) Establishment of thermophilic phytoplankton species in the North Sea: biological indicators of climatic changes? *ICES Journal of Marine Science* 55, 818–823.

Nehring, S. and Hesse, K.J. (2008) Invasive alien plants in marine protected areas: the *Spartina anglica* affair in the European Wadden Sea. *Biological Invasions* 10, 937–950.

Nichols, F.H., Thompson, J.K. and Schemel, L.E. (1990) Remarkable invasion of San Francisco

Bay (California, USA) by the Asian clam (*Potamocorbula amurensis*). II. Displacement of a former community. *Marine Ecology Progress Series* 66, 95–101.

Occhipinti-Ambrogi, A. (2003) Biological invasions as a component of global change in stressed marine ecosystems. *Marine Pollution Bulletin* 46, 542–551.

Ojaveer, H., Kotta, J., Pollumae, A., Pollupuu, M., Jaanus, A. and Vetemaa, M. (2011) Alien species in a brackish water temperate ecosystem: annual-scale dynamics in response to environmental variability. *Environmental Research* 111, 933–942.

Pancucci-Papadopoulou, M.A., Raitsos, D.E. and Corsini-Foka, M. (2012) Biological invasions and climatic warming: implications for south-eastern Aegean ecosystem functioning. *Journal of the Marine Biological Association of the United Kingdom* 92, 777–789.

Parker, I.M., Simberloff, D., Lonsdale, W.M., Goodell, K., Wonham, M., Kareiva, P.M., Williamson, M.H., Von Holle, B., Moyle, P.B., Byers, J.E. and Goldwasser, L. (1999) Impact: toward a framework for understanding the ecological effects of invaders. *Biological Invasions* 1, 3–19.

Petes, L.E., Menge, B.A. and Murphy, G.D. (2007) Environmental stress decreases survival, growth, and reproduction in New Zealand mussels. *Journal of Experimental Marine Biology and Ecology* 351, 83–91.

Rahel, F.J. and Olden, J.D. (2008) Assessing the effects of climate change on aquatic invasive species. *Conservation Biology* 22, 521–533.

Raitsos, D.E., Beaugrand, G., Georgopoulos, D., Zenetos, A., Pancucci-Papadopoulou, A.M., Theocharis, A. and Papathanassiou, E. (2010) Global climate change amplifies the entry of tropical species into the Eastern Mediterranean Sea. *Limnology and Oceanography* 55, 1478–1484.

Rivera, C.E. de, Steves, B.P., Fofonoff, P.W., Hines, A.H. and Ruiz, G.M. (2011) Potential for high-latitude marine invasions along western North America. *Diversity and Distributions* 17, 1198–1209.

Robinson, L.M., Elith, J., Hobday, A.J., Pearson, R.G., Kendall, B.E., Possingham, H.P. and Richardson, A.J. (2011) Pushing the limits in marine species distribution modeling: lessons from the land present challenges and opportunities. *Global Ecology and Biogeography* 20, 789–802.

Ruiz, G.M., Carlton, J.T., Grosholz, E.D. and Hines, A.H. (1997) Global invasions of marine and estuarine habitats by non-indigenous species: mechanisms, extent, and consequences. *American Zoologist* 37, 621–632.

Sabine, C.L., Feely, R.A., Gruber, N., Key, R.M., Lee, K., Bullister, J.L., Wanninkhof, R., Wong, C.S., Wallace, D.W.R. and Tilbrook, B. (2004) The oceanic sink for anthropogenic CO_2. *Science* 305, 367–371.

Sagarin, R.D., Barry, J.P., Gilman, S.E. and Baxter, C.H. (1999) Climate-related change in an intertidal community over short and long time scales. *Ecological Monographs* 69, 465–490.

Sakai, A.K., *et al.* (2001) The population biology of invasive species. *Annual Review of Ecology and Systematics* 32, 305–332.

Sala, O.E., *et al.* (2000) Global biodiversity scenarios for the year 2100. *Science* 287, 1770–1774.

Sangil, C., Sanson, M., Afonso-Carrillo, J., Herrera, R., Rodriguez, A., Martin-Garcia, L. and Diaz-Villa, T. (2012) Changes in subtidal assemblages in a scenario of warming: proliferations of ephemeral benthic algae in the Canary Islands (eastern Atlantic Ocean). *Marine Environmental Research* 77, 120–128.

Sareyka, J., Kraufvelin, P., Lenz, M., Lindström, M., Tollrian, R. and Wahl, M. (2011) Differences in stress tolerance and brood size between a non-indigenous and an indigenous gammarid in the northern Baltic Sea. *Marine Biology* 158, 2001–2008.

Saunders, M. and Metaxas, A. (2007) Temperature explains settlement patterns of the introduced bryozoan *Membranipora membranacea* in Nova Scotia, Canada. *Marine Ecology Progress Series* 344, 95–106.

Saunders, M.I., Metaxas, A. and Filgueira, R. (2010) Implications of warming temperatures for population outbreaks of a nonindigenous species (*Membranipora membranacea*, Bryozoa) in rocky subtidal ecosystems. *Limnology and Oceanography* 55, 1627–1642.

Schiel, D.R., Steinbeck, J.R. and Foster, M.S. (2004) Ten years of induced ocean warming causes comprehensive changes in marine benthic communities. *Ecology* 85, 1833–1839.

Schneider, K.R. (2008) Heat stress in the intertidal: comparing survival and growth of an invasive and native mussel under a variety of thermal conditions. *Biological Bulletin* 215, 253–264.

Schneider, K.R. and Helmuth, B. (2007) Spatial variability in habitat temperature may drive patterns of selection between an invasive and native mussel. *Marine Ecology Progress Series* 339, 157–167.

Shafer, D.J., Wyllie-Echeverria, S. and Sherman, T.D. (2008) The potential role of climate in the distributions and zonation of the introduced seagrass *Zostera japonica* in North America. *Aquatic Botany* 89, 297–302.

Smale, D.A. and Wernberg, T. (2012) Short-term *in situ* warming influences early development of sessile assemblages. *Marine Ecology Progress Series* 453, 129–136.

Smale, D.A., Wernberg, T., Peck, L.S. and Barnes, D.K.A. (2011) Turning on the heat: ecological response to simulated warming in the sea. *PLoS ONE* 6, e16050.

Somero, G.N. (2012) The physiology of global change: linking patterns to mechanisms. *Annual Review of Marine Science* 4, 39–61.

Sorte, C.J.B. (2010) Climate change, species invasions, and the composition of marine communities. PhD dissertation, University of California, Davis, California.

Sorte, C.J.B. and Stachowicz, J.J. (2011) Patterns and processes of compositional change in a California epibenthic community. *Marine Ecology Progress Series* 435, 63–74.

Sorte, C.J.B. and White, J.W. (2013) Competitive and demographic leverage points of community shifts under climate warming. *Proceedings of the Royal Society B* 280, 20130572.

Sorte, C.J.B., Williams, S.L. and Carlton, J.T. (2010a) Marine range shifts and species introductions: comparative spread rates and community impacts. *Global Ecology and Biogeography* 19, 303–316.

Sorte, C.J.B., Williams, S.L. and Zerebecki, R.A. (2010b) Ocean warming increases threat of invasive species in a marine community. *Ecology* 91, 2198–2204.

Sorte, C.J.B., Jones, S.J. and Miller, L.P. (2011) Geographic variation in temperature tolerance as an indicator of potential population responses to climate change. *Journal of Experimental Marine Biology and Ecology* 400, 209–217.

Sorte, C.J.B., Ibáñez, I., Blumenthal, D.M., Molinari, N.A., Miller, L.P., Grosholz, E.D., Diez, J.M., D'Antonio, C.M., Olden, J.D., Jones, S.J. and Dukes, J.S. (2013) Poised to prosper: a cross-system comparison of climate change impacts on native and non-native species performance. *Ecology Letters* 16, 261–270.

Stachowicz, J.J., Terwin, J.R., Whitlatch, R.B. and Osman, R.W. (2002) Linking climate change and biological invasions: ocean warming facilitates nonindigenous species invasions. *Proceedings of the National Academy of Sciences* 99, 15497–15500.

Steiner, S.C.C., Macfarlane, K.J., Price, L.M. and Willette, D.A. (2010) The distribution of seagrasses in Dominica, Lesser Antilles. *Revista de Biologia Tropical* 58, 89–98.

Sunday, J.M., Bates, A.E. and Dulvy, N.K. (2012) Thermal tolerance and the global redistribution of animals. *Nature Climate Change* doi:10.1038/nclimate1539.

Theoharides, K.A. and Dukes, J.S. (2007) Plant invasion across space and time: factors affecting nonindigenous species success during four stages of invasion. *New Phytologist* 176, 256–273.

Thieltges, D.W., Strasser, M., van Beusekom, J.E.E. and Reise, K. (2004) Too cold to prosper – winter mortality prevents population increase of the introduced American slipper limpet *Crepidula fornicate* in northern Europe. *Journal of Experimental Marine Biology and Ecology* 311, 375–391.

Thompson, R.C., Crowe, T.P. and Hawkins, S.J. (2002) Rocky intertidal communities: past environmental changes, present status and predictions for the next 25 years. *Environmental Conservation* 29, 168–191.

Thomsen, M.S., Wernberg, T., Tuya, F. and Silliman, B.R. (2009) Evidence for impacts of non-indigenous macroalgae: a meta-analysis of experimental field studies. *Journal of Phycology* 45, 812–819.

Urian, A.G., Hatle, J.D. and Gilg, M.R. (2011) Thermal constraints for range expansion of the invasive green mussel, *Perna viridis*, in the southeastern United States. *Journal of Experimental Zoology Part A – Ecological Genetics and Physiology* 315A, 12–21.

Valdizan, A., Beninger, P.G., Decottignies, P., Chantrel, M. and Cognie, B. (2011) Evidence that rising coastal seawater temperatures increase reproductive output of the invasive gastropod *Crepidula fornicata*. *Marine Ecology Progress Series* 438, 153–165.

Van der Velde, G., Leuven, R.S.E.W., Platvoet, D., Bacela, K., Huijbregts, M.A.J., Hendriks, H.W.M. and Kruijt, D. (2009) Environmental and morphological factors influencing predatory behaviour by invasive non-indigenous gammaridean species. *Biological Invasions* 11, 2043–2054.

van Kleunen, M., Weber, E. and Fischer, M. (2010) A meta-analysis of trait differences between invasive and non-invasive plant species. *Ecology Letters* 13, 235–245.

Williams, S.L. and Grosholz, E.D. (2008) The invasive species challenge in estuarine and coastal environments: marrying management and science. *Estuaries and Coasts* 31, 3–20.

Williams, S.L. and Smith, J.E. (2007) A global review of the distribution, taxonomy, and impacts of introduced seaweeds. *Annual Review of Ecology, Evolution, and Systematics* 38, 327–359.

Witte, S., Buschbaum, C., van Beusekom, J.E.E. and Reise, K. (2010) Does climatic warming explain why an introduced barnacle finally takes over after a lag of more than 50 years? *Biological Invasions* 12, 3579–3589.

Wonham, M.J. and Carlton, J.T. (2005) Trends in marine biological invasions at local and regional scales: the Northeast Pacific Ocean as a model system. *Biological Invasions* 7, 369–392.

Zerebecki, R.A. and Sorte, C.J.B. (2011) Temperature tolerance and stress proteins as mechanisms of invasive species success. *PLoS ONE* 6, e14806.

8 Ragweed in Eastern Europe

László Makra,[1] István Matyasovszky[2] and Áron József Deák[3]

[1]Department of Climatology and Landscape Ecology, University of Szeged, Szeged, Hungary; [2]Department of Meteorology, Eötvös Loránd University, Budapest, Hungary; [3]Department of Physical Geography and Geoinformatics, University of Szeged, Szeged, Hungary

Abstract

Ambrosia artemisiifolia, common ragweed, is an invasive plant species whose introduction and spread in Eastern Europe has resulted in enormous environmental and economic losses in agriculture and public health in recent decades. The aim of this chapter is: (i) to provide an overview on the origin and distribution of *Ambrosia* from North America to Europe, with special focus on Eastern Europe, particularly Hungary; and (ii) to identify and quantify those human-related factors on either a regional or global basis that may act to facilitate the spread and pollen production of this plant species. The chapter shows that socio-economic changes, particularly in agriculture following the fall of the Soviet Union, may be factors in contributing to the degree of soil disturbance necessary for ragweed establishment and spread. In addition, a temporal analysis is conducted of ragweed pollen characteristics and local meteorological factors from Szeged, Hungary (located in the biogeographical region of Hungary with the highest recorded ragweed pollen counts). This analysis demonstrates that most of the *A. artemisiifolia* pollen is from local (<100 km) sources on non-rainy days, but that recent anthropogenic changes in surface temperatures have likely lengthened both the growing season and the time of pollen exposure for common ragweed (i.e. delay in the onset of the autumn frost) in this location. Overall, both ongoing changes in land use and anthropogenic changes in climate are likely to continue to drive the spread of this invasive species, with negative consequences, particularly for allergic rhinitis, in Eastern Europe.

Introduction

With the intensification of global trade in the latter half of the 20th century, numerous invasive plant species were introduced to Europe. Among such species, perhaps the most egregious is common ragweed, *A. artemisiifolia*, a plant species with known detrimental effects as an agronomic weed but principally known globally as a source of allergenic pollen.

Allergic disorders represent an important group of chronic diseases globally, and economic costs can run into the billions of dollars (AAAAI, 2006; ASCIA, 2007). Aero-allergen exposure is associated with pollen exposure and the induction of two principal allergic diseases: allergic rhinitis (hay fever) and asthma. For much of the northern hemisphere, there are three distinct plant-based aeroallergen seasons: tree pollen in the spring; grass pollen in the early summer; and weed pollen, including ragweed (*Ambrosia* spp.), in the summer and autumn. Pollen from the genus *Ambrosia* includes *A. artemisiifolia* (short or common ragweed), *Ambrosia trifida* (giant ragweed), *Ambrosia psilostachya* (western ragweed) and *Ambrosia bidentata* (lanceleaf ragweed), and has long been acknowledged to be a significant cause of allergic diseases (Frenz, 2001). In fact, it has been suggested that *Ambrosia* may cause more seasonal allergic rhinitis than all other plants combined (Quest Diagnostics, 2011).

For Eastern Europe in recent decades, there has been a general recognition of the rapid spread of common ragweed, *A. artemisiifolia* (Csontos *et al.*, 2010). Aerobiological and allergological studies show that the pollen map of Europe is changing as a result (Kiss and Béres, 2006; D'Amato *et al.*, 2007), with subsequent effects on public health through increased exposure to bioallergens.

This chapter focuses on common ragweed, *A. artemisiifolia*, as the primary allergenic plant species of *Ambrosia* in Eastern Europe. The goal is to examine the potential linkages between the spread of this species and human activities, particularly those associated with environmental change. To that end, an overview is provided on the origin and distribution of *A. artemisiifolia* to Eastern Europe, particularly Hungary. In addition, meteorological and pollen characteristics associated with the city of Szeged, Hungary, are examined in more detail. This city is located in the geographical centre of the highest known ragweed pollen counts, with pollen counts approximately one order of magnitude higher than the rest of Europe (Makra *et al.*, 2004).

Impact on Public Health

Pollen grains from various plant taxa are frequently implicated in respiratory allergy symptoms. In Eastern Europe, pollen from *A. artemisiifolia* is the most important cause of allergy-associated respiratory complaints (Juhász, 1998). The Hungarian Great Plain, with the city of Szeged at its geographical centre, has among the highest diurnal and annual ragweed pollen counts per m^3 of air worldwide (Fig. 8.1) (Makra *et al.*, 2005). Clinical researchers recognize that this source of *A. artemisiifolia* is associated with the most serious and persistent clinical aspects of allergic rhinitis (Harsányi, 2009). In southern Hungary, *A. artemisiifolia* produces 44.1% of the total annual pollen production (Fig. 8.2) and is identified as the most important aeroallergen taxon in Hungary (Juhász and Juhász, 1997). Although *A. artemisiifolia* pollen production can change significantly with weather (Fig. 8.3), for this region overall, its annual pollen production is the highest of all taxa (Fig. 8.4).

Atmospheric pollen levels run parallel with ragweed distribution. When analysing the airborne pollen composition in Szeged in 1968, no ragweed pollen grains were identified (Simoncsics *et al.*, 1968). However, its concentration has increased dramatically since the early 1990s and has regularly reached values as high as 1000–2000 pollen grains per cubic metre of air on peak days (Makra *et al.*, 2004). Clinically, the threshold value of *Ambrosia* pollen concentration associated with allergic rhinitis for sensitive individuals is 20 pollen grains m^{-3} of air (Jäger, 1998). However, according to the Hungarian National Health Centre at least 30 pollen grains m^{-3} of air result in clinical symptoms (Basky, 2009). According to Juhász (1995), 50 pollen grains m^{-3} of air is deemed the threshold, above which 60–80% of ragweed-sensitive subjects display allergic symptoms. From the 1990s onwards, the southern part of Hungary and the northern part of Serbia are considered to be among those areas that are characterized by the highest pollen release in Europe. According to the available data, the highest daily European *A. artemisiifolia* pollen counts were recorded at Novi Sad in Serbia (3247 pollen grains m^{-3} of air in 2001) (Predrag Radišič, University of Novi Sad, Serbia, personal communication) and at Szeged in Hungary (2003 pollen grains m^{-3} of air in 1991) (Juhász, 1998). The highest annual total *Ambrosia* pollen counts were recorded at Novi Sad in Serbia (20,559 pollen grains m^{-3} of air in 2001) (Predrag Radišič, University of Novi Sad, Serbia, personal communication) and at Szeged in Hungary (17,242 pollen grains m^{-3} of air in 1994) (Juhász, 1998).

Over a 40-year period, the number of patients with registered allergic complaints

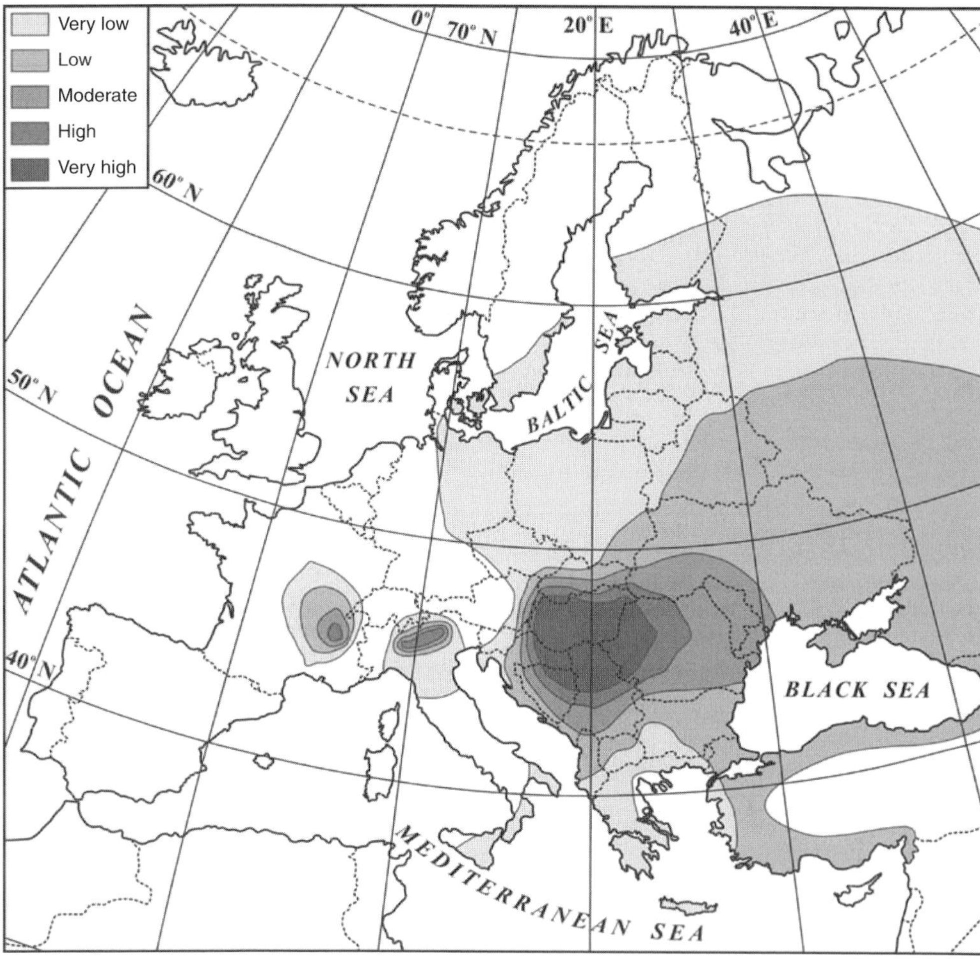

Fig. 8.1. Ragweed pollen distribution for the Szeged location in Europe, last 10 days of August (data from https://ean.polleninfo.eu/ean).

has doubled, while the number of cases of allergic asthma has quadrupled in southern Hungary (Makra *et al.*, 2005). Currently, 20% of the total population of Hungary suffers from pollen allergies, with 60–90% experiencing a specific allergy to *A. artemisiifolia* (Harsányi, 2009).

Impacts on Agriculture

In addition to its role in allergen exposure, *Ambrosia* spp. is recognized as a troublesome agronomic weed. Generally, *A. artemisiifolia* is associated with row crops and is a common weed problem in maize. *A. artemisiifolia* is considered less of a problem in solid and closed crop canopies such as wheat and rape fields.

A. artemisiifolia grows well in sunflower and fields with low crop cover, as well as on disturbed soils and uncovered land surfaces. Damp surfaces with grasses and closed substances provide adverse conditions (Béres *et al.*, 2005). Overall, *A. artemisiifolia* favours sandy or acidic soils, high April and annual precipitation and low May temperatures, but it dislikes soils containing

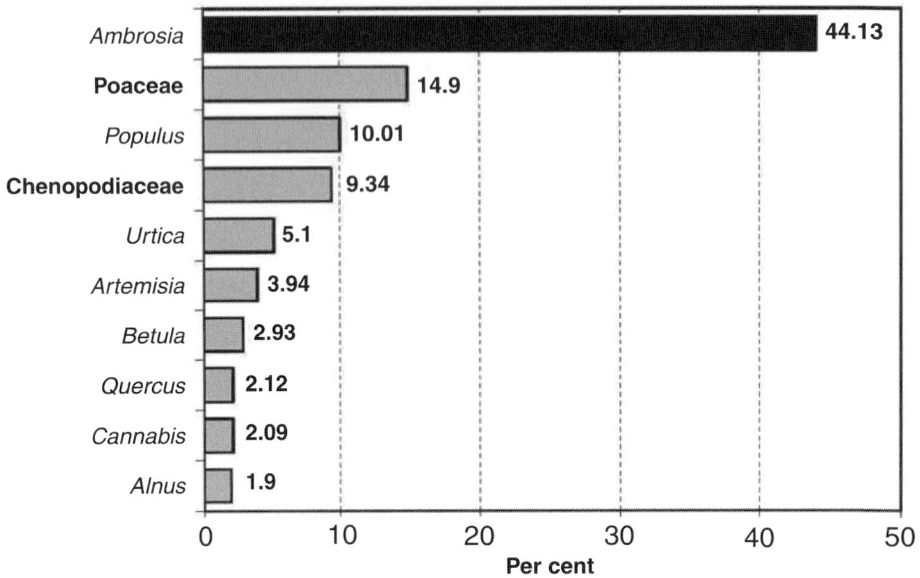

Fig. 8.2. Ratio of individual taxa from the total annual pollen count, per cent, Szeged, 1990–1996 (Juhász and Juhász, 1997).

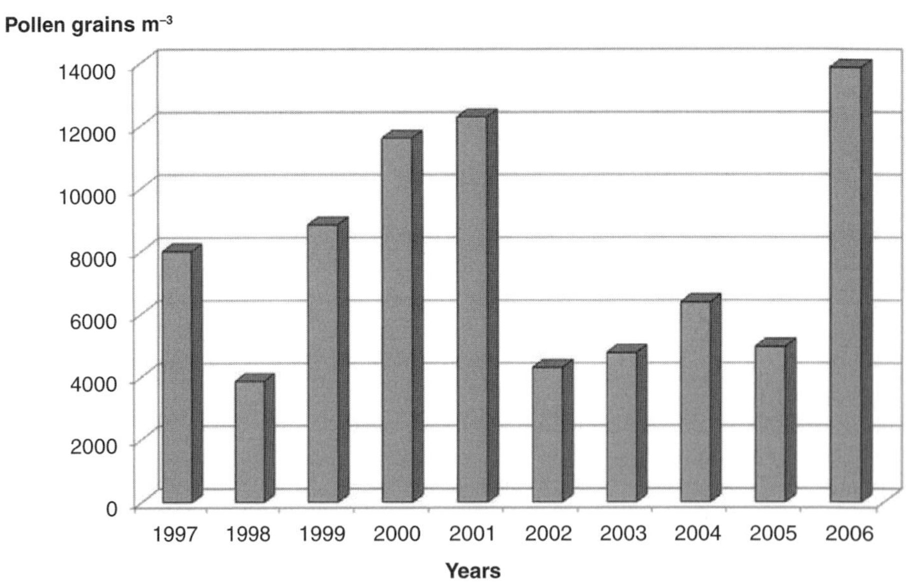

Fig. 8.3. Annual total counts of *Ambrosia* pollen from Szeged, Hungary, 1997–2006.

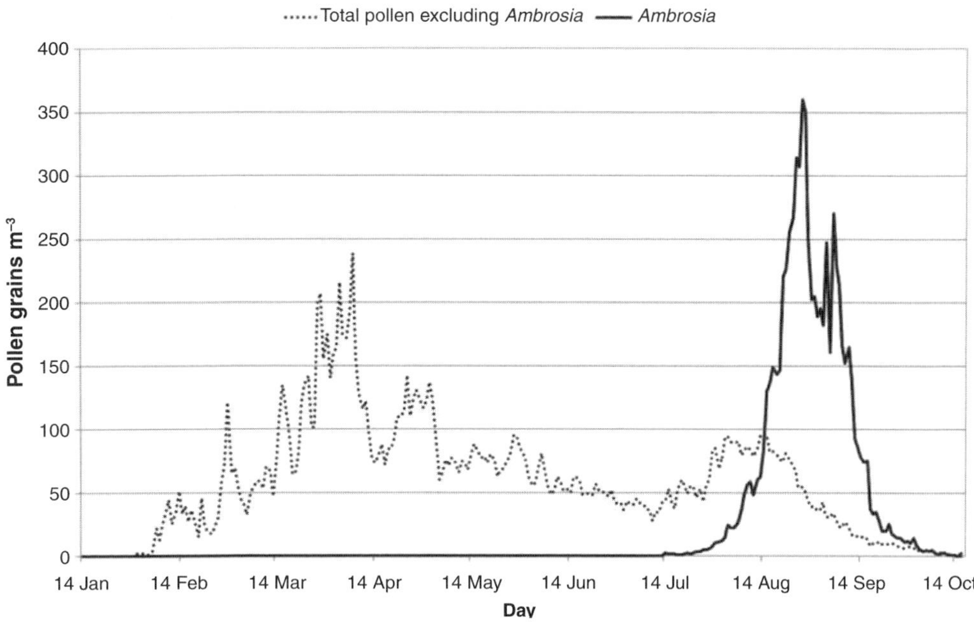

Fig. 8.4. Mean daily pollen concentrations in Szeged, 1 February to 31 October 1997–2006.

high concentrations of Na, K and Mn (Pinke *et al.*, 2011).

Chemical weed control occurs by a combination of herbicides that eradicate monocotyledonous and dicotyledonous plants. However, *Ambrosia* is less sensitive to herbicides than other weeds. Because of its prevalence and its herbicide tolerance, *Ambrosia* is one of the most-studied agronomic species (Béres *et al.*, 2005; Kazinczi *et al.*, 2008).

Economic Costs

Annual economic costs from ragweed in Hungary can reach €400–800 million (Mányoki *et al.*, 2011). These costs include the agronomic impact of ragweed on crop production and seed contamination, as well as medical expenses including sick leave, medicines, hospitalizations and the costs related to management and control.

Ambrosia: Origin and Distribution in Eastern Europe

A. artemisiifolia belongs to the phylum of angiosperms (Angiospermatophyta), to the class of Dicotyledonous plants (Dicotyledonopsida), to the order of Composites (Asterales), to the family of Daisies (Asteraceae), to the subfamily of disc florets (Tubuliflorae) and to the genus of ragweed (*Ambrosia* spp.) (Béres *et al.*, 2005). Among *Ambrosia* species, those most related to allergenic pollen production and health-related aspects include *A. artemisiifolia*, *A. psilostachya* and *A. trifida*. All of these species are native to North America.

The earliest described colonization of common ragweed (*A. artemisiifolia*) occurred in Dalmatia (Croatia) in 1842 (de Visiani, 1842; Béres *et al.*, 2005). In Western Europe, the first temporary colonization of *A. artemisiifolia* was reported from Brandenburg (Germany) in 1863 (Priszter,

1960). Although other American species have appeared in Europe, common ragweed is the most widely spread (Járai-Komlódi and Juhász, 1993).

Expansion of *A. artemisiifolia* in Europe began after the First World War (Comtois, 1998). Its expansion can be traced back to various European ports of entry; for example, Rijeka towards Croatia and Transdanubia (a region of Hungary), Trieste and Genoa towards northern Italy and Marseille towards the Rhône valley, France (Kiss and Béres, 2006; Kazinczi *et al.*, 2008). Overall, there appear to be three main introductory areas for Europe: the valley of the Rhône (France), northern Italy and the Carpathian Basin (Juhász, 1998). Gladieux *et al.* (2011) found that *A. artemisiifolia* from Eastern Europe did not originate from the earlier established French populations but came from other independent sources. This suggests separate introductions in Eastern and Western Europe, as well as divergent sources for these two invasions.

For Eastern Europe *in toto*, *Ambrosia* pollen distribution is wind driven. For example, pollen is likely to be transported from Hungary to eastern Austria during August and September, when south-eastern winds are predominant in the region (Jäger and Litschauer, 1998). The source region of *Ambrosia* in Slovakia is the Csallóköz (i.e. the plain of the Danube) and eastern Slovakia. Long-range transport can bring *Ambrosia* pollen to Poland predominantly from the Czech Republic, Slovakia, Hungary and the Ukraine (Stach *et al.*, 2007; Smith *et al.*, 2008; Kasprzyk *et al.*, 2011). Furthermore, the Pannonian Plain including Novi Sad and Ruma (Serbia) is a potential source region for *Ambrosia* pollen (Šikoparija *et al.*, 2009).

In Hungary, *A. artemisiifolia* was first detected at the beginning of the 20th century at Orsova (Jávorka, 1910), near the southern border of the country along the banks of the Danube. In the 1920s, it was found in the southern part of Transdanubia, and within 30 years it came to dominate the region. Its rapid spread is evident from national weed surveys. For example, in 1950 *A. artemisiifolia* was ranked 21st on the list of the most widespread weeds for wheat and maize, then 8th in 1970, 4th in 1988, and since the early 1990s, it has been ranked as the number one weed in the country (Béres, 2003).

Environmental and Socio-economic Factors Associated with Establishment and Biological Distribution of *A. artemisiifolia*

Once introduced, *A. artemisiifolia* can become established and spread if the new environment is similar to its native biogeographical range. For ragweed, this means that temperatures necessary for floral and seed development should exceed 1200 heating degree days (Comtois, 1998). *A. artemisiifolia* is adapted to a wide range of soil moisture, although it does not do well in flooded soils. In Eastern Europe, the Great Hungarian Plain in the Carpathian Basin and the southern part of the Ukraine and European Russia provide similar biophysical environments to those of North America. For southern Europe, the greater accumulation of heat without sufficient precipitation (e.g. the Mediterranean climate) may decrease pollen release; alternatively, cool temperatures below the floral optimum may result in unripened pollen grains and lack of pollen release (e.g. maritime or northern climates).

As a short-day plant, reductions in the light period below the diurnal maximum (21 June in the northern hemisphere) will result in floral initiation. In Hungary, the flowering of *A. artemisiifolia* starts in the second half of July and ends in October (or with the onset of frost). The peak pollen season occurs from mid-August to mid-September (Fig. 8.5). However, the timing and manner of pollination and the distance travelled by the pollen are highly dependent on meteorological parameters (temperature, wind speed, relative humidity and solar irradiance). As a consequence, the daily pollen exposure depends on locality, species and regional weather (Juhász and Juhász, 1997).

Aside from global trade, the establishment and spread of ragweed can be associated

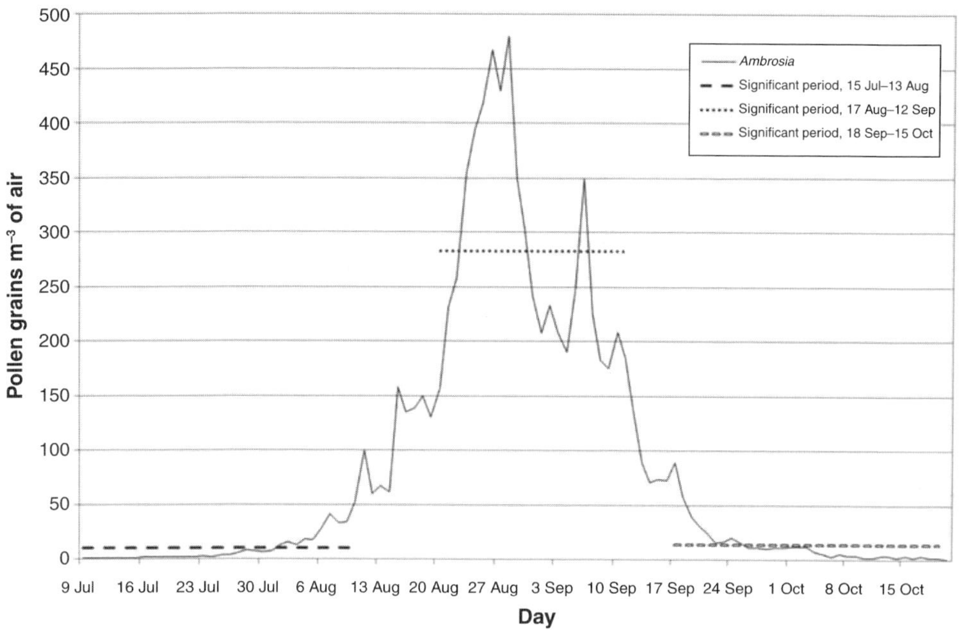

Fig. 8.5. Periods in which the daily mean *Ambrosia* pollen counts differ significantly from the mean of the whole period (Makra-test, 1% significance level) for Szeged, 15 July–15 October 1997–2006.

with major socio-economic transitions that increase the area of disturbed land (Bassett and Crompton, 1975; Comtois, 1998). For example, with the takeover of Hungary by the former USSR (soon after 1945), private agricultural fields were seized to create socialist-type cooperatives. Following the collapse of communism in 1989, agricultural fields belonging to cooperatives were divided into smaller parcels, with a greater emphasis on mechanization, particularly tillage and soil disturbance for weed management. Soil disturbance, in turn, may be a factor in eliciting ragweed seed germination (Bazzaz, 1970).

Anthropogenic Climate Change and *A. artemisiifolia*: Szeged, Hungary, as a Test Case

We examined Szeged, Hungary (46.25N; 20.10E), a medium-sized city (population ~170,000), recognized as the geographical centre of ragweed pollen production in eastern Europe (see Fig. 8.1), as a test case for interactions between the invasive species, *A. artemisiifolia*, and climate perturbations.

To determine potential changes in the season length of *A. artemisiifolia* as a function of recent temperature changes, we analysed the minimum temperature as a function of year beginning from 1977 through 2011. Because the reproductive cycle of *A. artemisiifolia* is known to be frost sensitive (e.g. it will continue to produce pollen until killed by frost), we quantified changes in the last frost date in the spring and the first frost date in the autumn over this period. These data (Fig. 8.6) indicated that, in recent decades, the last spring frost was arriving approximately 1 month earlier, and the first autumn frost was delayed by about 3 weeks. This indicates that the overall season length for exposure to *A. artemisiifolia* pollen has likely been extended for this region. This increase in season length is

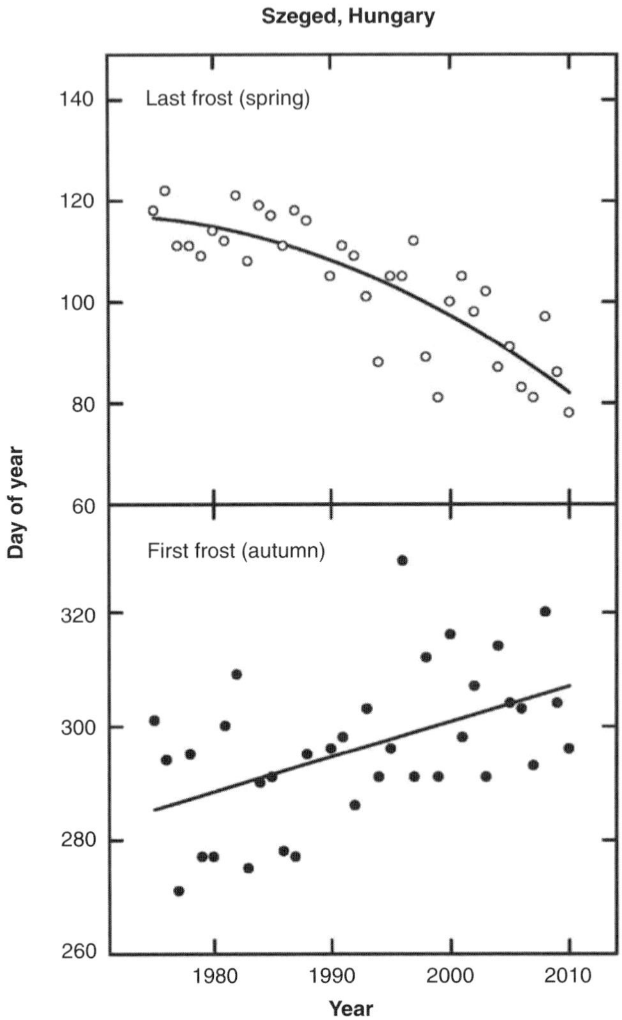

Fig. 8.6. Change in length (as day of year) for ragweed growing seasons from 1977 through 2011 as a function of the earlier onset of first spring frost (top graph) and delays in the time of first frost during the autumn, for Szeged, Hungary. Frost was determined as any day when the average minimal night-time temperature was <32°F. Data were determined as a function of simple regression. (Data are from: https://beaumont.tamu.edu/climaticdata/ViewWeatherDataPage.aspx?Type=Daily&index=HU_HU6&state=HU&county=CSONGRAD&station=SZEGED.)

consistent with recent observations regarding season length for common ragweed with increasing latitude in North America in relation to climate change (Ziska *et al.*, 2011). Higher levels of atmospheric CO_2 associated with urbanization (Ziska *et al.*, 2007) may also influence public health directly by stimulating the growth and pollen production of allergy-inducing species such as ragweed; however, urban CO_2 concentrations were not tested explicitly for Szeged.

To determine pollen concentration, we examined *A. artemisiifolia* (along with 19 other taxa) for Szeged. Pollen grains were collected using 7-day recording volumetric spore traps of the Hirst design (Hirst, 1952). The trap was placed on the roof of the building of the Faculty of Arts at the University of Szeged (20 m above ground level) in the centre of Szeged (Makra *et al.*, 2010). The identification and counting of pollen grains was conducted using light microscopy (Käpylä and Penttinen, 1981; Mandrioli, 2000). Counts were expressed as mean daily pollen concentrations (number of pollen grains m^{-3} of air). Daily pollen counts of the pollen season in Szeged (15 July–15 October) were analysed for the 10-year period 1997–2006 (Makra *et al.*, 2005). In our analysis, we found that the period 18 August–13 September was likely to have the highest level of pollen exposure and induced the subsequent development of allergic rhinitis among sensitized patients.

A linear trend analysis (Makra *et al.*, 2011) of the total annual pollen count (TAPC), annual peak pollen concentration (APC) and the start, end and duration of the pollen season were also analysed over this period. The pollen counts of *A. artemisiifolia* show a decrease in daily linear trends (Table 8.1). This decrease may be associated with a lack of available water and subsequent pollen release during the hottest summer period. Although the overall change in TAPC is not statistically significant, there is a weakly significant interaction concerning a later start of the pollen season. Such an interaction is consistent with the longer time analysis of climate-induced delays in the first autumn frost (Fig. 8.6).

How much of the *A. artemisiifolia* pollen collected reflects local meteorological conditions? Analysis of pollen transport indicated that if the daily cycle of pollen counts displayed a daytime peak, they were probably of local origin (Šikoparija *et al.*, 2009). As *Ambrosia* pollen in Szeged exhibits daily pollen peaks between 08:00 and 16:00 h (Juhász and Juhász, 1997), our results are consistent with Juhász and Juhász (1997), i.e. that pollen release for Szeged reflects medium-range transport, including local pollen dispersion, and hence reflects local meteorological factors.

Conclusions

In Eastern Europe, populations of an invasive plant species, common ragweed (*A. artemisiifolia*), have increased significantly since the 1970s. This species poses both an agronomic threat in its role as a weed limiting crop production, but is better known for its role in allergy-associated respiratory complaints. For Eastern Europe as a whole, following the invasion of *A. artemisiifolia*, Hungary, and in particular the Carpathian Basin, appears, in consequence, to have among the highest concentration of *A. artemisiifolia* pollen for this continent.

An analysis of *A. artemisiifolia* in the context of human activity indicates that social policies regarding land use since the collapse of the Soviet Union in the early 1990s may have influenced the establishment and spread of ragweed within Eastern Europe. In addition, a temporal examination of pollen trends in Szeged, Hungary, indicates that most pollen is transported from a medium or local distance and peaks during the period 18 August–13 September. Decadenal analysis of pollen trends is consistent with longer-term analysis of frost development and onset (longer growing

Table 8.1. Change in the total annual pollen count (TAPC) (pollen grains · m^{-3} of air $year^{-10}$), annual peak pollen concentration (APP) (pollen grains×m^{-3} of air $year^{-10}$), start, end and duration of the pollen season (days $year^{-10}$) for Ambrosia.

Taxon	TAPC[a]	TAPC[b]	APP[a]	Pollination season		
				Start	End	Duration
Ambrosia	229	−1170	230	14[c]	−9	−22

Notes: [a]Calculated by using linear trend; [b]calculated from daily linear trends; [c]denotes a significant change at the 10% level.

season) and a significant interaction with delay in the onset of the first autumn frost observed from 1996 to 2007. This observed temperature trend appears to be consistent with surface warming associated with anthropogenic climate change.

Overall, the influence of human activity, on both a regional and a global scale, appears to be associated with the spread and negative impacts of this invasive plant species for agronomic and human health considerations in Eastern Europe, particularly Hungary. Further spread and impact will depend, in large part, on recognition of this connection and whether appropriate control measures will be implemented by policy makers and the general public.

Acknowledgements

The authors would like to thank Gábor Motika (Environmental Conservancy Inspectorate, Szeged, Hungary) for providing the meteorological data for Szeged; Miklós Juhász for providing the pollen data for Szeged; Roland Draxler for his useful advice and consultations on the HYSPLIT model, version 4.8; and Zoltán Sümeghy for the digital mapping in Fig. 8.1. The authors gratefully acknowledge the NOAA Air Resources Laboratory (ARL) for the provision of the HYSPLIT transport and dispersion model and READY website (http://www.arl.noaa.gov/ready/hysplit4.html) used in this study. The European Union and the European Social Fund provided financial support for the project under grant agreements No TAMOP-4.2.1/B-09/1/KMR-2010-0003 and TAMOP-4.2.1/B-09/1/KONV-2010-0005.

References

AAAAI (American Academy of Allergy, Asthma and Immunology) (2006) *The Allergy Report: Science-based Findings on the Diagnosis and Treatment of Allergic Disorders, 1996–2005.* AAAAI, Milwaukee, Wisconsin.

ASCIA (Australian Society of Clinical Immunology and Allergy) (2007) *The Economic Impact of Allergic Disease in Australia: Not To Be Sneezed At.* CSIRO, Canberra.

Basky, Z.S. (2009) A Magyarországon őshonos levéltetvek hatása a parlagfû fejlődésére. (Effect of native aphids on the development of ragweed in Hungary.) *Növényvédelem* 45, 425–432.

Bassett, I.J. and Crompton, C.W. (1975) The biology of Canadian weeds. 11. *Ambrosia artemisiifolia* L. and *A. psilostachya* DC. *Canadian Journal of Plant Science* 55, 463–476.

Bazzaz, F.A. (1970) Secondary dormancy in the seeds of the common ragweed, *Ambrosia artemisiifolia. Bulletin of the Torrey Botanical Club* 97, 302–305.

Béres, I. (2003) Az ürömlevelű parlagfű (*Ambrosia artemisiifolia* L.) elterjedése, jelentősége és biológiája. (Distribution, importance and biology of common ragweed (*Ambrosia artemisiifolia* L.).) *Növényvédelem* 39, 293–302 (in Hungarian).

Béres, I., Novák, R., Hoffmanné Pathy, Z.S. and Kazinczi, G. (2005) Az ürömlevelű parlagfű (*Ambrosia artemisiifolia* L.) elterjedése, morfológiája, biológiája, jelentősége és a védekezés lehetőségei. (Distribution, morphology, biology and importance of common ragweed (*Ambrosia artemisiifolia* L.) and protection facilities.) *Gyomnövények, Gyomirtás* 6, 1–48.

Comtois, P. (1998) Ragweed (*Ambrosia* spp.): the phoenix of allergophytes. In: Spieksma, F.T. (ed.) *Ragweed in Europe.* Alk–Abelló A/S, Horsholm, Denmark, pp. 3–5.

Csontos, P., Vitalos, M., Barina, Z. and Levente, K. (2010) Early distribution and spread of *Ambrosia artemisiifolia* in Central and Eastern Europe. *Botanica Helvetica* 120, 75–78.

D'Amato, G., Cecchi, L., Bonini, S., Nunes, C., Annesi-Maesano, I., Behrendt, H., Liccardi, G., Popov, T. and van Cauwenberge, P. (2007) Allergenic pollen and pollen allergy in Europe. *Allergy* 62, 976–990.

Frenz, D.A. (2001) Interpreting atmospheric pollen counts for use in clinical allergy: allergic symptomology. *Annals of Allergy, Asthma, Immunology* 86, 150–158.

Gladieux, P., Giraud, T., Kiss, L., Genton, B.J., Jonot, O. and Shykoff, J.A. (2011) Distinct invasion sources of common ragweed (*Ambrosia artemisiifolia*) in Eastern and Western Europe. *Biological Invasions* 13, 933–944.

Harsányi, E. (2009) Parlagfû és allergia. (Ragweed and allergy.) *Növényvédelem,* 45, 454–458.

Hirst, J.M. (1952) An automatic volumetric spore trap. *Annals of Applied Biology* 39, 257–265.

Jäger, S. (1998) Global aspects of ragweed in Europe. In: Spieksma, F.Th.M. (ed.) *Ragweed in*

Europe. Satellite Symposium Proceedings of 6th International Congress on Aerobiology, Perugia, Italy. Alk–Abelló A/S, Horsholm, Denmark, pp. 6–10.

Jäger, S. and Litschauer, R. (1998) Ragweed (*Ambrosia*) in Austria. In: Spieksma, F.Th.M. (ed.) *Ragweed in Europe*. Satellite Symposium Proceedings of 6th International Congress on Aerobiology, Perugia, Italy. Alk–Abelló A/S, Horsholm, Denmark, pp. 22–26.

Járai-Komlódi, M. and Juhász, M. (1993) *Ambrosia elatior* (L.) in Hungary (1989–1990). *Aerobiology* 9, 75–78.

Jávorka, S. (1910) *Ambrosia artemisiifolia* (L.) in Hungary. *Botanikai Közlemények* 9, 303–333.

Juhász, I.E. and Juhász, M. (1997) Az aeroallergén növények pollenszórásának napi ritmusa a Dél-Alföld levegőjében. (Daily rhythm of pollen dispersal of aeroallergen plants in the air of the Southern Hungarian Plain.) Környezeti ártalmak és a légzőrendszer. (*Environmental Hazards and the Respiratory System*) 7, 28–34 (in Hungarian).

Juhász, M. (1995) New results of aeropalynological research in Southern Hungary. *Publications of the Regional Committee of the Hungarian Academy of Sciences* 5, 17–30.

Juhász, M. (1998) History of ragweed in Europe. In: Spieksma, F.Th.M. (ed.) *Ragweed in Europe*. Satellite Symposium Proceedings of 6th International Congress on Aerobiology, Perugia, Italy. Alk–Abelló A/S, Horsholm, Denmark, pp. 11–14.

Käpylä, M. and Penttinen, A. (1981) An evaluation of the microscopal counting methods of the tape in Hirst–Burkard pollen and spore trap. *Grana* 20, 131–141.

Kasprzyk, I., Myszkowska, D., Grewling, L., Stach, A., Šikoparija, B., Skjøth, C.A. and Smith, M. (2011) The occurrence of *Ambrosia* pollen in Rzeszów, Kraków and Poznań, Poland: investigation of trends and possible transport of *Ambrosia* pollen from Ukraine. *International Journal of Biometeorology* 55, 633–644.

Kazinczi, G., Béres, I., Novák, R., Bíró, K. and Pathy, Z.S. (2008) Common ragweed (*Ambrosia artemisiifolia*): a review with special regards to the results in Hungary. I. Taxonomy, origin and distribution, morphology, life cycle and reproduction strategy. *Herbologia* 9, 55–91.

Kiss, L. and Béres, I. (2006) Anthropogenic factors behind the recent population expansion of common ragweed (*Ambrosia artemisiifolia* L.) in Eastern Europe: is there a correlation with political transitions? *Journal of Biogeography* 33, 2156–2157.

Makra, L., Juhász, M., Borsos, E. and Béczi, R. (2004) Meteorological variables connected with airborne ragweed pollen in Southern Hungary. *International Journal of Biometeorology* 49, 37–47.

Makra, L., Juhász, M., Béczi, R. and Borsos, E. (2005) The history and impact of airborne *Ambrosia* (Asteraceae) pollen in Hungary. *Grana* 44, 57–64.

Makra L., Sánta, T., Matyasovszky, I., Damialis, A., Karatzas, K., Bergmann, K.C. and Vokou, D. (2010) Airborne pollen in three European cities: detection of atmospheric circulation pathways by applying three-dimensional clustering of backward trajectories. *Journal of Geophysical Research: Atmospheres* 115, 220–224.

Makra, L., Matyasovszky, I. and Deák, J.Á. (2011) Trends in the characteristics of allergenic pollen circulation in Central Europe based on the example of Szeged, Hungary. *Atmospheric Environment* 45, 6010–6018.

Mandrioli, P. (2000) Metodo di campionamento e conteggio dei granuli pollinici e delle spore fungine aerodisperse. (Method for sampling and counting of airborne pollen grains and fungal spores.) Consiglio Nazionale delle Ricerche, Istituto di Scienze dell'Atmosfera e dell'Oceano (National Research Council, Institute of Atmospheric and Oceanic Sciences) (ISAO) Manuscript, 11 p. (in Italian) (http://lnx.ilpolline.it/wp-content/uploads/2007/11/metodo.pdf, accessed 20 January 2014).

Mányoki, G., Apatini, D., Novák, E., Magyar, D., Bobvos, J., Bobvos, G., Málnási, T., Elekes, P. and Páldy, A. (2011) *Parlagfű – lakossági expozíció. Parlagfű helyzetkép és megoldási javaslatok az Aerobiológiai Hálózat mérései alapján és az OKI-AMO feldolgozásában.* (Ragweed – endangerment of the population. Distribution and pollen load of ragweed vs. suggestions for solution based on the measurements of the Aerobiological Network and the processing of OKI-AMO.) Országos Környezetegészségügyi Intézet, Egészséghatás Előrejelzés Főosztály, Aerobiológiai Monitorozási Osztály, kézirat (National Institute of Environmental Health, Health Effect Forecasting Department, Aerobiological Monitoring Group), Budapest.

Pinke, G., Karácsony, P., Czúcz, B. and Botta-Dukát, Z. (2011) Environmental and land-use variables determining the abundance of *Ambrosia artemisiifolia* in arable fields in Hungary. *Preslia* 83, 225–241.

Priszter, SZ. (1960) *Adventív gyomnövényeink terjedése*. (Adventive weeds.) Keszthelyi Mezőgazdasági Akadémia Kiadványai (in Hungarian). Agricultural Publishing Ltd, Budapest.

Quest Diagnostics (2011) Allergens Across America: The Largest Study of Allergy Testing in the United States (http://www.questdiagnostics.com/dms/Documents/Other/2011_QD_Allergy Report.pdf, accessed §4 January 2014), 54 pages.

Šikoparija, B., Smith, M., Skjøth, C.A., Radišič, P., Milkovska, S., Šimič, S. and Brandt, J. (2009) The Pannonian plain as a source of *Ambrosia* pollen in the Balkans. *International Journal of Biometeorology* 53, 263–272.

Simoncsics, P., Osváth, P. and Balázs, I. (1968) Qualitative analysis of pollen composition in the air. *Rheumatologia, Balneologia, Allergologia* 9, 117–119.

Smith, M., Skjøth, C.A., Myszkowska, D., Uruska, A., Puc, M., Stach, A., Balwierz, Z., Chlopek, K., Piotrowska, K., Kasprzyk, I. and Brandt, J. (2008) Long-range transport of *Ambrosia* pollen to Poland. *Agricultural and Forest Meteorology* 148, 1402–1411.

Stach, A., Smith, M., Skjøth, C.A. and Brandt, J. (2007) Examining *Ambrosia* pollen episodes at Poznan (Poland) using back-trajectory analysis. *International Journal of Biometeorology* 51, 275–286.

Visiani, R. de (1842) *Flora Dalmatica*. Vol. II. Hofmeister, Liepzig, Germany.

Ziska, L.H., George, K. and Frenz, D.A. (2007) Establishment and persistance of common ragweed (*Ambrosia artemisiifolia*) in disturbed soil as a function of an urban–rural macro-environment. *Global Change Biology* 13, 266–274.

Ziska, L., *et al.* (2011) Recent warming by latitude associated with increased length of ragweed pollen season in central North America. *Proceedings of the National Academy of Sciences* 108, 4248–4251.

9 Climate Change and Alien Species in South Africa

Ulrike M. Irlich, David M. Richardson, Sarah J. Davies and Steven L. Chown

Centre for Invasion Biology, Department of Botany and Zoology, Stellenbosch University, Matieland, South Africa

Abstract

South Africa has a long history of human-mediated introductions of species from all major taxonomic groups. Close to 9000 alien terrestrial plant species have been introduced, and all of the country's biomes have already been invaded. Invasive species are threatening the country's ecosystems in numerous ways, but the effect of climate change on these invasions is predicted to be complex and cascading and remains poorly understood. The relationship between climate and invasive species biology is well established, and there is no question that climate change will influence the ecology of invasive species significantly. If left unmanaged, these effects are expected to increase substantially. Besides terrestrial plants, numerous animals have also invaded the country's landscapes. South Africa's freshwater ecosystems have been invaded by both alien as well as extralimital introductions (indigenous species outside their historical extent of occurrence). The status of invasion in the marine environment remains poorly studied, and knowledge of the status of invasions and predictions regarding the impacts of climate change remain largely speculative. This chapter highlights the current status of invasions in South Africa and discusses some of the direct and indirect effects climate change is likely to have on these invasions.

Dedication

We dedicate this chapter to the memory of Phil Hockey, who had an enduring interest in avian range expansions and has made their study so much more straightforward through his significant contributions to ornithology.

Introduction

South Africa covers 1.22 million km^2, about 2% of the world's surface area, and has a coastline of about 2800 km. The country has a warm, temperate climate and is mostly arid, with large parts receiving less than 500 mm rainfall annually, especially the west, which is drier than the east (Fig. 9.1) (O'Brien, 1998; van Rensburg *et al.*, 2002). Nine major biomes (Desert, Succulent Karoo, Nama-Karoo, Fynbos, Albany Thicket, Grassland, Savannah, Forest and Indian Ocean Coastal Belt), each with distinct bioclimatic regimes, characterize the country (Mucina and Rutherford, 2006), which also includes three biodiversity hotspots (Cape Floristic Region, Maputaland–Pondoland–Albany and the Succulent Karoo). It is home to about 10% of the world's plant species, 13% of bird and mammals and 23% of marine inhabitants, with high levels of endemism in all groups (DEAT, 1997; EWT, 2002). South Africa also governs the Prince

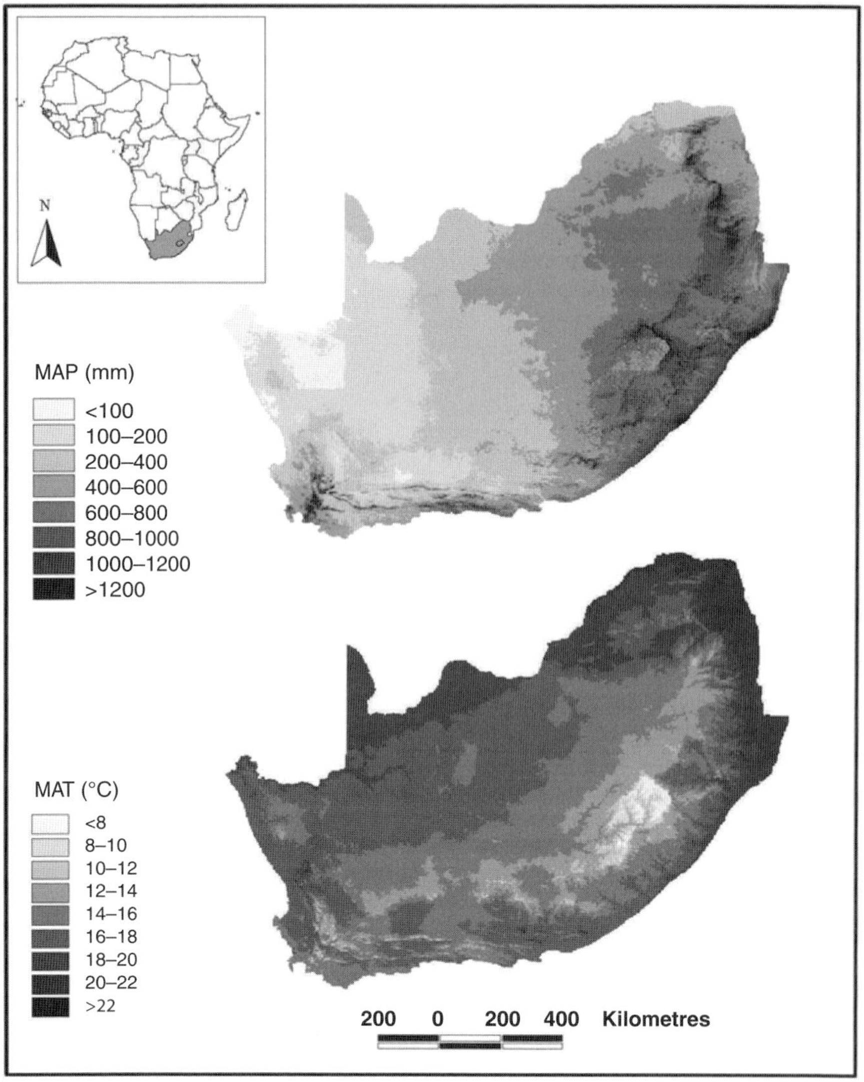

Fig. 9.1. Mean annual temperature (MAT) and precipitation (MAP) for South Africa. Data from the *South African Atlas of Agrohydrology and Climatology* (Schulze *et al.*, 1997; interpolated 1′ × 1′ data).

Edward Island archipelago in the sub-Antarctic, which lies approximately 1770 km south-east of Port Elizabeth, South Africa, and has a cool, moist climate characterized by two major biomes (Tundra and Polar Desert) and many distinct habitat types (Chown and Froneman, 2008).

South Africa has a long history of human-mediated introductions of organisms from all the major taxonomic groups (Richardson *et al.*, 2003). Cattle, sheep, goats, dogs and burweed were introduced approximately 2000 years ago as humans migrated southwards, but the arrival of European settlers in the mid-17th century led to a large increase in introductions (Macdonald *et al.*, 1986a). At present, all biomes are invaded, with the highest density of alien species occurring in the eastern parts of the country (Rouget *et al.*, 2004; Richardson *et al.*, 2005; Chown and Froneman, 2008; Spear and Chown, 2009). The prominence of alien trees and shrubs as major invaders of terrestrial systems is an unusual feature of South Africa's invasive biota; grasses and other herbaceous species dominate invasive floras in most other parts of the world (Richardson *et al.*, 1997).

Approximately 8750 alien terrestrial plant species from all growth forms are known to be present, and at least 180 of these are currently invasive (Table 9.1; Richardson *et al.*, 2011). At least 771 species of alien trees have been introduced to South Africa (von Breitenbach, 1990). The diversity of bioclimates in South Africa provides niches for species from tropical, temperate and Mediterranean climate regions. Most invasive alien plants from tropical regions are South American, while northern temperate species have originated mostly from Europe or Asia. Australia has

Table 9.1. Current estimates of the number of alien and invasive species (*sensu* Pyšek *et al.*, 2004) in South African terrestrial and aquatic habitats.

	Number of alien species	Number of invasive species	Sources
Terrestrial			
Plants[a]	8750	199	Macdonald *et al.*, 2003; Richardson *et al.*, 2011
Mammals	50	10	Richardson *et al.*, 2011
Birds	77	7	Dean, 2000; van Rensburg *et al.*, 2011
Reptiles	275	0	van Wilgen *et al.*, 2010
Amphibians	14	0	van Rensburg *et al.*, 2011; N.J. van Wilgen, Stellenbosch, 2010 personal communication
Arthropods[b]	305	Unknown	Lach *et al.*, 2002; Macdonald *et al.*, 2003
Molluscs	34	10	Herbert, 2010
Microbes	Unknown	Unknown	Pimentel *et al.*, 2001
Freshwater			
Plants	23	13	Hill, 2003; Macdonald *et al.*, 2003
Fish (freshwater)[c]	58	37	Macdonald *et al.*, 2003
Crayfish	4	4	de Moor, 2002
Molluscs	10	4	Appleton, 2003
Fish parasites/ diseases	8	8	Macdonald *et al.*, 2003
Marine[d]			
Plants/algae	6 (4)		Mead *et al.*, 2011
Crustaceans	22 (11)	2	Mead *et al.*, 2011
Insects	1		Mead *et al.*, 2011
Molluscs	12 (9)	1	Mead *et al.*, 2011
Fish	1		Mead *et al.*, 2011
Other	43 (15)	1	Mead *et al.*, 2011

Notes: [a]At least 771 tree species have been introduced to South Africa over the past 300 years (von Breitenbach, 1990); [b]numerous insect species were introduced as biocontrol agents; [c]33 extraregional (species from outside South Africa) and 25 extralimital (indigenous species outside their historical extent of occurrence); [d]number of cryptogenic species (species of unknown origin) in brackets.

been the largest contributor of southern temperate invasive plant species (Henderson, 2006). The most important aquatic weeds are from tropical South America (Henderson, 2006; Coetzee *et al.*, 2011). Many terrestrial invertebrate species have been introduced and have become invasive, including several that are pests of agriculture and which have substantial impacts on biodiversity, such as the Argentine ant, *Linepithema humile* (Annecke and Moran, 1982; Bond and Slingsby, 1984; Herbert, 2010).

Aquatic invertebrate invaders have not been well studied, but several molluscs are important, including *Aplexa marmorata*, *Lymnaea columella*, *Physa acuta* and *Tarebia granifera*, which are established in the freshwater ecosystems of the Kruger National Park (de Kock and Wolmarans, 2008). Alien freshwater fish have originated from South African river systems through inter-basin transfers (extralimital introductions; approximately 25 species) as well as extraregional introductions (species from outside South Africa; 33 species) (Table 9.1; Macdonald *et al.*, 2003).

In contrast to terrestrial plant invasions, far fewer terrestrial vertebrates have invaded South African ecosystems, and no extraregional amphibian or reptile species are currently considered invasive, although several species are established (van Wilgen *et al.*, 2008a; van Rensburg *et al.*, 2011). Few bird species have invaded, and those that have are commensals (Table 9.1; Dean, 2000). Among the mammals, commensal rodent species which pose substantial human health risks through their vector status are in a state of dynamic change: a recently discovered introduction, *Rattus tanezumi*, is widespread (Taylor *et al.*, 2008). Many indigenous mammal species have been moved intentionally outside of their historical ranges in South Africa. Such extralimital introductions have led to homogenization of the fauna (Spear and Chown, 2008, 2009). Species have also been introduced from outside South Africa (extraregional introductions) and, after the USA, South Africa is home to more alien ungulate species than any other country.

Assessing the Influence of Climate Change on Invasive Species

The impacts of invasive species on South African ecosystem services, biodiversity and the economy are multifaceted. Although these have not been quantified comprehensively, some countrywide assessments have been made, looking at all major taxonomic groups (van Wilgen *et al.*, 2001; Pimentel, 2002; van Wilgen and Richardson, 2009; van Rensburg *et al.*, 2011). The effect of climate change on these invasions is little understood. However, the extensive research conducted on invasive species elsewhere does enable several informed predictions regarding the possible impact and effect of climate change on invasive species to be made.

Direct effects

The most direct effect of a changing climate on invasive species will be through changes to distribution and life history as a consequence of altered patterns of precipitation and temperature and the increased frequency of extreme events, such as floods and droughts (Hobbs and Mooney, 2005). Kruger and Sekele (2012) found that since the 1960s, warm extreme temperatures have increased, while cold extreme temperatures have decreased, and this pattern has been observed across the country. For South Africa's interior, annual mean temperatures are expected to increase by 2.5–3.5°C by 2050, while the coastal regions could experience temperature increases of up to 2°C (Knoesen *et al.*, 2009). The increase in temperature will be accompanied by increased evaporation and the east to west rainfall gradient (Fig. 9.1) is likely to be exacerbated. A recent study investigated the precipitation trends over the past century and found no evidence of changes in rainfall patterns for large parts of the country, while only some areas showed increases or decreases in precipitation trends (Kruger, 2006). However, it is predicted that by 2100 the largest part of the country could experience increased rainfall, with only the west coast (Succulent Karoo and the western parts of the Fynbos biome) predicted to become more arid (Knoesen *et al.*, 2009). Projected rainfall anomalies include floods, which show no clear pattern, and short droughts, which are expected to decrease in frequency; current patterns of longer duration droughts are unlikely to change (Knoesen *et al.*, 2009). These direct effects of climate change are likely to drive substantial changes to the geographic ranges of both indigenous and alien species (Richardson *et al.*, 2000; Erasmus *et al.*, 2002; Rouget *et al.*, 2004; Huntley and Barnard, 2012). The range shift of species due to climate change is expected to be similar to natural dispersal, with leading edge range shifts, but long-distance dispersal and movement via corridors also needs to be considered (Tolley *et al.*, 2008; Wilson *et al.*, 2009). This is likely to be more important for alien species, especially those that are frequently dispersed by humans (Trakhtenbrot *et al.*, 2005). Following the changes in rainfall patterns, more alien species may appear in the eastern parts of the country (i.e. tropical species) than in the western arid parts (Rouget *et al.*, 2004).

Assessments of the effects of climate change on indigenous and invasive biota must also consider changes in ecosystem function that alter invasibility. For example, changes in the frequency and intensity of extreme events such as storms, fires, droughts and floods have the potential to increase ecosystem susceptibility to invasion (Hobbs and Mooney, 2005). Changing precipitation and rainfall regimes may also alter the demographic distribution of C_3 and C_4 plants, which will drive crucial changes in fire regimes (Wigley *et al.*, 2010; Higgins and Scheiter, 2012). Increases in carbon dioxide concentration ($[CO_2]$) are likely to have major effects on tree–grass dynamics in grasslands and savannahs, because of the different photosynthetic pathways used by the different growth forms. Differential responses of plant species to increased atmospheric $[CO_2]$ may be driving the expansion and densification of indigenous trees in South African grasslands (Bond and Midgley, 2012) and may alter the relative abundance of C_3 and C_4 plants radically in different biomes. This has profound implications for ecosystem functioning and invasibility (Richardson *et al.*, 2000), and such interactions are likely to cause a substantial reshuffling of invasive species in South Africa.

Indirect effects

Climate change will also affect introduced species indirectly by influencing the nature and intensity of human activities. Biological invasions are closely linked to human activities, particularly trade and the disturbance of natural ecosystems through changes in land use (Le Maitre *et al.*, 2004; Richardson *et al.*, 2005; Thuiller *et al.*, 2007; King and Tschinkel, 2008). The intentional introduction of species for human activities associated with recreation (fishing – Shelton *et al.*, 2008; hunting – Spear and Chown, 2009), the pet trade (van Wilgen *et al.*, 2010) and agriculture (forestry – Richardson, 1998; agriculture and horticulture – Le Maitre *et al.*, 2004) have contributed significantly to invasive species introductions. As humans adapt to novel climatic conditions, the donor and recipient regions of pathways, as well as the intensity with which these are used, may change (Galil *et al.*, 2007).

Changes in land use associated with increased urbanization and economic and infrastructure development are also likely to result in new dispersal corridors (e.g. roadside verges) for invasive species (Kalwij *et al.*, 2008). The intensification of land use may also increase nutrient deposition (e.g. nitrogen, phosphorus) in soil and water, enhancing invasion vulnerability (Richardson *et al.*, 2000; Coetzee *et al.*, 2009). Human settlements adjacent to protected areas may serve as sources of propagules of alien species, as well as drivers of disturbance that render areas inside protected areas more susceptible to invasion (Alston and Richardson, 2006; Foxcroft *et al.*, 2011; Jarošík *et al.*, 2011). Furthermore, in an attempt to curb climate change, carbon sequestration projects are being implemented, often in the form of plantations of fast-growing alien trees such as *Pinus* and *Eucalyptus* (Christie and Scholes, 1995), trees that have escaped the plantations and are invading the South African landscape. However, some sequestration projects are focusing on rehabilitating degraded rangelands using indigenous species. Such projects should be supported, as carbon storage in the long term has been found to be greater using indigenous than alien species (van Rooyen *et al.*, 2013). As climate change alters human activities, many introduced species may become 'change passengers', with greater exposure to new regions. Overall, human-mediated dispersal of invasive species is well recognized (Trakhtenbrot *et al.*, 2005), and the indirect effects of climate change as outlined above are likely to play a significant role in driving further invasions in South Africa.

A Synopsis of the Current Status of Invasion and the Predicted Effect of Climate Change on Invasive Species

Terrestrial

Current status

Invasive plant species now cover about 8%, or 10 million hectares (Mha), of South Africa's surface area (Le Maitre *et al.*, 2011), equating to approximately 1.813 million condensed hectares, calculated as the percentage invasion (density) as a proportion times the area of the polygon (Marais and Wannenburgh, 2008; van Wilgen *et al.*, 2012). Plant invaders pose a significant threat to South African biodiversity: they have already been implicated in the extinction of at least 58 plant species in the Cape Floristic Region and are threatening thousands more (Macdonald *et al.*, 2003; Gaertner *et al.*, 2009). Invasive plants are also negatively affecting other groups, such as birds (Allan *et al.*, 1997; Dean *et al.*, 2002) and several invertebrate groups (grasshoppers – Samways and Moore, 1991; ground-living invertebrates – Samways *et al.*, 1996; dung beetles – Steenkamp and Chown, 1996; spiders – van der Merwe *et al.*, 1996; dragonflies and damselflies – Samways and Taylor, 2004; Samways *et al.*, 2005).

Plant invasions also have a significant impact on South Africa's water resources. An estimated 7% of mean annual runoff is currently used by invasive plants, with an annual total of US$773 million in water-related losses (Le Maitre *et al.*, 2011). Invasive plants currently reduce the grazing potential by *c.*123,000 large livestock units, with an estimated unit-price estimate of US$329.47 per large livestock unit (Le Maitre *et al.*, 2011). If invasive species continue to spread and degrade grasslands, grazing potential could be reduced by up to 70% (van Wilgen *et al.*, 2008b). The cost of clearing invasive plant species across the country has been estimated at approximately US$1.2 billion (Pimentel, 2002). Currently, the Working for Water programme and a provincial programme in KwaZulu-Natal spend *c.*US$80 million per annum on the clearing and management of alien species. An estimated 42% of South Africa's arthropod agricultural pests are alien species which cause crop losses and impact food security to the value of approximately US$1 billion per annum, with alien weeds and pathogens adding a further US$3.3 billion in crop losses annually (Pimentel *et al.*, 2001). For South Africa alone, the total cost of invasive alien plant species has been estimated to be between US$867 million (Le Maitre *et al.*, 2011) and US$1 billion annually (Pimentel *et al.*, 2001). Many introduced species bring substantial economic benefits, but the environmental impacts of some of these species result in complex conflicts of interest. Considerable attention is being given to formulating trade-offs to resolve such conflicts; for example, in the case of forestry trees that also cause major problems as invasive species (van Wilgen *et al.*, 2011; van Wilgen and Richardson, 2012).

At least 50 mammal species have been introduced to South Africa, many intentionally for pest control, pets, food or as the source of other animal products, as well as for hunting and ornamental purposes (Table 9.1; van Rensburg *et al.*, 2011). In addition to domestic animals, such as goats, sheep and cattle, 38 ungulate species have also been introduced or translocated within the country, largely for the game ranching and hunting industries (Spear and Chown, 2009). At least 77 bird species have been introduced to South Africa, but only a few commensal species have established successfully (Table 9.1; Dean, 2000; van Rensburg *et al.*, 2011). Approximately 275 species of reptiles and at least 14 species of amphibians have been introduced to South Africa, mainly through the pet trade (van Wilgen *et al.*, 2010; N.J. van Wilgen, Stellenbosch, 2010, personal communication). Although some alien reptile species have established feral populations, none are currently invasive (Richardson *et al.*, 2011; van Rensburg *et al.*, 2011). Three amphibian species have established extralimital populations in the Western Cape and are invading (Measey and Davies, 2011). South African ecosystems appear resistant to invasion by extraregional alien amphibians and reptiles, but this may be an artefact of low propagule pressure and the short residence time of these species. Increasing rates of introduction and changing environmental conditions may well render ecosystems more susceptible to invasion.

Data on alien invertebrates in South Africa are limited. At least 24 alien spider species have been recorded (Macdonald *et al.*, 2003). Of the 34 mollusc species known to have been introduced, 28 are established and 10 of these are invasive (Table 9.1; Herbert, 2010). Of the 40 top crop pests in South Africa, about 17 species are alien (Pimentel, 2002). Some insect invaders with potentially detrimental effects on biodiversity include the varroa mite (*Varroa destructor*) and the Argentine ant (*L. humile*; Lach *et al.*, 2002). While not invasive, over 225 insect species have been introduced as biological control agents against insect pests and a further 56 species to control weeds (Lach *et al.*, 2002; Macdonald *et al.*, 2003; Klein, 2011). Little information is available on the status of alien microbes (viruses, bacteria, most fungi and most single-celled organisms) or their impact on the environment, human health and the economy (Table 9.1; Pimentel *et al.*, 2001).

Predictions

South Africa is expected to be affected significantly by climate change over the next few decades. Mean annual temperatures are expected to rise by 2.5–3.5°C by 2050 (Knoesen *et al.*, 2009). Large parts of the country are expected to experience more extreme rainfall events, which are likely to lead to the increased frequency as well as severity of floods. Predictions suggest an increase in shorter (1- to 2-year) droughts, whereas no change is predicted in longer (3 consecutive years or longer) drought events (Knoesen *et al.*, 2009). Raised levels of atmospheric [CO_2] have been shown to increase photosynthesis, plant water-use efficiency and biomass in many plant species (Bazzaz, 1990; Ziska, 2003), and both indigenous and alien woody plants will benefit from this CO_2 enrichment (Bond and Midgley, 2012; Higgins and Scheiter, 2012). The increased growth rate of woody species and their enhanced ability to survive fires is likely to drive a marked increase in the invasion of grasslands by woody plants, as is being observed with the range expansion of *Prosopis* species in North America (Polley *et al.*, 2002). Through this increased plant growth, the carbon:nitrogen ratio in the leaves may be modified, resulting in nitrogen becoming the limiting nutrient for herbivorous insects. Such altered plant–herbivore interactions could result in increased leaf herbivory and decreased insect growth, but ecosystem effects are variable (Coviella and Trumble, 1999; Lindroth, 2010). One consequence of rising CO_2 may well be a reduced ability of crop species to resist invasive herbivores (Zavala *et al.*, 2008), though such interactions have not been explored for the region. In the mesic parts of South Africa, increased water-use efficiency resulting from increased atmospheric [CO_2] could contribute to flooding events due to waterlogged soil, while in arid areas it could open up niches for alien species that require more water (Smith *et al.*, 2000; Betts *et al.*, 2007; but see Piao *et al.*, 2007).

The response of the *Prosopis* species complex to climatic factors in South Africa has been examined using bioclimatic modelling and by inferring the processes that have driven range changes in different parts of the invasive range (Richardson *et al*., 2000). *Prosopis* has invaded the arid and semi-arid areas of South Africa, covering 19% of the total invaded area, or an estimated 1.8 Mha (Marais and Wannenburgh, 2008), posing a significant threat to biodiversity (Steenkamp and Chown, 1996; Dean *et al*., 2002) and grazing resources (van Wilgen and Richardson, 2009). Warming alone is expected to decrease substantially the suitable area for this species, driving an eastward shift in range (Richardson *et al*., 2000). However, rainfall is predicted to increase across the area currently occupied by *Prosopis*; this could facilitate further spread in all directions, particularly to the currently arid western parts. Decreases in fire frequencies and increased grazing intensity will also alter spread by allowing this taxon to move eastwards and northwards, respectively (Richardson *et al*., 2000).

With increased urbanization and economic development, natural habitats are being transformed to meet the needs of a rapidly growing human population. Changes in urbanization and development associated with land use and land transformation in response to a changing climate are difficult to forecast for South Africa. Propagule pressure of alien species has been found to be closely related to human activities (Lockwood *et al*., 2005). Hence, it is likely that climate-driven changes to the spatial patterns of human activities will affect propagule availability significantly, exposing many new areas to invasion by alien species. In the absence of effective policy implementation, increased demands for goods and services may also therefore lead to heightened pressure from invasive species, as seen elsewhere (see Hulme *et al*., 2008).

Fresh water

Current status

South Africa's freshwater systems have been invaded by at least 58 alien fish species, 33 of which are extraregional introductions and the remaining 25 extralimital transfers (Table 9.1; Richardson *et al*., 2011). At least 37 of these alien species are considered detrimental to the aquatic habitats that they occupy. Predatory invasive fish reduce the abundance of indigenous fish and insects, and may cause local extinctions (Weyl and Lewis, 2006; Lowe *et al*., 2008; Shelton *et al*., 2008). Four freshwater crayfish have established following escape from aquaculture (de Moor, 2002). About ten freshwater gastropod species have been introduced, two of which are invasive and a further two undergoing range expansion (Appleton, 2003). At least eight fish parasites and diseases have also been introduced to South Africa (Macdonald *et al*., 2003). In addition to the aquatic animal invaders, South Africa has also been invaded by at least 21 aquatic and wetland plant species (Table 9.1; Henderson and Cilliers, 2002). These have caused extensive disruption to South Africa's freshwater systems by interrupting water flow, degrading water quality, increasing evapotranspiration and increasing erosion (Le Maitre *et al*., 2011; van Wilgen and de Lange, 2011).

The major pathways of introduction for South African freshwater invaders have been the aquarium trade, aquaculture and introductions of fish for sports and food, while most extralimital introductions are attributable to inter-basin transfers (Macdonald *et al*., 1986a; Richardson *et al*., 2011). The introduction and translocation of organisms often includes the transfer of their parasites, particularly in the case of fish. Furthermore, fish translocations have resulted in hybridization with closely related species; e.g. Nile tilapia (*Oreochromis niloticus*) hybridizes with other tilapia species (van der Waal and Bills, 2000; Canonico *et al*., 2005).

Predictions

Climate change will undoubtedly affect South Africa's aquatic ecosystems through increased temperatures and evaporation and through altered rainfall patterns. Increases in water temperature, together with decreased water flow, have been predicted for southern Africa. This could result in deterioration in water quality, affecting freshwater ecosystems and the utilization of water resources by humans (van Vliet *et al.*, 2013). Due to these changes, land use is likely to change in response to the changing climate. Urbanization, human population growth and economic development are increasing the pressure on South Africa's water supplies over much of the country. This will increase the demand for dam construction and inter-basin transfers, which will lead to more transfers of organisms between regions (both species native to South Africa and alien species) (Kolar and Lodge, 2000; Johnson *et al.*, 2008). The increased transformation of natural environments for agriculture will alter aquatic systems by increasing the runoff of agricultural effluent into rivers and dams, with impacts on water quality (de Villiers, 2007). The eutrophication of freshwater systems often makes ecosystems less suitable for indigenous species and more suitable for alien species, as eutrophication acts as a disturbance event and some alien species might be better adapted to nutrient-rich environments than indigenous species (Byers, 2002). More intensive agriculture may also alter water temperature (Kolar and Lodge, 2000), which may reduce habitat quality for freshwater species such as dragonflies and damselflies (Samways and Taylor, 2004). However, increased evaporation and increased aridity in the western parts of the country will decrease annual runoff and river flow. The predicted increases in the frequency and severity of precipitation events across the country will have marked effects on freshwater systems (Knoesen *et al.*, 2009).

Some aquatic weeds have been brought under control successfully through the introduction of biological control agents (Coetzee *et al.*, 2011). The effect of climate change on biocontrol agents has been poorly investigated. However, variable levels of success in controlling water hyacinth (*Eichhornia crassipes*) have been ascribed to different climatic conditions among invaded sites, pointing to the effect climate may have on biocontrol agents (Coetzee *et al.*, 2011). In the case of the water hyacinth and its biocontrol agents, climate change could increase the effectiveness of the agents if conditions become more favourable and enhance the reproductive success of the insects.

Large physiological tolerance ranges of the four alien crayfish species (de Moor, 2002) indicate that these species could potentially spread across large areas and will not be much affected by climate change. This could prove problematic, as freshwater molluscs and crayfish act as intermediate hosts for parasites such as lung fluke and bilharzias, affecting human health (de Kock and Wolmarans, 1998; de Moor, 2002). The potential effects of climate change on other aquatic invaders have not been assessed, and research is required to understand the full range of possible outcomes.

Marine

Current status

Alien marine organisms have arrived in South Africa mainly in the ballast water of ships and attached to ships' hulls as fouling organisms, with fewer than ten introductions associated with mariculture (Mead *et al.*, 2011). South Africa's marine habitats support at least 83 alien species and 37 cryptogenic species (species of unknown origin; Mead *et al.*, 2011), a drastic increase from the lists of 22 alien and 18 cryptogenic species published in 2009 (Griffiths *et al.*, 2009). The lack of knowledge of species distributions and life-history characteristics complicates the task of making informed predictions regarding the potential effects of climate change on indigenous and alien biodiversity. Furthermore, the number of species that are invasive and the potentially environmental and economic impacts of the invaders remains poorly documented. South Africa has 259 estuaries, all of which are extremely vulnerable ecologically. They are being affected by marine and freshwater invasions (Robinson *et al.*, 2005), as well as by invasions along riparian zones, which affect the freshwater discharge into estuaries (Richardson and van Wilgen, 2004).

Predictions

There is limited knowledge of indigenous and alien biodiversity along the South African coastline and a lack of data on the species distributions and life-history characteristics of most species. Despite recent advances (e.g. Mead *et al.*, 2011), the number of species that are invasive, as well as the potential environmental and economic impacts of the invaders, are not well known. Predictions regarding the responses of marine species to climate change are therefore largely speculative.

The effects of climate change on the marine environment are predicted to be complex and cascading. Between 1985 and 2006, sea surface temperatures of the Agulhas Current have increased 0.7°C per decade, while increases of 1.1°C per decade have been noted at 500 m depth (Rouault *et al.*, 2009). Sea surface temperatures are expected to increase by a further 1–3°C before 2050 (Clark, 2006). Sea levels have risen 10–15 mm over the past century and are expected to rise an additional 25 cm by 2050. The frequency and severity of storms will increase; this, in combination with sea level rise, will affect the intertidal zone negatively through coastal erosion, as seen in KwaZulu-Natal in 2007 (Smith *et al.*, 2007). Atmospheric [CO_2] heavily influences the marine environment, and the acidity of surface waters will decrease by 0.4 pH units by the end of the century, accompanied by a decline in calcium carbonate saturation levels (Clark, 2006). Increases in ultraviolet radiation are expected to affect photosynthetic rates negatively. The changes are expected to affect the indigenous and alien biota profoundly, either individually or in combination.

Globalization together with climate change will add new trade paths and alter existing ones, across the oceans as well as across land (Galil *et al.*, 2007). These will aid the movement of alien species in ballast water, as well as hull-fouling organisms. Furthermore, it has been found that the pollution of harbours by heavy metals affects the indigenous biota negatively and increases the dominance of invasive species (Piola and Johnston, 2008).

Changes in precipitation patterns across the country may result in changes in runoff, and subsequently affect the marine and estuarine environments (Clark *et al.*, 2000). The greatest potential effect on South Africa's marine biota due to climate change may be from changes in pressure systems, which are expected to manipulate large-scale oceanographic processes, particularly the upwelling associated with the Benguela region (Clark, 2006). All these climatic alterations will disturb marine environments, making them more susceptible to invasion. Invasive species from lower latitudes are likely to arrive in South African oceanic systems, and warm-water species are likely to become more abundant (Carlton, 2000; Drinkwater *et al.*, 2010). Consequently, an overall decline in cold-water indigenous species can be expected.

The response of alien marine organisms to climate change has not been investigated specifically in South Africa, but some predictions can be made by relating the physiological tolerances of a given species to predicted temperature shifts. For example, *Balanus glandula*, a marine invader with a preference for cool, temperate waters, is currently distributed along the western coast, and its range along the south coast is limited by the warmer waters of the Agulhas region. This species could become less abundant with increased ocean temperatures, as it only tolerates temperatures below 17°C (Laird and Griffiths, 2008). The invasion of the European green crab (*Carcinus maenas*) is currently limited to intertidal sites with low wave exposure (Griffiths *et al.*, 2009). Changes in storm frequency and severity resulting in increased wave action could induce local extinction of this species. The Mediterranean mussel (*Mytilus galloprovincialis*) affects local mussel diversity negatively by displacing indigenous species on the west coast of South Africa. It seems to prefer cool, temperate conditions and is limited by subtropical environments (Branch and Steffani, 2004), but it is capable of surviving high sea temperatures (Rajagopal *et al.*, 2005), and the temperature range for spawning and larval survival is relatively broad (Chícharo and Chicharo, 2000). Wave action has been shown to cause higher mortality in *M. galloprovincialis*, and thus its invasiveness could decline with increasing wave action as a result of climate change (Zardi *et al.*, 2008).

Southern Ocean islands

Marion Island, the larger of the two islands in the Prince Edward group, is permanently occupied by a research team, whereas Prince Edward Island is visited only occasionally. The islands have a cool, moist climate, generally very stable, but temperature has increased steadily over the past few centuries and precipitation has declined (Smith and Steenkamp, 1990; Smith, 2002; le Roux and McGeoch, 2008). The Prince Edward Islands have been invaded by numerous alien species, the numbers of which are correlated with the numbers of human visitors during the last two centuries (Chown *et al.*, 1998). Because of the much greater number of visitors to Marion, this island has more alien species. Introductions to these islands have been both accidental and intentional. After the impact of the unintentional introduction of the feral house mouse (*Mus musculus domesticus*) on Marion Island was noted, the domestic cat (*Felis catus*) was introduced to alleviate the problem. However, the cats became feral, killing thousands of seabirds, resulting in the active removal of the cats from the island (Chown and Froneman, 2008). The islands have been invaded by at least 13 species of vascular plants (three on Prince Edward) and 18 species of terrestrial invertebrates (one on Prince Edward) (Frenot *et al.*, 2005).

Global environmental change, especially rising temperatures, ocean acidification and changes in sea-ice distribution are of immediate conservation concern for the Prince Edward Islands (Chown *et al.*, 2012). With climate change, the islands are expected to become warmer and drier. This could open up niches for new alien species to occupy the islands and assist in the spread of established aliens (Frenot *et al.*, 2005; Chown *et al.*, 2012).

New Threats

Climate change poses a major challenge for the management of alien species in South Africa on many fronts. Besides rearranging the distribution and abundance of species that are already invasive in the country, changing climates will alter the performance of other alien species already in the country but which are not invasive – either because current conditions are suboptimal for them or because they have not had enough time to invade. Dealing with ‘invasion debt’ (the latent expansion of exotic species populations) must be part of a comprehensive strategy for the long-term management of alien species. Nel *et al.* (2004) identified 85 emerging plant invaders – species that currently have little impact on natural or semi-natural ecosystems in South Africa but which are known to have invasive tendencies. Many of these species are likely to flourish under a changing climate, and management and control strategies need to be responsive to these dynamics.

The changing needs of humans will drive the need for new introductions. Interest in biofuels is growing due to global economic trends and the need to reduce dependence on fossil fuels. Although the cultivation of alien plants for the production of biofuels may, in some cases, have clear economic benefits, there are many environmental costs that need to be considered, including the problem that many of the species most suitable for biofuel production are notorious invasive species (Raghu *et al.*, 2006; Blanchard *et al.*, 2011). Similarly, invasions have resulted and can be expected where plants are cultivated in carbon sequestration projects (Christie and Scholes, 1995; de Wit *et al.*, 2001).

Invasion meltdown, the reshuffling and disassembly of communities, may result from novel interactions among species, invasive and indigenous, resulting sometimes in beneficial, but largely detrimental, effects on species or entire ecosystems

(Simberloff and Von Holle, 1999; O'Dowd *et al.*, 2003). It is possible that more species indigenous to South Africa may become invasive in parts of the country outside their natural ranges as a result of ongoing human-mediated translocation (Spear and Chown, 2009) or climatic shifts. However, differentiating between unassisted change (i.e. climate change-related only) from directly human-assisted change (i.e. alien invasive species) is a challenge for invasion biology. While the enemy release hypothesis (the ability to thrive in a new environment free of the species' natural enemy) might hold for some species, a recent meta-analysis revealed that not all alien species benefited (Chun *et al.*, 2010). Numerous freshwater fish have already spread through inter-basin transfers and now threaten the indigenous biota of these aquatic systems (Macdonald *et al.*, 2003). The hadeda ibis (*Bostrychia hagedash* – Macdonald *et al.*, 1986b) and the red-billed quelea (*Quelea quelea* – Oschadleus and Underhill, 2006) have expanded their distributional ranges in South Africa as a response to human activity. Dynamic range shifts of several other bird species have been attributed to a combination of land-use change and climate change (Hockey *et al.*, 2011). The guttural toad (*Amietophrynus gutturalis*), painted reed frog (*Hyperolius marmoratus*) and African clawed toad (*Xenopus laevis*) have been moved within South Africa and are now considered invasive in the Western Cape (Measey and Davies, 2011). However, distinguishing between climate-induced range expansion and human-assisted translocation is not always straightforward (e.g. Tolley *et al.*, 2008; Hockey *et al.*, 2011), and is likely to become more complicated in future.

In the face of climate change, indigenous species may have to be translocated for conservation purposes (Mueller and Hellmann, 2008; Richardson *et al.*, 2009). Conservation planning must take climate change into account, establishing appropriate corridors between conservation areas, protecting areas of evolutionary significance, expanding transfrontier approaches, practising managed relocation and meeting the challenges posed by invasive species.

Prevention and Management

The prevention and management of biological invasions forms an integral part of South African policy, legislation and government action aimed at protecting the country's biodiversity. Most legislation of direct relevance to the management of biological invasions is incorporated within the Conservation of Agricultural Resources Act (No 43 of 1983), the National Environmental Management: Protected Areas Act (No 57 of 2003) and the National Environmental Management: Biodiversity Act (No 10 of 2004), although many other legal instruments have clauses that can be invoked in this regard (Richardson *et al.*, 2003). Management interventions have been implemented both nationally (Working for Water at US\$66 million year^{-1}), provincially (KwaZulu-Natal Invasive Alien Species Programme at US\$13 million year^{-1}) and regionally (e.g. the Invasive Alien Species Strategy for the Greater Cape Floristic Region). Research and capacity-building initiatives include the creation of the DST-NRF Centre of Excellence for Invasion Biology (http://academic.sun.ac.za/cib/), a partnership between government and the tertiary education sector. Biological control of invasive species has proven successful for controlling several invasive species (Coetzee *et al.*, 2011; Klein, 2011). Currently, between 1.6 and 2.6% of the total of the Working for Water budget is allocated to biological control, substantially less than is spent on chemical and mechanical control (just under 20% and 80%, respectively) (Le Maitre *et al.*, 2011; van Wilgen and de Lange, 2011). The effective implementation of legislation, creating sufficient capacity to undertake the extension and identification work required to prevent new invasions, and the research capacity to keep pace with changing circumstances are all crucial ingredients of an effective strategy to reduce the rates and impacts of biological invasions in the future.

Merging species distribution models (also called niche models) with mechanistic models has potential for predicting the outcomes of complex interactions involving alien species and changing environmental conditions. This has been done for the invasive tree, *Schinus molle*, in South Africa (Richardson *et al.*, 2010). Although the overall range of *S. molle* is predicted to shrink in South Africa with predicted climate change, the response of the species is likely to be fundamentally different in different biomes. For example, in areas where it is already highly invasive (notably around Kimberley in the Northern Cape), the considerable propagule pressure that already exists, and which will persist, will facilitate persistence and perhaps even drive further spread, even if climate conditions become marginal. In other areas, the changing flow dynamics of rivers (which create microsites for establishment and facilitate dispersal) are likely to shape invasion dynamics in ways that cannot be accommodated in standard species distribution models (Richardson *et al.*, 2010). Ongoing developments in combining mechanistic models with species distribution models could facilitate better predictions.

Overall, the identification of new and emerging invaders, preventing their introduction and managing populations if prevention fails will depend on an informed and aware public, and on science-based policies that take note of the many issues discussed in this chapter. The increased availability of global databases of invasive species is improving our ability to act in preventing further introductions of high-risk species and in calculating and managing the invasion debt. A complete catalogue of all alien species, including those used for agriculture, forestry, horticulture and other purposes, is needed urgently to identify potentially invasive species among already introduced taxa. Proactive early detection and eradication can substantially reduce the costs of control and damage caused by these species (van Wilgen and Richardson, 2009). An important component of mitigation is the effective regulation of the introduction and movement of species. South African legislation to curb such movement is already in place within the National Environmental Management: Biodiversity Act (No 10 of 2004), and the implementation thereof now needs to be addressed. Work will soon start on a National Invasive Species Strategy and Action Plan.

Summary

South African ecosystems are being affected negatively by a wide range of invasive species. Alien invasions and their effects on terrestrial and freshwater systems are currently better understood than marine alien invasions; these are the most prevalent and urgent problems to biodiversity and food security. However, the lack of knowledge and understanding of marine invasions cannot be ignored. It is also clear that numerous potential and ongoing interactions exist between invasive species and aspects of anthropogenic climate change. Our understanding of many of these interactions is conceptual and theoretical, and more illustrative than definitive. None the less, the relationship between climate and invasive species biology is well established, and there is no question that climate change will significantly influence the ecology of biological invasions in this region. For these reasons, several preventive and management efforts could be adapted to help manage changes to system function and services resulting from the interactions between climate change and invasion.

Acknowledgements

This work was funded by the DST-NRF Centre of Excellence for Invasion Biology (C•I•B) and the Working for Water Programme through their collaborative research project on 'Research for Integrated Management of Invasive Alien Species'.

References

Allan, D.G., Harrison, J.A., Navarro, R.A., van Wilgen, B.W. and Thompson, M.W. (1997) The impact of commercial afforestation on bird populations in Mpumalanga Province, South Africa – insights from bird-atlas data. *Biological Conservation* 79, 173–185.

Alston, K.P. and Richardson, D.M. (2006) The roles of habitat features, disturbance, and distance from putative source populations in structuring alien plant invasions at the urban/wildland interface on the Cape Peninsula, South Africa. *Biological Conservation* 132, 183–198.

Annecke, D.P. and Moran, V.C. (1982) *Insects and Mites of Cultivated Plants in South Africa*. Butterworths, Durban, South Africa.

Appleton, C.C. (2003) Alien and invasive fresh water Gastropoda in South Africa. *African Journal of Aquatic Science* 28, 69–81.

Bazzaz, F.A. (1990) The response of natural ecosystems to the rising global CO_2 levels. *Annual Review of Ecology and Systematics* 21, 167–196.

Betts, R.A., Boucher, O., Collins, M., Cox, P.M., Falloon, P.D., Gedney, N., Hemming, D.L., Huntingford, C., Jones, C.D., Sexton, D.M.H. and Webb, M.J. (2007) Projected increase in continental runoff due to plant responses to increasing carbon dioxide. *Nature* 448, 1037–1041.

Blanchard, R., Richardson, D.M., O'Farrell, P. and von Maltitz, G. (2011) Biofuels and biodiversity in South Africa. *South African Journal of Science* 107, 19–26.

Bond, W.J. and Midgley, G.F. (2012) Carbon dioxide and the uneasy interactions of trees and savannah grasses. *Philosophical Transactions of the Royal Society B Biological Sciences* 367, 601–612.

Bond, W.J. and Slingsby, P. (1984) Collapse of an ant–plant mutualism: the Argentine ant (*Iridomyrmex humilis*) and myrmecochorous Proteaceae. *Ecology* 65, 1031–1037.

Branch, G.M. and Steffani, C.N. (2004) Can we predict the effects of alien species? A case-history of the invasion of South Africa by *Mytilus galloprovincialis* (Lamarck). *Journal of Experimental Marine Biology and Ecology* 300, 189–215.

Byers, J.E. (2002) Impact of non-indigenous species on natives enhanced by anthropogenic alteration of selection regimes. *Oikos* 97, 449–458.

Canonico, G.C., Arthington, A., McCrary, J.K. and Thieme, M.L. (2005) The effects of introduced tilapias on native biodiversity. *Aquatic Conservation: Marine and Freshwater Ecosystems* 15, 463–483.

Carlton, J.T. (2000) Global change and biological invasions in the oceans. In: Mooney, H.A. and Hobbs R.J. (eds) *Invasive Species in a Changing World*. Island Press, Washington, DC, pp. 31–53.

Chícharo, L.M.Z. and Chicharo, M.A. (2000) Estimation of the life history parameters of *Mytilus galloprovincialis* (Lamarck) larvae in a coastal lagoon (Ria Formosa – south Portugal). *Journal of Experimental Marine Biology and Ecology* 243, 81–94.

Chown, S.L. and Froneman, P.W. (eds) (2008) *The Prince Edward Islands. Land–Sea Interactions in a Changing Ecosystem*. Sun Media, Stellenbosch, South Africa.

Chown, S.L., Gremmen, N.J.M. and Gaston, K.J. (1998) Ecological biogeography of southern ocean islands: species–area relationships, human impacts, and conservation. *American Naturalist* 152, 562–575.

Chown, S.L., *et al.* (2012) Challenges to the future conservation of the Antarctic. *Science* 337, 158–159.

Christie, S.I. and Scholes, R.J. (1995) Carbon sequestration in Eucalyptus and Pine plantations in South Africa. *Environmental Monitoring and Assessments* 38, 231–241.

Chun, Y.L., van Kleunen, M. and Dawson, W. (2010) The role of enemy release, tolerance and resistance in plant invasions: linking damage to performance. *Ecology Letters* 13, 937–946.

Clark, B.M. (2006) Climate change: a looming challenge for fisheries management in southern Africa. *Marine Policy* 30, 84–95.

Clark, B.M., Steffani, C.N., Young, S., Richardson, A.J. and Lombard, A.T. (2000) The effects of climate change on marine biodiversity in South Africa. Report prepared for the Foundation for Research Development, South African Country Study on Climate Change, Vulnerability and Adaptation Assessment, Marine Biodiversity Section, Pretoria.

Coetzee, J.A., Hill, M.P. and Schlange, D. (2009) Potential spread of the invasive plant *Hydrilla verticillata* in South Africa based on anthropogenic spread and climate suitability. *Biological Invasions* 11, 801–812.

Coetzee, J.A., Hill, M.P., Byrne, M.J. and Bownes, A. (2011) A review of the biological control programmes on *Eichhornia crassipes* (C. Mart.) Solms (Pontederiaceae), *Salvinia molesta* D.S. Mitch. (Salviniaceae), *Pistia stratiotes* L. (Araceae), *Myriophyllum aquaticum* (Vell.) Verdc. (Haloragaceae) and *Azolla filiculoides* Lam. (Azollaceae) in South Africa. *African Entomology* 19, 451–468.

Coviella, C.E. and Trumble, J.T. (1999) Effects of elevated atmospheric carbon dioxide on insect–plant interactions. *Conservation Biology* 13, 700–712.

de Kock, K.N. and Wolmarans, C.T. (1998) A re-evaluation of the occurrence of freshwater molluscs in the Kruger National Park. *Koedoe* 41, 1–8.

de Kock, K.N. and Wolmarans, C.T. (2008) Invasive

alien freshwater snail species in the Kruger National Park, South Africa. *Koedoe* 50, 49–53.

de Moor, I. (2002) Potential impacts of alien freshwater crayfish in South Africa. *African Journal of Aquatic Science* 27, 125–139.

de Villiers, S. (2007) The deteriorating nutrient status of the Berg River, South Africa. *Water SA* 33, 659–664.

de Wit, M.P., Crookes, D.J. and van Wilgen, B.W. (2001) Conflicts of interest in environmental management: estimating the costs and benefits of a tree invasion. *Biological Invasions* 3, 167–178.

Dean, W.R.J. (2000) Alien birds in southern Africa: what factors determine success? *South African Journal of Science* 96, 9–14.

Dean, W.R.J., Anderson, M.D., Milton, S.J. and Anderson, T.A. (2002) Avian assemblages in native *Acacia* and alien *Prosopis* drainage line woodland in the Kalahari, South Africa. *Journal of Arid Environments* 51, 1–19.

DEAT (1997) White paper on the conservation and sustainable use of South Africa's biological diversity. Department of Environmental Affairs and Tourism, Pretoria.

Drinkwater, K.F., Beaugrand, G., Kaeriyama, M., Kim, S., Ottersen, G., Perry, R.I., Pörtner, H.-O., Polovina, J.J. and Takasuka, A. (2010) On the processes linking climate to ecosystem changes. *Journal of Marine Systems* 79, 374–388.

Erasmus, B.F.N., van Jaarsveld, A.S., Chown, S.L., Kshatriya, M. and Wessels, K.J. (2002) Vulnerability of South African animal taxa to climate change. *Global Change Biology* 8, 679–693.

EWT (Endangered Wildlife Trust) (2002) *The Biodiversity of South Africa 2002: Indicators, Trends and Human Impacts.* Struik, Cape Town.

Foxcroft, L.C., Jarošík, V., Pyšek, P., Richardson, D.M. and Rouget, M. (2011) Protected-area boundaries as filters of plant invasions. *Conservation Biology* 25, 400–405.

Frenot, Y., Chown, S.L., Whinam, J., Selkirk, P.M., Convey, P., Skotnicki, M. and Bergstrom, D.M. (2005) Biological invasions in the Antarctic: extent, impacts and implications. *Biological Reviews* 80, 45–72.

Gaertner, M., Den Breeÿen, A., Hui, C. and Richardson, D.M. (2009) Impacts of alien plant invasions on species richness in Mediterranean-type ecosystems: a meta-analysis. *Progress in Physical Geography* 33, 319–338.

Galil, B.S., Nehring, S. and Panov, V. (2007) Waterways as invasion highways – impact of global climate change and globalization. In: Nentwig, W. (ed.) *Biological Invasions.* Springer-Verlag, Berlin, pp. 59–74.

Griffiths, C.L., Robinson, T.B. and Mead, A. (2009) The status and distribution of marine alien species in South Africa. In: Rilov, G. and Crooks, J.A. (eds) *Biological Invasions in Marine Ecosystems.* Springer Verlag, Heidelberg, Germany, pp. 393–407.

Henderson, L. (2006) Comparisons of invasive plants in southern Africa originating from southern temperate, northern temperate and tropical regions. *Bothalia* 36, 201–222.

Henderson, L. and Cilliers, C.J. (2002) *Invasive Aquatic Plants: A Guide to the Identification of the Most Important and Potentially Dangerous Invasive Aquatic and Wetland Plants in South Africa.* Agricultural Research Council, Pretoria.

Herbert, D.G. (2010) *The Introduced Terrestrial Mollusca of South Africa.* SANBI Biodiversity Series 15. South African National Biodiversity Institute, Pretoria.

Higgins, S.I. and Scheiter, S. (2012) Atmospheric CO_2 forces abrupt vegetation shifts locally, but not globally. *Nature* 488, 209–212.

Hill, M.P. (2003) The impact and control of alien aquatic vegetation in South African aquatic ecosystems. *African Journal of Aquatic Science* 28, 19–24.

Hobbs, R.J. and Mooney, H.A. (2005) Invasive species in a changing world: the interactions between global change and invasives. In: Mooney, H.A., Mack, R.N., McNeely, J.A., Neville, L.E., Schei, P.J. and Waage, J.K. (eds) *Invasive Alien Species: A New Synthesis.* Island Press, Washington, DC, pp. 310–331.

Hockey, P.A.R., Sirami, C., Ridley, A.R., Midgley, G.F. and Babiker, H.A. (2011) Interrogating recent range changes in South African birds: confounding signals from land use and climate change present a challenge for attribution. *Diversity and Distributions* 17, 254–261.

Hulme, P.E., *et al.* (2008) Grasping at the routes of biological invasions: a framework for integrating pathways into policy. *Journal of Applied Ecology* 45, 403–414.

Huntley, B. and Barnard, P. (2012) Potential impacts of climatic change on southern African birds of fynbos and grassland biodiversity hotspots. *Diversity and Distributions* 18, 769–781.

Jarošík, V., Pyšek, P., Foxcroft, L.C., Richardson, D.M., Rouget, M. and MacFadyen, S. (2011) Predicting incursion of plant invaders into Kruger National Park, South Africa: the interplay of general drivers and species-specific factors. *PloS One* 6, e28711.

Johnson, P.T.J., Olden, J.D. and van der Zanden, M.J. (2008) Dam invaders: impoundments facilitate biological invasions into freshwaters. *Frontiers in Ecology and the Environment* 6, 357–363.

Kalwij, J.M., Robertson, M.P. and van Rensburg, B.J. (2008) Human activity facilitates altitudinal expansion of exotic plants along a road in montane grassland, South Africa. *Applied Vegetation Science* 11, 491–498.

King, J.R. and Tschinkel, W.R. (2008) Experimental evidence that human impacts drive fire ant invasions and ecological change. *Proceedings of the National Academy of Sciences* 105, 20339–20343.

Klein, H. (2011) A catalogue of the insects, mites and pathogens that have been used or rejected, or are under consideration, for the biological control of invasive alien plants in South Africa. *African Entomology* 19, 515–549.

Knoesen, D., Schulze, R.E., Pringle, C., Summerton, M., Dickens, C. and Kunz, R. (2009) Water for the future: impacts of climate change on water resources in the Orange-Senqu River basin. Report to NeWater, a project funded under the Sixth Research Framework of the European Union. Institute of Natural Resources, Pietermaritzburg, South Africa.

Kolar, C.S. and Lodge, D.M. (2000) Freshwater nonindigenous species: interactions with other global changes. In: Mooney, H.A. and Hobbs, R.J. (eds) *Invasive Species in a Changing World.* Island Press, Washington, DC, pp. 3–30.

Kruger, A.C. (2006) Observed trends in daily precipitation indices in South Africa: 1910–2004. *International Journal of Climatology* 26, 2275–2285.

Kruger, A.C. and Sekele, S.S. (2012) Trends in extreme temperature indices in South Africa: 1962–2009. *International Journal of Climatology* doi:10.1002/joc.3455.

Lach, L., Picker, M.D., Colville, J.F., Allsopp, M.H. and Griffiths, C.L. (2002) Alien invertebrate animals in South Africa. In: Pimentel, D. (ed.) *Biological Invasions: Economic and Environmental Costs of Alien Plant, Animal, and Microbe Species.* CRC Press, Boca Raton, Florida, pp. 267–282.

Laird, M.C. and Griffiths, C.L. (2008) Present distribution and abundance of the introduced barnacle *Balanus glandula* Darwin in South Africa. *African Journal of Marine Science* 30, 93–100.

Le Maitre, D.C., Richardson, D.M. and Chapman, R.A. (2004) Alien plant invasions in South Africa: driving forces and the human dimension. *South African Journal of Science* 100, 103–112.

Le Maitre, D.C., de Lange, W.J., Richardson, D.M., Wise, R.M. and van Wilgen, B.W. (2011) The economic consequences of the environmental impacts of alien plant invasions in South Africa. In: Pimentel, D. (ed.) *Biological Invasions: Economic and Environmental Costs of Alien Plant, Animal, and Microbe Species.* CRC Press, New York, pp. 295–323.

le Roux, P.C. and McGeoch, M.A. (2008) Changes in climate extremes, variability and signature on sub-Antarctic Marion Island. *Climatic Change* 86, 309–329.

Lindroth, R.L. (2010) Impacts of elevated atmospheric CO_2 and O_3 on forests: phytochemistry, trophic interactions, and ecosystem dynamics. *Journal of Chemical Ecology* 36, 2–21.

Lockwood, J.L., Cassey, P. and Blackburn, T.M. (2005) The role of propagule pressure in explaining species invasions. *Trends in Ecology and Evolution* 20, 223–228.

Lowe, S.R., Woodford, D.J., Impson, N.D. and Day, J.A. (2008) The impact of invasive fish and invasive riparian plants on the invertebrate fauna of the Rondegat River, Cape Floristic Region, South Africa. *African Journal of Aquatic Science* 33, 51–62.

Macdonald, I.A.W., Kruger, F.J. and Ferrar, A.A. (eds) (1986a) *The Ecology and Management of Biological Invasions in Southern Africa.* Oxford University Press, Cape Town.

Macdonald, I.A.W., Richardson, D.M. and Powrie, F.J. (1986b) Range expansion of the hadeda ibis *Bostrychia hagedash* in southern Africa. *South African Journal of Zoology* 21, 331–342.

Macdonald, I.A.W., Reaser, J.K., Bright, C., Neville, L.E., Howard, G.W., Murphy, S.J. and Preston, G. (eds) (2003) *Invasive Alien Species in Southern Africa: National Reports and Directory of Resources.* Global Invasive Species Programme, Cape Town.

Marais, C. and Wannenburgh, A.M. (2008) Restoration of water resources (natural capital) through the clearing of invasive alien plants from riparian areas in South Africa – costs and water benefits. *South African Journal of Botany* 74, 526–537.

Mead, A., Carlton, J.T., Griffiths, C.L. and Rius, M. (2011) Revealing the scale of marine bioinvasions in developing regions. *Biological Invasions* 13, 1991–2008.

Measey, G.J. and Davies, S.J. (2011) Struggling against domestic exotics at the southern end of Africa. *FrogLog* 97, 28–30.

Mucina, L. and Rutherford, M.C. (2006) *The Vegetation of South Africa, Lesotho and Swaziland.* Strelitzia 19. South African National Biodiversity Institute, Pretoria.

Mueller, J.M. and Hellmann, J.J. (2008) An assessment of invasion risk from assisted migration. *Conservation Biology* 22, 562–567.

Nel, J.L., Richardson, D.M., Rouget, M., Mgidi,

T.N., Mdzeke, N., Le Maitre, D.C., van Wilgen, B.W., Schonegevel, L., Henderson, L. and Neser, S. (2004) A proposed classification of invasive alien plant species in South Africa: towards prioritizing species and areas for management action. *South African Journal of Science* 100, 53–64.

O'Brien, E.M. (1998) Water–energy dynamics, climate, and prediction of woody plant species richness: an interim general model. *Journal of Biogeography* 25, 379–398.

O'Dowd, D.J., Green, P.T. and Lake, P.S. (2003) Invasion 'meltdown' on an oceanic island. *Ecology Letters* 6, 812–817.

Oschadleus, H.D. and Underhill, L.G. (2006) Range expansion of the red-billed quelea, *Quelea quelea*, into the Western Cape, South Africa. *South African Journal of Science* 102, 12–13.

Piao, S., Friedlingstein, P., Ciais, P., de Noblet-Ducoudré, N., Labat, D. and Zaehle, S. (2007) Changes in climate and land use have a larger direct impact than rising CO_2 on global river runoff trends. *Proceedings of the National Academy of Sciences* 104, 15242–15247.

Pimentel, D. (ed.) (2002) *Biological Invasions: Economic and Environmental Costs of Alien Plant, Animal, and Microbe Species*. CRC Press, Boca Raton, Florida.

Pimentel, D., McNair, S., Janecka, J., Wightman, J., Simmonds, C., O'Connell, C., Wong, E., Russel, L., Zern, J., Aquino, T. and Tsomondo, T. (2001) Economic and environmental threats of alien plant, animal, and microbe invasions. *Agriculture, Ecosystems and Environment* 84, 1–20.

Piola, R.F. and Johnston, E.L. (2008) Pollution reduces native diversity and increases invader dominance in marine hard-substrate communities. *Diversity and Distributions* 14, 329–342.

Polley, H.W., Johnson, H.B. and Tischler, C.R. (2002) Woody invasion of grasslands: evidence that CO_2 enrichment indirectly promotes establishment of *Prosopis glandulosa*. *Plant Ecology* 164, 85–94.

Pyšek, P., Richardson, D.M., Rejmánek, M., Webster, G.L., Williamson, M. and Kirschner, J. (2004) Alien plants in checklists and floras: towards better communication between taxonomists and ecologists. *Taxon* 53, 131–143.

Raghu, S., Anderson, R.C., Daehler, C.C., Davis, A.S., Wiedenmann, R.N., Simberloff, D. and Mack, R.N. (2006) Adding biofuels to the invasive species fire? *Science* 313, 1742–1742.

Rajagopal, S., van der Velde, G., Jansen, J., van der Gaag, M., Atsma, G., Janssen-Mommen, J.P.M., Polman, H. and Jenner, H.A. (2005) Thermal tolerance of the invasive oyster *Crassostrea gigas*: feasibility of heat treatment as an antifouling option. *Water Research* 39, 4335–4342.

Richardson, D.M. (1998) Forestry trees as invasive aliens. *Conservation Biology* 12, 18–26.

Richardson, D.M. and van Wilgen, B.W. (2004) Invasive alien plants in South Africa: how well do we understand the ecological impacts? *South African Journal of Science* 100, 45–52.

Richardson, D.M., Macdonald, I.A.W., Hoffmann, J.H. and Henderson, L. (1997) Alien plant invasions. In: Cowling, R.M., Richardson, D.M. and Pierce, S.M. (eds) *The Vegetation of Southern Africa.* Cambridge University Press, Cambridge, UK, pp. 535–570.

Richardson, D.M., Bond, W.J., Dean, W.R.J., Higgins, S.I., Midgley, G.F., Milton, S.J., Powrie, L.W., Rutherford, M.C., Samways, M.J. and Schulze, R.E. (2000) Invasive alien species and global change: a South African perspective. In: Mooney, H.A. and Hobbs, R.J. (eds) *Invasive Species in a Changing World.* Island Press, Washington, DC, pp. 303–349.

Richardson, D.M., Cambray, J.A., Chapman, R.A., Dean, W.R.J., Griffiths, C.L., Le Maitre, D.C., Newton, D.J. and Winstanley, T.J. (2003) Vectors and pathways of biological invasions in South Africa – past, present, and future. In: Ruiz, G.M. and Carlton, J.T. (eds) *Invasive Species: Vectors and Management Strategies.* Island Press, Washington, DC, pp. 292–349.

Richardson, D.M., Rouget, M., Ralston, S.J., Cowling, R.M., van Rensburg, B.J. and Thuiller, W. (2005) Species richness of alien plants in South Africa: environmental correlates and the relationship with indigenous plant species richness. *Écoscience* 12, 391–402.

Richardson, D.M., *et al.* (2009) Multidimensional evaluation of managed relocation. *Proceedings of the National Academy of Sciences* 106, 9721–9724.

Richardson, D.M., Iponga, D.M., Roura-Pascual, N., Krug, R.M., Milton, S.J., Hughes, G.O. and Thuiller, W. (2010) Accommodating scenarios of climate change and management in modelling the distribution of the invasive tree *Schinus molle* in South Africa. *Ecography* 33, 1049–1061.

Richardson, D.M., Wilson, J.R.U., Weyl, O.L.F. and Griffiths, C.L. (2011) South Africa: invasions. In: Simberloff, D. and Rejmánek, M. (eds) *Encyclopedia of Biological Invasions.* University of California Press, Berkeley, California, pp. 643–651.

Robinson, T.B., Griffiths, C.L., McQuaid, C. and Rius, M. (2005) Marine alien species of South Africa – status and impacts. *African Journal of Marine Science* 27, 297–306.

Rouault, M., Penven, P. and Pohl, B. (2009) Warming in the Agulhas Current system since the 1980's. *Geophysical Research Letters* 36, L12602, doi:10.1029/2009GL037987.

Rouget, M., Richardson, D.M., Nel, J.L., Le Maitre, D.C., Egoh, B. and Mgidi, T. (2004) Mapping the potential ranges of major plant invaders in South Africa, Lesotho and Swaziland using climatic suitability. *Diversity and Distributions* 10, 475–484.

Samways, M.J. and Moore, S.D. (1991) Influence of exotic conifer patches on grasshopper (Orthoptera) assemblages in a grassland matrix at a recreational resort, Natal, South Africa. *Biological Conservation* 57, 117–137.

Samways, M.J. and Taylor, S. (2004) Impacts of invasive alien plants on Red-Listed South African dragonflies (Odonata). *South African Journal of Science* 100, 78–80.

Samways, M.J., Caldwell, P.M. and Osborn, R. (1996) Ground-living invertebrate assemblages in native, planted and invasive vegetation in South Africa. *Agriculture, Ecosystems and Environment* 59, 19–32.

Samways, M.J., Taylor, S. and Tarboton, W. (2005) Extinction reprieve following alien removal. *Conservation Biology* 19, 1329–1330.

Schulze, R.E., Maharaj, M., Lynch, S.D., Howe, B.J. and Melvil-Thompson, B. (1997) *South African Atlas of Agrohydrology and Climatology*. Beta 1.002. Computing Centre for Water Research, Pietermaritzburg, South Africa.

Shelton, J.M., Day, J.A. and Griffiths, C.L. (2008) Influence of largemouth bass, *Micropterus salmoides*, on abundance and habitat selection of Cape galaxias, *Galaxias zebratus*, in a mountain stream in the Cape Floristic Region, South Africa. *African Journal of Aquatic Science* 33, 201–210.

Simberloff, D. and Von Holle, B. (1999) Positive interactions of nonindigenous species: invasional meltdown? *Biological Invasions* 1, 21–32.

Smith, A.M., Guastella, L.A., Bundy, S.C. and Mather, A.A. (2007) Combined marine storm and Saros spring high tide erosion events along the KwaZulu-Natal coast in March 2007. *South African Journal of Science* 103, 274–276.

Smith, S.D., Huxman, T.E., Zitzer, S.F., Charlet, T.N., Housman, D.C., Coleman, J.S., Fenstermaker, L.K., Seemann, J.R. and Nowak, R.S. (2000) Elevated CO_2 increases productivity and invasive species success in an arid ecosystem. *Nature* 408, 79–81.

Smith, V.R. (2002) Climate change in the sub-Antarctic: an illustration from Marion Island. *Climatic Change* 52, 345–357.

Smith, V.R. and Steenkamp, M. (1990) Climatic change and its ecological implications at a subantarctic island. *Oecologia* 85, 14–24.

Spear, D. and Chown, S.L. (2008) Taxonomic homogenization in ungulates: patterns and mechanisms at local and global scales. *Journal of Biogeography* 35, 1962–1975.

Spear, D. and Chown, S.L. (2009) The extent and impacts of ungulate translocations: South Africa in a global context. *Biological Conservation* 142, 353–363.

Steenkamp, H.E. and Chown, S.L. (1996) Influence of dense stands of an exotic tree, *Prosopis glandulosa* Benson, on a savanna dung beetle (Coleoptera: Scarabaeinae) assemblage in southern Africa. *Biological Conservation* 78, 305–311.

Taylor, P.J., Arntzen, L., Hayter, M., Iles, M., Frean, J. and Belmain, S. (2008) Understanding and managing sanitary risks due to rodent zoonoses in an African city: beyond the Boston Model. *Integrative Zoology* 3, 38–50.

Thuiller, W., Richardson, D.M. and Midgley, G.F. (2007) Will climate change promote alien species invasions? In: Nentwig, W. (ed.) *Biological Invasions*. Springer Verlag, Heidelberg, Germany, pp. 197–211.

Tolley, K.A., Davies, S.J. and Chown, S.L. (2008) Deconstructing a controversial local range expansion: conservation biogeography of the painted reed frog (*Hyperolius marmoratus*) in South Africa. *Diversity and Distributions* 14, 400–411.

Trakhtenbrot, A., Nathan, R., Perry, G. and Richardson, D.M. (2005) The importance of long-distance dispersal in biodiversity conservation. *Diversity and Distributions* 11, 173–181.

van der Merwe, M.R., Dippenaar-Schoeman, A.S. and Scholtz, C.H. (1996) Diversity of ground-living spiders at Ngome State Forest, Kwazulu/Natal: a comparative survey in indigenous forest and pine plantations. *African Journal of Ecology* 34, 342–350.

van der Waal, B.C.W. and Bills, R. (2000) *Oreochromis niloticus* (Teleostei: Cichlidae) now in the Limpopo River system. *South African Journal of Science* 96, 47–48.

van Rensburg, B.J., Chown, S.L. and Gaston, K.J. (2002) Species richness, environmental correlates, and spatial scale: a test using South African birds. *American Naturalist* 159, 566–577.

van Rensburg, B.J., Chimimba, C.T., Davies, S.J., Peacock, D.S., Spear, D., van Wilgen, N.J. and Weyl, O.L.F. (2011) Invasive vertebrates of South Africa. In: Pimentel, D. (ed.) *Biological*

Invasions: Economic and Environmental Costs of Alien Plant, Animal, and Microbe Species. CRC Press, New York, pp. 325–378.

van Rooyen, M.W., van Rooyen, N. and Stoffberg, G.H. (2013) Carbon sequestration potential of post-mining reforestation activities on the KwaZulu-Natal coast, South Africa. *Forestry* 86, 211–223.

van Vliet, M.T.H., Franssen, W.H.P., Yearsley, J.R., Ludwig, F. Haddeland, I., Lettenmaier, D.P. and Kabat, P. (2013) Global river discharge and water temperature under climate change. *Global Environmental Change* (http://dx.doi.org/10.1016/j.gloenvcha.2012.11.002, accessed 15 January 2014).

van Wilgen, B.W. and de Lange, W.J. (2011) The costs and benefits of biological control of invasive alien plants in South Africa. *African Entomology* 19, 504–514.

van Wilgen, B.W. and Richardson, D.M. (2009) Current and future consequences of invasion by alien species: a case study from South Africa. In: Perrings, C., Mooney, H.A. and Williamson, M. (eds) *Bioinvasions and Globalization.* Oxford University Press, Oxford, UK, pp. 183–201.

van Wilgen, B.W. and Richardson, D.M. (2012) Three centuries of managing introduced conifers in South Africa: benefits, impacts, changing perceptions and conflict resolution. *Journal of Environmental Management* 106, 56–68.

van Wilgen, B.W., Richardson, D.M., Le Maitre, D.C., Marais, C. and Magadlela, D. (2001) The economic consequences of alien plant invasions: examples of impacts and approaches to sustainable management in South Africa. *Environment, Development and Sustainability* 3, 145–168.

van Wilgen, B.W., Reyers, B., Le Maitre, D.C., Richardson, D.M. and Schonegevel, L. (2008b) A biome-scale assessment of the impact of invasive alien plants on ecosystem services in South Africa. *Journal of Environmental Management* 89, 336–349.

van Wilgen, B.W., Dyer, C., Hoffmann, J.H., Ivey, P., Le Maitre, D.C., Moore, J.L., Richardson, D.M., Rouget, M., Wannenburgh, A. and Wilson, J.R.U. (2011) National-scale strategic approaches for managing introduced plants: insights from Australian acacias in South Africa. *Diversity and Distributions* 17, 1060–1075.

van Wilgen, B.W., Forsyth, G.G., Le Maitre, D.C., Wannenburgh, A., Kotzé, J.D.F., van den Berg, E. and Henderson, L. (2012) An assessment of the effectiveness of a large, national-scale invasive alien plant control strategy in South Africa. *Biological Conservation* 148, 28–38.

van Wilgen, N.J., Richardson, D.M. and Baard, E.H.W. (2008a) Alien reptiles and amphibians in South Africa: towards a pragmatic management strategy. *South African Journal of Science* 104, 13–20.

van Wilgen, N.J., Wilson, J.R.U., Elith, J., Wintle, B. and Richardson, D.M. (2010) Alien invaders and reptile traders: what drives the live animal trade in South Africa? *Animal Conservation* 13 (Suppl. 1), 24–32.

von Breitenbach, F. (1990) *National List of Introduced Trees.* Dendrological Foundation, Pretoria.

Weyl, O.L.F. and Lewis, H. (2006) First record of predation by the alien invasive freshwater fish *Micropterus salmoides* L. (Centrarchidae) on migrating estuarine fishes in South Africa. *African Zoology* 41, 294–296.

Wigley, B.J., Bond, W.J. and Hoffman, M.T. (2010) Thicket expansion in a South African savanna under divergent land use: local vs. global drivers? *Global Change Biology* 16, 964–976.

Wilson, J.R.U., Dormontt, E.E., Prentis, P.J., Lowe, A.J. and Richardson, D.M. (2009) Something in the way you move: dispersal pathways affect invasion success. *Trends in Ecology and Evolution* 24, 136–144.

Zardi, G.I., Nicastro, K.R., McQuaid, C.D. and Erlandsson, J. (2008) Sand and wave induced mortality in invasive (*Mytilus galloprovincialis*) and indigenous (*Perna perna*) mussels. *Marine Biology* 153, 853–858.

Zavala, J.A., Casteel, C.L., DeLucia, E.H. and Berenbaum, M.R. (2008) Anthropogenic increase in carbon dioxide compromises plant defense against invasive insects. *Proceedings of the National Academy of Sciences* 105, 5129–5133.

Ziska, L.H. (2003) Evaluation of the growth response of six invasive species to past, present and future atmospheric carbon dioxide. *Journal of Experimental Botany* 54, 395–404.

10 Climate Change and 'Alien Species in National Parks': Revisited

Thomas J. Stohlgren,[1] Jessica R. Resnik[2] and Glenn E. Plumb[2]

[1]US Geological Survey, Fort Collins Science Center, Fort Collins, Colorado, USA; [2]National Park Service, Biological Resource Management Division, Fort Collins, Colorado, USA

Abstract

Many units comprising the National Park Service system in the USA were established to protect native species and historical landscapes and scenery, and to provide public enjoyment of the same, as long as the natural area remained 'unimpaired for future generations'. However, a growing human population and a global economy have spawned a 40-fold increase in global trade and transportation since 1950. This created what may turn out to be the most significant challenge to Park Service management: the invasion of alien (i.e. exotic, non-native) plants, animals and diseases into 'protected areas'. Climate change may compound the problem by interacting with other existing ecosystem stressors to affect the distribution, abundance and impact of invasive species. Scientists' warnings have not been entirely ignored, but harmful invasions may be outpacing effective managerial responses. This chapter revisits a paper published in 1995 by D.B. Houston and E.G. Schreiner entitled 'Alien Species in National Parks: Drawing Lines in Space and Time', to demonstrate that despite efforts to contain alien plants, animals and diseases, the number and abundance of conspicuous invaders may be increasing. New and spreading alien species are directly affecting native species abundance, their genetic integrity and the management of adjacent lands, and human health. While climate change may add an additional or interacting threat to park resources, we found that at park scales and over short periods (17 years in this case), it has become increasingly difficult to achieve the primary mission of the National Park Service, for example protecting native species and ecosystems for present and future generations, due to the continuing invasions of alien organisms largely resulting from increased trade and transportation.

Introduction

Because of their beauty, uniqueness and historical value, the National Park Service (Park Service) was created to protect natural resources and cultural features (National Park Service Organic Act 1916; 16 U.S.C. 1 2 3, and 4). The legislation is quite clear. The National Park Service is directed:

> ... to conserve the scenery and the natural and historic objects and the wild life therein and to provide for the enjoyment of the same in such manner and by such means as will leave them unimpaired for the enjoyment of future generations.

However, a brief history of invasive species management in national parks, adapted from Houston and Schreiner (1995) and Drees (2004) and expanded here, shows how, each year, the National Park Service may be falling further behind in its primary obligation.

About 80 years ago, Wright *et al.* (1932) suggested that alien animals posed a significant threat to national parks. They identified the purposeful introduction of game birds and fish, the accidental release of alien species and the human-caused range extensions of alien species as inappropriate for areas set aside to preserve natural conditions.

In the 1960s, two influential reports also asserted that the introduction of alien species was inappropriate for areas set aside to preserve natural resources (Leopold *et al.*, 1963; Robbins, 1963). In 1968, National Park Service policy stated that 'nonnative species may not be introduced into natural areas. Where they have become established or threaten invasion of a natural area, an appropriate management plan should be developed to control them, where feasible...' and that 'nonnative species of plants and animals will be eliminated where it is possible to do so by approved methods which will preserve wilderness qualities' (NPS, 1968). By the 1990s, scientists urged policy makers more strongly on the potential for alien species to displace native species and change so-called 'protected' landscapes (Usher *et al.*, 1988; Vitousek, 1990; Mack *et al.*, 2000; Pimentel *et al.*, 2000; Myers and Bazely, 2003). More recently, environmental changes, including climate change and invasive species, have been recognized by the National Park System Advisory Board (2012) as requiring urgent re-examination of Park Service resource management and stewardship principles as described in the Leopold report (Leopold *et al.*, 1963). *Revisiting Leopold* (NPS Advisory Board, 2012) urged integration of the precautionary principle, whereby stewardship decisions should be 'conservative in allowing actions and activities that may heighten impairment of park resources' and avoid 'actions and activities that may irreversibly impact park resources and systems'.

The notable paper by Houston and Schreiner (1995) entitled 'Alien Species in National Parks: Drawing Lines in Space and Time', drew attention to a dozen or more conspicuous invaders including: eastern brook trout (*Salvelinus fontinalis*) in Yellowstone National Park (NP) and Olympic NP; feral goats (*Capra hircus*) and eucalyptus trees (*Eucalyptus deglupta*) in Maui's Haleakala NP; European wild boars (*Sus scrofa*) in Great Smoky Mountains NP; free-ranging burros (*Equus asinus*) in Grand Canyon NP and Bandelier National Monument; bison (*Bison bison*) in Wrangell-St Elias NP; horses (*Equus caballus*) in Ozark National Scenic Riverways; and mountain goats (*Oreamnos americanus*) in Olympic NP. The purpose of the Houston and Schreiner (1995) paper was not to conduct an exhaustive survey of invasive species in National Park Service units but to provide a few case histories in which determination of what was considered an alien species was resolved by the courts. The paper demonstrated that: 'Improved conservation will require revisiting agency policies and definitions, including those concerned with alien species' (p. 208).

In 2006, the National Park Service clarified its policy regarding invasive species:

> All exotic plant and animal species that are not maintained to meet an identified park purpose will be managed—up to and including eradication—if (1) control is prudent and feasible, and (2) the exotic species:
>
> - interferes with natural processes and the perpetuation of natural features, native species or natural habitats, or
> - disrupts the genetic integrity of native species, or
> - disrupts the accurate presentation of a cultural landscape, or damages cultural resources, or
> - significantly hampers the management of park or adjacent lands, or
> - poses a public health hazard as advised by the U.S. Public Health Service (which includes the Centers for Disease Control and the NPS public health program), or
> - creates a hazard to public safety.
>
> (NPS, 2006; Section 4.4.4.2 Removal of Exotic Species Already Present)

Presently, intrepid but modest efforts continue to be made by the National Park Service in containing harmful invasive species. For example, Exotic Plant Management Teams continue to control invasive weeds in many parks. Many parks have programmes to contain or eradicate at least some alien animal species. Ongoing programmes in Yellowstone NP are intended to protect native cutthroat trout (*Oncorhynchus clarki* spp.) from invasive lake trout (*Salvelinus namaycush*). Grand Canyon NP continues to remove invasive burros. Channel Islands NP eradicated alien ungulates from Santa Cruz and Santa Rosa islands. Several parks are also removing some alien fish to restore native fish populations.

This chapter's objective is to revisit the paper by Houston and Schreiner (1995) to ask three questions: (i) how much progress has been made in protecting National Park Service units from harmful invasive species? (ii) are new or conspicuous invasions outpacing containment efforts? and (iii) in light of climate change and trade concerns, what approaches should the National Park Service consider in addressing invasive species?

Some Progress in Protecting the Parks

We know National Park Service officials were aware of hundreds of alien species in parks, and that they were attempting to contain many species, when the Houston and Schreiner (1995) paper was published. Great Smoky Mountains NP implemented control actions for the feral hog in 1959; it may be the longest-running invasive animal programme in the National Park Service. However, the first service-wide report providing a spatial assessment of alien plant infestations was produced 9 years later. Drees (2004) reported that, of the 34 million ha (Mha) managed by the National Park Service, 1.1 Mha were infested by alien plants and animals. Many of the alien animals, including feral hogs and goats, hemlock woolly adelgid, New Zealand mudsnail and African oryx, and the West Nile virus, were rampant in the Park System. Some staff of the National Park Service unofficially estimated that staffing and funding for a viable invasive species programme might cost US$80 million year^{-1}, but that level of funding far outstripped the agency's capacity. For example, in 2003, Exotic Plant Management Teams received US$2.8 million for weed control in national parks (Drees, 2004). While scientists at the time argued that invasive species science and management remained fragmentary and underfunded relative to the scope of the problem (Vitousek, 1990; Mack *et al.*, 2000; Pimentel *et al.*, 2000, Myers and Bazely, 2003), some Park Service officials remained optimistic. 'If we stay the course, management of invasive species in parks is within our grasp' (Drees, 2004).

Alien plant species

It is difficult to assess the progress made in containing alien plant species in the National Parks. Currently, there are 17 Exotic Plant Management Teams spread over 243 parks, with over 700 invasive plant species targeted for containment. Preliminary results (http://www.ipcconnect.com/forestglen/Swearingen.pdf) suggest that between 2001 and 2010, 43,760 ha (108,135 acres) were treated of the 102,955 ha (254,407 acres) identified as infested. However, estimates of gross infested areas (any hectare containing a target plant species) increased from 457,544 ha (1,130,617 acres) to 1,092,600 ha (~2.7 million acres) infested over that same period, due to better surveys and continued invasions. Potential target species have increased from 700 to 5600 alien plant species. The National Park Service reports the number of 'treated acres', but it is unknown whether these areas are reinvaded by the same or other invasive plants species a few years later. Despite significant progress reported by the Exotic Plant Management Teams, concern remains whether management efforts are keeping pace with recurring plant invasions and new invaders. International trade and travel continue to be the ultimate drivers of invasive species introductions (Levine and D'Antonio, 2003, Lodge *et al.*, 2006). Whether or not climate change enhances the success of invasive species, however, depends on the species in question and their effects on ecosystem interactions; recent climate change is linked to both increases and decreases in population size (Dukes and Mooney, 1999; McCarty, 2001).

An alarming study by Allen *et al.* (2009) reported 20,305 alien plant species infestations, with 3756 unique alien plants, totalling 7.3 Mha in 218 national parks. Note that the new figure of 7.3 Mha of alien plant infestations greatly exceeded the 2004 estimate of 1.1 Mha infested by alien plants *and* animals reported by Drees (2004). Allen *et al.* (2009) also reported that many of the 243 parks within the Exotic Plant Management Team network had inadequate data, or mapped data for only a small fraction of their invasive species. Thus, the reported 7.3 Mha of parklands infested with alien plants is likely underestimated. Many national parks reported hundreds of alien plant species. Chesapeake and Ohio Canal National Historic Park reported 483 alien plant species. Alien plant species richness was associated significantly with the richness of rare plant species and native plant richness, and roads and trails may be important sources of harmful invasive species (Gelbard and Belnap, 2003; Allen *et al.*, 2009).

Other scientists showed that plant invasions directly affected fire regimes, including the frequency, intensity, extent, type and seasonality of fire, described in Brooks *et al.* (2004). The authors demonstrate the threat of an 'invasive plant–fire regime cycle' developing, where increased fire is caused by, and promotes, invasions. Altered precipitation and temperature patterns resulting from climate change can also be expected to change fire regimes, further facilitating the spread of some invasive species (López-Hoffman and Quijada-Mascareñas, 2012). Another recent example comes from Saguaro NP in Arizona. Saguaro cacti are now severely threatened by the invasion of buffelgrass (*Pennisetum ciliare*), a fire-prone perennial bunchgrass from Africa. Just in the past decade, buffelgrass has spread rapidly across the park and greatly threatens the Sonoran Desert ecosystem, where fire historically played a minor role (Esque *et al.*, 2004, 2006).

And the alien plant species keep coming. In Grand Canyon NP, careful, long-term monitoring has revealed a continued increase in alien plant species (Fig. 10.1). Five alien plant species were thought to arrive in 2011, based on field surveys in previously well-surveyed areas. Cutleaf

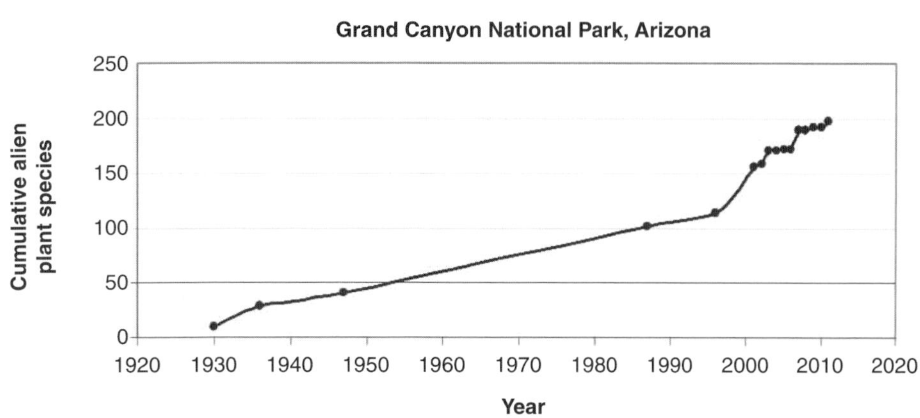

Fig. 10.1. Cumulative exotic plant species reported in Grand Canyon National Park over time. (Data from Lori J. Makarick, NPS; for more information see Stohlgren *et al.*, 2013b.)

vipergrass (*Scorzonera lacinata*) had only one previous specimen collected in Arizona, but based on its rapid spread in other states, it is likely to keep expanding its range (Stohlgren *et al.*, 2013b).

Alien animal species

An assessment of the status of invasive animals in national parks was in the early stages of development at the time of writing. However, there are numerous examples of efforts by the agency to control invasive animals in its parks since the Houston and Schreiner (1995) paper. In a recent 5-year period, the National Park Service has spent between US$8 and US$12 million year^{-1} to control invasive animal species in parks (NPS Reports to Congress FY07–FY11). It is important to recognize that enabling legislation specific to each park establishes its management mandates and priorities. Consequently, the National Park Service essentially manages alien animals of two types: appealing alien animals (e.g. feral horses at Ozark National Scenic Riverways) and non-appealing alien animals (e.g. feral hogs wherever they persist). While the categories are convenient, they are an oversimplification. The point has likely been made in earlier chapters of this book but it is important to reiterate that invasive species can be either native or alien; and alien animals can include species that may be native to one area of the USA but not native to another area (e.g. mountain goats in Olympic NP). Thus, careful consideration must be given to the park and the species in question when deciding on management approaches.

Let us revisit briefly the case studies examined by Houston and Schreiner (1995; Table 10.1). While we understand that Houston and Schreiner (1995) did not intend to be comprehensive, some comparisons to the present day may still be instructive. For example, their paper specifically mentioned eastern brook trout (*Salvelinus fontinalis*) in Yellowstone and Olympic NPs. Despite some progress to control this fish in Yellowstone NP, populations of eastern brook trout are still thriving in both parks. A recent review of National Park Service data (Table 10.2) lists that particular alien fish species in 33 park units, and climate change could allow further range expansion as temperatures warm in the coldest streams, where native bull trout (*Salvelinus confluentus*) currently have a competitive edge (Rieman *et al.*, 1997; Rieman and Isaak, 2007). Houston and Schreiner (1995) mention European wild boars (*Sus scrofa*) in Great Smoky Mountains NP; they currently occur in 29 park units (Table 10.2). Although Houston and Schreiner (1995) did not include a case study involving alien birds, five of the most frequently reported alien animals in national park units are birds (Table 10.2).

It is not all doom and gloom – there are some success stories. To name a few, alien ungulates have been eradicated successfully from Channel Islands NP, and Pinnacles National Monument is on track to being certifiably 'feral hog free'. Yellowstone NP finalized native fish conservation plans in 2011, calling for the removal of alien trout from its waterways. Hawai'i Volcanoes NP is close to finalizing a management plan for non-native ungulates, and Big Bend NP is drafting a management plan for all alien species in the park. Since 2007, at least 28 management plans incorporating invasive species issues have initiated the National Environmental Policy Act process; and the Service has conducted over 30 technical assistance projects relating to invasive animal management in parks. Cumulatively, the National Park Service is now attempting to address invasive animal species across the spectrum of taxa: amphibians, birds, fish, invertebrates (aquatic and terrestrial), mammals and reptiles.

Despite some progress, much work is left to be done. A subset of alien animals in national parks may deserve special attention – major predators. In the eyes of many native freshwater invertebrates, invasive alien fish in many headwater lakes and streams in national parks are apex predators. Most of these populations were stocked by agencies tasked to protect native species, or by recreationists (Marnell, 1995; Dunham

Table 10.1. Updated status of selected invasive species from Houston and Schreiner (1995).

Species/Park noted by Houston and Schreiner (1995)	Status now	Additional notable invaders since 1995
Mountain goats (*Oreamnos americanus*) in Olympic NP (source: Patti Happe, Wildlife Branch Chief, personal communication, 19 June 2012)	The population persists in the Park. The population is growing after years of low and stable numbers. Lack of air support has made removal of the goats difficult. One visitor was fatally wounded by a goat (lawsuit is pending).	Barred owls (*Strix varia*), first documented in 1985 on the Olympic Peninsula, are heavily outcompeting the rare northern spotted owl (*Strix occidentalis caurina*). Deer lice (*Bovicola tibialis*) have blasted through the black-tailed deer (*Odocoileus hemionus sitkensis*) population. Coyotes (*Canis latrans*), not native to the Park, have expanded following wolf (*Canis lupus*) extirpation and habitat change – and they prey on the endemic (and state listed) Olympic marmots (*Marmota olympus*), causing an alarming population decline.
Eucalyptus globulus in Haleakala NP (source: L. Loope, USGS Research Ecologists, personal communication, 21 February 2012)	Still a significant weed but is under reasonable control.	The most serious weeds in the Park (all three in Kipahulu Valley) now are probably *Clidemia hirta*, *Hedychium gardnerianum* and *Psidium cattleianum*. The first two of these have increased drastically since 1992. An eminent threat is *Miconia calvescens*, which has proved very expensive to contain outside the Park (*c.*US$1 million year^{-1} from various sources); biological control is desperately needed.
Feral horses (*Equus ferus caballus*) in Ozark National Scenic Riverways (source: Victoria M. Grant, Natural Resource Program Manager, personal communication, 20 June 2012)	The feral horses are persisting. They were the subject of an amendment to the Park's enabling legislation, identifying them as park resources and providing for their management. Thus, they were actually 'retained' (within certain caveats) per NPS legislation (Ref. P.L. 104-333, 12 November 1996).	Park staff are working to control feral hogs (*Sus scrofa*), which are likely new or expansions of illegally locally-released groups. Chytrid fungus is an invasive disease that has been documented here on a federally endangered aquatic salamander, the Ozark hellbender (*Cryptobranchus alleganiensis*).
Bison (*Bison bison*) in Wrangell-St Elias NP (source: Eric Veich, Resource Manager, personal communication, 17 June 2012)	The bison introduced in 1980 still persist. There has been no attempt to remove the animals, despite their recognition as an alien species.	White sweet clover (*Melilotus alba*) is the highest priority invasive plant in Wrangell-St Elias NP.
Feral burros (*Equus asinus*) in Grand Canyon NP (source: Brian Healy, Fisheries Biologist, GCNP, personal communication, July 2012)	Feral burros have effectively been removed. During control projects in 2011, crews removed 664,811 individual non-native plants from nearly 6500 infested acres. To date, more than 290,000 tamarisk trees have been removed from at least 130 tributaries (http://www.nps.gov/grca/naturescience/cynsk-v24.htm).	About a dozen species of alien plants, and New Zealand mudsnails, have become established in the Park since 1995. Alien rainbow trout have become self-sustaining in the Lees Ferry reach. Native fish were found in the stomachs of 18% of the alien brown trout and 5% of rainbow trout, indicating native fish populations may be impacted by non-native fish via predation.
Feral burros (*Equus asinus*) in Bandelier National Monument (source: Craig Allen, USGS, personal communication, 22 February 2012)	Feral burros no longer occur in the Monument.	Chytrid fungus (*Batrachochytrium dendrobatidis*; Chytridiomycota) was documented near the Monument in the endemic Jemez Mountains Salamander (*Plethodon neomexicanus*).

Table 10.2. Most commonly reported alien animals in US National Parks. (Source: NPSpecies data set; acquired by G. Plumb, NPS, February 2012.)

Scientific name	Common name	Number of Parks
Sturnus vulgaris	European starling	219
Passer domesticus	House sparrow	195
Columba livia	Rock dove	161
Mus musculus	House mouse	118
Felis catus	Domestic cat	87
Carpodacus mexicanus	House finch	85
Phasianus colchicus	Common pheasant	79
Cyprinus carpio	Common carp	73
Rattus norvegicus	Norway rat	70
Canis familiaris	Domestic dog	57
Micropterus salmoides	Largemouth black bass	57
Lepomis macrochirus	Bluegill	52
Salmo trutta	Brown trout	52
Oncorhynchus mykiss	Rainbow trout	49
Rana catesbeiana	American bullfrog	46
Lepomis cyanellus	Green sunfish	39
Meleagris gallopavo	Wild turkey	38
Rattus rattus	Black rat	36
Alectoris chukar	Chukar partridge	33
Micropterus dolomieu	Smallmouth bass	33
Salvelinus fontinalis	Eastern brook trout	33
Sus scrofa	Wild boar/domestic pig	29
Vulpes vulpes	Red fox	28
Gambusia affinis	Mosquitofish	27
Streptopelia decaocto	Eurasian collared dove	25
Ictalurus punctatus	Channel catfish	24
Carassius auratus	Goldfish	22
Cygnus olor	Mute swan	21
Ambloplites rupestris	Rock bass	20
Pimephales promelas	Fathead minnow	20
Bos taurus	Cattle/bovine	19
Didelphis virginiana	Virginia opossum	19
Molothrus ater	Brown-headed cowbird	19
Perdix perdix	Grey partridge	18
Pomoxis nigromaculatus	Black crappie	18
Equus caballus	Horse	17
Notemigonus crysoleucas	Golden shiner	17
Pomoxis annularis	White crappie	16
Trachemys scripta elegans	Red-eared slider	16
Bubulcus ibis	Cattle egret	15
Myocastor coypus	Nutria	14
Trachemys scripta	Pond slider	14
Callipepla californica	California quail	13
Morone saxatilis	Striped bass	13
Pimephales notatus	Bluntnose minnow	13
Perca flavescens	Yellow perch	12
Ameiurus melas	Black bullhead	11
Ameiurus natalis	Yellow bullhead	11
Ctenopharyngodon idella	Grass carp	11
Cyprinella lutrensis	Red shiner	11
Lepomis megalotis	Creek perch	11

et al., 2004). Alien fish compete directly with native fish species and consume native invertebrates and amphibians, and their introductions have been accompanied by many undesirable alien pathogens, including water mould (*Saprolegnia ferax*) and whirling disease (Dunham *et al.*, 2004). In addition, alien fish have been implicated in assisting the spread of alien bullfrogs (*Rana* (*Lithobates*) *catesbeiana*), which also decimate native amphibians in parks (Adams *et al.*, 2003).

Asian carp (*Hypophthalmic molitrix*) threaten the Great Lakes. Snakehead fish (*Channa argus*) in Chesapeake Bay and lionfish (*Pterois* spp.) in south Florida and Caribbean parks are additional examples of aquatic predators affecting native fisheries. In national parks in Hawai'i, ground-nesting birds have been decimated by introduced Norway rats (*Rattus norvegicus*), small Indian mongooses (*Herpestes auropunctatus*), wild house cats (*Felis catus*) and pet dogs (*Canis familiaris*; Vitousek *et al.*, 1987). The introduction of alien predators in national parks creates the potential for a millennial history of evolution to be wiped out in perhaps less than a few hundred years. But these predators may pale in comparison to the Burmese python (*Python molurus bivittatus*) that has recently become well established in the Florida Everglades (Fig. 10.2).

A recent paper by Dorcas *et al.* (2012) showed that small mammal abundance and diversity in Everglades NP were declining at alarming rates. For example, road surveys totalling 56,971 km from 2003 to 2011 showed a 99.3% decrease in the frequency of raccoon observations, a 98.9% decline in opossums, an 87.5% decrease in bobcats and the complete absence of rabbits in recent surveys. Twenty-five species of birds have also been documented as part of the Burmese python diet, including four species of special concern in Florida and one federally endangered species (Dove *et al.*, 2011). While the number of pythons removed each year has increased dramatically, only a small fraction of the python

Fig. 10.2. Drawing a new line in space and time with an invasive Burmese python in Everglades National Park, Florida, USA. (NPS photo by Dan DeGross; used with permission.)

population has likely been captured. Burmese pythons, which can reach 5.5 m in length and over 80 kg in weight, have taken prey as big as white-tailed deer and alligators. There is renewed concern for the federally endangered Key deer (*Odocoileus virginianus clavium*), which lives only in the Florida Keys, south of the Everglades. Until recently, saltwater was thought to act as a natural barrier to Burmese python dispersal, but recent research explicitly cautions against considering saltwater alone as a barrier (Hart *et al.*, 2012).

Because the game has changed, so must our approach to management. The dramatic and precipitous impact of Burmese pythons on the Everglades ecosystem may have provided the catalyst. In general, prevention, detection and control efforts for invasive animals have developed in an ad hoc fashion at both the State and Federal level over the past decade, tending to address each new species as a single issue. These efforts have been as efficient as limited funding would allow, and have culminated in the formation of grassroots but formalized interagency groups such as the Everglades Cooperative Invasive Species Management Area (CISMA). The Everglades CISMA, created partly in response to Burmese pythons, is a promising approach and one that the National Park Service actively participates in and is considering using as a prototype for other invasive animal management efforts.

On a broader scale, the NPS has yet to develop fully a service-wide organizational approach to invasive animal management. Consequently, parks seek and employ a variety of best management solutions at the local level, albeit often without strategic coherence or outcomes. This piecemeal approach may not provide a cost-effective service-wide strategy for managing invasive animals. Recognizing this, the agency has begun the process of developing an institutional framework that will deploy people strategically, efficiently and effectively, and funding to address the highest priorities among the growing suite of invasive animal issues.

Alien forest pathogens

Specific survey information on the number, extent and effects of alien forest pathogens is lacking for the national parks. More than 450 alien forest insects are established in the USA (Aukema *et al.*, 2011), and the national parks are not immune to their ecological effects. Forest composition in most national parks in the eastern USA has been altered radically by Dutch elm disease (*Ceratocystis ulmi*), the Asian gypsy moth (*Lymantria dispar*), chestnut blight (*Cryphonectria parasitica*), the balsam woolly adelgid (*Adelges piceae*), the Emerald ash borer (*Agrilus planipennis*) and many other species of alien forest pests (Lovett *et al.*, 2006). Likewise, the forest composition of national parks in the western USA has likely been altered by white pine blister rust (*Cronartium ribicola*), sudden oak death (*Phytophthora ramorum*) and other alien pathogens (Wittenberg and Cock, 2001). Climate directly influences the survival and spread of forest insects and pathogens and affects the competitors and natural enemies that regulate the abundance of potential pests and pathogens (Dale *et al.*, 2001). Enormous uncertainty remains when modelling the potential impacts of forest pathogens, (and invasive species in general) resulting from climate change. Common causes for the uncertainty include inadequate data; the complex, incompletely understood and unstable relationships involved; and the rapid pace of climate change relative to the adaptability of many species (Simberloff, 2000; Hansen *et al.*, 2001; Dukes *et al.*, 2009).

In national parks, large-scale forest restoration to the forest composition at the time the National Park Service was established, let alone the time of first contact with Europeans, is probably not yet feasible. However, an emphasis on prevention and early detection of future invaders may be feasible. Targeting the nun moth (*Lymantria monacha*) from north-west forests, or the Asian long-horned beetle (*Anoplophora glabripennis*) from mid-western and eastern forests, or other harmful insects, nematodes and fungi may ultimately prove wise.

Many forests are also affected by a steady influx of competing tree species from other countries which may harbour hitchhiking pathogens. Common tree invaders include Norway maple (*Acer platanoides*), tree-of-heaven (*Ailanthus altissima*), Brazilian peppertree (*Schinus terebinthifolius*), Myrica (*Myrica faya*) and tropical ash (*Fraxinus uhdei*), black locust (*Robinia pseudoacacia*), chinaberry tree (*Melia azedarach*), mimosa (*Albizia julibrissin*), royal paulownia (*Paulownia tomentosa*), white poplar (*Populus alba*) and tallow tree (*Triadica sebifera*). Compounding the problem is that, like most introduced woody species in the USA (Reichard and White, 2001), many of these trees are commercially available, and many of their seeds are dispersed by wind and birds far beyond their intended garden.

Alien animal pathogens

Diseases affecting all wild or domestic animals around the globe are tracked by the World Organisation for Animal Health (OIE). The National Park Service, along with other federal agencies, relies on the OIE, the US Department of Agriculture, the US Geological Survey's National Wildlife Health Center and the Center for Disease Control to stay abreast of alien animal pathogen occurrence in the USA. Recognizing the importance of wildlife health as a component of the interactive 'One Health' paradigm, the National Park Service established a Wildlife Health Program to guide strategic approaches to achieving the agency's mission fully (see http://nature.nps.gov/biology/wildlifehealth/index.cfm). For example, the rapid decline of amphibians attributed to chytrid disease, or *Chytridiomycosis*, has attracted attention (Fellers *et al.*, 2001; Kilpatrick *et al.*, 2010). Caused by the chytrid fungus, this disease has been implicated in significant frog die-offs in the Sierra Nevada of California, including Sequoia and Kings Canyon NPs, although a complex of bacterial pathogens (e.g. *Aeromonas hydrophila*) may be partly responsible (Bradford, 1991). Feral hogs are widely recognized as vectors for pathogens infectious to both animals and humans (Meng *et al.*, 2009). Foot-and-mouth disease, classical swine fever (or swine cholera) and African swine fever would threaten domestic animal production; foot-and-mouth disease could also impact native wild ruminant populations. All of these diseases would be extremely difficult to eradicate if introduced to feral hogs. If feral hog populations grow unchecked, so does the likelihood for transmission of the pathogens they carry.

An unintended consequence of stocking alien fish in streams and lakes in and adjacent to national parks has been the ubiquitous spread of whirling disease (Allen and Bergersen, 2002). Whirling disease is a parasite that affects a fish's ability to feed normally, resulting in starvation and death. The infected fish is also more vulnerable to predators. Spores formed in infected fish are released into the water when the fish dies and decomposes, where the spores are hosted by tubiflex worms in the sediments for up to 30 years. In this case, the worm kills the fish! In addition, the microscopic spores are transported among lakes and streams via mud on boats and recreationists' waders, and by birds and other animals. It is no wonder that this European disease has spread to at least 25 states since the 1950s (http://whirlingdisease.montana.edu/). Allen and Bergersen (2002) found that over 300 sites in 14 of 15 major drainages in Colorado tested positive for the parasite. The effects of whirling disease may paralyse native fish restoration projects. For example, whirling disease in Yellowstone Lake in Yellowstone NP affects native cutthroat trout (*O. clarki* spp.) but not the invasive lake trout (*S. namaycush*).

There are many wildlife diseases transmitted from alien domestic stock (see Table 10.2) to native ungulates in national parks. The two most famous examples include the lungworm–pneumonia complex transmission from domestic sheep to native bighorn sheep, and bovine brucellosis (*Brucella abortus*) transmitted from cattle to bison. In the first case, the lungworm–pneumonia complex affects the abundance and distribution (and recovery efforts) of the native bighorn sheep greatly, restricting the bighorn sheep to a fraction of their previous range (Aguirre and Starkey, 1994). Bovine brucellosis is an alien bacterial disease of ungulates, causing abortions in infected bison and elk in the Greater Yellowstone area (Plumb *et al.*, 2009). Reported first in 1917, bison and elk in the Yellowstone region maintain the disease (Meagher and Meyer, 1994). Carnivores are exposed to the disease when feeding on infected animals (Meagher and Meyer, 1994). Wildlife in national parks is now seen as a reservoir and as 'spillover' vectors of the disease back to domestic cattle, where efforts must now be made to protect the livestock industry.

An emerging wildlife disease in national parks may be White-nosed Bat Syndrome, a disease of hibernating North American bats caused by the fungus, *Geomyces destructans*. Since first observed in the winter of 2007, the disease has caused unprecedented mortality in several species of bats in the eastern USA. It may spread rapidly across the continent (Blehert *et al.*, 2009; Flory *et al.*, 2012).

Threats to human health

Alien species such as feral hogs, feral horses and cattle are commonly reported in national parks and contribute to the spread of *Escherichia coli* bacteria. In a recent study in Ozark National Scenic Riverways, Missouri, about half of *E. coli* isolates were assigned to cattle and horses (Davis and Barr, 2006). Feral hogs were not specifically examined in that study. However, feral hogs are vectors for several other diseases transmissible to humans. Swine brucellosis (*Brucella suis* biovars 1 and 3) is zoonotic and occurs in feral hogs in the USA. Type A influenza in swine is an emerging concern, primarily due to its zoonotic potential. Other foreign animal diseases transmissible to humans that have been documented in wild hogs outside the USA include Hepatitis E and Japanese encephalitis virus (Meng *et al.*, 2009).

Plague, the infectious disease caused by the bacterium, *Yersinia pestis*, was introduced inadvertently and spread into the western USA (Antolin *et al.*, 2002). Since it was introduced to the USA in the early 1900s, plague has been reported in all continental states west of the 100th meridian, and in at least 76 vertebrate species, including most rodents (Barnes, 1993). In many natural areas in the Rocky Mountains, Black Hills and Central Grasslands, prairie dogs (*Cynomys* spp.) are a common vector of fleas, with the distinct potential for zoonotic transmission of plague to humans. Urbanization adjacent to parks increases the threat of disease transmission. Aside from the obvious human health concern is that many domestic pets and native mammal species are potential carriers. In addition, periodic collapse of prairie dogs makes the restoration of the rare black-footed ferret (*Mustela nigripes*) difficult (Antolin *et al.*, 2002).

West Nile virus, a mosquito-borne flavivirus, was first detected in New York in 1999 (Asnis *et al.*, 1999). It spread rapidly across the continent, arriving in California in 2003 (Reisen *et al.*, 2004). Data from the California Department of Public Health Dead Bird Surveillance Program (cited in Hull *et al.*, 2010) reported that 26 species tested positive for West Nile virus RNA during the summers of 2003–2007 (soon after it arrived). West Nile virus has been detected in dead birds of at least 326 species, causing significant mortality in species in the families Accipitridae, Strigidae and Corvidae (http://www.cdc.gov/westnile/index.html). Human cases, via mosquito bites in the first 10 years of invasion, have resulted in 1100 deaths (http://www.cdc.gov/westnile/index.html). Virtually every national park in the conterminous USA, aside from the Channel Islands off the coast of California, probably harbours the West Nile virus.

Native and alien mosquitoes in national parks can transmit additional diseases to park visitors, including cholera, dengue fever and malaria. The number of documented human cases of disease transmission from mosquitoes to humans in national parks, at present, may be small, for reasons including: (i) relatively short visits; (ii) the illness may occur after the visit; or (iii) poor reporting programmes. Increased trade and transportation, increasing numbers of international visitors and increases in both the number of alien diseases and alien vectors of diseases may provide significant challenges for the National Park Service in the future. For example, bird-biting mosquitoes in the *Culex pipiens* complex are becoming better adapted to human-altered environments (Farajollahi *et al.*, 2011). This complex of mosquitoes has a broad, global distribution, interact with highly mobile hosts (i.e. birds) and carry multiple diseases. As some NPS units include extensive landscapes that serve as refugia of biodiversity, and may not be suitable for broad-scale mosquito control and bird control programmes, they may also become refugia for an increased number of alien organisms, including diseases and pathogens.

Emerging diseases specific to national parks are difficult to pin down because of the spatial variability and boom-and-bust patterns of epidemiology, frequent mutations of the disease organism and the changing biological responses of the hosts. One potential (and potentially horrifying) emerging disease is Asian bird flu, the influenza A (H5N1) virus. This disease, which drew global attention in 2005, can be spread globally by migratory birds and locally by mechanical means such as soil relocated by vehicle tyres (Webster *et al.*, 2007). The virus is now considered endemic in many parts of Indonesia and in some parts of China, Egypt and Nigeria (http://www.fao.org/avianflu/en/qanda.html#4). While this disease is a global issue, it may be of particular concern to the National Park Service because many species involved in the bird–mammal transmission of the virus are found in high concentrations in the parks. Many species of birds can carry the virus, and ubiquitous mammals including pigs, seals, whales, mink, ferrets and domestic cats (Webster *et al.*, 2007). Most cases reported to date occurred by bird–human transmission, but human–human transmission would be a game changer, particularly for containing the disease.

Climate Change and Alien Invasive Species: Uncertain Challenges

Over the past 17 years, albeit a short time, mean monthly temperature and precipitation from stations near Everglades NP, Hawai'i Volcanoes NP and Yellowstone Lake have shown no obvious trends (Fig. 10.3). While climate change might add an additional or interacting threat to park resources, we found that at park scales and over short periods of time (17 years in this case), with or without changes in climate, it might become increasingly difficult to achieve the primary mission of the National Park Service, for example, protecting native species and ecosystems for present and future generations, due to the continuing invasions of alien organisms resulting largely from increased trade and transportation. The invasions continue. We believe that the urgent and compelling alien species issues in national parks may be more related to trade and transportation, and purposeful and accidental introductions, than to short-term climate trends (Fig. 10.4). For example, in a study of potential climate effects on a dozen highly invasive plant species in the USA, Holcombe *et al.* (2010) found that habitat suitability was expected to increase for half of the plant species, while habitat suitability decreased for the remaining species. Evangelista *et al.* (2011) reported a similar story for forest pests in the intermountain west; half the species benefited, half did not. In a study of two alien grasses in Hakalau Forest National Wildlife Refuge in Hawai'i, one grass was projected to migrate out of the refuge, while the other grass was projected to spread throughout the refuge (D. Barnett, NEON Inc, personal communication, October 2012). Drought, severe cold and extreme events harm many alien species – and current models cannot provide accurate, complex climate scenarios at park scales over short time frames (Pielke *et al.*, 2012). Cautious use of climate projections may aid in selecting priority species to control, or to assess potential changes in local habitat suitability, but immigration from other

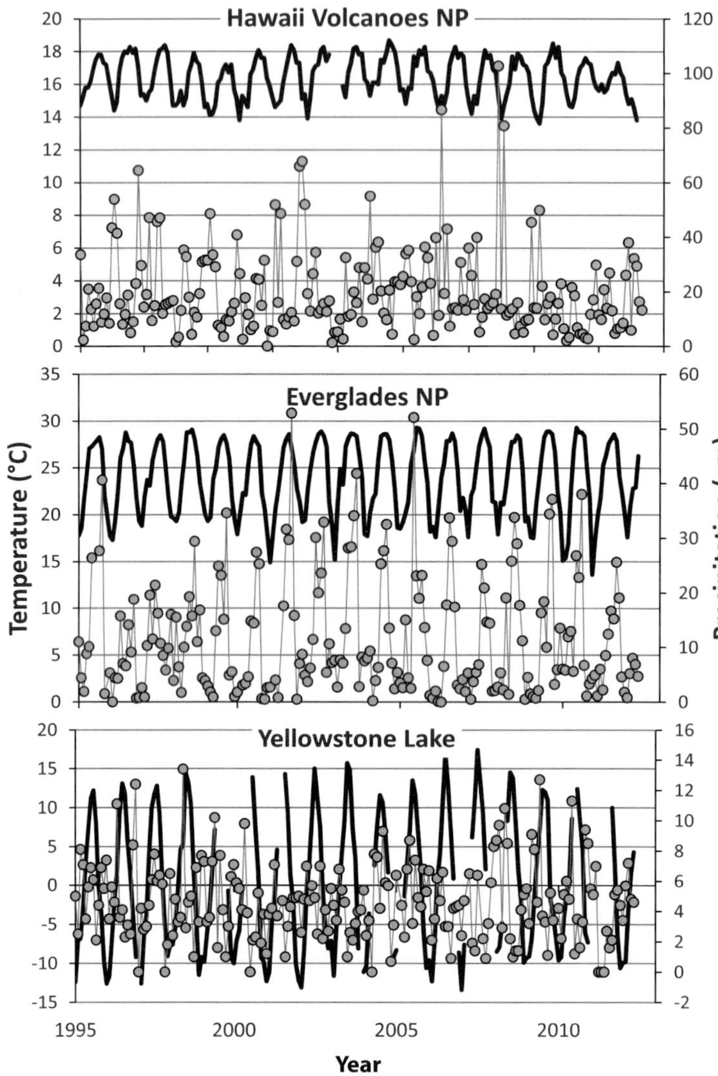

Fig. 10.3. Data on monthly temperature (dark line) and precipitation (lighter line) from the Global Historical Climatology Network (monthly summaries) data set for Everglades NP (Naples, Florida, Station COOP086078), Hawaii Volcanoes NP (Station COOP511303) and Yellowstone NP (Lake Yellowstone Station COOP485345) from the study period (1995–2012). Missing data were left blank.

sources (i.e. trade and transportation) may spread invaders at a much faster rate than climate change (Stohlgren *et al.*, 2013a).

As plant communities respond to changes in climate, their ranges may shift, expand or contract; and there will likely be corresponding shifts in the distributions of the animals that depend on them. This would be the case for all species, whether native or alien. However, it may be immaterial that distributions of alien species may be associated negatively or positively with climate change (Holcombe *et al.*, 2010). Alien species reach the parks in many ways, but the primary and proximate cause of invasions is widely recognized as trade and

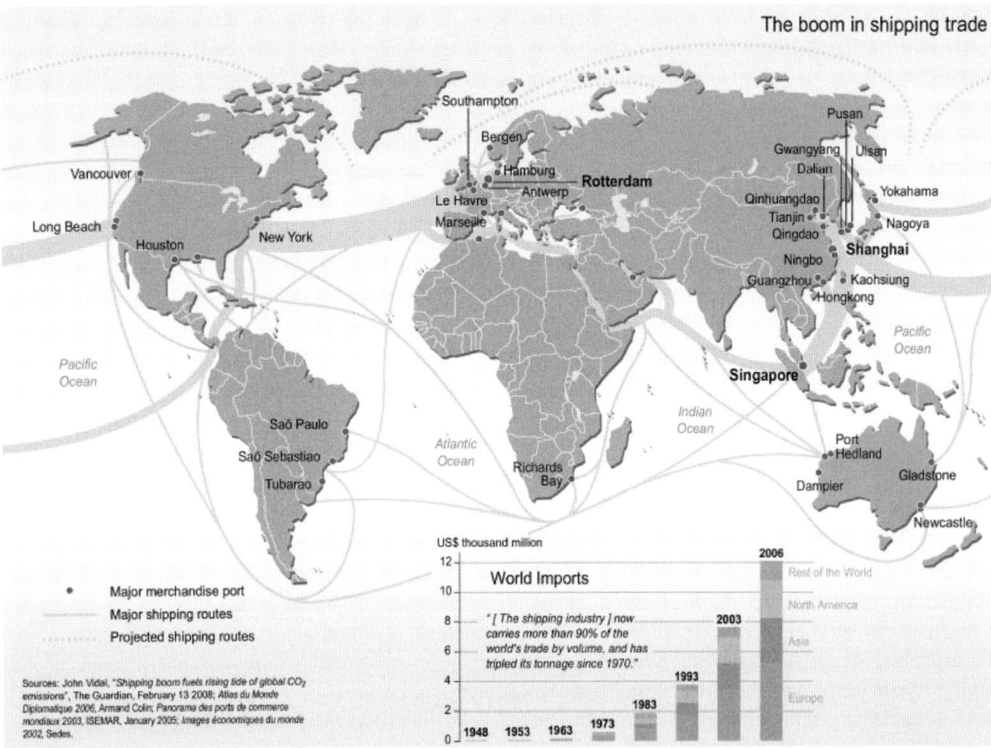

Fig. 10.4. The boom in the shipping trade, used with permission from the website http://maps.grida.no/go/graphic/the-boom-in-shipping-trade1, and based on John Vidal, 'Shipping boom fuels rising tide of global CO2 emissions', *The Guardian*, 13 February 2008; *Atlas du Monde Diplomatique, 2006*, Armand Colin; *Panorama des ports de commerce mondiaux 2003*, ISEMAR, January 2005; Images économiques du monde 2002, Sedes.

transportation (Mooney and Cleland, 2001; Levine and D'Antonio, 2003). According to US Fish and Wildlife Service records, legal wildlife shipments into the USA between 1999 and 2010 comprised over 2.8 billion individual alien animals, representing at least 4200 different species from over 150 countries (Romagosa, in press).

How many alien animals entering the USA through illegal trade are obviously not reported? Intentional and unintentional introductions make it increasingly difficult to protect national parks (and all protected areas) from the invasion of harmful plants, animals and diseases. Global trade has increased 40-fold in the past 60 years (Fig. 10.4). Trade and transportation are linked directly to plant species invasions, with about half of the most widely distributed plant species in the USA being from other countries (Stohlgren *et al.*, 2011). It is typical in national parks to see common plant species from Europe and Asia (e.g. tamarisk, thistles, annual grasses, composites) dominating landscapes. Thus, external forces (globalization and trade) are combining with internal forces (transportation networks and park visitation) to increase greatly the dispersal of alien propagules. Urbanization adjacent to the parks, combined with landscaping and the nursery trade (Reichard and White, 2001), may increasingly add to the number and genetic diversity of alien plant invaders in US national parks.

Intentional stocking of alien fish in and adjacent to parks has filled their waters with alien fish, whirling disease and other organisms (Fuller *et al.*, 1999). Hunting enthusiasts, including state departments of game and fish, have intentionally released dozens of alien animals now found in national parks (Table 10.2). Intentional releases of animals sold in the pet trade industry now fill parks' lands, air and water. Unintentional introductions by campers, boaters, fisherpersons and sightseers are commonplace; and there will be new opportunities as a result of current climate trends. Loss of Arctic sea ice may lead to a Northwest Passage that will cut travel time substantially for some ships, thus affecting the survival rates of alien organisms hitchhiking on ships' hulls and in ballast water (Hellman *et al.*, 2008; Pyke *et al.*, 2008). Each year, it becomes more obvious that designating areas as 'protected' does not protect them from biological invasions. Studies in central Europe have reached similar conclusions (Pyšek *et al.*, 2002, 2003). Economists in Australia have demonstrated that increasing inspections for harmful invaders at the ports of entry may save money in the long run (McAusland and Costello, 2004). It may behove the National Park Service to adopt a global strategy to reduce the costs of future invasions.

Invasions are likely outpacing containment efforts. There are plenty more conspicuous invaders in the parks today than occurred in 1995, when the Houston and Schreiner paper was published. There is no question that Burmese pythons, Africanized honeybees, zebra mussels, New Zealand mudsnails, West Nile virus, sudden oak death, white-nose bat syndrome and dozens of newly introduced plant species are flourishing (Fig. 10.1). However, it is likely impossible to determine unequivocally the rate of increase in invasions in all national parks over time, due to variable monitoring programmes (Allen *et al.*, 2009) and the cryptic nature of small invaders, especially pathogens and diseases. However, based on the ease with which recreational boats carry tiny alien aquatic organisms, birds and wind carry alien plant seeds and bats transmit diseases hundreds of kilometres (Flory *et al.*, 2012), it may be safe to assume that the invasions will keep coming.

It may be time to 'draw another line in the sand' – this time with a live Burmese python (Fig. 10.2). Few scientists paid attention to the pleas of Houston and Schreiner – the paper had been cited only 20 times at the last check (21 January 2014). General papers on the problems created by alien species have been cited hundreds or thousands of times (e.g. Mack *et al.*, 2000). It is time for the National Park Service to develop strategically a synthesis report similar to the 1963 Leopold and Robbins reports that focus specifically on issues related to invasive species. The synthesis would need to provide the Park Service with detailed options for effectively evaluating and prioritizing multiple threats to park resources (including native plant and wildlife species) from invasive species within the prevailing context of land-use change, contaminants, recreationists as vectors and climate change.

Potential Approaches to Consider

An ounce of prevention

When Ben Franklin suggested that, 'An ounce of prevention is worth a pound of cure', he might have underestimated the weight of the invasive species challenge for park managers. Welcoming millions of visitors to the national parks invites hitchhiking organisms. Plant seeds may be hiding in their car tyres and boots. New Zealand mudsnails, whirling disease, zebra mussels and aquatic weeds may be lurking in boats, fishing waders and live-bait containers. Tree insects and diseases may be transported by firewood. Pets might be harbouring zoonotic diseases. Many parks and recreation areas with aquatic resources now have boat-washing stations and inspection services. Some parks wash construction vehicles before they can enter. More attention may be given to the alien house mouse (*Mus musculus*), found in 118 parks, and the alien Norway rat (*Rattus norvegicus*), found in 79 park units, when they are linked to a potential epidemic of hantavirus or plague. Educating the visiting public is an important first step in the process of protecting the national parks from invaders, but many of the hitchhiking organisms have alternative means of transportation by native species, wind and other means of dispersal. Regardless, the environmental and economic costs of species invasions often far outweigh the costs associated with prevention (Simberloff, 1997). Prevention remains the most effective and cost-efficient method for combating invasive species (Pyke *et al.*, 2008). We need to be realistic though – it may take much more than an ounce of prevention.

A risk assessment approach to biological invasions

Early warning systems may require a 'risk assessment' approach to biological invasions, with climate variables well represented among predictor variables (Stohlgren and Schnase, 2006; Jarnevich and Stohlgren, 2009). Modelling the potential distributions of harmful species with Geographic Information Systems is an important component of an early warning system (Jarnevich *et al.*, 2010). Such models are often data intensive, meaning that a large number of field observations may be required inside and adjacent to the national park for best results. Developing and training teams of park visitors as 'citizen scientists' may increase awareness and help contain invasive species problems (Newman *et al.*, 2010).

Operating in tandem with an early warning system needs to be an early response system. For extremely harmful invasive species, especially those with inherently low detection probabilities such as snakes, control efforts must be early and swift (Rejmánek and Pitcairn, 2002; Kraus, 2009; Willson *et al.*, 2011). A strategy similar to the one used to contain wildfires may be useful. Such a strategy might include inter-agency centres and specifically trained crews for battling harmful alien plants, animals and diseases. However, some new invasions will likely remain undetected, and a long-term commitment to invasive species containment will likely be necessary.

Triage

It may be important to revise current science and management priorities based on a 'triage approach' to internal and external threats. The Park Service might begin by asking, 'What are the proximate causes of native population declines, degradation of habitat quality, and species extirpations?' In many cases, the answer will lead to invasive species. Another important question to ask is, 'What invasive alien organisms are finding refuge in the national parks, and do they threaten human health inside and outside the parks?' National parks may find it increasingly challenging to maintain native species populations (i.e. maintain wildlife refugia) and provide a relatively safe environment for the public, when the same wildlife the agency protects may harbour an increasing load of harmful infectious diseases (e.g. plague, West Nile virus, mosquito-borne diseases). A study by Taylor *et al.* (2001) found that nearly 75% of emerging pathogens were zoonotic. As we look ahead, another question must be considered, 'How will the agency prevent national parks from increasingly becoming refugia for harmful alien species?' A triage approach might also entail prioritizing resources by the type and severity of threat posed by individual species. As a consequence, we may need to begin evaluating critically whether there are some invasive species we will just have to 'put on the back burner' because it is not currently feasible to eradicate or control them once their populations are too large or geographically widespread. We might need to tolerate, at least temporarily, species that have minimal deleterious impacts on ecosystem functions in order to focus management resources on more harmful species.

Scenario planning

Scenario building offers a way to examine systematically possible future and cascading consequences that are complex and uncertain (Chermack *et al.*, 2001; Peterson *et al.*, 2003). The National Park Service uses scenario planning to assess the vulnerability of park resources to climate change, and it is in the beginning stages of applying such an approach to assess management options for Burmese pythons at its three parks in southern Florida (Everglades NP, Biscayne NP and Big Cypress National Preserve). Indeed, a coupled natural–human systems model (Machlis *et al.*, 1997) that was adapted and applied by the US Department of the Interior to the Deepwater Horizon oil spill (Machlis and McNutt, 2010) may prove to be a meaningful approach to assess the emergent properties of invasive species' interactions with climate change. Senior Park Service staff were members of the strategic sciences working group conducting that assessment.

The human–natural systems approach explicitly addresses complex interactions and feedback within and between human and natural systems and has the potential to reveal new patterns or processes that might initially be overlooked by decision makers (Liu *et al.*, 2007; Machlis and McNutt, 2010). Applying this approach to the management challenge presented by a top-down invasive predator such as the Burmese python could present decision makers with a series of scenarios that are adaptable to new information; identify new monitoring, management capacity or policy needs; and outline intervention points for reducing further negative impacts and/or increase the resilience and effectiveness of management actions. The limitations of scenario building include constraints due to available expert opinion, information and lack of theory (Peterson *et al.*, 2003).

Conclusions

Control costs climb as new harmful invasive species establish and existing invaders spread and reproduce (Rejmánek and Pitcairn, 2002). Many parks have integrated pest management programmes to help contain alien termites, cockroaches and rodents, while regions are coordinating Exotic Plant Management Teams to control major weed infestations. However, even back in 2004, the National Park Service admitted that 'the funding needs for a viable invasive species program is a far cry from the amount allocated'. Because of the uncertainty of the impacts of climate change at local scales, it is hard to conceptualize how to address its more nebulous impacts on species invasions when more immediate threats such as trade and transportation are more tangible. It seems clear that invasive species management within the National Park Service needs to develop a servicewide, strategic management approach that considers both the immediate and long-term threats. There is compelling evidence that invasive species issues, costs and consequences have continued to grow in national parks since Houston and Schreiner drew their line in the sand. Indeed, it appears that the beachhead of natural resource integrity in parks is being actively eroded at rates likely not less than projected coastline demise under certain climate change scenarios. It is time to do more than just draw a new line in the sand.

References

Adams, M.J., Pearl, C.A. and Bury, R.B. (2003) Indirect facilitation of an anuran invasion by nonnative fishes. *Ecology Letters* 6, 343–351.

Aguirre, A.A. and Starkey, E.E. (1994) Wildlife disease in US national parks: historical and coevolutionary perspectives. *Conservation Biology* 8, 654–661, doi: 10.1046/j.1523-1739.1994.08030654.x

Allen, J.A., Brown, C.S. and Stohlgren, T.J. (2009) Non-native plant invasions of the United States National Parks. *Biological Invasions* doi: 10.1007/s10530-008-9376-1.

Allen, M.B. and Bergersen, E.P. (2002) Factors influencing the distribution of *Myxobolus cerebralis*, the causative agent of whirling disease in the Cache La Poudre River, Colorado. *Diseases of Aquatic Organisms* 49, 51–60.

Antolin, M.F., Gober, P., Luce, B., Biggins, D.E., Van Pelt, W.E., Seery, D.B., Lockhart, M. and Ball, M. (2002) The influence of sylvatic plague on North American wildlife at the landscape level, with special emphasis on black-footed ferret and prairie dog conservation. *Transactions of North America Wildlife Natural Resource Conservation* 67, 104–127.

Asnis, D.S., Conetta, R., Teixeira, A.A., Waldman, G. and Sampson, B.A. (1999) Outbreak of West Nile-like viral encephalitis, New York 1999. *Morbidity and Mortality Weekly Report* 48, 845–849.

Aukema, J.E., Leung, B., Kovacs, K., Chivers, C., Britton, K.O., Englin, J., Frankel, S.J., Haight, R.G., Holmes, T.P., Liebhold, A.M., McCullough, D.G. and Von Holle, B. (2011) Economic impacts of non-native forest insects in the Continental United States. *PLoS ONE* 6, e24587, doi:10.1371/journal.pone.0024587.

Barnes, A.M. (1993) A review of plague and its relevance to prairie dog populations and the black-footed ferret. In: Oldemeyer, J.L., Biggins, D.E. and Miller, B. (eds) *Proceedings of the Symposium on the Management of Prairie Dog Complexes for the Reintroduction of the Black-footed Ferret.* USDI Biological Report 13. Department of the Interior, Washington, DC, pp. 28–37.

Blehert, D.S., *et al.* (2009) Bat white-nose syndrome: an emerging fungal pathogen? *Science* 323, 227.

Bradford, D.F. (1991) Mass mortality and extinction in a high elevation population of *Rana muscosa. Journal of Herpetology* 25, 369–377.

Brooks, M.L., D'Antonio, C.M., Richardson, D.M, Grace, J.B., Keeley, J.E., DiTomaso, J.M., Hobbs, R.J, Pellant, M. and Pyke, D. (2004) Effects of invasive alien plants on fire regimes. *BioScience* 54, 677–688.

Chermack, T.J., Lynham, S.A. and Ruona, W.E.A. (2001) A review of scenario planning literature. *Futures Research Quarterly* 7, 7–32.

Dale, V.H., *et al.* (2001) Climate change and forest disturbances. *BioScience* 51, 723–734.

Davis, J.V. and Barr, M.N. (2006) Assessment of possible sources of microbiological contamination in the water column and streambed sediment of the Jacks Fork, Ozark National Scenic Riverways, Missouri – Phase III. US Scientific Investigations Report 06-5161. US Geological Survey, Reston, Virginia, pp. 1–32.

Dorcas, M.E., Willson, J.D., Reed, R.N., Snow, R.W., Rochford, M.R., Miller, M.A., Meshaka, W.E. Jr, Andreadis, P.T., Mazzotti, F.J., Romagosa, C.M. and Hart, K.M. (2012) Severe mammal declines coincide with proliferation of invasive Burmese pythons in Everglades National Park. *Proceedings of the National Academy of Sciences* 109, 2418–2422.

Dove, C.J., Snow, R.W., Rochford, M.R. and Mazzotti, F.J. (2011) Birds consumed by the invasive Burmese python (*Python moolurus bivittatus*) in Everglades National Park, Florida, USA. *Wilson Journal of Ornithology* 123, 126–131.

Drees, L. (2004) A retrospective on NPS invasive species policy and management. *Park Science* 22 (Fall 2004), 1–8, ISSN 1090-9966.

Dukes, J.S. and Mooney, H.A. (1999) Does global change increase the success of biological invaders? *Trends in Ecology and Evolution* 14, 135–139.

Dukes, J.S., *et al.* (2009) Responses of insect pests, pathogens, and invasive plant species to climate change in the forests of northeastern North America: What can we predict? *Canadian Journal of Forest Research* 39, 231–248.

Dunham, J.B., Pilliod, D.S. and Young, M.K. (2004) Assessing the consequences of nonnative trout in headwater ecosystems in western North America. *Fisheries* 29, 18–26.

Esque, T.C., Schwalbe, C.R., Haines, D.F. and Halvorson, W.L. (2004) Saguaros under siege: invasive species and fire. Special Issue *Desert Plants* 20, 49–55.

Esque, T.C., Schwalbe, C.R., Lissow, J.A., Haines, D.F., Foster, D. and Garnett, M.C. (2006) Buffelgrass fuel loads in Saguaro National Park, Arizona, increase fire danger and threaten native species. *Park Science* 24, 33–37.

Evangelista, P.H., Kumar, S., Stohlgren, T.J. and Young, N.E. (2011) Assessing forest vulnerability and the potential distribution of pine beetles under current and future climate scenarios in the Interior West of the US. *Forest Ecology and Management* 262, 307–316.

Farajollahi, A., Fonseca, D.M., Kramer, L.D. and Kilpatrick, A.M. (2011) 'Bird biting' mosquitoes and human disease: a review of the role of *Culex pipiens* complex mosquitoes in epidemiology. *Infection, Genetics and Evolution* 11, 1577–1585.

Fellers, G.M., Green, D.E. and Longcore, J.E. (2001) Oral chytridiomycosis in the mountain yellow-legged frog (*Rana muscosa*). *Copeia* 2001, 945–953.

Flory, A.R., Kumar, S., Stohlgren, T.J. and Cryan, P. (2012) Environmental conditions associated with bat white-nose syndrome in the northeastern United States. *Journal of Applied Ecology* 49, 680–689.

Fuller, P.L., Nico, L.G. and Williams, J.D. (1999) *Nonindigenous Fishes Introduced into Inland Waters of the United States.* Special Publication 27. American Fisheries Society, Bethesda, Maryland, 613 pp.

Gelbard, J.L. and Belnap, J. (2003) Roads as conduits for exotic plant invasions in a semiarid landscape. *Conservation Biology* 17, 420–432.

Hansen, A.J., Neilson, R.P., Dale, V.H., Flather, C.H., Iverson, L.R., Currie, D.J., Shafer, S., Cook, R. and Bartlein, P.J. (2001) Global change in forests: responses of species, communities, and biomes. *BioScience* 51, 765–779.

Hart, K.M., Schofield, P.J. and Gregoire, D.R. (2012) Experimentally derived salinity tolerance of hatchling Burmese pythons (*Python molurus bivittatus*) from the Everlades, Florida (USA). *Journal of Experimental Marine Biology and Ecology* 413, 56–59.

Hellmann, J.J., Byers, J.E., Bierwagen, B.G. and Dukes, J.S. (2008) Five potential consequences of climate change for invasive species. *Conservation Biology* 22, 534–543.

Holcombe, T.R., Stohlgren, T.J. and Jarnevich, C.S. (2010) From points to forecasts: predicting invasive species habitat suitability in the near term. *Diversity* 2, 738–767.

Houston, D.B. and Schreiner, E.G. (1995) Alien species in National Parks: drawing lines in space and time. *Conservation Biology* 9, 204–209.

Hull, J.M., Keane, J.J., Tell, L. and Ernest, H.B. (2010) West Nile virus antibody surveillance in three Sierra Nevada raptors of conservation concern. *Condor* 112, 168–172.

Jarnevich, C.S. and Stohlgren, T.J. (2009) Near term climate projections for invasive species distributions. *Biological Invasions* 11, 1373–1379.

Jarnevich, C.S., Holcombe, T., Barnett, D.T. Stohlgren, T.J. and Kartesz, J.T. (2010) Forecasting weed distributions using climate data: a GIS early warning tool. *Invasive Plant Science and Management* 3, 365–375.

Kilpatrick, A.M., Briggs, C.J. and Daszak, P. (2010) The ecology and impact of chytridiomycosis: an emerging disease of amphibians. *Trends in Ecology and Evolution* 25, 109–118.

Kraus, F. (2009) *Alien Reptiles and Amphibians: A Scientific Compendium and Analysis.* Springer, New York.

Leopold, A.S., Cain, S.A., Cottam, C.M., Gabrielson, I.N. and Kimball, T.L. (1963) Wildlife management in the national parks: advisory board on wildlife management appointed by Secretary of the Interior. *Transactions of the North American Wildlife and Natural Resources Conference* 28, 29–44.

Levine, J.M. and D'Antonio, C.M. (2003) Forecasting biological invasions with increasing international trade. *Conservation Biology* 17, 322–326.

Liu, J., *et al.* (2007) Complexity of coupled human and natural systems. *Science* 317, 1513–1516.

Lodge, D.M., Williams, S., MacIsaac, H.J., Hayes, K.R., Leung, B., Reichard, S., Mack, R.N., Moyle, P.B., Smith, M., Andow, D.A., Carlton, J.T. and McMichael, A. (2006) Biological invasions: recommendations for U.S. policy and management. *Ecological Applications* 16, 2035–2054.

López-Hoffman, L. and Quijada-Mascareñas, A. (2012) Madrean Sky Islands, North America. In: Hilty, J.A., Chester, C. and Cross, M.S. (eds) *Climate and Conservation.* Island Press, Washington, DC, pp. 217–226.

Lovett, G.M., Canham, C.D., Arthur, M.A., Weathers, K.C. and Fitzhugh, R.D. (2006) Forest ecosystem responses to exotic pests and pathogens in eastern North America. *Bioscience* 56, 395–405.

McAusland, C. and Costello, C. (2004) Avoiding invasives: trade-related policies for controlling unintentional exotic species introductions. *Journal of Environmental Economics and Management* 48, 954–977.

McCarty, J.P. (2001) Ecological consequences of recent climate change. *Conservation Biology* 15, 321–331.

Machlis, G.E. and McNutt, M.K. (2010) Scenario-building for the Deepwater Horizon oil spill. *Science* 329, 1018–1019.

Machlis, G.E., Force, J.E. and Burch, W.R. (1997) The human ecosystem part I: the human ecosystem as an organizing concept in ecosystem management. *Society and Natural Resources* 10, 347–367.

Mack, R.N., Simberloff, D., Lonsdale, W.M., Evans, H., Clout, M. and Bazzaz, F.A. (2000) Biotic invasions: causes, epidemiology, global consequences, and control. *Ecological Applications* 10, 689–710.

Marnell, L.F. (1995) Cutthroat trout in Glacier National Park, Montana. In: LaRoe, E.T., Farris, G.S., Puckett, C.S., Doran, P.D. and Mac, M.J. (eds) *Our Living Resources, 1995: A Report to the Nation on the Distribution, Abundance, and Health of U.S. Plants, Animals, and Ecosystems.* National Biological Service, Washington, DC, pp. 153–154.

Meagher, M. and Meyer, M.E. (1994) On the origin of brucellosis in bison of Yellowstone National Park: a review. *Conservation Biology* 8, 645–653.

Meng, X.J., Lindsay, D.S. and Sriranganathan, N. (2009) Wild boars as sources for infectious diseases in livestock and humans. *Philosophical Transactions of the Royal Society of Biological Sciences* 364, 2697–2707.

Mooney, H.A. and Cleland, E.E. (2001) The evolutionary impact of invasive species. *Proceedings of the National Academy of Sciences* 98, 5446–5451.

Myers, J.H. and Bazely, D.R. (2003) *Ecology and Control of Introduced Plants.* Cambridge University Press, Cambridge, UK, 313 pp.

Newman, G., Laituri, M., Graham, Stohlgren, T.J., Moore, J.C. and Holfelder, K. (2010) Teaching citizen science skills online: implications for invasive species training programs. *Applied Environmental Education and Communication* 9, 276–286.

NPS (National Park Service) (1968) *Administrative Policies for Natural Areas of the National Park System*. US Government Printing Office 16–21. Washington, DC.

NPS (National Park Service) (2006) *National Park Policy*. Superintendent of Documents, US Government Printing Office, Washington, DC.

NPS (National Park System) Advisory Board (2012) *Revisiting Leopold: Resource Stewardship in the National Parks*. National Park System Advisory Board, Washington, DC.

Peterson, G.D., Cumming, G.S. and Carpenter, S.R. (2003) Scenario planning: a tool for conservation in an uncertain world. *Conservation Biology* 17, 358–366.

Pielke, R.A. Sr, Wilby, R., Niyogi, D., Hossain, D., Dairuku, F., Adegoke, K., Kallos, J., Seastedt, G. and Suding, T. (2012) Dealing with complexity and extreme events using a bottom-up, resource-based vulnerability perspective. In: Sharma, A.S., Bunde, A., Dimri, V.P. and Baker, N. (eds) *Extreme Events and Natural Hazards: The Complexity Perspective. Geophysical Monographs*, Vol 196. AGU, Washington, DC, pp. 345–359, doi:10.1029/2011GM001086.

Pimentel, D., Lach, L., Zuniga, R. and Morrison, D. (2000) Environmental and economic costs of nonindigenous species in the United States. *BioScience* 50, 53–65.

Plumb, G.E., White, P.J., Coughenour, M. and Wallen, R. (2009) Carrying capacity, migration and dispersal in Yellowstone bison. *Biological Conservation* 142, 2377–2387.

Pyke, C.R., Thomas, R., Porter, R.D., Hellmann, J.J., Dukes, J.S., Lodge, D.M. and Chavarria, G. (2008) Current practices and future opportunities for policy on climate change and invasive species. *Conservation Biology* 22, 585–592.

Pyšek, P.V., Jarošík, V. and Kučera, T. (2002) Patterns of invasions in temperate nature reserves. *Biological Conservation* 104, 13–24.

Pyšek, P.V., Jarošík, V. and Kučera, T. (2003) Inclusion of native and alien species in temperate nature reserves: an historical study from Central Europe. *Conservation Biology* 17, 1414–1424.

Reichard, S.H. and White, P. (2001) Horticulture as a pathway of invasive plant introductions in the United States. *BioScience* 51, 103–113.

Reisen, W., Lothrop, H., Chiles, R., Madon, M.B., Cossen, C., Woods, L., Husted, S., Kramer, V.L. and Edman, J.D. (2004) Invasion of California by West Nile virus. *Emerging Infectious Diseases* 10, 1369–1378.

Rejmánek, M. and Pitcairn, M.J. (2002) When is eradication of exotic pest plants a realistic goal? In: Veitch, C.R. and Clout, M.N. (eds) *Turning the Tide: The Eradication of Invasive Species*. IUCN SSC Invasive Species Specialist Group, Gland, Switzerland, and Cambridge, UK, pp. 249–253.

Rieman, B.E. and Isaak, D. (2007) Anticipated climate warming effects on bull trout habitats and populations across the interior Columbia River basin. *Transactions of the American Fisheries Society* 136, 1552–1565.

Rieman, B.E., Lee, D.C. and Thurow, R.F. (1997) Distribution, status, and likely future trends of bull trout within the Columbia River and Klamath River basins. *North American Journal of Fisheries Management* 17, 1111–1125.

Robbins, W.J. (Chairman) (1963) *A Report by the Advisory Committee to the National Park Service on Research*. National Academy of Sciences, National Research Council, Washington, DC.

Romagosa, C.M. (in press) Global trade in live vertebrates and the contribution to biological invasions. In: Keller, R.P., Cadotte, M. and Sandiford, G. (eds) *Invasive Species in a Globalized World*. University of Chicago Press, Illinois.

Simberloff, D. (1997) The biology of invasions. In: Simberloff, D., Schmitz, D. and Brown, T. (eds) *Strangers in Paradise: Impact and Management of Nonindigenous Species in Florida*. Island Press, Washington, DC.

Simberloff, D. (2000) Global climate change and introduced species in United States forests. *Science of the Total Environment* 262, 253–261.

Stohlgren, T.J. and Schnase, J.L. (2006) Risk analysis for biological hazards: what we need to know about invasive species. *Risk Analysis* 26, 163–173.

Stohlgren, T.J., *et al.* (2011) Widespread plant species: natives vs. aliens in our changing world. *Biological Invasions* 13, 1931–1944, doi:10.1007/s10530-011-0024-9.

Stohlgren, T.J., *et al.* (2013a) Globalization effects on common plant species. In: Levin, S.A. (ed.) *Encyclopedia of Biodiversity*, 2nd edn, Vol 3. Academic Press, Princeton, New Jersey, and Waltham, Massachusetts, pp. 700–706.

Stohlgren, T.J., Loope, L.L. and Makarick, L.J. (2013b) Plant species invasions in the United States National Parks. In: Foxcroft, L., Richardson, D., Pysek, P. and Genoves, A. (eds)

Alien Plant Invasions in Protected Areas: A Global Assessment. Springer, New York, pp. 267–283.

Taylor, L.H., Latham, S.M. and Woolhouse, M.E.J. (2001) Risk factors for human disease emergence. *Philosophical Transactions of the Royal Society B: Biological Sciences* 356, 983–989.

Usher, M.B., Kruger, F.J., Macdonald, I.A.W., Loope, L.L. and Brockie, R.E. (1988) The ecology of biological invasions into nature reserves: an introduction. *Biological Conservation* 44, 1–8.

Vitousek, P.M. (1990) Biological invasions and ecosystem processes: towards integration of population biology and ecosystem studies. *OIKOS* 57, 7–13.

Vitousek, P.M., Loope, L.L. and Stone, C.P. (1987) Introduced species in Hawaii: biological effects and opportunities for ecological research. *Trends in Ecology and Evolution* 2, 224–227.

Webster, R.G., Guan, Y., Poon, M., Krauss, L., Webby, S., Govorkovai, R. and Peiris, E. (2007) The spread of the H5N1 bird flu epidemic in Asia in 2004. *Infectious Diseases from Nature: Mechanisms of Viral Emergence and Persistence* 2005, 117–129, doi:10.1007/3-211-29981-5_10.

Willson, J.D., Dorcas, M.E. and Snow, R.W. (2011) Identifying plausible scenarios for the establishment of invasive Burmese pythons (*Python molurus*) in southern Florida. *Biological Invasions* 13, 1493–1504.

Wittenberg, R. and Cock, M.J.W. (eds) (2001) *Invasive Alien Species: A Toolkit of Best Prevention and Management Practices.* CAB International, Wallingford, UK, 228 pp.

Wright, G.M., Dixon, J.S. and Thompson, B.H. (1932) Fauna of the national parks of the United States: a preliminary survey of faunal relations in national parks. *Fauna series 1.* US National Park Service, Washington, DC.

11 Invasive Plants in a Rapidly Changing Climate: An Australian Perspective

Bruce L. Webber,[1,2] Rieks D. van Klinken[3] and John K. Scott[1]

[1]*CSIRO Ecosystem Sciences and Climate Adaptation Flagship, Wembley, Western Australia, Australia;* [2]*School of Plant Biology, The University of Western Australia, Crawley, Western Australia, Australia;* [3]*CSIRO Ecosystem Sciences, Brisbane, Queensland, Australia*

Abstract

Invasive plants in Australia are currently among the greatest threats to native biological diversity and a significant cost to agriculture. The impact of climate change is projected to be particularly significant for Australia and will add to and change the impact of invasive plants on natural and managed ecosystems. Australia is a large country incorporating a wide variety of climates that often experience significant disturbance events, is likely to undergo significant land-use change with climate change and has a long history of some 26,000 plant species introductions and about 2700 established alien plant species. This combination of features makes it a particularly pertinent place to explore the effect of climate change on invasive plants. Here, the chapter begins by providing background context on plant introductions, climate change and the policy underpinning management relevant for invasive plants in Australia. Then, the existing literature linking climate change and invasive plants is reviewed, focusing on climate change drivers and shifting species distributions, before looking at how the management of invasive plants will be particularly influenced by climate change. The chapter concludes by highlighting five areas where greater research effort is required and where Australian insight is likely to improve best practice management significantly, with global applicability for invasive plants in a changing climate.

Introduction

Invasive plants in Australia are currently among the greatest threats to native biological diversity and a significant cost to agriculture. The economic impact on agriculture alone is estimated at in excess of AUS$4 billion year^{-1} (Sinden *et al.*, 2004) and the impact on the environment, while unquantified, is regarded as significant. The prevention of further invasions through weed risk assessment and the management of existing invaders has been the focus of considerable policy, research and management effort. Yet despite this effort, there has been a mismatch between the perceived magnitude of the threat from invasive plants in Australia, relative to the resources invested to adequately address the problem.

Rapid change in the global climate both adds to and changes the impact of invasive plants by altering the dynamics of ecosystem function, ecosystem services and land use,

 Invasive Species and Global Climate Change (eds L.H. Ziska and J.S. Dukes)

among other drivers of change (Hellmann *et al.*, 2008; Walther *et al.*, 2009). It is clear that with future climate change, the management of invasive plants is only going to become more challenging. Greater uncertainty for the projecting possible threats from invasive species, combined with conceptual and policy frameworks that are becoming increasingly untenable, is likely to make it more difficult to find robust solutions to underpin management responses. For example, current policy-driven actions for climate change adaptation are not often whole-of-system based and have not adequately addressed trade-offs between benefits and threats (e.g. Webber *et al.*, 2011a).

Like many parts of the world, Australia's climate has already changed in recent decades. The annual mean maximum daily temperatures have increased by 0.75°C since 1910, as have daily mean and minimum temperatures (0.9 and 1.1°C, respectively; CSIRO and Bureau of Meteorology, 2012). There is a trend towards increased spring and summer monsoonal precipitation across the north, higher than normal precipitation across the centre and decreased precipitation in the south (CSIRO and Bureau of Meteorology, 2012). The most marked example of change is the 15% reduction of precipitation in south-western Australia since the 1970s, leading to up to a 45% reduction in runoff into major water storage dams (Climate Commission, 2011).

The Australian continent covers a broad range of climates, and the projected impact of future climate change scenarios reflects that diversity (Steffen *et al.*, 2009; CSIRO and Bureau of Meteorology, 2012). Maximum and minimum temperatures are likely to increase in all regions and seasons. The consensus across scenarios indicates a decrease in precipitation in the temperate regions of southern Australia, with considerable variability in the south-east and the strongest declines in the south-west. The direction of precipitation change in tropical areas is unclear, but it is likely that there will be fewer, yet more intense, tropical cyclones. Inland desert regions are likely to have longer droughts. Overall, weather events are likely to become less predictable, more frequent and more intense (Steffen *et al.*, 2009; CSIRO and Bureau of Meteorology, 2012).

There are five reasons that make Australia a particularly pertinent continent in which to explore the effect of climate change on invasive plants. First, Australia has a long history of plant introductions. Some 26,000 plant species have been introduced to Australia by humans (Randall, 2007), significantly more than the 20,000 or so extant species of native flowering plants (Chapman, 2009). Although some plant introductions into Australia took place thousands of years ago (Bean, 2007), the majority of these introductions have occurred in the last 230 years since European settlement. About 2700 of these introduced species are currently considered potential invaders (naturalized or invasive; Groves *et al.*, 2003; Randall, 2007) and approximately 30% (798) of these potential invaders are now of serious economic and environmental concern (Groves *et al.*, 2003). Second, the majority of Australian terrestrial ecosystems are characterized by frequent disturbance events (e.g. fire, cyclones) that are projected to increase in intensity or frequency under climate change projections. Third, many agricultural areas, both grazing and cropping, are established in regions of marginal quality. With increasing stresses driven by climate change, the pressure to abandon previously managed areas or to switch to new (possibly invasive) crops will be relatively high. Fourth, Australia has significant variation in the distribution of settlements and population density, with urban ecosystems and vast, little-managed wilderness regions having strongly contrasting requirements for the management of invasive species. Last, the Australian mainland is a continental island of some 7.6 million km^2, and together with the nearby islands, has considerable internal variation in current (and projected future) climate, as well as climatic similarities between disjunct regions. Australia, therefore, has both the potential for considerable internal movement of plants into areas where they are not native, as well as the geographical isolation to be a world leader in minimizing any future species introductions.

Here, an overview is provided of the interaction between invasive plants and climate change within an Australian context. More specifically, this chapter will summarize relevant management and research to date, focus on significant drivers of change, outline the implications of climate change for invasive species management and provide insight into adaptation options and open research questions for managing invasive plants under climate change in Australia. Management responses for invasive species need to consider the potentially profound effect of global climate change, and this chapter provides a synthesis of the Australian research that is underpinning such responses. (Additional insight into the management of invasive species with climate change is provided in Parts III and IV of this volume). This chapter focuses on: (i) terrestrial plants, but acknowledges that many issues are equally applicable to invasive animals and pathogens; (ii) recognizes that the global climate always changes, but uses the term 'climate change' to refer to rapid anthropogenic climate change; (iii) attaches no implication of impact (either positive or negative) to the term 'invasion' (*sensu* Richardson *et al.*, 2011); (iv) assumes both natives and non-natives can be invasive (i.e. by expanding rapidly or shifting their range) or weeds (not wanted where they are found), or both; and (v) does not confer non-native status automatically to anthropogenic plant introductions (*sensu* Webber and Scott, 2012; see also Trudgen *et al.*, 2012).

The Australian Policy Framework for Invasive Plants

In Australia, the federal government has overall responsibility for national borders and quarantine, while the state governments have responsibility for overall management of invasive species. However, this state-level management is often facilitated by federal funding: individual states can have regional quarantine barriers and individual landholders can have ultimate management responsibility. The Australian Weed Strategy (Australian Weeds Committee, 2007) and the agreement on 32 'Weeds of National Significance' (WoNS; Australian Weeds Committee, 2013) has provided overall national policy direction for weed issues, and each state has developed its own matching strategies. This approach to weed management has flowed down to regional bodies, such as Catchment Councils, Natural Resource Management Regions and local governments. Most of these strategies now contain a climate change objective. For example, in the Australian Weed Strategy there is an objective (1.4) to 'implement weed risk management practices to respond to climate change' (Australian Weeds Committee, 2007).

National climate change mitigation policies encourage the planting of vegetation for carbon capture (carbon farming). The planting of invasive species is explicitly excluded, based on national lists of weed species (Commonwealth of Australia, 2010b). While the national climate change adaptation strategy (Commonwealth of Australia, 2010a) does not mention invasive organisms, it is recognized that native biological diversity is threatened by climate change and that this is exacerbated by the threat of invasive plants (Steffen *et al.*, 2009). The mitigation and adaptation responses of individual states to invasive organisms and climate change are patchy, from little emphasis in some states, to detailed assessments in South Australia (Kriticos *et al.*, 2010) and Victoria (Steel *et al.*, 2008). Some assessments are being made by local government (e.g. Pillora, 2010) for regions where weeds are recognized for their potential to change their distribution and then subsequently cause altered fire regimes and loss of biodiversity. However, most of the various policies do not address the issue of the management of range-shifting native species that become invasive beyond their historical range, either through migration or translocation. For example, *Pittosporum undulatum* is native to parts of south-east Australia, but its spread in areas outside its historical range (Howell, 2003) is a 'potentially threatening process' under the Flora and Fauna Guarantee Act 1988 in Victoria. Potential solutions to this conundrum exist. For example, Webber and Scott (2012) proposed a framework based on potential dispersal distance to help guide the management of range-shifting species, both native and non-native, within a climate change context.

Ultimately, good policy is guided by locally relevant research. The link between climate change and invasive plants in Australia received relatively early attention in the papers of Sutherst *et al.* (1996) and O'Donnell and Adkins (1996), but these were followed by comparatively few papers over the next 10 years (Fig. 11.1). It is only recently that there has been more of an attempt at an overall synthesis of research to date (e.g. Low, 2008) and an increasing research effort specifically on Australian issues. This change has been largely facilitated by government policy (e.g. the Australian Weeds Strategy) and funding opportunities (e.g. the Defeating the Weeds Menace programme and the National Weeds and Productivity Research Program). Despite an increasingly detailed framework for managing invasive species under climate change since Low (2008), there has been no national synthesis on invasive plants and climate change in Australia to ensure this framework is as relevant as possible to local management issues.

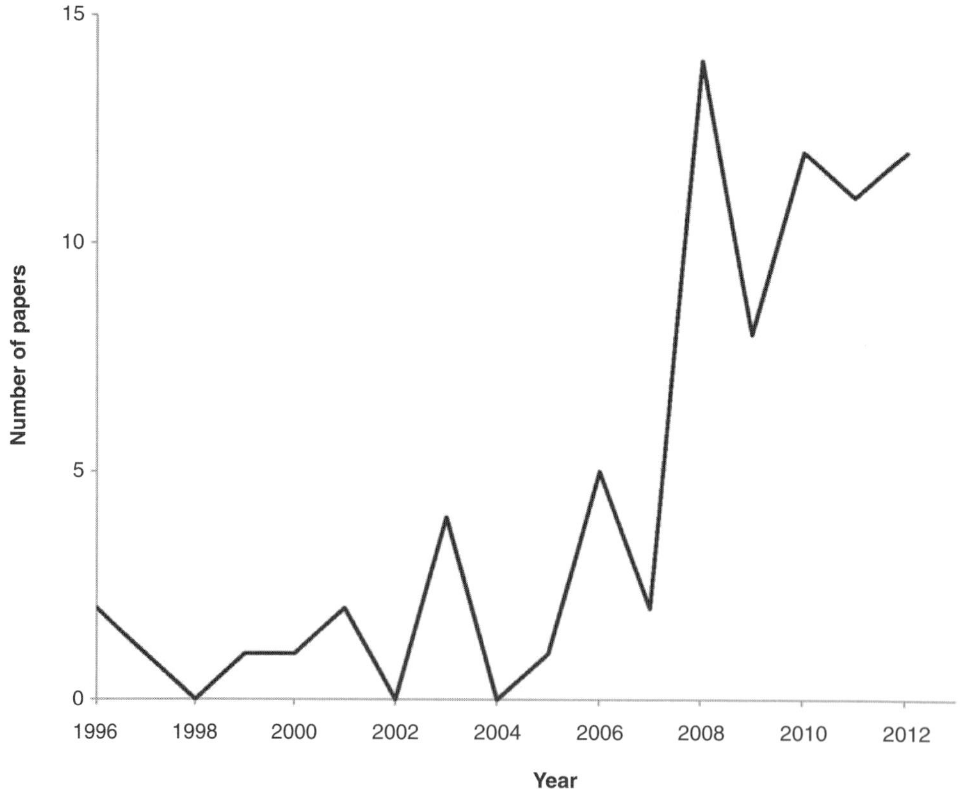

Fig. 11.1. Number of papers on climate change and invasive plants focusing on Australia published between 1996 and 2012 and listed on Web of Science. The search terms used on 25 November 2012 were: Topic = (climate change) AND (invas* AND plant OR weed)) AND Address = (Australia). Search results were subsequently checked for relevance by scanning abstracts and introductions, as required.

The Changing Face of Plant Invasions in Australia

The impact of climate change on plant invasions are manifested directly through abiotic changes such as altered temperature and precipitation regimes and increasing concentrations of carbon dioxide (CO_2) in the atmosphere. Climate change can also influence invasions indirectly via change in human behaviour (e.g. changes in introduction pathways and land use), altered natural disturbance regimes and altered biotic interactions, including biological control. The possible outcomes, when rapid climate change is layered on to the multiple other drivers influencing plant invasions, remain similar to that of classical invasion biology with respect to changed distributions, ecosystem impacts and management issues. However, it is critical that adaptation responses to plant invasions take into account climate change drivers, including their interaction with other drivers, in order to deliver an appropriate management response. If not, solutions may not remain relevant and effective into the future. Much of the research undertaken in Australia on climate change and plant invasions can be partitioned into either exploring how climate change drivers affect the fitness of invasive plants or understanding projected changes in the distribution of invasive populations.

Changing drivers

In contrast to the relatively large body of work done overseas on the impact of climate change drivers on invasive species, Australian research has been somewhat limited. The research that does exist has focused on a number of factors that are altering the dynamics of plant invasions via complex interactions and multiple feedback loops between the plants and the ecosystems in which they exist. These drivers can be grouped broadly into those that act globally, such as temperature, precipitation and atmospheric CO_2, and those regional drivers that are locally scaled, such as extreme weather, land use and disturbance regimes.

Globally scaled drivers

Higher global temperatures and altered precipitation patterns are likely to have significant effects on plant invasion dynamics, but there are few studies exploring the direct impacts of these drivers on Australian plant invasions. Mason *et al*. (2012) demonstrated that water stress affects the degree of competition between native plants and *Chrysanthemoides monilifera* subsp. *rotundata*, possibly allowing for increased dominance of this non-native invader with further climate change. As a component of mechanistic niche modelling with the CLIMEX package, a number of studies have explored growth rate responses to temperature and precipitation regimes for invasive plants in controlled condition experiments (e.g. van Klinken *et al*., 2009; Michael *et al*., 2012a,b).

Many parts of Australia, however, are characterized by climate that is sometimes exceptionally variable between years, as well as at decadal timescales (Power *et al*., 1999). The challenge is, therefore, to distinguish the effects of rapid climate change within the context of a naturally variable climate (Potter and Chiew, 2011). The recruitment of invasive plants in Australia can be highly sensitive to how precipitation is distributed within and between years (Lawes, 2004; Perring and Hovenden, 2012). Multi-decadal precipitation patterns are important ecologically in Australia, for example in determining tree survival (Fensham and Holman, 1999) and mass recruitment events (Harrington, 1991; van Klinken and Campbell, 2009). Population projections of the long-lived invasive shrub, *Prosopis* sp. (Mesquite), in Western Australia used a 100-year historical climate record to account for multi-scale variability in climate (precipitation; J. Pichancourt and R.D. van Klinken, personal communication). In this study, projected climate change was assumed to alter the relative proportion of wet and dry years, but not the underlying historically observed intra- and inter-annual pattern of precipitation. The underlying multi-decadal climate pattern did influence the strength of the effect of projected climate change, although it was not as strong as expected, due to strong regulation by biological control agents.

In one of the earliest studies on the effect of increasing and/or elevated atmospheric CO_2 (eCO_2) on invasive plants, O'Donnell and Adkins (1996, 2001) demonstrated that the non-native *Avena fatua* had improved growth and seed production under eCO_2 conditions. Hovenden and colleagues examined Tasmanian grassland ecosystems and found that eCO_2 and temperature did not change the flowering patterns of invasive grasses, but increased temperature had a negative impact on plant growth rates (Williams *et al.*, 2007a; Hovenden *et al.*, 2008a,b; Hovenden and Williams, 2010); indicating that climate change is not always beneficial for weedy plants. Adkins *et al.* (2010) noted that competitiveness was reduced under eCO_2 (in glasshouse pot trials) for two pasture grasses, *Setaria incrassata* (C_4, non-native) and *Astrebla squarrosa* (C_4, native), relative to the non-native *Parthenium hysterophorus* (C_3–C_4 intermediate; Moore *et al.*, 1987). In another comparison, the C_3 non-native *Vulpia myuros* was always able to suppress the native C_3 grass, *Austradanthonia eriantha*, under conditions of differing initial sizes, but this relationship was not changed by increased temperature and eCO_2 (Hely and Roxburgh, 2005). In contrast, Manea and Leishman (2011) compared 14 native–non-native species pairs (eight C_3 and six C_4) and found reduced competitive effects of the native species under eCO_2, suggesting that the relative success of non-native species might be increased. These contrasting responses to eCO_2 for both C_3 and C_4 non-native grasses suggests that competitive ability and resulting community composition may be context specific, as also found by Leishman *et al.* (1999) and Tooth and Leishman (2013).

Locally scaled drivers

Extreme weather events act locally and are arguably more influential on plant distributions than climate averages. For invasive plants, these events may alter their competitive dynamics with native congeners or change their impact on the seasonal cycles of agricultural production. Although we are not aware of any work in Australia directly exploring the possible influence of climate change-related temperature extremes or precipitation frequency and intensity (i.e. floods and droughts) on plant invasions, invasive plants are known to benefit from flooding events (11 WoNS listed species, including *Tamarix aphylla*, *Salix* spp.; Florentine and Westbrooke, 2005; Low, 2008). Furthermore, the potential distribution of the invasive wetland species, *Phyla canescens* (Lippia), is sensitive to altered flooding regimes that are expected to result directly from both climate change and the anthropogenic responses to climate change (Murray *et al.*, 2012a).

Locally scaled natural disturbance regimes, such as bushfires and cyclones, are significant components of the Australian landscape. They are projected to have altered regimes, in terms of frequency, timing and intensity, under future climates and can act directly on invasive species. For example, cyclones can facilitate the spread of invasive plants into new regions by decreasing the resilience of once-intact forest (Murphy *et al.*, 2008, 2011). Disturbance events can also act indirectly and interactively with other elements of climate change. Bushfires, a major disturbance component of some Australian landscapes, are likely to be further influenced by the interruption of natural burning cycles as land management practices respond to shifting human populations and climate change adaptation practices. Bushfire intensity and frequency can also be altered by invasive plant species themselves. For example, invasive alien grasses have already been observed to increase fuel loads, influencing the frequency, intensity and extent of fires (e.g. *Andropogon gayanus*; Rossiter *et al.*, 2003), and climate change is likely to enhance existing interactions between flammable invasives and ecosystem change (Steffen *et al.*, 2009).

Anthropogenic disturbance, such as land-use change, is also likely to be a significant driver for invasive plants as agriculture, conservation practices and human populations respond to climate change. For example, the main driver for change to invasive weeds in agricultural situations will be changed agronomic practices as farmers implement climate change adaptation measures. The result of agriculture being replaced by pastoralism under a drier climate change scenario in south-eastern Australia is expected to result in the wetland invader, *P. canescens*, becoming less likely to invade new areas but more difficult to manage once established (Murray *et al.*, 2012a). In a drying and warmer environment across much of Australia, the type of agronomic changes likely include: (i) a shift to pastoralism; (ii) opportunistic cropping to take advantage of years with good rain; (iii) the introduction of new crop and pasture species better able to tolerate future conditions; and (iv) changes in sown crops (e.g. increased production of canola in previously wetter areas). These changes and the inevitable development of new cropping regions (in particular in the north of Australia, aided by irrigation) will see a new set of weeds that require ongoing management and investment of resources. New herbicide resistance issues are one of many potential problems that would require careful consideration.

Shifting distributions

Despite a wide range of potential drivers mediating the response of invasive plants to climate change, most Australian research in this field has been on how climate change can influence shifts in plant distributions. The widespread popularity of modelling these distributions with a variety of methods (e.g. niche, species distribution, bioclimatic and mechanistic modelling) has resulted in many of these models being used to project various representations of potential climatic or habitat suitability with future climate scenarios. Helped by two large studies (Scott *et al.*, 2008; Wilson *et al.*, 2011), there are now projected distributions for future climate scenarios in Australia for 153 non-native, invasive or weedy species from 55 families (Table 11.1).

In an early example, Kriticos *et al.* (2003a) concluded that climate change could potentially increase the Australian distribution of *Acacia nilotica*, due to the plant's ability to invade drier sites further inland and to spread polewards (south) due to increased temperatures. Scott *et al.* (2008) modelled 41 'sleeper' and 'alert' list species (Table 11.1) and found an overall continental trend toward a polewards (southern) movement for model centroids projected to 2070, relative to current climate (Fig. 11.2). Sleeper and alert species, by definition, have very limited distributions in Australia but are recognized as a threat to agriculture and the environment, respectively (Australian Weeds Committee, 2007). An analysis modelling 72 species (WoNS, alert species and others; Wilson *et al.*, 2011; Table 11.1) also concluded that the south-west and south-east of Australia were likely to become 'invasion hotspots' (O'Donnell *et al.*, 2012).

Projections from this modelling do, however, have to be framed and interpreted cautiously (Venette *et al.*, 2010). Unfortunately, a significant proportion of models that explore distributions for future climate scenarios in Australia have used inappropriately applied techniques, have not quality controlled the input data properly or have not presented enough results to adequately interpret the projections. It is a frequently stated, but often overlooked, point that correlative species modelling methods (e.g. MaxEnt, Climatch, Biomod ensemble models) are inherently risky for extrapolating into novel climates (Elith and Leathwick, 2009; Elith *et al.*, 2010). Yet novel climates are exactly the situations commonly involved when modelling invasive species (van Klinken *et al.*, 2009; Webber *et al.*, 2012) and with future climate scenarios (Williams *et al.*, 2007b). Even if regions of extrapolation are revealed to the end user (e.g. MESS maps; Elith *et al.*, 2010), the primary utility of using models to identify novel climates at risk of invasion is limited with correlative techniques (Sutherst and Bourne, 2009; Rodda *et al.*, 2011; Webber *et al.*, 2011b). Some of these limitations can often be addressed through properly validated models using more mechanistic approaches (e.g. Kearney and Porter, 2009; van Klinken

Table 11.1. Studies modelling the distribution of invasive species in Australia that include projections of future climate change scenarios. The inclusion of models in this table implies nothing about their quality or how useful they might be for understanding species invasions with climate change.

Species	Family	Common name	Category[a]	Modelling software (Reference)
Acacia catechu (L.f.) Willd.	Fabaceae	Cutch tree	Alert	CLIMEX (Scott *et al.*, 2008), MaxEnt (Wilson *et al.*, 2011)
Acacia cyclops A.Cunn. ex G. Don	Fabaceae	Red-eyed wattle	Native weed	CLIMEX, MaxEnt, Boosted Regression Trees (Webber *et al.*, 2011b)
Acacia farnesiana (L.) Willd	Fabaceae	Mimosa bush	Potential weed in Victoria	BIOCLIM within ANUCLIM (Steel *et al.*, 2008)
Acacia karroo Hayne	Fabaceae	Karroo thorn	Alert	CLIMEX (Scott *et al.*, 2008; Kriticos *et al.*, 2010), MaxEnt (Wilson *et al.*, 2011)
Acacia nilotica (L.) Delile ssp. *indica* (Benth.) Brenan	Fabaceae	Prickly acacia	WoNS	CLIMEX (Kriticos *et al.*, 2000, 2003a, 2006; Kriticos, 2008), MaxEnt (Wilson *et al.*, 2011)
Acacia pycnantha Benth.	Fabaceae	Golden wattle	Native weed	CLIMEX, MaxEnt, Boosted Regression Trees (Webber *et al.*, 2011b)
Acetosa vesicaria (L.) A. Löve	Polygonaceae	Bladder dock	Potential weed in Victoria	BIOCIM within ANUCLIM (Steel *et al.*, 2008)
Aeschynomene paniculata Willd. ex Vogel	Fabaceae	Pannicle jointvetch	Sleeper	CLIMEX (Scott *et al.*, 2008)
Alternanthera philoxeroides Griseb.	Amaranthaceae	Alligator weed	WoNS	MaxEnt (Wilson *et al.*, 2011)
Andropogon gayanus Kunth.	Poaceae	Gamba grass	New WoNS	MaxEnt (Wilson *et al.*, 2011)
Annona glabra L.	Annonaceae	Pond apple	WoNS	MaxEnt (Wilson *et al.*, 2011)
Anredera cordifolia (Ten.) Steenis	Basellaceae	Madeira vine	New WoNS, Candidate WoNS 2000, non-native vine	MaxEnt (Gallagher *et al.*, 2010; Wilson *et al.*, 2011)
Argemone ochroleuca Sweet	Papaveraceae	Mexican poppy	Candidate WoNS 2000	MaxEnt (Wilson *et al.*, 2011)
Asparagus aethiopicus L.	Asparagaceae	Basket asparagus	New WoNS, potential weed in Victoria	BIOCLIM within ANUCLIM (Steel *et al.*, 2008)
Asparagus asparagoides (L.) Druce	Asparagaceae	Bridal creeper	WoNS	CLIMEX (Kriticos *et al.*, 2010), MaxEnt (Wilson *et al.*, 2011)
Asparagus declinatus L.	Asparagaceae	Bridal veil	New WoNS	CLIMEX (Kriticos *et al.*, 2010), MaxEnt (Wilson *et al.*, 2011)

Asystasia gangetica (L.) T.Anders. ssp. *micrantha* (Nees) Ensermu	Acanthaceae	Chinese violet	Alert and sleeper	CLIMEX (Scott *et al.*, 2008), MaxEnt (Wilson *et al.*, 2011)
Baccharis halimifolia L.	Asteraceae	Groundsel bush		CLIMEX (Sims-Chilton *et al.*, 2010)
Baccharis pingraea DC.	Asteraceae	Chilquilla	Sleeper	CLIMEX (Scott *et al.*, 2008)
Barleria prionitis L.	Acanthaceae	Barleria or porcupine flower	Alert	CLIMEX (Scott *et al.*, 2008), MaxEnt (Wilson *et al.*, 2011)
Bassia scoparia (L.) A.J. Scott	Chenopodiaceae	Kochia	Alert, Candidate WoNS 2000	CLIMEX (Scott *et al.*, 2008), MaxEnt (Wilson *et al.*, 2011)
Bidens pilosa L.	Asteraceae	Cobbler's pegs	Potential weed in Victoria	BIOCLIM within ANUCLIM (Steel *et al.*, 2008)
Billardiera heterophylla (Lindl.) L. Cayzer & Crisp	Pittosporaceae	Bluebell creeper	Potential weed in Victoria	BIOCLIM within ANUCLIM (Steel *et al.*, 2008)
Brillantaisia lamium (Nees) Benth.	Acanthaceae		Sleeper	CLIMEX (Scott *et al.*, 2008)
Bryophyllum tubiflorum Harv. and hybrids	Crassulaceae	Mother of millions	Candidate WoNS 2000	MaxEnt (Wilson *et al.*, 2011)
Cabomba caroliniana Gray	Cabombaceae	Cabomba	WoNS	MaxEnt (Wilson *et al.*, 2011)
Calluna vulgaris (L.) Hull	Ericaceae	Heather	Alert	CLIMEX (Scott *et al.*, 2008), MaxEnt (Wilson *et al.*, 2011)
Calotropis procera (Aiton) W.T. Aiton	Apocynaceae	Calotrope	Candidate WoNS 2000	MaxEnt (Wilson *et al.*, 2011)
Cenchrus ciliaris L.	Poaceae	Buffelgrass	Potential weed in Victoria	BIOCLIM within ANUCLIM (Steel *et al.*, 2008), MaxEnt (Wilson *et al.*, 2011)
Celtis sinensis Pers.	Cannabaceae	Chinese elm	Candidate WoNS 2000	MaxEnt (Wilson *et al.*, 2011)
Centaurea eriophora L.	Asteraceae	Mallee cockspur	Sleeper	CLIMEX (Scott *et al.*, 2008)
Chloris truncata R. Br.	Poaceae	Windmill grass	Crop weed, native weed	CLIMEX (Michael *et al.*, 2012a)
Chromolaena odorata (L.) King & Robinson	Asteraceae	Siam weed	Alert	CLIMEX (Kriticos *et al.*, 2006; Kriticos, 2008; Scott *et al.*, 2008), MaxEnt (Wilson *et al.*, 2011)
Chrysanthemoides monilifera (L.) Norl	Asteraceae	Bitou bush, bone-seed	WoNS	MaxEnt (Wilson *et al.*, 2011)
Cortaderia selloana (Schult. & Schult. f.) Asch. & P. Graebn.	Poaceae	Pampas grass	Candidate WoNS 2000	MaxEnt (Wilson *et al.*, 2011; Gallagher *et al.*, 2012)
Cotoneaster glaucophyllus Franch.	Rosaceae	Cotoneaster	Potential weed in Victoria	BIOCLIM within ANUCLIM (Steel *et al.*, 2008)
Crupina vulgaris Pers. ex Cass.	Asteraceae	Common crupina	Sleeper	CLIMEX (Scott *et al.*, 2008)

Continued

Table 11.1. Continued

Species	Family	Common name	Category[a]	Modelling software (Reference)
Cryptostegia grandiflora Roxb. ex R. Br.	Apocynaceae	Rubber vine	WoNS	CLIMEX (Kriticos *et al.*, 2000, 2003b; Kriticos, 2008), MaxEnt (Wilson *et al.*, 2011)
Cuscuta campestris Yunck.	Cuscutaceae	Golden dodder	Candidate WoNS 2000	MaxEnt (Wilson *et al.*, 2011)
Cuscuta suaveolens Ser.	Cuscutaceae	Chilean dodder	Sleeper	CLIMEX (Scott *et al.*, 2008)
Cynoglossum creticum Mill.	Boraginaceae	Blue hound's tongue	Alert	CLIMEX (Scott *et al.*, 2008), MaxEnt (Wilson *et al.*, 2011)
Cyperus teneristolon Mattf. & Kük.	Cyperaceae	Cyperus	Alert	CLIMEX (Scott *et al.*, 2008), MaxEnt (Wilson *et al.*, 2011)
Cytisus multiflorus (L'Hér.) Sweet	Fabaceae	White Spanish broom	Alert	CLIMEX (Scott *et al.*, 2008), MaxEnt (Wilson *et al.*, 2011)
Cytisus scoparius (L.) Link	Fabaceae	Scotch broom	New WoNS	CLIMEX (Potter *et al.*, 2008, 2009; Kriticos *et al.*, 2010), MaxEnt (Wilson *et al.*, 2011)
Dittrichia viscose (L.) Greuter	Asteraceae	False yellowhead	Alert	CLIMEX (Scott *et al.*, 2008), MaxEnt (Wilson *et al.*, 2011)
Echium plantagineum L.	Boraginaceae	Paterson's curse	Candidate WoNS 2000, Potential weed in Victoria	BIOCLIM within ANUCLIM (Steel *et al.*, 2008), MaxEnt (Wilson *et al.*, 2011)
Eichhornia crassipes (C. Mart.) Solms	Pontederiaceae	Water hyacinth	New WoNS	MaxEnt (Wilson *et al.*, 2011)
Eleocharis parodii Barros	Cyperaceae	Parodi spike rush	Sleeper	CLIMEX (Scott *et al.*, 2008)
Elephantopus mollis Kunth	Asteraceae	Tobacco weed	Candidate WoNS 2000	MaxEnt (Wilson *et al.*, 2011)
Equisetum spp.	Equisetaceae	Horsetails	Alert	MaxEnt (Wilson *et al.*, 2011)
Equisetum arvense L.	Equisetaceae	Horsetail	Alert	CLIMEX (Scott *et al.*, 2008)
Eragrostis curvula (Schrad.) Nees	Poaceae	African love grass	Candidate WoNS 2000	MaxEnt (Wilson *et al.*, 2011; Gallagher *et al.*, 2012)
Eremophila sturtii R. Br.	Myoporaceae	Narrow-leaf emu bush	Potential weed in Victoria	BIOCLIM within ANUCLIM (Steel *et al.*, 2008)
Erica lusitanica Rudolphi	Ericaceae	Spanish heath	Candidate WoNS 2000	MaxEnt (Wilson *et al.*, 2011)
Euphorbia paralias L.	Euphorbiaceae	Sea spurge	Candidate WoNS 2000	MaxEnt (Wilson *et al.*, 2011)
Euphorbia terracina L.	Euphorbiaceae	Terracina spurge	Potential weed in Victoria	BIOCLIM within ANUCLIM (Steel *et al.*, 2008)

Froelichia floridana (Nutt.) Moq.	Amaranthaceae	Snakecotton	Sleeper	CLIMEX (Scott *et al*., 2008)
Genista monspessulana (L.) L.A.S. Johnson	Fabaceae	Broom	New WoNS, Candidate WoNS 2000	MaxEnt (Wilson *et al*., 2011)
Gleditsia triacanthos L.	Fabaceae	Honey locust	Candidate WoNS 2000	MaxEnt (Wilson *et al*., 2011)
Gmelina elliptica Sm.	Lamiaceae	Badhara bush	Sleeper	Climatch (Baker, 2010), CLIMEX (Scott *et al*., 2008)
Gomphocarpus fruticosus (L.) W.T. Aiton	Apocynaceae	Narrow leaf cotton bush	Candidate WoNS 2000	MaxEnt (Wilson *et al*., 2011)
Gymnocoronis spilanthoides (D. Don.) DC.	Asteraceae	Senegal tea plant	Alert	CLIMEX (Scott *et al*., 2008), MaxEnt (Wilson *et al*., 2011)
Heliotropium amplexicaule Vahl	Boraginaceae	Blue heliotrope	Potential weed in Victoria	BIOCLIM within ANUCLIM (Steel *et al*., 2008)
Hieracium aurantiacum L.	Asteraceae	Orange hawkweed	Alert and sleeper	Climatch (Baker, 2010), CLIMEX (Scott *et al*., 2008), BIOMOD ensemble (8 correlative models) (Beaumont *et al*., 2009a,b), MaxEnt (Wilson *et al*., 2011)
Hieracium murorum L.	Asteraceae	Hawkweed		BIOMOD ensemble (8 correlative models) (Beaumont *et al*., 2009a,b)
Hieracium pilosella L.	Asteraceae	Hawkweed		BIOMOD ensemble (8 correlative models) (Beaumont *et al*., 2009a,b)
Hordeum glaucum Steud.	Poaceae	Blue barley-grass	Potential weed in Victoria	BIOCLIM within ANUCLIM (Steel *et al*., 2008)
Hydrocotyle ranunculoides L.f.	Apiaceae	Hydrocotyl	Candidate WoNS 2000	MaxEnt (Wilson *et al*., 2011)
Hymenachne amplexicaulis (Rudge) Nees	Poaceae	Hymenachne	WoNS	MaxEnt (Wilson *et al*., 2011)
Hyparrhenia hirta (L.) Stapf	Poaceae	Coolatai grass		CLIMEX (Chejara *et al*., 2010; Kriticos *et al*., 2010), MaxEnt (Wilson *et al*., 2011)
Hypericum perforatum L.	Hypericaceae	St John's wort	Candidate WoNS 2000	MaxEnt (Wilson *et al*., 2011)
Hypericum tetrapterum Fries	Clusiaceae	Square-stalked St John's wort	Sleeper	CLIMEX (Scott *et al*., 2008)
Hyptis suaveolens (L.) Poit.	Lamiaceae	Hyptis	Candidate WoNS 2000	MaxEnt (Wilson *et al*., 2011)
Ipomoea indica (Burm.f.) Merr.	Convolvulaceae	Blue morning glory	Non-native vine	MaxEnt (Gallagher *et al*., 2010)

Continued

Table 11.1. Continued

Species	Family	Common name	Category[a]	Modelling software (Reference)
Ipomoea purpurea (L.) Roth	Convolvulaceae	Purple morning glory	Non-native vine	MaxEnt (Gallagher *et al.*, 2010)
Jatropha gossypifolia L.	Euphorbiaceae	Bellyache bush	New WoNS, Candidate WoNS 2000	MaxEnt (Wilson *et al.*, 2011)
Koelreuteria elegans (Seem.) A.C. Sm. ssp. *formosana* (Hayata) F.G. Mey.	Sapinaceae	Chinese rain tree	Alert	CLIMEX (Scott *et al.*, 2008), MaxEnt (Wilson *et al.*, 2011)
Lachenalia reflexa Thunb.	Hyacinthaceae	Yellow soldier	Alert	CLIMEX (Scott *et al.*, 2008), MaxEnt (Wilson *et al.*, 2011)
Lagarosiphon major (Ridl.) Moss	Hydrocharitaceae	Lagarosiphon	Alert	CLIMEX (Scott *et al.*, 2008), MaxEnt (Wilson *et al.*, 2011)
Lantana camara L.	Verbenaceae	Lantana	WoNS, potential weed in Victoria	BIOCLIM within ANUCLIM (Steel *et al.*, 2008), CLIMEX (Taylor *et al.*, 2012), MaxEnt (Wilson *et al.*, 2011)
Lantana montevidensis (Spreng.) Briq.	Verbenaceae	Creeping lantana	Candidate WoNS 2000	MaxEnt (Wilson *et al.*, 2011)
Leycesteria formosa Wall.	Caprifoliaceae	Himalayan honey-suckle	Potential weed in Victoria	BIOCLIM within ANUCLIM (Steel *et al.*, 2008)
Ligustrum lucidum Aiton	Oleaceae	Broad-leaved privet	Candidate WoNS 2000	MaxEnt (Wilson *et al.*, 2011)
Ligustrum sinense Lour.	Oleaceae	Small-leaved privet	Candidate WoNS 2000, potential weed in Victoria	BIOCLIM within ANUCLIM (Steel *et al.*, 2008), MaxEnt (Wilson *et al.*, 2011)
Lycium ferocissimum Miers.	Solanaceae	African boxthorn	New WoNS	CLIMEX (Kriticos *et al.*, 2010), MaxEnt (Wilson *et al.*, 2011)
Macfadyena unguis-cati (L.) A.H. Gentry	Bignoniaceae	Cat's claw creeper	New WoNS, Candidate WoNS 2000	MaxEnt (Wilson *et al.*, 2011)
Medicago laciniata (L.) Mill.	Fabaceae	Cut-leaf medic	Potential weed in Victoria	BIOCLIM within ANUCLIM (Steel *et al.*, 2008)
Melaleuca quinquenervia (Cav.) S.T. Blake	Myrtaceae	Broad-leaved paper-bark		CLIMEX (Watt *et al.*, 2009)
Mimosa pigra L.	Fabaceae	Mimosa	WoNS	MaxEnt (Wilson *et al.*, 2011)
Nassella charruana (Arechav.) Barkworth	Poaceae	Lobed needle grass	Alert and sleeper	CLIMEX (Scott *et al.*, 2008), MaxEnt (Wilson *et al.*, 2011)
Nassella hyalina (Nees) Barkworth	Poaceae	Cane needle grass	Alert	CLIMEX (Scott *et al.*, 2008), MaxEnt (Wilson *et al.*, 2011; Gallagher *et al.*, 2012)

Nassella neesiana (Trin. & Rupr.) Barkworth	Poaceae	Chilean needle grass	Potential weed in Victoria, WoNS	BIOCLIM within ANUCLIM (Steel *et al.*, 2008), CLIMEX (Kriticos *et al.*, 2010), MaxEnt (Wilson *et al.*, 2011; Gallagher *et al.*, 2012)
Nassella trichotoma (Nees) Hack. ex Arechav.	Poaceae	Serrated tussock	Potential weed in Victoria, WoNS	BIOCLIM within ANUCLIM (Steel *et al.*, 2008), CLIMEX (Watt *et al.*, 2011; Kriticos *et al.*, 2012), MaxEnt (Wilson *et al.*, 2011; Gallagher *et al.*, 2012)
Oenanthe pimpinelloides L.	Apiaceae	Meadow parsley	Sleeper	CLIMEX (Scott *et al.*, 2008)
Olea europaea L. ssp. *europaea*	Oleaceae	Wild olive		CLIMEX (Kriticos *et al.*, 2010)
Onopordum acanthium L.	Asteraceae	Scotch thistle	Candidate WoNS 2000	MaxEnt (Wilson *et al.*, 2011)
Onopordum acaulon L.	Asteraceae	Stemless thistle	Candidate WoNS 2000	MaxEnt (Wilson *et al.*, 2011)
Onopordum illycirum L.	Asteraceae	Illyrian thistle	Candidate WoNS 2000	MaxEnt (Wilson *et al.*, 2011)
Onopordum tauricum Wild.	Asteraceae	Taurian thistle	Sleeper	CLIMEX (Scott *et al.*, 2008)
Opuntia robusta J.C. Wendl. ex Pfeiff.	Cactaceae	Wheel cactus	New WoNS	CLIMEX (Kriticos *et al.*, 2010)
Orobanche minor Sm.	Orobanchaceae	Broomrape	Candidate WoNS 2000	MaxEnt (Wilson *et al.*, 2011)
Parkinsonia aculeata L.	Caesalpinaceae	Parkinsonia	WoNS	CLIMEX (van Klinken *et al.*, 2009), MaxEnt (Wilson *et al.*, 2011)
Parthenium hysterophorus L.	Asteraceae	Parthenium weed	WoNS	MaxEnt (Wilson *et al.*, 2011)
Passiflora suberosa L.	Passifloraceae	Corky passionflower	Potential weed in Victoria	BIOCLIM within ANUCLIM (Steel *et al.*, 2008)
Passiflora subpeltata Ortega	Passifloraceae	White passionflower	Non-native vine	MaxEnt (Gallagher *et al.*, 2010)
Pelargonium alchemilloides (L.) L'Hérit.	Geraniaceae	Geranium	Alert	CLIMEX (Scott *et al.*, 2008), MaxEnt (Wilson *et al.*, 2011)
Pennisetum polystachion (L.) Schultes	Poaceae	Mission grass	Candidate WoNS 2000	MaxEnt (Wilson *et al.*, 2011; Gallagher *et al.*, 2012)
Pereskia aculeata Miller	Cactaceae	Leaf cactus	Alert	CLIMEX (Scott *et al.*, 2008), MaxEnt (Wilson *et al.*, 2011)
Phyla canescens (Kunth) Greene	Verbenaceae	Lippia	Candidate WoNS 2000	Bayesian Belief Network with CLIMEX (Murray *et al.*, 2012a), MaxEnt (Wilson *et al.*, 2011)
Piptochaetium montevidense (Spreng.) Parodi	Poaceae	Uruguayan rice grass	Alert and sleeper	CLIMEX (Scott *et al.*, 2008), MaxEnt (Wilson *et al.*, 2011; Gallagher *et al.*, 2012)
Polygala myrtifolia L.	Polygalaceae	Myrtleleaf milkwort	Candidate WoNS 2000	MaxEnt (Wilson *et al.*, 2011)

Continued

Table 11.1. Continued

Species	Family	Common name	Category[a]	Modelling software (Reference)
Praxelis clematidea (Griseb.) R.B. King & H. Rob.	Asteraceae	Praxelis	Alert	CLIMEX (Scott *et al*., 2008), MaxEnt (Wilson *et al*., 2011)
Prosopis pallida (Humb. & Bonpl. ex Willd.) Kunth	Fabaceae	Mesquite	Potential weed in Victoria	BIOCLIM within ANUCLIM (Steel *et al*., 2008)
Prosopis spp.	Fabaceae	Mesquite	WoNS	MaxEnt (Wilson *et al*., 2011)
Rapistrum rugosum (L.) All.	Brassicaceae	Turnip weed	Crop weed	CLIMEX (Michael *et al*., 2012c)
Reseda luteola L.	Resedaceae	Wild mignonette	Candidate WoNS 2000	MaxEnt (Wilson *et al*., 2011)
Retama raetam (Forsk.) Webb	Fabaceae	White weeping broom	Alert	CLIMEX (Scott *et al*., 2008; Kriticos *et al*., 2010), MaxEnt (Wilson *et al*., 2011)
Rorippa sylvestris (L.) Besser	Brassicaceae	Yellow creeping cress	Sleeper	CLIMEX (Scott *et al*., 2008)
Rubus anglocandicans A. Newton	Rosaceae	Blackberry	WoNS	CLIMEX (Kriticos *et al*., 2010)
Rubus fruticosus L. agg.	Rosaceae	Blackberry	Potential weed in Victoria, WoNS	BIOCLIM within ANUCLIM (Steel *et al*., 2008), MaxEnt (Wilson *et al*., 2011)
Salix spp.	Salicaceae	Willows	WoNS	MaxEnt (Wilson *et al*., 2011)
Salvinia molesta D.S. Mitch.	Salviniaceae	Salvinia	WoNS	MaxEnt (Wilson *et al*., 2011)
Schinus terebinthifolia Raddi	Anacardiaceae	Brazilian pepper	Candidate WoNS 2000	MaxEnt (Wilson *et al*., 2011)
Senecio glastifolius L.	Asteraceae	Holly leaved senecio	Alert	CLIMEX (Scott *et al*., 2008), MaxEnt (Wilson *et al*., 2011)
Senecio jacobaea L.	Asteraceae	Ragwort	Candidate WoNS 2000, potential weed in Victoria	BIOCLIM within ANUCLIM (Steel *et al*., 2008), MaxEnt (Wilson *et al*., 2011)
Senecio madagascariensis Poir.	Asteraceae	Fireweed	New WoNS, Candidate WoNS 2000	MaxEnt (Wilson *et al*., 2011)
Senna obtusifolia (L.) H.S. Irwin & Barneby	Fabaceae	Sicklepod	Candidate WoNS 2000	MaxEnt (Wilson *et al*., 2011)
Senna tora (L.) Roxb.	Fabaceae	Sicklepod	Candidate WoNS 2000	MaxEnt (Wilson *et al*., 2011)
Sida rhombifolia L.	Malvaceae	Paddy's lucerne	Potential weed in Victoria	BIOCLIM within ANUCLIM (Steel *et al*., 2008), MaxEnt (Wilson *et al*., 2011)
Solanum elaeagnifolium Cav.	Solanaceae	Sliver leaf nightshade	New WoNS	CLIMEX (Kriticos *et al*., 2010), MaxEnt (Wilson *et al*., 2011)
Solanum hoplopetalum Bitter & Summerh.	Solanaceae	Afghan thistle	Crop weed, native weed	CLIMEX (Michael *et al*., 2012b)
Solanum seaforthianum Andrews	Solanaceae	Climbing nightshade	Non-native vine	MaxEnt (Gallagher *et al*., 2010)
Spartina anglica C.E. Hubb.	Poaceae	Common cordgrass	Candidate WoNS 2000	MaxEnt (Wilson *et al*., 2011)

Sporobolus africanus (Poir.) Robyns & Tournay	Poaceae	Parramatta grass	Candidate WoNS 2000	MaxEnt (Wilson *et al.*, 2011; Gallagher *et al.*, 2012)
Sporobolus natalensis (Steud.) Th. Dur. & Schinz	Poaceae	Giant rat's tail grass	Candidate WoNS 2000	MaxEnt (Wilson *et al.*, 2011; Gallagher *et al.*, 2012)
Sporobolus pyramidalis Beauv.	Poaceae	Parramatta grass	Candidate WoNS 2000	MaxEnt (Wilson *et al.*, 2011; Gallagher *et al.*, 2012)
Stachytarpheta jamaicensis (L.) Vahl	Verbenaceae	Snake weed	Candidate WoNS 2000	MaxEnt (Wilson *et al.*, 2011)
Stachytarpheta mutibilis (Jacq.) Vahl	Verbenaceae	Snake weed	Candidate WoNS 2000	MaxEnt (Wilson *et al.*, 2011)
Stevia ovata Willd.	Asteraceae	Candyleaf		CLIMEX (Murphy *et al.*, 2009)
Tamarix aphylla (L.) Karst.	Tamaricaceae	Athel pine	Potential weed in Victoria, WoNS	BIOCLIM within ANUCLIM (Steel *et al.*, 2008), MaxEnt (Wilson *et al.*, 2011)
Themeda quadrivalvis (L.) Kuntze	Poaceae	Grader grass	Candidate WoNS 2000	MaxEnt (Wilson *et al.*, 2011; Gallagher *et al.*, 2012)
Thunbergia grandiflora (Roxb. ex Rottler) Roxb.	Acanthaceae	Blue thunbergia	Candidate WoNS 2000	MaxEnt (Wilson *et al.*, 2011)
Thunbergia laurifolia Lindl.	Acanthaceae	Laurel clock vine	Alert	CLIMEX (Scott *et al.*, 2008), MaxEnt (Wilson *et al.*, 2011)
Tipuana tipu (Benth.) Kuntze	Fabaceae	Rosewood or tipuana tree	Alert	CLIMEX (Scott *et al.*, 2008), MaxEnt (Wilson *et al.*, 2011)
Trianoptiles solitaria (C.B. Clarke) Levyns	Cyperaceae	Subterranean Cape sedge	Alert	CLIMEX (Scott *et al.*, 2008), MaxEnt (Wilson *et al.*, 2011)
Ulex europaeus L.	Fabaceae	Gorse	WoNS	MaxEnt (Wilson *et al.*, 2011)
Urochloa mutica (Forssk.) T.Q. Nguyen	Poaceae	Para grass		MaxEnt (Wilson *et al.*, 2011)
Watsonia spp.	Iridaceae	Watsonia	Candidate WoNS 2000	MaxEnt (Wilson *et al.*, 2011)
Xanthium occidentale Bertol	Asteraceae	Noogoora burr	Candidate WoNS 2000	MaxEnt (Wilson *et al.*, 2011)
Xanthium spinosum L.	Asteraceae	Bathurst burr	Candidate WoNS 2000, potential weed in Victoria	BIOCLIM within ANUCLIM (Steel *et al.*, 2008), MaxEnt (Wilson *et al.*, 2011)
Zantedeschia aethiopica (L.) Spreng.	Araceae	Arum lily	Candidate WoNS 2000	CLIMEX (Kriticos *et al.*, 2010), MaxEnt (Wilson *et al.*, 2011)
Ziziyphus mauritiana Lam.	Rhamnaceae	Chinese apple	Candidate WoNS 2000	MaxEnt (Wilson *et al.*, 2011)

Notes: [a]'WoNS' = Weed of National Significance. Initially there was a candidate list (Candidate WoNS 2000), from which 20 WoNS were selected. A new selection process led to an additional 12 WoNS being declared in 2012. Details on these species and the selection process can be found at http://www.weeds.org.au/. 'Sleeper' and 'alert' species are those currently with a restricted distribution that are predicted to be important future weeds. Background context on these species can be found in Scott *et al.* (2008) and http://www.environment.gov.au/biodiversity/invasive/weeds/weeds/lists/sleeper.html and at http://www.environment.gov.au/biodiversity/invasive/weeds/weeds/lists/alert.html. Steel *et al.* (2008) listed a set of species considered to be potential weeds in Victoria. 'Crop weeds' are weeds of grains crops. 'Native weeds' are species indigenous to Australia that are unwanted plants where they are found. 'Non-native vines' refers to the focal species in Gallagher *et al.* (2010).

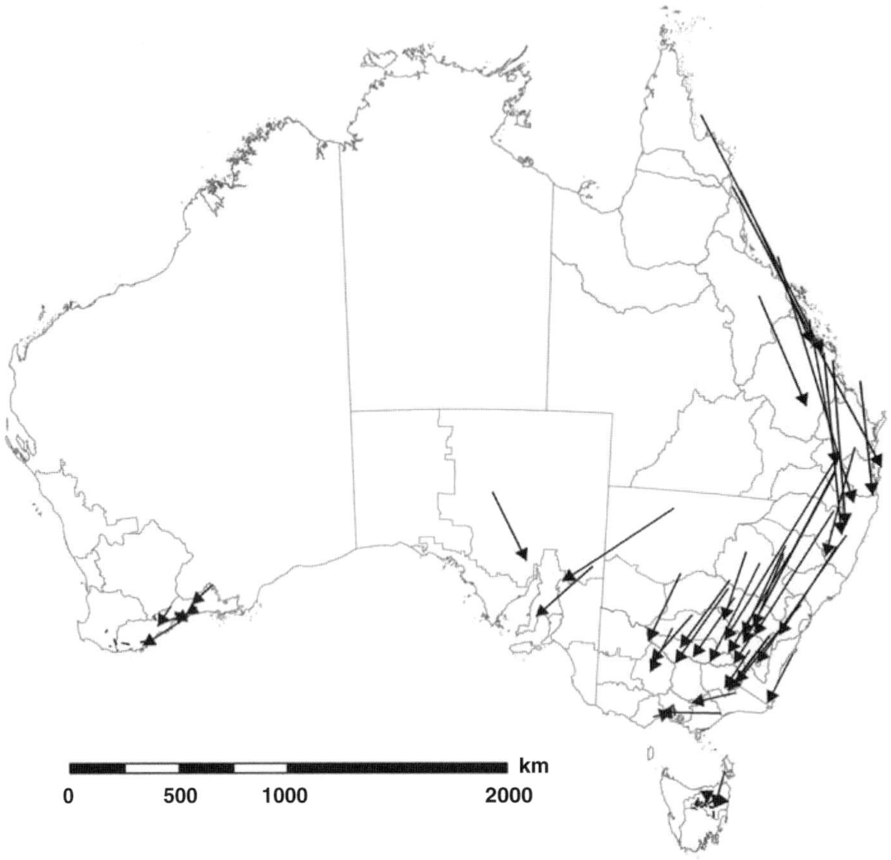

Fig. 11.2. Trajectories of non-native plants on the Australian 'sleeper and alert list' with projected climate change for 2070. State boundaries and Natural Resource Management (NRM) regions are indicated (grey lines). The base of each arrow is the centre of the current modelled distribution, the arrow tip is the shifted centre of the distribution when projected under the Echam 3 climate change scenario. (Figure modified from Scott *et al.*, 2008, with permission.)

et al., 2009). In fact, these techniques (e.g. CLIMEX) have been applied more widely in Australia to modelling invasive plants with future climate scenarios than anywhere else in the world (Table 11.1; 44% of the 152 species).

Although the choice of taxa for modelling studies has been influenced strongly by national priority lists (Scott *et al.*, 2008; Wilson *et al.*, 2011), distinct knowledge gaps remain. First, there is a distinct lack of modelling done on species in tropical Australia relative to the number of invasive plants in these regions, particularly in the west. While this trend may reflect the underlying geographical bias in research into invasive species, it may also mask different or more nuanced trends in the directional shift of suitable habitat for invasive species in Australia (Fig. 11.2). Second, a number of very important agricultural weeds have not been modelled (e.g. *Raphanus raphanistrum*, *Lolium rigidum*) because of the focus on invasive plant species in natural environments over agricultural weeds. Last, few models have been done for invasive plants that are native to other regions of Australia (but see *Chloris truncata*; Michael *et al.*, 2012a) or invasive native weeds (but see *Solanum hoplopetalum*;

Michael *et al.*, 2012b). We caution, however, that a call for more modelling must be balanced against a need also to understand other important aspects of plant invasions in a changing climate, such as ecosystem interactions and impact, and for any new modelling research to apply best practice techniques to their research questions.

There is an increasing recognition that climate change can influence the potential distribution of invasive plants in ways that are not typically captured in the modelling of species distributions with bioclimatic covariates based largely on average conditions. Such drivers include altered fire or flood regimes, changing land-use patterns or altered competitive and trophic interactions (Hellmann *et al.*, 2008; Scoble and Lowe, 2010). Taking this step, however, to include additional modelling of these drivers can enhance significantly the utility of research output for management recommendations and on-ground action.

Recent modelling in Australia has sought to address this issue by combining bioclimatic modelling with other environmental factors that influence the establishment and persistence of invasive plant populations (Michael *et al.*, 2012a; Murray *et al.*, 2012a; Smith *et al.*, 2012). The Murray Darling Basin in south-eastern Australia, for example, is likely to see fundamental changes in land use and altered flood regimes with future climate change, driven by interactions between a reduction in overall precipitation, different precipitation patterns and anthropogenic responses to changing water availability (Colloff and Baldwin, 2010; Murray *et al.*, 2012a). Such changes are very likely to impact the potential distribution of invasive plants. For example, modelling indicates that the climatically suitable area for the wetland invader, *P. canescens*, would be reduced in the modelled region (Queensland Murray Darling Basin) under the most likely drier scenario (Murray *et al.*, 2012a; Fig. 11.3). Suitable area would be further reduced by the expected decrease in area and frequency of flooding, irrespective of land-use changes (Fig. 11.3e). However, existing clonal populations of *P. canescens* would persist if the region remained under grazing, as they are not dependent on flooding regimes (Murray *et al.*, 2012a).

Managing Invasive Plants in a Changing Climate

With an inherently variable climate and invasive species often spread across wide climatic gradients, it would be easy to assume that Australia might be well positioned to integrate adaptation to climate change into 'business as usual' management of invasive species. For many aspects of such management, climate change is just another element acting in an additive or interactive way with existing drivers. Indeed, other aspects of global change, such as demographic change, land-use change and global connectivity, may well prove to be more significant factors influencing the future threat posed by invasive plants to native biodiversity and agriculture. Independent of other factors, climate change is important enough to be consolidated as an essential component of invasive species research and management in Australia. Otherwise, for example, poorly considered general adaptation responses to climate change, such as landscape corridors, may inadvertently exacerbate the threat posed by invasive plants to Australian ecosystems.

Invasive plants can impact profoundly the environmental, economic and social values that we seek to protect. There is an emerging emphasis on ensuring that the objective of management is to minimize any negative impacts, rather than to focus on the invasive species per se (Hulme *et al.*, 2013). None the less, most climate change research on invasive plants up to now has focused on understanding the changes in realized or potential distributions, rather than on how the consequences of invasion might change with climate change. It is sometimes implicitly assumed that if the distribution of invasive plants changes with climate change, then this will result in a concomitant change in impact. This may indeed be the case: for example, if it allows a non-native species to invade and impact

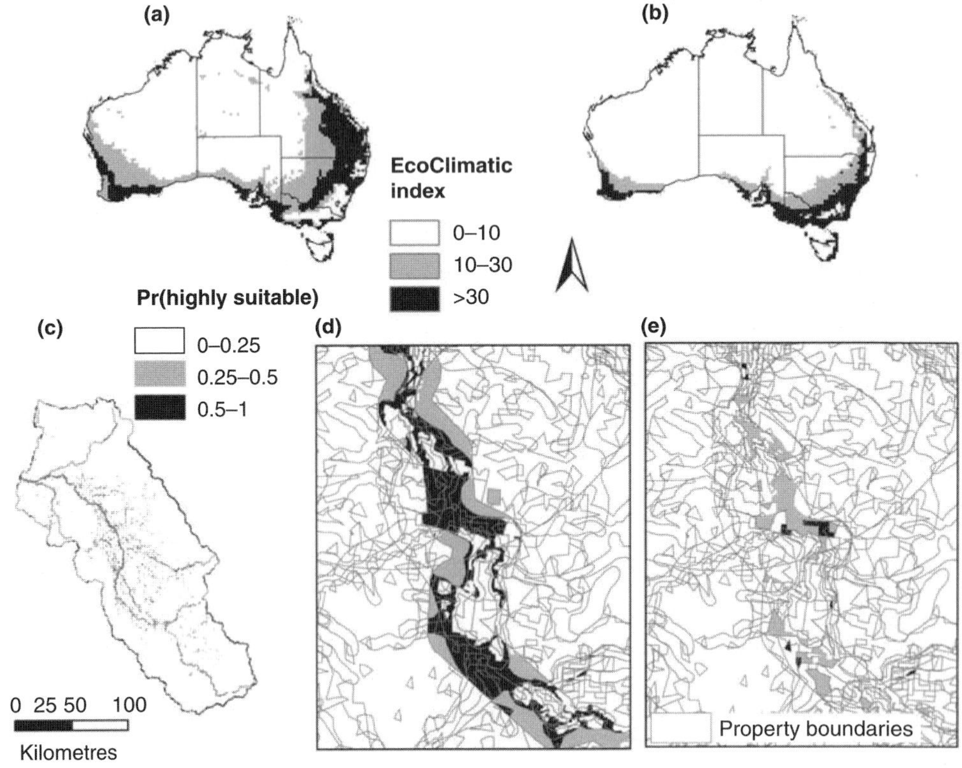

Fig. 11.3. The projected effect of climate change on the distribution of an invasive wetland plant, *Phyla canescens* (Lippia), in Australia when considering only bioclimatic factors (a,b) and when including other determinants of habitat suitability such as land use, soil type, herbaceous cover and flooding regimes (c–e). Current climatic conditions are shown at different spatial resolutions (a,c,d) and compared against a generally hotter, drier climate change scenario (HADCM2, 2070 climate scenario) (b,e) with an associated reduction in flooding frequency and extent (e). (Figure modified from Murray *et al.*, 2012a, with permission.)

new areas. It may even exacerbate the impacts: for example, if it allows the invader to shift into remnant areas of high biodiversity value. However, the distribution of at-risk values can also change with climate change. Land use, for example, is expected to change dramatically across the Murray Darling Basin under certain climate change scenarios (Jiang and Grafton, 2012). If climate change resulted in increased precipitation, it would lead to more intensive agriculture in the region, and therefore less economic impact from *P. canescens*, even though this invasive species should thrive under wetter conditions (Murray *et al.*, 2012a). Conversely, a drying, hotter basin might make the region more susceptible to a wide range of new agricultural weeds, but this would be of little consequence in parts of the basin that become largely unsuited to intensive agriculture (Jiang and Grafton, 2012).

Contextualising the possible responses of invasive plants to climate change within broader management, adaptation and mitigation challenges is a priority for Australia. We have identified four areas where management issues for invasive plants in Australia will be particularly influenced by, or need to consider carefully, climate change.

Dispersal and landscape connectivity

An unfortunate paradox of climate change adaptation for the conservation of native species is that improving landscape connectivity to facilitate the tracking of similar climates for native plants also increases the risk of non-native plants becoming more invasive (Procheş *et al.*, 2005). With significant capital investment and planning, Australia intends to build a national network of wildlife corridors as part of an adaptation response for climate change by improving environmental connectivity for native flora and fauna (Commonwealth of Australia, 2012). It is somewhat ironic that the clearest measure of success for such corridors is that native plants will expand rapidly or shift their range (i.e. become invasive). The national strategy on wildlife corridors recognizes that these corridors could also become pathways for invasive non-native plants and pests (Commonwealth of Australia, 2012). This is to be managed by corridor design and management plans for non-natives. Many of the major existing corridor initiatives – Gondwana Link, the Great Eastern Ranges Initiative, Habitat 141°, NatureLinks, Trans-Australia Eco-Link, Tasmanian Midlandscapes (Commonwealth of Australia, 2012) – have climate change as a major design consideration (Whitten *et al.*, 2011). Some corridors are laid out in an east–west direction, whereas both the modelling of non-native species (Fig. 11.2) and the few native species examples (Michael *et al.*, 2012a,b) point to a north–south direction of movement. One north–south corridor is the Great Eastern Ranges Initiative (Whitten *et al.*, 2011). This corridor corresponds closely to the projected trajectory of invasive non-native species in the same area (Fig. 11.2). Adaptation responses to manage invasive non-native plants in corridors include east–west quarantine boundaries at natural or artificial restriction points (e.g. the urban area of Perth, Western Australia) and convenient political boundaries (e.g. the Queensland–New South Wales border). It is clear that careful, nuanced thought needs to be given to what outcomes will be viewed as 'successful', for both donor and recipient communities, involving range-shifting species, both native and non-native.

Biological control

Australia is a world leader in the use of biological control agents, mainly arthropods and pathogens, for invasive plant control. A recent review found that over 90 invasive plant species had been targeted in Australia using some 242 agents (Julien *et al.*, 2012). Biological control currently provides permanent and effective control for some major invasive plant species. Many of these biological control agents were selected because they came from regions with climates that matched the climate of the target weed-infested areas of Australia. There has, however, been little research on what happens to the agent–target association with climate change.

Climate change is likely to affect biological control in the same manner as it will affect any arthropod– or pathogen–plant interaction. Increased temperatures, changes to precipitation and higher CO_2 concentrations can all act, directly or indirectly, to influence the effectiveness of existing biological control programmes. For example, plant responses under eCO_2 that may affect their insect agents directly include increased growth rates, increased carbon to nitrogen ratio (C:N), decreased nitrogen (N) concentration and changes to secondary chemistry (Stiling and Cornelissen, 2007; DeLucia *et al.*, 2012). However, there have been very few Australian-focused studies to indicate likely direction of change(s), either positive or negative, for biological control.

One of two biological control examples that have been studied in the context of climate change in Australia is *Dialectica scalariella*, a leaf miner on *Echium plantagineum* (Johns and Hughes, 2002). Elevated CO_2 increased biomass, reduced leaf N and increased C:N in the host plant, but the growth response of the leaf miner was related mostly to higher temperatures and not CO_2 concentration. However, *D. scalariella* had poorer survival and lower weight when developed under higher temperatures and with eCO_2, possibly because the increased development rate was compromised by the poorer quality foliage (Johns and Hughes, 2002). Another example are the chrysomelids, *Octotoma championi* and *Octotoma scabripennis*, which are leaf miners as larvae and leaf feeders as adults on the widespread invasive shrub, *Lantana camara*. The plant responded to eCO_2 as described for *E. plantagineum* above, but had premature leaf loss at higher temperatures (Johns *et al.*, 2003). This effect, however, was reduced by higher CO_2 concentrations, which might have benefited insects that fed inside leaves (e.g. leaf miners). In feeding trials, adult beetles showed a preference for leaves grown at ambient CO_2 conditions (Johns *et al.*, 2003).

Carbon dioxide plays multiple roles in insect behaviour, as shown by the above two examples. In addition, eCO_2 may affect the physiology of the insect's CO_2 sensing system (Guerenstein and Hildebrand, 2008). This may be particularly relevant to biological control in Australia, in relation to one of the best-known examples of successful biological control worldwide: *Cactoblastis cactorum* and its control of opuntioid cacti. To detect oviposition sites on *Opuntia stricta* (i.e. healthy leaf pads), the female moth uses the CO_2 gradient generated by the plants crassulacean acid metabolism (CAM) photosynthesis mechanism (Stange *et al.*, 1995). The resulting larvae hatch and consume the cactus, leading to control of the invasion. While the moth's sensors are adapted to pre-industrial atmospheric CO_2, free-air CO_2 enrichment experiments show that likely future CO_2 levels are outside the range that tested moths can detect (Stange *et al.*, 1995; Osmond *et al.*, 2008). Ninety years after its introduction in 1840, the invasion of *Opuntia* cacti in Queensland and New South Wales covered more than 25 million ha, and at its peak was expanding at *c.*50 ha h^{-1}. With the cacti still scattered at low levels across the old invasion area, any large-scale failure of the *Cactoblastis* biocontrol agent due to rising atmospheric CO_2 concentrations could lead to a re-invasion that would be on the scale of a national disaster for Australia. There is an urgent need to assess this risk, and potentially the risk of other biocontrol agents decoupling from their hosts, and put in place suitable adaptation procedures to maintain the effective management, biological control or otherwise, of these invasive plants.

Herbicides (see also Chapter 18)

Herbicides are an integral part of crop production in Australia and are used increasingly to control non-agricultural weeds. There are, however, few Australian studies that specifically examine the role of climate change on the effectiveness of herbicides. In a recent glasshouse experiment, four invasive exotic C_4 grass species (*Chloris gayana*, *Eragrostis curvula*, *Paspalum dilatatum* and *Sporobolus indicus*) were grown under eCO_2 conditions and treated with the recommended levels of glyphosate (Manea *et al.*, 2011). Three of the species showed increased tolerance of the herbicide, thus indicating that dosage levels would need to be increased to remain effective under climate change.

Such results support similar insight from overseas research on herbicides and various aspects of projected climate change. One difference for how these trends influence the on-ground management of weeds in Australia is that Australian cropping regions already have some of the world's highest incidences of herbicide resistance (Heap, 2012). This resistance is extensive over many herbicide groups and includes the most widely used herbicide, glyphosate (Powles and Yu, 2010). It would seem that Australia has much to gain from recent calls to move beyond a dominant focus on herbicides to a more integrated approach to controlling agricultural weeds (e.g. Murray *et al.*, 2012b).

The main lesson we learn from the evolution of herbicide resistance is that agricultural weeds evolve quickly; eight consecutive applications of a herbicide is sufficient to evolve resistance (Powles *et al.*, 1997). Consequently, it may be reasonable to assume that weeds, at least those with short generation times, could evolve rapidly in response to other environmental stressors, such as climate change. If this were the case, it is highly likely that the major annual agricultural weed problems of today (e.g. wild radish, annual ryegrass) will remain significant problems for Australian farmers in the future, assuming current agronomic practices continue.

Arrivals new and old: a welcome addition in a changing climate or an additional threat?

Most new invasive species in Australia were originally garden plants (Groves *et al.*, 2003), and at least 606 Australian plant species have become invasive in Australia outside of their native range (Randall, 2007). With projected reductions in rainfall across much of Australia, a likely adaptation response to climate change, for both agricultural and horticultural reasons, will be to grow more drought-resistant Australian and non-native plant species. Altered planting patterns of non-native species due to, or anticipating, climate change could increase the risk of unwanted invasions from new introductions or plants already in the country (Low, 2008). For example, non-native alpine species have been planted in revegetation and amenity areas and subsequently have spread into native vegetation (McDougall *et al.*, 2005). Planting non-native alpine species in amenity areas as an adaptation response to climate change represents an additional threat to potential refugia for natives, particularly in the treeless vegetation of the Australian Alps (McDougall *et al.*, 2005). A more appropriate climate change adaptation response in this situation would be to use local native species for revegetation and amenity plantings (McDougall *et al.*, 2011a,b).

There has been a long history of planting non-native grasses introduced into Australia for pasture improvement, with limited net benefit (Lonsdale, 1994). A similar introduction history of beneficial pasture and crop legumes has also occurred (Emms *et al.*, 2005; Cook and Dias, 2006). Both examples have resulted in substantial unintended negative impacts from subsequent invasions. There are appropriate risk assessments now in place for new pasture introductions in Australia (Stone *et al.*, 2008; Nichols *et al.*, 2012), but these have yet to take climate change into account. The global push for biofuels for climate change mitigation has not escaped Australia either (Low and Booth, 2008; Low *et al.*, 2011), and the same conflict of interest applies between what is viewed as a successful biofuel species and a high-risk invasive plant (Raghu *et al.*, 2006; Murphy *et al.*, 2012). A recent proposal to move potential pasture legumes (*Cullen* spp.) from their native range in eastern Australia to Western Australia (Bennett *et al.*, 2011), and continued prospecting for new rangeland legumes in South Africa (Howieson *et al.*, 2008), highlights the ongoing need to consider the full range of risks and impacts associated with species translocations as an adaptation to climate change (Webber *et al.*, 2011a).

Summary and Future Research Needs

There is little doubt that invasive plants represent a significant threat to natural environments and impact on efficient agricultural production systems in Australia. While our knowledge on how to manage invasive species in current climates has improved tremendously in recent decades, climate change adds further uncertainty and new challenges. Climate change, therefore, must become a core consideration for current research on and management of invasive plants if outcomes are to remain relevant and effective into the future. Research in Australia appears to be particularly well placed to inform best practice management with global applicability for invasive plants in a changing climate. Unfortunately, Australia has had a recent history of stop–start funding for research on invasive plants, leading to short-term projects often involving the application of results from earlier studies (McFadyen,

2012). A new approach is needed to funding the research and management of invasive plants in Australia with a national long-term focus. Recognizing that others have previously drawn attention to priorities for future work on climate change and invasive species (e.g. Hellmann *et al.*, 2008; Walther *et al.*, 2009; Murray *et al.*, 2012b), we conclude by highlighting five areas of interest where a greater research effort is required to tackle issues of global importance, but with particular relevance for Australia:

- **Addressing geographical bias:** We have shown that invasive plants of the north and west of Australia remain disproportionately understudied. Although large parts of northern Australia have a relatively low diversity of non-native plants (*c*.5% of the total flora; J.K. Scott, unpublished data), the abundance of non-natives can be very high in some areas. As McFadyen (2012) points out, there is also little research capacity in northern Australia. Yet this area is sometimes considered the future location for agricultural expansion, because its precipitation is expected to remain relatively constant or increase, in contrast to the drying south. If agricultural expansion in the north is to be pursued, research will need to focus on invasion pathways, containment and dispersal mechanisms in these tropical areas and risk assessments for potential new invaders, both new arrivals and existing species responding to local change. It is possible that shifts of suitable climates for invasive species with climate change might be detected that differ to the polewards shifts documented up to now. Furthermore, the relatively short distances between the northern edge of the Australian continent and the islands of the Indonesian archipelago (including Timor Leste) and New Guinea mean we may need to assess what species may be tolerated if they migrate naturally from their current native range in these islands into newly suitable regions in Australia.
- **Getting the measure of climate change impact:** It is essential that we improve our understanding of how climate change and climate change adaptation is altering the impact of invasive plants on the environment, economy and society. To achieve this, a better understanding of and ability to quantify all measures of impact is required (e.g. Simberloff *et al.*, 2013). This includes the combined effect of climate change on invasive plants and the values which they impact (such as the distribution and value of land use or biodiversity). We must be mindful of considering the potential positive impacts of certain invasive non-native species (Grice *et al.*, 2012), but caution that the inherently subjective nature of quantifying the range of impacts for any given invasion may become more polarized with future climate change. For example, ecosystem services may be compromised under climate change if native species are impacted seriously, but their function could possibly be replaced or maintained by appropriate non-native species.
- **Improving confidence and certainty in the face of uncertainty:** With so many global change drivers impacting on invasive species (e.g. landscape change, disturbance and climate change), we need to improve our ability to detect the fingerprint of climate change over and above 'business as usual' invasion dynamics. Investing further in existing long-term monitoring sites, as well as establishing new ones, will be an important step in this process. Such detailed monitoring could make it easier to detect an evolutionary response to climate change and could include research on propagule dispersal, invasion risks, pathways and impacts, including methodologies for refugia-scale quarantine and eradication of invasive non-native species. Indeed, invasion will become one of the main threats to refugia protecting biological diversity from climate change that may contain the last vestige of native species. The modelling

of species distributions is likely to remain an important tool for understanding invasion risk with climate change, but improving confidence in the results is a high priority to inform management and policy makers better. A more rigorous focus on model choice and interpretation, improved modelling frameworks, better model validation and more attention to the framing of results may help to address these needs (Kriticos *et al.*, 2012).

- **Advancing concepts and value systems:** Climate change has the potential to change people's perceptions of what is an invasive plant, or even what is non-native, because so many other changes may be occurring, including an increase in the prevalence of novel ecosystems. These sociological and cultural aspects need to be measured and understood, so that future policy development and management of invasive plants, both native and non-native, meets community expectations. Concordant with these expectations is the need to understand the most appropriate means of community education on the issues of native and non-native plants that are shifting their range. In situations of change, it may be that some invasive non-native plants will become useful, and research needs to be perceptive to this possibility. Ecosystem restoration faces a major challenge: what is the restoration objective in the face of climate change and what is the role in restoration of invasive native species? Non-native plants, both those already introduced and those yet to be introduced, may have the potential to dominate restored ecosystems. Which species should stay and which managed for removal will be a sensitive process requiring dynamic baselines that explicitly recognize that rapid climate change will drive significant recombination of existing ecosystems.
- **Revising management needs:** All aspects of biological control, in particular existing control successes currently controlling invasive plants, need re-assessment and adaptation strategies put in place to manage risks due to climate change. Herbicides need critical assessment of their role under climate change, because of their importance to Australian agriculture and management of invasive plants in the environment. While its relative isolation and broad oceanic boundaries may represent a seemingly robust barrier to the movement of many species, careful consideration will have to be given to factoring in climate change to national biosecurity issues. Australia faces an increasing diversity and (often unintended) effectiveness of artificial dispersal vectors for non-native plants, as well as invasive non-native species in the wider region that may be able to expand their distributions naturally over existing barriers to track suitable climates. Australia will thus need to pay closer attention to the biosecurity standards of their less developed neighbouring countries, where effective management strategies and biosecurity priorities are likely to be very different. New forms of invasive plant management, underpinned by revised conceptual and policy frameworks, may be required under climate change, especially for novel plant communities. The need for managing weeds that represent a serious threat to the environment and agriculture will continue, but the focal species, regions and priorities are likely to be different to those that currently dominate research and management effort.

Acknowledgements

We thank the CSIRO Climate Adaptation Flagship for their support and Trevor Booth, Matt Colloff and two anonymous reviewers for very useful feedback on earlier versions of this manuscript.

References

Adkins, S.W., O'Donnell, C., Khan, N., Thi, N., Shabbir, A., Dhileepan, K., George, D. and Navie, S. (2010) Parthenium weed (*Parthenium hysterophorus* L.) research in Australia: new management possibilities. In: Zydenbos, S.M. (ed.) *Proceedings of the 17th Australasian Weeds Conference*. New Zealand Plant Protection Society, Christchurch, New Zealand, pp. 120–123.

Australian Weeds Committee (2007) The Australian Weeds Strategy (http://www.environment.gov.au/biodiversity/invasive/weeds/publications/strategies/pubs/weed-strategy.pdf, accessed 21 January 2014).

Australian Weeds Committee (2013) Weeds of National Significance (http://www.weeds.org.au/WoNS/, accessed 21 January 2014).

Baker, J. (2010) Climate change – managing the waking sleepers. In: Zydenbos, S.M. (ed.) *Proceedings of the 17th Australasian Weeds Conference*. New Zealand Plant Protection Society, Christchurch, New Zealand, pp. 202–204.

Bean, A.R. (2007) A new system for determining which plant species are indigenous in Australia. *Australian Systematic Botany* 20, 1–43.

Beaumont, L.J., Gallagher, R.V., Thuiller, W., Downey, P.O., Leishman, M.R. and Hughes, L. (2009a) Different climatic envelopes among invasive populations may lead to under-estimations of current and future biological invasions. *Diversity and Distributions* 15, 409–420.

Beaumont, L.J., Gallagher, R.V., Downey, P.O., Thuiller, W., Leishman, M.R. and Hughes, L. (2009b) Modelling the impact of *Hieracium* spp. on protected areas in Australia under future climates. *Ecography* 32, 757–764.

Bennett, R.G., Ryan, M.H., Colmer, T.D. and Real, D. (2011) Prioritisation of novel pasture species for use in water-limited agriculture: a case study of Cullen in the Western Australian wheatbelt. *Genetic Resources and Crop Evolution* 58, 83–100.

Chapman, A.D. (2009) *Numbers of Living Species in Australia and the World*, 2nd edn. Report for the Australian Biological Resources Study, Canberra, 78 pp.

Chejara, V.K., Kriticos, D.J., Kristiansen, P., Sindel, B.M., Whalley, R.D.B. and Nadolny, C. (2010) The current and future potential geographical distribution of *Hyparrhenia hirta*. *Weed Research* 50, 174–184.

Climate Commission (2011) *The Critical Decade; Western Australian Climate Change Impacts.* Department of Climate Change and Energy Efficiency, Commonwealth of Australia, Canberra, 16 pp.

Colloff, M.J. and Baldwin, D.S. (2010) Resilience of floodplain ecosystems in a semi-arid environment. *Rangeland Journal* 32, 305–314.

Commonwealth of Australia (2010a) Adapting to climate change in Australia, an Australian Government position paper (http://www.climatechange.gov.au/sites/climatechange/files/documents/03_2013/adapt-climate-change-position-paper.pdf, accessed 4 January 2014).

Commonwealth of Australia (2010b) Planting a known weed species (http://www.climatechange.gov.au/government/initiatives/carbon-farming-initiative/activities-eligible-excluded/excluded-activities-negative-list/weeds.aspx, accessed 21 January 2014).

Commonwealth of Australia (2012) *National Wildlife Corridors Plan: A Framework for Landscape-scale Conservation*. Department of Sustainability, Environment, Water, Population and Communities, Canberra, 53 pp.

Cook, G.D. and Dias, L. (2006) It was no accident: deliberate plant introductions by Australian government agencies during the 20th century. *Australian Journal of Botany* 54, 601–625.

CSIRO and Bureau of Meteorology (2012) *State of the Climate 2012*. CSIRO and Bureau of Meteorology, Commonwealth of Australia, Canberra, 12 pp.

DeLucia, E.H., Nabity, P.D., Zavala, J.A. and Berenbaum, M.R. (2012) Climate change: resetting plant–insect interactions. *Plant Physiology* 160, 1677–1685.

Elith, J. and Leathwick, J.R. (2009) Species distribution models: ecological explanation and prediction across space and time. *Annual Review of Ecology, Evolution and Systematics* 40, 677–697.

Elith, J., Kearney, M. and Phillips, S. (2010) The art of modelling range-shifting species. *Methods in Ecology and Evolution* 1, 330–342.

Emms, J., Virtue, J.G., Preston, C. and Bellotti, W.D. (2005) Legumes in temperate Australia: a survey of naturalisation and impact in natural ecosystems. *Biological Conservation* 125, 323–333.

Fensham, R.J. and Holman, J.E. (1999) Temporal and spatial patterns in drought-related tree dieback in Australian savanna. *Journal of Applied Ecology* 36, 1035–1050.

Florentine, S.K. and Westbrooke, M.E. (2005) Invasion of the noxious weed *Nicotiana glauca*. *Journal of Arid Environments* 60, 531–545.

Gallagher, R.V., Hughes, L., Leishman, M.R. and Wilson, P.D. (2010) Predicted impact of exotic vines on an endangered ecological community under future climate change. *Biological Invasions* 12, 4049–4063.

Gallagher, R.V., Englert Duurmsa, D., O'Donnell, J., Wilson, P.D., Downey, P.O., Hughes, L. and Leishman, M.R. (2012) The grass may not always be greener: projected reductions in climatically suitable habitat for exotic grasses under future climates. *Biological Invasions* 15, 961–975.

Grice, A.C., Friedel, M.H., Marshall, N.A. and Van Klinken, R.D. (2012) Tackling contentious invasive plant species: a case study of buffel grass in Australia. *Environmental Management* 49, 285–294.

Groves, R.H., *et al.* (2003) *Weed Categories for Natural and Agricultural Ecosystem Management.* Bureau of Rural Sciences, Canberra.

Guerenstein, P.G. and Hildebrand, J.G. (2008) Roles and effects of environmental carbon dioxide in insect life. *Annual Review of Entomology* 53, 161–178.

Harrington, G.N. (1991) Effects of soil moisture on shrub seedling survival in a semi-arid grassland. *Ecology* 72, 1138–1149.

Heap, I.M. (2012) International survey of herbicide resistant weeds (http://www.weedscience.org, accessed 21 January 2014).

Hellmann, J.J., Byers, J.E., Bierwagen, B.G. and Dukes, J.S. (2008) Five potential consequences of climate change for invasive species. *Conservation Biology* 22, 534–543.

Hely, S.E.L. and Roxburgh, S.H. (2005) The interactive effects of elevated CO_2, temperature and initial size on growth and competition between a native C_3 and an invasive C_3 grass. *Plant Ecology* 177, 85–98.

Hovenden, M.J. and Williams, A.L. (2010) The impacts of rising CO_2 concentrations on Australian terrestrial species and ecosystems. *Austral Ecology* 35, 665–684.

Hovenden, M.J., Wills, K.E., Vander Schoor, J.K., Williams, A.L. and Newton, P.C.D. (2008a) Flowering phenology in a species-rich temperate grassland is sensitive to warming but not elevated CO_2. *New Phytologist* 178, 815–822.

Hovenden, M.J., Williams, A.L., Pedersen, J.K., Vander Schoor, J.K. and Wills, K.E. (2008b) Elevated CO_2 and warming impacts on flowering phenology in a southern Australian grassland are related to flowering time but not growth form, origin or longevity. *Australian Journal of Botany* 56, 630–643.

Howell, J. (2003) *Pittosporum undulatum* as a case study for native species that change range – how to avoid inappropriate responses? *Cunninghamia* 8, 153–155.

Howieson, J.G., Yates, R.J., Foster, K.J., Real, D. and Besier, R.B. (2008) Prospects for the future use of legumes. In: Dilworth, M.J., James, E.K., Sprent, J.I. and Newton, W.E. (eds) *Nitrogen-fixing Legumunous Symbioses.* Springer, Dordrecht, the Netherlands, pp. 363–393.

Hulme, P.E., Pyšek, P., Jarošík, V., Pergl, J., Schaffner, U. and Vilà, M. (2013) Bias and error in understanding plant invasion impacts. *Trends in Ecology and Evolution* doi:10.1016/j.tree.2012.10.010.

Jiang, Q. and Grafton, R.Q. (2012) Economic effects of climate change in the Murray-Darling Basin, Australia. *Agricultural Systems* 110, 10–16.

Johns, C.V. and Hughes, L. (2002) Interactive effects of elevated CO_2 and temperature on the leaf-miner *Dialectica scalariella* Zeller (Lepidoptera: Gracillariidae) in Paterson's Curse, *Echium plantagineum* (Boraginaceae). *Global Change Biology* 8, 142–152.

Johns, C.V., Beaumont, L.J. and Hughes, L. (2003) Effects of elevated CO_2 and temperature on development and consumption rates of *Octotoma championi* and *O. scabripennis* feeding on *Lantana camara. Entomologia Experimentalis et Applicata* 108, 169–178.

Julien, M., McFadyen, R. and Cullen, J. (eds) (2012) *Biological Control of Weeds in Australia.* CSIRO Publishing, Melbourne, Australia.

Kearney, M. and Porter, W. (2009) Mechanistic niche modelling: combining physiological and spatial data to predict species' ranges. *Ecology Letters* 12, 334–350.

Kriticos, D.J. (2008) Weeds getting steamed up in the tropics: climate change effects on ecosystem water use efficiency and woody plant range. In: Van Klinken, R.D., Osten, V.A., Panetta, F.D. and Scanlan, J.C. (eds) *Proceedings of the 16th Australian Weeds Conference.* Queensland Weeds Society, Cairns, Australia, pp. 506–509.

Kriticos, D.J., Sutherst, R.W. and Adkins, S.W. (2000) *Exotic Woody Weeds: Use of Simulation Models to Predict Future Distributions under Climate Change.* Rural Industries Research and Development Corporation, Canberra, 1–43 pp.

Kriticos, D.J., Sutherst, R.W., Brown, J.R., Adkins, S.W. and Maywald, G.F. (2003a) Climate change and the potential distribution of an invasive alien plant: *Acacia nilotica* ssp. *indica* in Australia. *Journal of Applied Ecology* 40, 111–124.

Kriticos, D.J., Sutherst, R.W., Brown, J.R., Adkins, S.A. and Maywald, G.F. (2003b) Climate change and biotic invasions: a case history of a tropical woody vine. *Biological Invasions* 5, 145–165.

Kriticos, D.J., Alexander, N.S. and Kolomeitz, S.M. (2006) Predicting the potential geographic distribution of weeds in 2080. In: Preston, C., Watts, J.H. and Crossman, N.D. (eds) *Proceedings of the 15th Australian Weeds Conference*. Weed Management Society of South Australia, Torrens Park, Australia, pp. 27–34.

Kriticos, D.J., Crossman, N.D., Ota, N. and Scott, J.K. (2010) Climate change and invasive plants in South Australia (http://www.csiro.au/Portals/Publications/Research--Reports/South-Australia-weeds-and-climate-change.aspx, accessed 6 February 20104).

Kriticos, D.J., Le Maitre, D.C. and Webber, B.L. (2012) Essential elements of discourse for advancing the modelling of species' current and potential distributions. *Journal of Biogeography* 40, 608–611.

Lawes, R. (2004) Temporal fluctuations of *Parkinsonia aculeata* L. seedling germination and growth in three Australian rangelands habitats. In: Sindel, B.M. and Johnson, S.B. (eds) *Proceedings of the 14th Australian Weeds Conference*. Weed Society of New South Wales, Sydney, Australia, pp. 164–167.

Leishman, M.R., Sanbrooke, K.J. and Woodfin, R.M. (1999) The effects of elevated CO_2 and light environment on growth and reproductive performance of four annual species. *New Phytologist* 144, 455–462.

Lonsdale, W.M. (1994) Inviting trouble – introduced pasture species in northern Australia. *Australian Journal of Ecology* 19, 345–354.

Low, T. (2008) *Climate Change and Invasive Species: A Review of Interactions November 2006 Workshop Report*. Biological Diversity Advisory Committee, Canberra, 50 pp.

Low, T. and Booth, C. (2008) The Weedy Truth About Biofuels (http://www.invasives.org.au/documents/file/reports/isc_biofuels_revised_march08.pdf, accessed 21 January 2014).

Low, T., Booth, C. and Sheppard, A. (2011) Weedy biofuels: what can be done? *Current Opinion in Environmental Sustainability* 3, 55–59.

McDougall, K.L., Morgan, J.W., Walsh, N.G. and Williams, R.J. (2005) Plant invasions in treeless vegetation of the Australian Alps. *Perspectives in Plant Ecology Evolution and Systematics* 7, 159–171.

McDougall, K.L., Alexander, J.M., Haider, S., Pauchard, A., Walsh, N.G. and Kueffer, C. (2011a) Alien flora of mountains: global comparisons for the development of local preventive measures against plant invasions. *Diversity and Distributions* 17, 103–111.

McDougall, K.L., Khuroo, A.A., Loope, L.L., Parks, C.G., Pauchard, A., Reshi, Z.A., Rushworth, I. and Kueffer, C. (2011b) Plant invasions in mountains: global lessons for better management. *Mountain Research and Development* 31, 380–387.

McFadyen, R.E.C. (2012) Food security for a 9 billion population: more R&D for weed control will be critical. In: Eldershaw, V. (ed.) *Proceedings of the 18th Australasian Weeds Conference*. Weed Society of Victoria, Australia, pp. 306–309.

Manea, A. and Leishman, M.R. (2011) Competitive interactions between native and invasive exotic plant species are altered under elevated carbon dioxide. *Oecologia* 165, 735–744.

Manea, A., Leishman, M.R. and Downey, P.O. (2011) Exotic C_4 grasses have increased tolerance to glyphosate under elevated carbon dioxide. *Weed Science* 59, 28–36.

Mason, T.J., French, K. and Russell, K. (2012) Are competitive effects of native species on an invader mediated by water availability? *Journal of Vegetation Science* 23, 657–666.

Michael, P.J., Yeoh, P.B. and Scott, J.K. (2012a) Potential distribution of the Australian native *Chloris truncata* based on modelling both the successful and failed global introductions. *PLoS ONE* 7, e42140.

Michael, P.J., Yeoh, P.B. and Scott, J.K. (2012b) The current and future projected distribution of *Solanum hoplopetalum* (Solanaceae): an indigenous weed of the south-western Australian grain belt. *Australian Journal of Botany* 60, 128–135.

Michael, P.J., Yeoh, P.B. and Scott, J.K. (2012c) Potential distribution of *Rapistrum rugosum* (turnip weed). In: Eldershaw, V. (ed.) *Proceedings of the 18th Australasian Weeds Conference*. Weed Society of Victoria, Australia, pp. 369–372.

Moore, B.D., Franceschi, V.R., Cheng, S., Wu, J. and Ku, M.S.B. (1987) Photosynthetic characteristics of the C_3–C_4 intermediate *Parthenium hysterophorus*. *Plant Physiology* 85, 984–989.

Murphy, H., Metcalfe, D., Kriticos, D. and Lawson, T. (2009) Pro-active management of weeds based on their potential distributions under changing climatic conditions. In: Kinnear, S. and Baker, T. (eds) *Proceedings of the 10th Queensland Weeds Symposium*. Weed Society of Queensland, Australia, pp. 61–67.

Murphy, H.T., Metcalfe, D.J., Bradford, M.G., Ford, A.F., Galway, K.E., Sydes, T.A. and Westcott,

D.J. (2008) Recruitment dynamics of invasive species in rainforest habitats following Cyclone Larry. *Austral Ecology* 33, 495–502.

Murphy, H.T., Metcalfe, D.J., Bradford, M.G. and Ford, A.J. (2011) *Weed Response to Cyclones in the Wet Tropics Rainforests. Impacts and Adaptation.* Publication No 11/010. Australian Government Rural Industries Research and Development Corporation, Barton, ACT, Australia, 10 pp.

Murphy, H.T., *et al.* (2012) A common view of the opportunities, challenges, and research actions for *Pongamia* in Australia. *Bioenergy Research* 5, 778–800.

Murray, J.V., Stokes, K.E. and van Klinken, R.D. (2012a) Predicting the potential distribution of a riparian invasive plant: the effects of changing climate, flood regimes and land-use patterns. *Global Change Biology* 18, 1738–1753.

Murray, J.V., Lehnhoff, E.A., Neve, P., Poggio, S.L. and Webber, B.L. (2012b) 'Raising the bar': improving the standard and utility of weed and invasive plant research. *New Phytologist* 196, 678–680.

Nichols, P.G.H., Revell, C.K., Humphries, A.W., Howie, J.H., Hall, E.J., Sandral, G.A., Ghamkhar, K. and Harris, C.A. (2012) Temperate pasture legumes in Australia – their history, current use, and future prospects. *Crop and Pasture Science* 63, 691–725.

O'Donnell, C.C. and Adkins, S.W. (1996) The response of wild oats to climate change. In: Shepherd, R.C.H. (ed.) *Proceedings of the 11th Australian Weeds Conference.* Weed Science Society of Victoria, Australia, pp. 101–104.

O'Donnell, C.C. and Adkins, S.W. (2001) Wild oat and climate change: the effect of CO_2 concentration, temperature, and water deficit on the growth and development of wild oat in monoculture. *Weed Science* 49, 694–702.

O'Donnell, J., Gallagher, R.V., Wilson, P.D., Downey, P.O., Hughes, L. and Leishman, M.R. (2012) Invasion hotspots for non-native plants in Australia under current and future climates. *Global Change Biology* 18, 617–629.

Osmond, B., Neales, T. and Stange, G. (2008) Curiosity and context revisited: crassulacean acid metabolism in the Anthropocene. *Journal of Experimental Botany* 59, 1489–1502.

Perring, M.P. and Hovenden, M.J. (2012) Seedling survivorship of temperate grassland perennials is remarkably resistant to projected changes in rainfall. *Australian Journal of Botany* 60, 328–339.

Pillora, S. (2010) *Australian Local Government and Climate Change.* University of Technology, Sydney, Australia.

Potter, K.J.B., Kriticos, D.J. and Leriche, A. (2008) Climate change impacts on Scotch broom in Australia. In: van Klinken, R.D., Osten, V.A., Panetta, F.D. and Scanlan, J.C. (eds) *Proceedings of the 16th Australian Weeds Conference.* Queensland Weeds Society, Brisbane, Queensland, Australia, pp. 522–524.

Potter, K.J.B., Kriticos, D.J., Watt, M.S. and Leriche, A. (2009) The current and future potential distribution of *Cytisus scoparius*: a weed of pastoral systems, natural ecosystems and plantation forestry. *Weed Research* 49, 271–282.

Potter, N.J. and Chiew, F.H.S. (2011) An investigation into changes in climate characteristics causing the recent very low runoff in the southern Murray-Darling Basin using rainfall-runoff models, *Water Resources Research* 47, W00G10, doi: 10.1029/2010WR010333.

Power, S., Tseitkin, F., Mehta, V., Lavery, B., Torok, S. and Holbrook, N. (1999) Decadal climate variability in Australia during the twentieth century. *International Journal of Climatology* 19, 169–184.

Powles, S.B. and Yu, Q. (2010) Evolution in action: plants resistant to herbicides. *Annual Review of Plant Biology* 61, 317–347.

Powles, S.B., Preston, C., Bryan, I.B. and Jutsum, A.R. (1997) Herbicide resistance: impact and management. *Advances in Agronomy* 58, 57–93.

Procheş, Ş., Wilson, J.R.U., Veldtman, R., Kalwij, J.M., Richardson, D.M. and Chown, S.L. (2005) Landscape corridors: possible dangers? *Science* 310, 779–780.

Raghu, S., Anderson, R.C., Daehler, C.C., Davis, A.S., Wiedenmann, R.N., Simberloff, D. and Mack, R.N. (2006) Adding biofuels to the invasive species fire? *Science* 313, 1742.

Randall, R.P. (2007) *The Introduced Flora of Australia and its Weed Status.* CRC for Australian Weed Management, Adelaide, Australia.

Richardson, D.M., Pyšek, P. and Carlton, J.T. (2011) A compendium of essential concepts and terminology in invasion ecology. In: Richardson, D.M. (ed.) *Fifty Years of Invasion Ecology: The Legacy of Charles Elton.* Blackwell Publishing Ltd, Chichester, UK, pp. 409–420.

Rodda, G.H., Jarnevich, C.S. and Reed, R.N. (2011) Challenges in identifying sites climatically matched to the native ranges of animal invaders. *PLOS One* 6, 1–18.

Rossiter, N.A., Setterfield, S.A., Douglas, M.M. and Hutley, L.B. (2003) Testing the grass–fire cycle: alien grass invasion in the tropical savannas of northern Australia. *Diversity and Distributions* 9, 169–176.

Scoble, J. and Lowe, A.J. (2010) A case for incorporating phylogeography and landscape genetics into species distribution modelling approaches to improve climate adaptation and conservation planning. *Diversity and Distributions* 16, 343–353.

Scott, J.K., Batchelor, K.L., Ota, N. and Yeoh, P.B. (2008) Overview of modelling climate change impacts on sleeper and alert weeds in Australia (http://www.csiro.au/en/Outcomes/Food-and-Agriculture/Sleeper-Alert-Weeds.aspx, accessed 21 January 2014).

Simberloff, D., *et al.* (2013) Impacts of biological invasions: what's what and the way forward. *Trends in Ecology and Evolution* 28, 58–66.

Sims-Chilton, N.M., Zalucki, M.P. and Buckley, Y.M. (2010) Long-term climate effects are confounded with the biological control programme against the invasive weed *Baccharis halimifolia* in Australia. *Biological Invasions* 12, 3145–3155.

Sinden, J., Jones, R., Hester, S., Odom, D., Kalisch, D., James, R. and Cacho, O. (2004) The economic impact of weeds in Australia. *CRC for Australian Weed Management, Technical Series* 8, 1–55.

Smith, C., van Klinken, R.D., Seabrook, L. and McAlpine, C. (2012) Estimating the influence of land management change on weed invasion potential using expert knowledge. *Diversity and Distributions* 18, 818–831.

Stange, G., Monro, J., Stowe, S. and Osmond, C.B. (1995) The CO_2 sense of the moth *Cactoblastis cactorum* and its probably role in the biological control of the CAM plant *Opuntia stricta*. *Oecologia* 102, 341–352.

Steel, J., Kohout, M. and Newell, G. (2008) Climate change and potential distribution of weeds. Whither the weeds under climate change? (http://www.climatechange.vic.gov.au/__data/assets/pdf_file/0007/73249/Whithertheweedsunderclimatechange2008v1.pdf, accessed 21 January 2014).

Steffen, W., Burbidge, A.A., Hughes, L., Kitching, R., Lindenmayer, D., Musgrave, W., Stafford Smith, M. and Werner, P.A. (2009) *Australia's Biodiversity and Climate Change: A Strategic Assessment of the Vulnerability of Australia's Biodiversity to Climate Change. A Report to the Natural Resource Management Ministerial Council Commissioned by the Australian Government.* CSIRO Publishing, Canberra, 289 pp.

Stiling, P. and Cornelissen, T. (2007) How does elevated carbon dioxide (CO_2) affect plant–herbivore interactions? A field experiment and meta-analysis of CO_2-mediated changes on plant chemistry and herbivore performance. *Global Change Biology* 13, 1823–1842.

Stone, L.M., Byrne, M. and Virtue, J.G. (2008) An environmental weed risk assessment model for Australian forage improvement programs. *Australian Journal of Experimental Agriculture* 48, 568–574.

Sutherst, R.W. and Bourne, A.S. (2009) Modelling non-equilibrium distributions of invasive species: a tale of two modelling paradigms. *Biological Invasions* 11, 1231–1237.

Sutherst, R.W., Yonow, T., Chakraborty, S., O'Donnell, C. and White, N. (1996) A generic approach to defining impacts of climate change on pests, weeds and diseases in Australasia. In: Bouma, W.J., Pearman, G.I. and Manning, M.R. (eds) *Greenhouse: Coping with Climate Change.* CSIRO Publishing, Collingwood, Victoria, Australia, pp. 281–307.

Taylor, S., Kumar, L., Reid, N. and Kriticos, D.J. (2012) Climate change and the potential distribution of an invasive shrub, *Lantana camara* L. *PLoS ONE* 7, e35565.

Tooth, I.M. and Leishman, M.R. (2013) Post-fire resprouting responses of native and exotic grasses from Cumberland Plain Woodland (Sydney, Australia) under elevated carbon dioxide. *Austral Ecology*, 38, 1–10.

Trudgen, M.S., Webber, B.L. and Scott, J.K. (2012) Human-mediated introduction of *Livistona* palms into central Australia: conservation and management implications. *Proceedings of the Royal Society B: Biological Sciences* 279, 4115–4117.

van Klinken, R.D. and Campbell, S. (2009) Australian weeds series: *Prosopis* species. In: Panetta, F.D. (ed.) *The Biology of Australian Weed,* Volume 3. R.G. and F.J. Richardson, Melbourne, Australia, pp. 238–273.

van Klinken, R.D., Lawson, B.E. and Zalucki, M.P. (2009) Predicting invasions in Australia by a Neotropical shrub under climate change: the challenge of novel climates and parameter estimation. *Global Ecology and Biogeography* 18, 688–700.

Venette, R.C., *et al.* (2010) Pest risk maps for invasive alien species: a roadmap for improvement. *Bioscience* 60, 349–362.

Walther, G.-R., *et al.* (2009) Alien species in a warmer world: risks and opportunities. *Trends in Ecology and Evolution* 24, 686–693.

Watt, M.S., Kriticos, D.J. and Manning, L.K. (2009) The current and future potential distribution of *Melaleuca quinquenervia. Weed Research* 49, 381–390.

Watt, M.S., Kriticos, D.J., Lamoureaux, S.L. and Bourdôt, G.W. (2011) Climate change and the potential global distribution of serrated tussock (*Nassella trichotoma*). *Weed Science* 59(4), 538–545.

Webber, B.L. and Scott, J.K. (2012) Rapid global change: implications for defining natives and aliens. *Global Ecology and Biogeography* 21, 305–311.

Webber, B.L., Scott, J.K. and Didham, R.K. (2011a) Translocation or bust! A new acclimatization agenda for the 21st century? *Trends in Ecology and Evolution* 26, 495–496.

Webber, B.L., Yates, C.J., Le Maitre, D.C., Scott, J.K., Kriticos, D.J., Ota, N., McNeill, A., Le Roux, J.J. and Midgley, G.F. (2011b) Modelling horses for novel climate courses: insights from projecting potential distributions of native and alien Australian acacias with correlative and mechanistic models. *Diversity and Distributions* 17, 978–1000.

Webber, B.L., Le Maitre, D.C. and Kriticos, D.J. (2012) Comment on 'Climatic niche shifts are rare among terrestrial plant invaders'. *Science*, 338, 193.

Whitten, S.M., Freudenberger, D., Wyborn, C., Doerr, V. and Doerr, E. (2011) *A Compendium of Existing and Planned Australian Wildlife Corridor Projects and Initiatives, and Case Study Analysis of Operational Experience.* CSIRO Ecosystem Sciences, Canberra, 152 pp.

Williams, A.L., Wills, K.E., Janes, J.K., Vander Schoor, J.K., Newton, P.C.D. and Hovenden, M.J. (2007a) Warming and free-air CO_2 enrichment alter demographics in four co-occurring grassland species. *New Phytologist* 176, 365–374.

Williams, J.W., Jackson, S.T. and Kutzbach, J.E. (2007b) Projected distributions of novel and disappearing climates by 2100 AD. *Proceedings of the National Academy of Sciences* 104, 5738–5742.

Wilson, P.D., Downey, P.O., Gallagher, R.V., O'Donnell, J., Leishman, M.R. and Hughes, L. (2011) *Modelling Climate Suitability for Exotic Plants in Australia Under Future Climate. Final Report on the Potential Impact of Climate Change on the Distribution of National Priority Weeds in Australia.* Macquarie University and New South Wales Office of Environment and Heritage, Sydney, Australia, 122 pp.

12 Invasive Species of China and their Responses to Climate Change

Bo Li, Shujuan Wei, Hui Li, Qiang Yang and Meng Lu

Coastal Ecosystems Research Station of Yangtze River Estuary, Ministry of Education Key Laboratory for Biodiversity Science and Ecological Engineering, The Institute of Biodiversity Science, Fudan University, Shanghai, PR China

Abstract

China is a rapidly developing country with the largest share of the world's population. The extent of human activity, combined with diverse climates and landscapes, may allow for greater risk of biological invasions. To date, at least 529 invasive species have been identified, including 270 species of higher plants, 198 species of animals and 61 species of microbes, resulting in an estimated annual economic loss of US$18.9 billion. Evidence to date suggests that different components of climate change (i.e. temperature, altered precipitation, extreme weather events and rising CO_2 concentration) have already influenced the biology of invasive species in China. The influence of climate change may increase the negative economic or environmental consequences of some invasives; however, others may become disadvantaged. Although we are still at the earliest stages of understanding the consequences of climate change on invasive species biology, China is one of the countries that may be affected most dramatically by invasive species. Overall, understanding the interactions between climate change and invasive species biology is an important scientific challenge, but one in which Chinese scientists can make significant contributions. The inclusion of the invasive species issue in some of China's ongoing projects on global change could begin to address critical research needs in this area.

Introduction

China, considered here to include the mainland, Taiwan, Hong Kong and Macao, is the third largest country geographically, spanning about 50 degrees of latitude, or 5500 km, from south to north, and stretching 60 degrees of longitude, or 5200 km, from east to west. It has complex topography and diverse landscapes covering five climatic zones, i.e. tropical, subtropical, warm temperate, temperate and cold temperate. However, 80% of China's area is classified as subtropical and temperate (SEPA, 1998). The complex topography and climate allow for biological diversity. This is reflected in species diversity, with approximately 31,000 vascular plant species (*c.*12% of the world's total flora; FoCEC, 2004) and 6300 vertebrate species (*c.*10% of the global total; Wang and Fang, 2003).

China also has 1.4 billion people, accounting for 20% of the world's population, and is a rapidly developing country, meaning that human activities such as land use and trade will facilitate biological invasions. Such introductions (for agricultural purposes) began during the Han Dynasty, more than 2100 years ago (Xie *et al.*, 2001). Hence,

  Invasive Species and Global Climate Change (eds L.H. Ziska and J.S. Dukes)

these natural and anthropogenic factors have resulted in an accumulation of invasive species from different biogeographical regions of the world. In recent years, these invasive species have become recognized as serious environmental and economic threats to both native (e.g. forests) and managed ecosystems (e.g. agroecosystems). This chapter provides an overview of both the biology and distribution of these invasives and offers a context for how anthropogenic climate change may be exacerbating their impact. The chapter will also highlight a number of research needs that will be essential if China is to control and manage the impact of invasive species with increased climatic uncertainty.

Invasive Species Diversity and Geographic Distributions

A comprehensive understanding of invasive species diversity is critical to manage biological invasions (Silva and Smith, 2004; Khuroo *et al.*, 2007). Although historical databases on invasive species are available for many countries and areas, less is known about invasive species in Asian countries (Wu *et al.*, 2004). Two national-scale invasive species inventories have been conducted in China: the first was done from 2001 to 2003 (Xu *et al.*, 2006) and the other from 2008 to 2010 (Ding *et al.*, 2011). However, a complete list of invasive species is still lacking. Based on different sources, 529 invasive alien species have been identified, including 270 species of higher plants (Weber and Li, 2008), 198 species of animals and 61 species of microbes (Yang *et al.*, 2009). Of these, 50 species are on the IUCN's list of 100 of the world's worst invasive species. However, it is believed that the actual number of invasive species in China may be greater than this (Ju *et al.*, 2012). We have compiled a partial list of 25 of the most harmful invasive species in China, 20 of which have been identified by Wan *et al.* (2009) (Table 12.1).

The 270 invasive plant species of China are represented by 59 families and 177 genera, of which 79.6% of the species belong to dictolyedons and 20.4% to monocotyledons (Weber *et al.*, 2008). The families Asteraceae, Poaceae and Brassicaceae are the most abundant taxa of invasive plants, accounting for 40.6% of all species (Weber *et al.*, 2008). However, in Taiwan, most alien plant species belong to the Amaranthaceae, Convolvulaceae and Onagraceae (Wu *et al.*, 2004). The most common taxa of invasive animals in China include insects, especially invasive agricultural pests from the Coleoptera, Homoptera, Diptera and Lepidoptera, and fish. In relation to invasive microbes of China, plant pathogenic fungi are the most dominant, followed by plant viruses, fish and shrimp viruses and plant pathogenic bacteria (Xu and Qiang, 2011).

The distribution and density of invasive species decrease from the south-eastern coast to the north-western regions of China. At present, the largest number of invasive species occurs in China's coastal areas and in Yunnan province (Yang *et al.*, 2002; Weber *et al.*, 2008; Ding *et al.*, 2011). South-eastern China is heavily populated and economically well developed, and has convenient transport networks. Such networks create opportunities for invasive species to become established and to spread.

Viewed nationally, the occurrence of invasive species is correlated positively with population (Weber *et al.*, 2008; Wang *et al.*, 2010a) and road density (Weber and Li, 2008). However, we recognize that not all regions in China have been categorized equally with respect to invasive species demography, which could lead to sampling bias (Williams *et al.*, 2002). Nevertheless, socio-economic factors such as human land use, economic development and trade of animal and plant germplasm are considered significant factors that drive biological invasions.

Geographic Origins of Invasives

The vulnerability of an ecosystem is determined, in part, by the overlap with the biogeographic range of the organism being introduced (Williamson, 1996). Because of the biogeographic diversity of China, it is a

Table 12.1. Twenty-five of the most harmful invasive species of China.

Name	Origin	Distribution in China	Ecosystems damaged	Economic loss (US$ million)[a]
Fungi				
Verticillium dahliae Kleb.	USA	All cotton-planted areas	Agroecosystems	94.2
Xanthomonas oryzae pv. *oryzicola* (Fang) Swings	Philippines?	Central, south and north China	Agroecosystems	81.8
Higher plants				
Alternanthera philoxeroides	South America	All provinces south of 45°N and east of 97°E	Agroecosystems and wetlands	94.6
Ambrosia artemisiifolia L.	North America	Most of the provinces	Agroecosystems and human health	397.9
Eichhornia crassipes (Mart.) Solms	South America	17 southern provinces	Wetlands	15.8
Eupatorium adenophorum Spreng.	Mexico	South China	Forests	155.9
Mikania micrantha (L.) Kunth.	Central and South America	South China	Forests	N/A
Solidago canadensis L.	North America	Central, south and east China	Agroecosystems and wastelands	N/A
Spartina alterniflora	North America	East coast	Salt marshes	86.0
Nematodes				
Bursaphelenchus xylophilus Steiner & Buhrer	USA	South-east China, Guizhou and Sichuan	Forests	29.3
Radopholus similis (Cobb) Thorne	South-east Asia	Central and south China	Agroecosystems	16.8
Insects				
Bactrocera dorsalis Hendel	South-east Asia	Coastal provinces, south-west and central China	Agroecosystems	N/A
Bemisia tabaci Gennadius	India	All provinces south of 35°N and east of 100°E	Agroecosystems	353.9
Brontispa longissima Gestro	Tropical Asia	South China	Agroecosystems	N/A
Dendroctonus valens LeConte	North America	Shanxi, Shan'xi, Hebei, Henan	Forests	103.0
Hemiberlesia pitysophila Takagi	Japan	South China and Jiangxi	Forests	85.7
Laspeyresia pomonella L.	Europe	Xinjiang and Gansu	Agroecosystems	9.8
Leptinotarsa decemlineata Kraatz	USA	Xinjiang	Agroecosystems	1.7
Liriomyza sativae Blanchard	South America	All provinces except Tibet, Qinghai and Heilongjiang	Agroecosystems	500.3
Lissorhoptrus oryzophilus Kuschel	North America	Central and east China	Agroecosystems	90.7
Pectinophora gossypiella Saund.	India	All cotton-planted areas except north-west China	Agroecosystems	96.9
Solenopsis invicta Buren	South America	South China	Human health	N/A
Crustacea				
Procambarus clarkii Girard	South USA and north Mexico	All provinces except Tibet	Wetlands	N/A
Molluscs				
Pomacea canaliculata Lamarck	South America	Provinces south of 35°N	Agroecosystems and wetlands	16.8
Mammals				
Rattus norvegicus Berkenhout	South-east Asia	All over China	Human health and agroecosystems	62.8

Note: [a]The data on economic loss were extracted from Ding *et al.* (2004).

suitable habitat for numerous invasive species. At present, the Americas are the largest source of alien germplasm and are estimated to have contributed 54% of the invasive species, with 30% and 20% from North and South America, respectively (Yang *et al.*, 2009). Other regions, specifically Asia, Europe, Africa and Oceania, have contributed 20%, 17%, 6% and 3%, respectively (Yang *et al.*, 2009). The larger contributions of the Americas reflect, in part, the greater exchange of goods between this region and China. Unlike mainland China, the southern islands of Taiwan, Hong Kong and Macao have a greater portion of invasive species originating from South America, as these regions have warmer climates (Wang *et al.*, 2004; Wu *et al.*, 2004; Li *et al.*, 2007).

Habitats and Impacts of Invasive Species

In China, all provinces and almost all ecosystems now harbour invasive species. As in other regions/countries, the degree of infestation by invasive species is proportional to the extent of human activities. Thus, urban, manmade and coastal ecosystems are subject to more invasions than rural, natural and inland ecosystems. For example, 43.3% of invasive species are found in arable fields and botanical gardens, 34% in constructed areas, 18.6% in coastal ecosystems, 17.4% in orchid farms and plant nurseries, 16.6% in forests, 15.8% in grasslands and 14.7% in inland waters (Yang *et al.*, 2009). The number of invasive species present can be ranked (in decreasing order): roadsides including railways, wastelands and other disturbed grounds; agricultural fields; grasslands including pastures and grassy slopes; tree plantations including afforested lands; lakeshores and riparian habitats; natural forests and forest margins; coastal areas; and swamps and marshes (Weber *et al.*, 2008).

Invasive species are associated with considerable economic and environmental losses globally. For example, in the USA, the British Isles, Australia, Europe, South Africa, India and Brazil, the introduction and establishment of alien organisms cost more than US$300 billion year^{-1} due to damage and control (Pimentel, 2011). For China, invasive species damage and control efforts are estimated at US$18.9 billion, with US$3.1 billion in direct losses and US$15.7 billion in indirect losses (Ding *et al.*, 2004) (Table 12.2).

Estimates of economic loss may be conservative because full information is often lacking. For example, early estimates of wetland loss from *Spartina* spp. were US$86 million across China (Ding *et al.*, 2004), but a more recent assessment showed that in Fujian province alone, *S. alterniflora* caused a total loss of US$251 million in 2006 (direct: US$128 million; indirect: US$123 million) (Wang, 2010).

Direct, rather than indirect, losses resulting from invasive species may be easier to quantify. For example, invasive species may reduce native biodiversity such as *S. alterniflora* invading salt marshes on Chinese coasts. The occurrence of *S. alterniflora* in turn can lead, indirectly, to a loss of naïve biodiversity and a reduction in ecosystem services. However, estimating the exact economic damage resulting from a reduction in these services is difficult (see Costanza *et al.*, 1997).

Table 12.2. Direct and indirect losses caused by invasive species in China. (Data from Ding *et al.*, 2004.)

Sources of loss	Economic loss (US$ million)
Direct economic loss	*3134*
Agriculture	1,927
Forestry	345
Stockbreeding	156
Fisheries	97
Transport and storage	134
Environment, public facilities	14
Human health	461
Indirect costs	*15,746*
Forest ecosystems	2,559
Agroecosystems	1,834
Grasslands	414
Wetlands	10,937
Lawns	2

Overall, it is clear that, at present, the number and impact of invasive species pose a significant challenge to China's economic and environmental interests. Rapid economic development will continue to contribute to the introduction of new, potentially invasive germplasm; but it is the role of human-induced climate change on exacerbating the impact of new or existing alien species that deserves additional attention.

Biological Invasions and Climate Change in China

Climate is an important determinant of species distributions; consequently, human-induced changes in climate (e.g. climate change or global warming) are likely to affect various aspects of biological invasions, including species introduction pathways, range and impact (Hellmann *et al.*, 2008). At present, there has been relatively little work focusing on the interactions between climate change and biological invasions in China, despite their potential significance and relevance. Here, we review the studies that have been conducted to date regarding the effects of rising temperature, altered precipitation regimes, extreme climatic events and CO_2 enrichment on invasive species in China.

Rising temperature

Rising air temperatures are one of the major components of global climate change. In the past 50 years, the surface air temperature of China has increased, with the greatest rate of increase occurring since the 1980s (Fig. 12.1; Liu *et al.*, 2004); however, the degree of increase is dependent on region (Fig. 12.2; Xue *et al.*, 2012). For example, the temperature has increased by 2°C in north-eastern China and the Xinjiang region, but only by 0.5°C in the north-west, middle and south of China (Zhai *et al.*, 1999; Liu, 2009).

Increasing temperature, as a major component of global climate change, will have far-reaching consequences for ecosystem diversity and functioning, including invasive biodiversity. For invasive plant species, rising temperature may alter their phenologies and prolong their growing season, with consequences for establishment, colonization and spread. Hence, as temperatures warm, invasive species are likely to migrate to higher elevations or

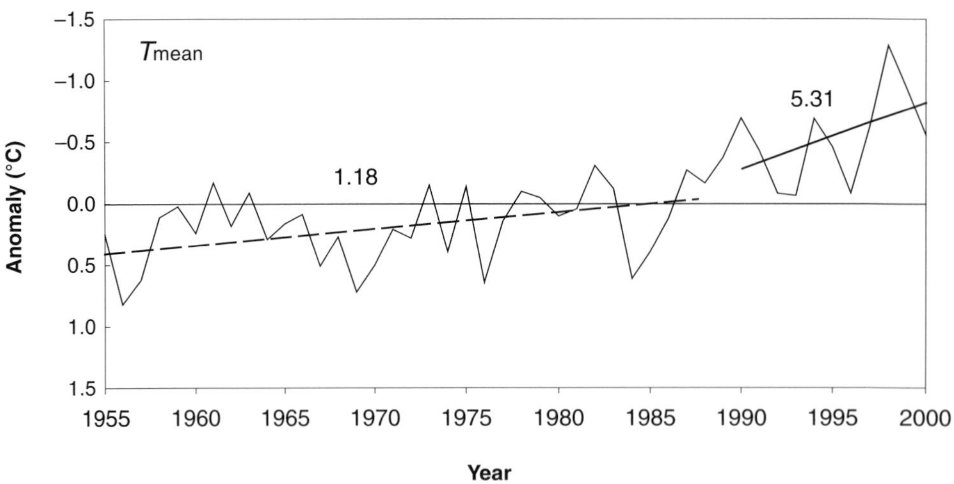

Fig. 12.1. Linear trends and anomalies from annual means of daily temperature (T_{mean}) for China (1955–2000). Linear trends: 1955–1989 (dashed line) and 1990–2000 (solid line). Values along the trend lines are the rate of change [°C (100 y)$^{-1}$]. (After Liu *et al.*, 2004.)

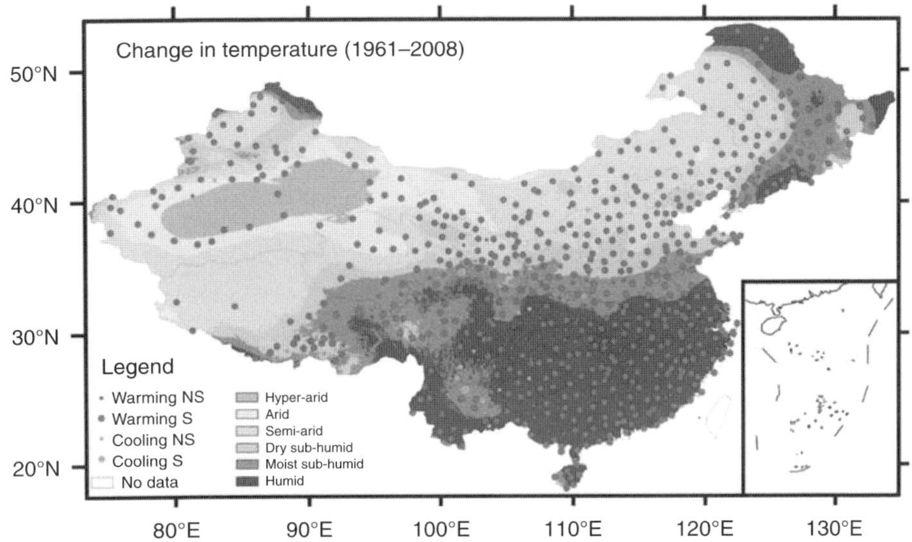

Fig. 12.2. Distributions of climate types and recorded changes in annual mean temperature in China. NS = non-significant; S = significant. (From Xue *et al.*, 2012.)

towards the poles (Hellmann *et al.*, 2008; Walther *et al.*, 2009; Ziska *et al.*, 2011).

For China overall, there are only a few relevant studies regarding temperature responses at the individual species level (Table 12.3). Spanish needles (*Bidens pilosa* L.) increased its height, total biomass, leaf area and shoot:root ratio all in response to increased temperature (Wang *et al.*, 2011a). Palmate-leaved morning glory (*Ipomoea cairica*) showed similar growth responses to increased temperature; interestingly, its allelopathic effects on other plants were also stimulated (Wang *et al.*, 2011b). Other studies have shown that rising temperature can enhance photosynthesis, increase the activity of stress-related antioxidative enzymes, enhance seed germination and stimulate the biomass of some invasive plants (e.g. Ni *et al.*, 2005; Zhang *et al.*, 2006; Lu *et al.*, 2007; Ren *et al.*, 2008; He *et al.*, 2012). These results suggest that rising temperature could favour the establishment and success of invasive plants. However, at present, there are insufficient data to confirm such a hypothesis (Lortie *et al.*, 2007).

Invasive animals, particularly invasive insects, are also sensitive to temperature changes. For example, European corn borer (*Ostrinia nubilalis* Hubner) and North American pine bark beetles have migrated to more northern and higher elevation regions in response to rising winter temperatures (Porter, 1995; Logan and Powell, 2001). In China, climate warming could facilitate similar migrations and range expansions of invasive insects/arthropods such as red imported fire ant (*Solenopsis invicta*), American serpentine leaf miner (*Liriomyza trifolii*), and golden apple snail (*Pomacea canaliculata*) (Lv *et al.*, 2011; Wu *et al.*, 2011).

Increased temperature may facilitate the spread of these invasives throughout China. For example, the snail, *P. canaliculata*, was introduced into China in 1981 and is currently recorded in 13 provinces of southern China. *Pomacea* has become a key intermediate host for the rat lungworm, *Angiostrongylus cantonensis*, a pathogen that causes angiostrongyliasis. *Pomacea*'s range is determined by mean January temperature and snail generation intensity (number of generations per year), which is determined by the accumulated effective thermal energy and developmental threshold temperature (Lv *et al.*, 2011). Models predict that the infested area will increase by 56.9% by the 2030s relative to the current range, which would lead to a doubled occurrence of *A. cantonensis* (Lv *et al.*, 2011). Thus, the interplay between climate change and invasive species for rat lungworm may have serious consequences for human health.

Rising temperature can also enhance invasive animal colonization rates. For example, Huang *et al.* (2011) found that the establishment rates of invasive insects in mainland China was correlated with the increase in mean annual surface air temperature for the period of 1900–2005. They extrapolated that a 1°C increase in mean annual surface temperature could lead to a 0.5 species-per-year increase in the establishment rate of invasive insects (Table 12.3). This relationship between rising temperature and increasing establishment rate still held true even after accounting for increases in international trade for the same period (Huang *et al.*, 2011).

Rising temperature may also change the synchrony of insect development and host phenology (Visser and Holleman, 2001; Thomson and Benton, 2007). Synergy between host and insect could occur with

Table 12.3. Studies examining the effects of rising temperature on invasive species in China.

Species/taxon	Parameters	Conclusion	Reference
Animals			
Invasive insects in China	Establishment rate	Establishment rate increases with warming.	Huang *et al.*, 2011
Bemisia tabaci B-biotype	Survival rate, longevity and fecundity	Warming climate favours the population outbreak of this species through increasing survival rate, longevity and fecundity.	Xie *et al.*, 2011
Pomacea canaliculata	Range	Range size increases with warming.	Lv *et al.*, 2011
Plants			
Alternanthera philoxeroides	Photosynthesis	Photosynthetic capacity and competitive ability are greater at elevated temperature.	Zhang *et al.*, 2006
	Growth and vegetative reproduction	Warming benefits distal ramets rather than proximal ramets through increasing biomass and the numbers of distal ramets.	Li *et al.*, 2012a
Bidens pilosa	Seed germination, seedling biomass allocation, allelopathic effect	Warming may favour plant invasion through increasing plant growth, biomass allocation to stem and leaves and enhancing allelopathic effects.	Wang *et al.*, 2011a
Eupatorium adenophorum	Mortality, plant height, total biomass, biomass allocation	Warming may promote invasions through increasing its growth and stress tolerance.	He *et al.*, 2012
Ipomoea cairica	Seed germination, growth rates, shoot biomass, root biomass, leachate phytotoxicity	Warming may increase its invasiveness through increasing shoot biomass, decreasing root biomass and enhancing phytotoxic effects.	Wang *et al.*, 2011b
Spartina alterniflora	Aboveground biomass, stem density, CH_4 emission	Inundation does not affect aboveground biomass and stem density greatly, but stimulates CH_4 production and emission.	Ding *et al.*, 2010

higher temperatures, increasing the survival rate, longevity and fecundity of *Bemisia tabaci* biotype B, which may favour population outbreaks of this species in China (Xie *et al.*, 2011). However, asynchrony between host and insect could have negative impacts on the fecundity of invasive insects and could potentially decrease their populations and reduce outbreak risks if food shortages occur.

Overall, the initial evidence obtained from a limited number of studies shows that rising temperature may increase the prevalence of invasive species within China. However, it is too early to offer a generalization about how climate change affects invasive species biology for China per se, given the lack of sufficient data from different climatic zones.

Altered precipitation regimes

Unlike projections for climate warming, which have a relatively consistent trend globally, projections for altered precipitation regimes have strong spatial and temporal heterogeneity and greater uncertainty (New *et al.*, 2001; Lana *et al.*, 2004; Xoplaki *et al.*, 2004). The IPCC's (2007) synthesis showed that precipitation increased significantly in eastern North America and South America, northern Europe and central Asia, but decreased in the Sahel, the Mediterranean and southern Africa. Such heterogeneities make predictions of the consequences of altered precipitation regimes for ecosystems uncertain.

Currently, observations from meteorological stations indicate that precipitation in China is quite variable, both spatially and temporally (Figs 12.3 and 12.4; Zhai *et al.*, 2005; Xue *et al.*, 2012). Southern coastal China has abundant precipitation, while middle, western and northern China is dry. Overall, there is a precipitation gradient from north to south in China, with the Yangtze River as an approximate dividing line. Although variable by region (Zhai *et al.*, 1999; Liu, 2009), the overall trend from 1961 to 2008 indicates a decrease in precipitation for China as a whole.

Precipitation and water availability play an important role in the growth, development and distribution of plants and animals. In arid and semi-arid ecosystems, higher

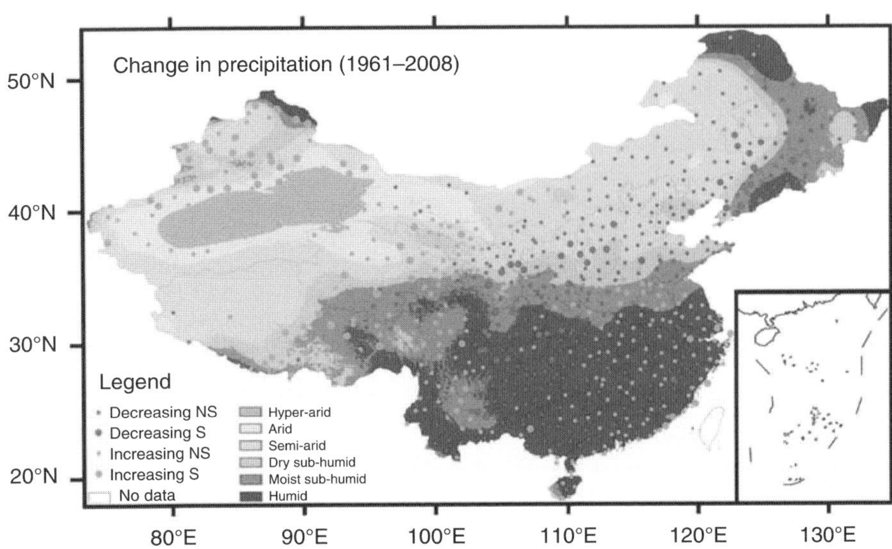

Fig. 12.3. Distributions of climate types and recorded changes in annual mean precipitation in China. NS = non-significant; S = significant. (From Xue *et al.*, 2012.)

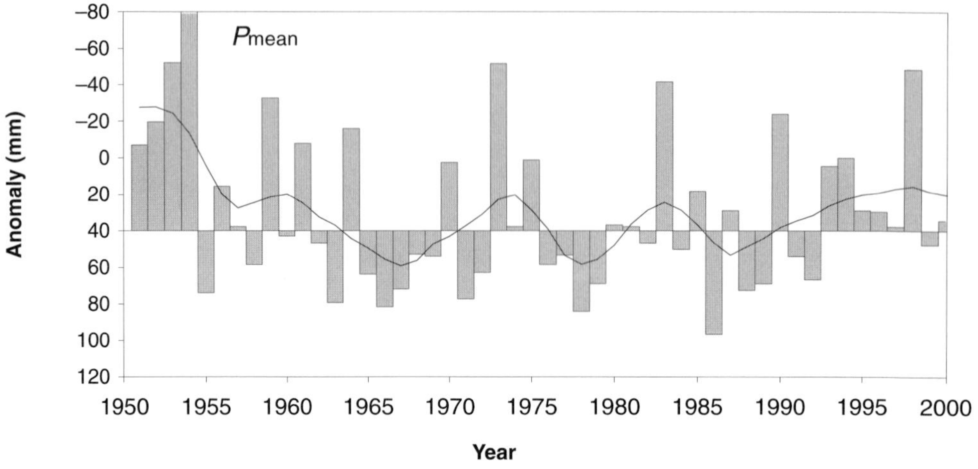

Fig. 12.4. Area-averaged time series of annual precipitation anomaly (in mm) for mainland China (1951–2000). Bars represent annual values and solid curves are moving averages calculated from an 11-year binomial filter. (After Zhai *et al.*, 2005.)

precipitation increases the abundance of some invasive plants (Hobbs and Mooney, 1991; Dukes and Mooney, 1999; Smith *et al.*, 2000). Although most studies have indicated that increased precipitation could facilitate alien species invasion, there are also some exceptions. For example, dry conditions were helpful for the distribution of invasive *Tamarix* (Vilà *et al.*, 2007).

In China, a handful of studies have examined the responses of invasive plants to altered rainfall patterns. Crofton weed (*Eupatorium adenophorum*) is a very aggressive plant invader in south-western China whose reproduction and dispersal are constrained by increased precipitation. Hence, in humid areas, *E. adenophorum* might spread more slowly; however, *E. adenophorum* could spread much more quickly and is predicted to become invasive in drier areas of this region (He *et al.*, 2011). *Praxelis clematidea* may also shift its range with altered precipitation, spreading from southern to central China (Qiu *et al.*, 2011).

An integrative analysis by Feng *et al.* (2010) indicated that the number of invasive species was correlated positively with the mean annual and spatial variation of precipitation. Another synthesis by Wu *et al.* (2010) showed a significant relationship between the invasion index and climatic factors, and suggested that the intensity of plant invasion declined with decreasing temperature and precipitation across China.

Relative humidity (RH) also affects insect development (Wu *et al.*, 2011), and there are interspecific differences in sensitivity to RH. Therefore, altered precipitation might be either beneficial or harmful to an insect, depending on its moisture requirement. For example, the suitable relative humidity for *Bactrocera dorsalis* ranges from 60% to 80%: when air humidity is less than 60% or higher than 80%, insect growth and development is inhibited (Alyokhin *et al.*, 2001). For the invasive insects, fire ant (*S. invicta*), banana moth (*Opogona sacchari*) and *Matsucoccus matsumurae*, higher air humidity could be helpful to their spread (Wu *et al.*, 2011). A study with *Dendroctonus valens* indicated that decreased precipitation or drought influenced the host, *Pinus tabulaeformis*, significantly, which in turn facilitated the infestation of the insect (Wang *et al.*, 2007). Another investigation with chestnut phylloxerid (*Moritziella castaneivora*) demonstrated that mean annual precipitation ranging from 500 to 1600 mm could benefit chestnut phylloxerid's invasion in China (Wang *et al.*, 2010b).

Excessive precipitation could also inhibit insect growth. For example, excessive precipitation constrains the reproduction of the banana moth (*O. sacchari*) by affecting its breeding and decreasing egg hatchability (Du *et al.*, 2006). Moreover, some invasive insects prefer drier conditions. For example, western flower thrips (*Frankliniella occidentalis*) are invading Yunnan province, Weihe River basin, Huaihe River basin and the upper reaches of the Yangtze River, where the climate is relatively dry. It is anticipated that it will be unable to survive in the lower reaches of the Yangtze River, where it is quite humid in the growing season (Wu *et al.*, 2011). *B. tabaci*, whose adults lay more eggs under conditions of lower humidity, would be affected adversely under wetter conditions (Chen, 2006).

In summary, altered precipitation regimes will influence the establishment and spread of invasive insects in China, but the nature of such effects is likely to be species specific. Given the current lack of available data, the distribution of current or new invasive species as a function of precipitation will be difficult to quantify.

Extreme weather events

The intensity and frequency of extreme weather events; e.g. cyclones, heatwaves, minimum temperature, precipitation are projected to increase with climate change (Jentsch and Beierkuhnlein, 2008; Diez *et al.*, 2012). This is consistent with Chinese meteorological observations since the mid-20th century. That is, extreme cold weather events have become less severe, whereas high temperature extremes have intensified. The frequency and intensity of extreme weather events related to precipitation have increased significantly in the lower reaches of the Yangtze River and south-eastern China but have decreased in northern, north-eastern and south-western China, whereas the frequency of drought has increased significantly all over the country, particularly in northern and north-eastern China (Ren *et al.*, 2010; Li *et al.*, 2012b). However, typhoons and tropical cyclones have become less frequent in China. In addition, the frequency of dust storm events in northern China has decreased significantly over recent years (Ren *et al.*, 2010).

Extreme weather events, as intense natural disturbances, strongly influence ecosystems (Jentsch and Beierkuhnlein, 2008). Studies that have examined the effects of extreme weather events on biological invasions (e.g. Diez *et al.*, 2012) suggest that such events could facilitate biological invasions through several mechanisms.

First, extreme weather can act as an enhanced dispersing force for propagules, increasing opportunities for invasive species' transport and expansion (Diez *et al.*, 2012). For example, floods often occur during the summer in southern China and can transport propagules quickly throughout a region. Flooding can divide and transport fragments of *Alternanthera philoxeroides*, which can affect its local abundance (Lu and Ding, 2010). In the Yangtze River delta, flooding can also enhance the dispersal of *Eichhornia crassipes* collected from other watersheds. Moreover, *A. philoxeroides* often works synergistically with *E. crassipes* in the freshwater wetlands in the Yangtze delta, causing 'invasional meltdown' (Simberloff and Von Holle, 1999; Gao and Li, 2004).

Second, physical disturbances caused by extreme weather can create resource and space opportunities (Davis *et al.*, 2000) that facilitate the colonization, growth and spread of invasive species (Diez *et al.*, 2012). For example, hurricanes destroy existing vegetation, creating spatial gaps for plants to colonize. In 2005, Hurricane Katrina in the Gulf of Mexico, USA, accelerated Chinese tallow tree (*Triadica sebifera*) invasion in Louisiana and adjacent areas (Middleton, 2009, Li *et al.*, 2013). In the coastal wetlands in the Yangtze estuary, frequent flooding and land reclamation has restricted an endemic plant, *Scirpus mariqueter*, to a narrow belt in the salt marshes while, conversely, the alien invader, *S. alterniflora*, can take advantage of such disturbances, colonizing the empty niches and spreading rapidly.

Finally, extreme weather events can shift the competitive balance between invasive and native species, potentially putting native species or existing alien species at a competitive disadvantage (Diez *et al.*, 2012). As a result, at the community level, differences among species' responses can alter species demography and ecosystem functioning. For example, *A. philoxeroides*, a ubiquitous invasive weed in China capable of surviving extreme drought conditions, was selected for during the severe drought that occurred in the middle and lower reaches of the Yangtze River in early summer, 2011. As a result, *A. philoxeroides* became more regionally abundant in the freshwater lakes and shallow rivers (personal observation). Interestingly, a comparison of native and invasive species of the same genus (*Wedelia chinensis*, native; *W. trilobata*, invasive) indicated that the invasive species was less susceptible to extreme high temperature. This hints that the invasive *W. trilobata* may become more aggressive with the extreme hot weather that might occur in the future (Song *et al.*, 2010). Similarly, over recent years, the continuous drought has decreased native host (*Pinus*) trees' resistance to the red turpentine beetle (*D. valens*) and favoured the invasive nematode, consequently increasing the damage caused by the pest (Wang *et al.*, 2007; Yao *et al.*, 2008).

Extreme weather events are not consistently advantageous to invasive species selection (Diez *et al.*, 2012). Like native species, some invasive species have been found to be affected negatively by extreme weather. For example, *Acacia mearnsii* is a shallow-rooted tree introduced in Yunnan province. Significant dieback was observed for this species relative to native trees during the drought of 2009–2010. Another example of the detrimental effects of climate change on invasive species biology is the destruction of shell beds in the Yellow River by extreme storms. Such beds trap sediment and allow the establishment of the invasive plant species, *S. alterniflora*. The removal of these beds by storms can lead to the decline of invading *S. alterniflora* populations (Guan, 2009). The case of *S. alterniflora* demonstrates that extreme weather events could allow for the recolonization of native species (Li *et al.*, 2009).

Invasive animal establishment and spread could also be limited by extreme weather events. For example, extreme low temperatures inhibit the development of *Bactrocera correcta* individuals and increase the adults' mortality (Dou *et al.*, 2011). Similarly, low temperature can increase the pupae mortality of *L. huidobrensis* and *L. sativae* (Chen and Kang, 2002). Extreme high temperature can also decrease significantly the egg hatching and survival rate of *Mononychellus tanajioa* (Lu *et al.*, 2011), decrease the fecundity of *B. tabaci* biotype B and *Trialeurodes vaporariorum* (Cui *et al.*, 2008; Yu and Wan, 2009) and the survival rate of *Corythucha ciliata* (Ju *et al.*, 2011, 2013).

Overall, the data to date suggest that extreme weather events can influence invasive species strongly, either positively or negatively. However, at present, insufficient evidence regarding consistent patterns of susceptibility or life-cycle vulnerability to extreme weather events makes general predictions of native versus invasive species competitive outcomes difficult.

CO_2 enrichment

The global increase in atmospheric CO_2 concentration (+23% since 1970) is one of the most important components of global change. This is because CO_2 is the primary source of carbon for all plants; and approximately 95% of all plant species lack optimal amounts of CO_2 for photosynthesis. Consequently, this 'fertilization effect' will directly affect plant growth and function globally, while animals will be affected indirectly through their host plants. Rising CO_2 can also impact on plants indirectly through altering other environmental factors such as temperature, moisture and nutrient availability (Dukes and Mooney, 1999; Bradley *et al.*, 2010).

Studies on invasive plants in China indicated that increasing CO_2 could promote the growth and development of invasive relative to native plants (Song *et al.*, 2009; Table 12.4). For instance, elevated CO_2 concentration enhances *E. crassipes* growth

Table 12.4. Studies examining the effects of elevated CO_2 on biological invasion in China.

Species/taxon	CO_2 concentration range used (ppm)	Parameters	Conclusion	Reference
Animals				
Bemisia tabaci biotype B	375–750	Sizes of egg, pupa, adult; host selection behaviour	Enriched CO_2 concentration has had no effect on the size of eggs, pupa, adults of both sexes in the second and third generations of *B. tabaci*, while *B. tabaci* prefers to select and lay eggs on cottons grown at elevated CO_2 concentration.	Wang *et al.*, 2011c
Bemisia tabaci biotype Q	375–750	Population abundance	Rising CO_2 concentration may favour *B. tabaci* more than its sympatric competitor, *Aphis gossypii*, through increasing the population abundance of *B. tabaci.*	Li *et al.*, 2011
C_3 plants				
Eichhornia crassipes	380–800	Fresh weight, RGR	Elevated CO_2 enhances plant growth, but shoots are more stimulated than roots.	Liu *et al.*, 2010
Galinsoga parviflora	380–700	SLA, leaf perimeter and leaf length	SLA, leaf perimeter and leaf length decreased, while the area and perimeter of leaf stomata increased.	Qi *et al.*, 2009
Mikania micrantha *Wedelia trilobata* *Ipomoea cairica*	400–700	Photosynthesis, biomass, SLA	All three invasive species increased their dry biomass and net photosynthesis significantly and performed better than their native species under elevated CO_2.	Song *et al.*, 2009
Lantana camara	5~1500	Net photosynthesis, water-use efficiency	Its net photosynthesis and water-use efficiency increased with increasing CO_2 concentration.	Zhu and Ma, 2009
C_4 plants				
Spartina anglica	50~1200	Net photosynthesis, transpiration	Transpiration decreases while photosynthetic rate and water-use efficiency increases with increasing CO_2 concentration.	Shi *et al.*, 2009
Spartina alterniflora	50~1500	Net photosynthesis, dark respiration rate	Net photosynthesis increases but dark respiration rate decreased with increasing CO_2 concentration.	Liang *et al.*, 2006

Notes: RGR = relative growth rate; SLA = specific leaf area.

and shoot:root ratio compared with the control group (Liu *et al.*, 2010). Another comparative study using three invasive species, *Mikania micrantha*, *W. trilobata* and *I. cairica*, has shown that all three species increased their net photosynthesis and dry biomass significantly under elevated CO_2 and performed better than the native species (Song *et al.*, 2009). In addition, net photosynthesis and water-use efficiency of *Lantana camara* (Zhu and Ma, 2009) and *Spartina anglica* (Shi *et al.*, 2009) increased significantly, while the dark respiration rate of *S. alterniflora* (Liang *et al.*, 2006) and transpiration (Shi *et al.*, 2009) of *S. anglica* decreased with increasing CO_2 concentration.

Because of the difference in photosynthetic pathways, in theory increasing atmospheric CO_2 concentration would favour the growth of C_3 plants over C_4 plants. However, as Dukes (2000) points out, it seems premature to make any generalizations as to which group (C_3 or C_4 plants) will be more prevalent in the future under elevated atmospheric CO_2 concentration since the response between C_3 and C_4 species may be modified by temperature (e.g. Alberto *et al.*, 1996).

At the ecosystem level, there are a number of implications with respect to CO_2 and the establishment and competitive abilities of invasive plants. Although data are limited, it has been suggested that, on average, invasive species may show a stronger response to both recent and projected changes in carbon dioxide relative to other plant species within a community (Ziska, 2010). This prediction is consistent with recent data that compare the CO_2 response of three invasive species from China and their native, co-occurring plant species (Song *et al.*, 2009).

Elevated CO_2 concentration can affect animals' physiology and behaviour indirectly (Stange, 1997; Abrell *et al.*, 2005; Guerenstein and Hildebrand, 2008) by altering the composition of the host plant. It has been shown, for example, that elevated CO_2 can reduce the nutritional quality of plants, due, in part, to a reduction in N content, protein and changes in phenolic concentration (Cotrufo *et al.*, 1998; Agrell *et al.*, 2000; Hartley *et al.*, 2000; Johns and Hughes, 2002).

To date, most studies of animal responses to elevated CO_2 concentration have focused on insects. In China, invasive insects are the largest group among invasive animals, accounting for nearly half of all invasive animal species (Ding *et al.*, 2010). The response of insects to CO_2-induced changes in the plant host can be mixed, i.e. species specific. Leaf chewers such as cotton bollworm exhibited decreased fecundity, survivorship and developmental rate at elevated CO_2 concentration (Wu *et al.*, 2006). However, sap-sucking insects such as aphids can show a positive response (Sun *et al.*, 2010).

Responses can also vary intraspecifically. For example, for whiteflies, which are very aggressive alien pests in China, a multi-generation study (three successive generations) found that enriched CO_2 concentration had no effect on the size of eggs, pupa or adults of either sex in the second and third generations of *B. tabaci* biotype B, while *B. tabaci* biotype Q preferred to select and lay eggs on cotton plants grown at the elevated CO_2 concentration (Wang *et al.*, 2011c). Similarly, elevated CO_2 concentrations may benefit the Q biotype of *B. tabaci* by increasing its abundance relative to its sympatric competitor, *Aphis gossypii* (Li *et al.*, 2011). These data, while preliminary, suggest that whiteflies may do more harm to crops in China as atmospheric CO_2 increases.

Research Needs

China, with its huge population, diverse landscapes and intensive trade, may be particularly vulnerable not only to climate change but also to a potential synergy between climate change and the occurrence and severity of biological invasions.

Recently, China has increased its efforts to understand global change science and mitigate and/or adapt its potential effects on natural and societal systems. For example, the Ministry of Science and Technology of China strategically initiated a long-term Key Program on Global Change

Research in 2009, with a large economic investment (a half billion Yuan each year) to elucidate better the various topics and impact of global climate change. Although this investment represents a large commitment of the Chinese government and the research community, to our knowledge little effort is currently directed toward the interactions between climate change and biological invasions. Yet, as is clear globally, the impact of invasive species on economic losses and habitat destruction is considerable. If climate change and/or rising CO_2 exacerbate these losses, as could occur, understanding and managing this new threat will be imperative to maintain China's natural and agricultural resources. Overall, we feel that while progress is being made, relevant research on the links between climate, CO_2 and invasive species is underrepresented in China, given the significance and relevance of such research to national resources overall.

What research, then, is most imperative? A global perspective on this topic by Hellmann and colleagues (2008) identified several key research gaps that also need to be addressed in China. Here, we extend the framework of Hellmann *et al.* (2008) and emphasize five issues that need to be addressed in future. Fortunately, many of these can fit into the existing research framework of the national programme on global change research.

Climate is the most important abiotic factor that determines ecosystem functioning (Chapin *et al.*, 2011). How future climate change will alter the vulnerability of ecosystems to biological invasions should be a research priority for Chinese scientists. Vulnerable ecosystem types include temperate grasslands, inland and coastal wetlands and northern semi-arid ecosystems (Li *et al.*, 2005). In turn, as these ecosystems become invaded, how will this feedback on their ability to function as an economic (e.g. watersheds for agriculture) or environmental (e.g. bacterial decomposition in wetlands) resource?

Climate change and CO_2 will affect both native and introduced species. However, at present, invasive species are believed to respond more rapidly to sudden environmental change, including climatic change; consequently, they could become more dominant in a given system. Although a number of studies have examined the plight of native species, little effort has focused on the temporal and spatial dynamics of invading populations under climate change. For major invasive species in China, studies will be required to quantify climate change and CO_2 impacts on their growth, reproduction, behaviour and population dynamics, as well as their competitive interactions with native biota in order to understand changes in their biology better; particularly in regard to shifts in their potential biogeographic range.

Potential invasive plants are important components of many Chinese botanical gardens. The possible escape of these species, particularly those introduced from warmer or tropical climates, represents a potential threat to Chinese landscapes, given current warming trends. For example, two noxious species, *A. philoxeroides* and *E. crassipes*, previously restricted to managed ecosystems, for example gardens in southern China, are now colonizing almost all the freshwater bodies in the south. A warming climate could, in turn, favour their spread to more northern regions. Identifying such species, and their potential spread, should be an important priority.

The impact of biological invasions on native ecosystem functioning has been one of the key issues in biological invasion science. Detrimental impacts on ecosystem functionality, in turn, make biological invasion one of the most pressing global environmental problems (Sala *et al.*, 2000). Assessing the impact of invasives on the functionality of native ecosystems (e.g. bamboo forests) and managed systems (rice paddies) will be necessary to determine the long-term sustainability of these systems with the interactive effects of climate change and existing invasive species in China.

Mitigation of climate change is an obvious strategy to reduce any potential impacts of invasive species. At present, however, measures have been designed separately in assessing climate change and biological invasions. In order to minimize the interactive effects of biological invasions and climate change, an integrated approach needs to be developed when risk assessments or control measures are designed for invasive species. For example, a large portion of China's invasive plants come from South America, and a warmer China may provide more opportunities for the alien plants native to this region. Therefore, quarantine measures need to take into account changes in the likelihood of successful introductions from particular regions.

Conclusions

China is now facing biological invasions by an array of alien species (at least 529 recognized invasive species to date), and more invasive species are likely to accompany future economic development. Existing evidence, while limited, already suggests that different components of climate change (e.g. climate warming, altered precipitation patterns, extreme weather events and elevated CO_2 concentration) may have influenced already the establishment and spread of invasive species. Ongoing changes in climate and rising CO_2 levels, in turn, may exacerbate future and current biological invasions in China. In that context we hope that in this chapter we have: (i) drawn attention to the potential synergy of climate and invasive species as an underappreciated aspect of global change in China, capable of significant economic and environmental impacts; and (ii) advocated successfully for additional resources to begin to predict, prevent and manage these impacts.

Acknowledgements

We thank Dr Jeff Dukes and an anonymous reviewer for their constructive comments on the earlier version of this chapter, and Dr Lewis H. Ziska for his professional editing of our manuscript, which improved its quality and readability greatly. This work was supported financially by the National Basic Research Program of China (Grant No 2013CB430404), the National Science Foundation of China (Grant No 30930019) and the Science and Technology Commission of Shanghai Municipality (Grants No 10JC1400700 and 12XD1400500).

References

Abrell, L., Guerenstein, P.G., Mechaber, W.L., Stange, G., Christensen, T.A., Nakanishi, K. and Hildebrand, J.G. (2005) Effect of elevated atmospheric CO_2 on oviposition behavior in *Manduca sexta* moths. *Global Change Biology* 11, 1272–1282.

Agrell, J., McDonald, E.P. and Lindroth, R.L. (2000) Effects of CO_2 and light on tree phytochemistry and insect performance. *Oikos* 88, 259–272.

Alberto, A.M., Ziska, L.H., Cervancia, C.R. and Manalo, P.A. (1996) The influence of increasing carbon dioxide and temperature on competitive interactions between a C_3 crop, rice (*Oryza sativa*), and a C_4 weed (*Echinochloa glabrescens*). *Australian Journal of Plant Physiology* 23, 795–802.

Alyokhin, A.V., Mille, C., Messing, R.H. and Duan, J.J. (2001) Selection of pupation habitats by oriental fruit fly larvae in the laboratory. *Journal of Insect Behavior* 14, 57–67.

Bradley, B.A., Blumenthal, D.M., Wilcove, D.S. and Ziska, L.H. (2010) Predicting plant invasions in an era of global change. *Trends in Ecology and Evolution* 25, 310–318.

Chapin, F.S. III, Matson, P.A. and Vitousek, P.M. (2011) *Principle of Terrestrial Ecosystem Ecology*, 2nd edn. Springer-Verlag, New York.

Chen, B. and Kang, L. (2002) Cold hardiness and supercooling capacity in the pea leafminer *Liriomyza huidobrensis*. *Cryoletters* 23, 173–182.

Chen, Y.H. (2006) The impact and prevention of alien invasive pest, *Bemisia tabaci*. *Anhui Agricultural Science Bulletin* 12, 133–134.

Costanza, R., *et al.* (1997) The value of the world's ecosystem services and natural capital. *Nature* 387, 253–260.

Cotrufo, M.F., Ineson, P. and Scott, A. (1998) Elevated CO_2 reduces the nitrogen concentration of plant tissues. *Global Change Biology* 4, 43–54.

Cui, X.H., Wan, F.H., Xie, M. and Liu, T.X. (2008) Effects of heat shock on survival and reproduction of two whitefly species, *Trialeurodes vaporariorum* and *Bemisia tabaci* biotype B. *Journal of Insect Science* 8, 24.

Davis, M.A., Grime, J.P. and Thompson, K. (2000) Fluctuating resources in plant communities: a general theory of invisibility. *Journal of Ecology* 88, 528–534.

Diez, J.M., *et al.* (2012) Will extreme climatic events facilitate biological invasions? *Frontiers in Ecology and the Environment* 10, 249–257.

Ding, H., Li, M.Y. and Xu, H.G. (2004) Assessing economic costs of invasive exotic species in China. In: Xu, H.G., Wang, J.M., Qiang, S. and Wang, C.Y. (eds) *Alien Species Invasion, Biosafety and Genetic Resources.* Science Press, Beijing, pp. 78–128.

Ding, H., Xu, H.G., Qiang, S., Meng, L., Han, Z.M., Miao, J.L., Sun, B.S., Sun, H.Y., Huang, C., Lei, C.F. and Le, Z.F. (2011) Status quo and trends of biological invasion into China. *Journal of Ecology and Rural Environment* 27, 35–41.

Ding, W., Zhang, Y. and Cai, Z. (2010) Impact of permanent inundation on methane emissions from a *Spartina alterniflora* coastal salt marsh. *Atmospheric Environment* 44, 3894–3900.

Dou, Q.C., Liu, X.F., Yao, W.F. and Ye, H. (2011) Tolerance of adult guava fruit fly (Diptera: Tephritidae) to low temperature. *Southwest China Journal of Agricultural Science* 24, 1771–1774.

Du, Y.Z., Ju, R.T., Zheng, F.S., Gong, W.R. and Diao, C.Y. (2006) Effects of the environmental factors on the development and survival of banana moth, *Opogona sacchari* (Bojer) (Lepidoptera: Hieroxestidae). *Acta Phytophylacica Sinica* 33, 11–15.

Dukes J.S. (2000) Will the increasing atmospheric CO_2 concentration affect the success of invasive species? In: Mooney, H.A. and Hobbs, R.J. (eds) *Invasive Species in a Changing World.* Island Press, Washington, DC, pp. 95–113.

Dukes, J.S. and Mooney, H.A. (1999) Does global change increase the success of biological invaders? *Trends in Ecology and Evolution* 14, 135–139.

Feng, J.M., Dong, X.D. and Xu, C.D. (2010) Spatial patterns of species diversity of alien invasive plants in China and their relationship with native plants. *Journal of Southwest University* (Natural Science Edition) 32, 50–57.

FoCEC (Flora of China Editorial Committee) (2004) *Flora of China*, Vol 1. Science Press, Beijing.

Gao, L. and Li, B. (2004) The study of a specious invasive plant, water hyacinth (*Eichhornia crassipes*): achievements and challenges. *Chinese Journal of Plant Ecology* 28, 735–752.

Guan, D.M. (2009) *Spartina Salt Marsh Ecosystem in the Coastal Wetlands of China.* Ocean Press, Beijing.

Guerenstein, P.G. and Hildebrand. J.G. (2008) Roles and effects of environmental carbon dioxide in insect life. *Annual Review of Entomology* 53, 161–178.

Hartley, S.E., Jones, C.G., Couper, G.C., and Jones, T.H. (2000) Biosynthesis of plant phenolic compounds in elevated atmospheric CO_2. *Global Change Biology* 6, 497–506.

He, W.M., Li, J.J. and Peng, P.H. (2012) A congeneric comparison shows that experimental warming enhances the growth of invasive *Eupatorium adenophorum. PloS One* 7, e35681.

He, Y.L., Zhang, L.Y. and Guo, Z.F. (2011) Effects of precipitation on germination and emergence of *Eupatorium adenophorum* Spreng. *Jiangsu Agricultural Sciences* 39, 145–148.

Hellmann, J.J., Byers, J.E., Bierwagen, B.G. and Dukes, J.S. (2008) Five potential consequences of climate change for invasive species. *Conservation Biology* 22, 534–543.

Hobbs, R.J. and Mooney, H.A. (1991) Effects of rainfall variability and gopher disturbance on serpentine annual grassland dynamics. *Ecology* 72, 59–68.

Huang, D.C., Haack, R.A. and Zhang, R.Z. (2011) Does global warming increase establishment rates of invasive alien species? A centurial time series analysis. *PloS One* 6, e24733.

IPCC (Intergovernmental Panel on Climate Change) (2007) *Climate Change 2007: The Physical Science Basis. Contribution of Working Group I to the Fourth Assessment Report of the IPCC.* Cambridge University Press, Cambridge, UK.

Jentsch, A. and Beierkuhnlein, C. (2008) Research frontiers in climate change: effects of extreme meteorological events on ecosystems. *Comptes Rendus Geoscience* 340, 621–628.

Johns, C.V. and Hughes, A. (2002) Interactive effects of elevated CO_2 and temperature on the leaf-miner *Dialectica scalariella* Zeller (Lepidoptera: Gracillariidae) in Paterson's Curse, *Echium plantagineum* (Boraginaceae). *Global Change Biology* 8, 142–152.

Ju, R.T., Chen, G.B., Wang, F. and Li, B. (2011) Effects of heat shock, heat exposure pattern, and heat hardening on survival of the sycamore lace bug, *Corythucha ciliata. Entomologia Experimentalis et Applicata* 141, 168–177.

Ju, R.T., Li, H., Shih, C.J. and Li, B. (2012) Progress of biological invasions research in China over the last decade. *Biodiversity Science* 20, 581–611.

Ju, R.T., Wang, F., Zhou, X.H. and Li, B. (2013) Critical thermal tolerance to extreme high temperatures in a new invasive lace bug, *Corythucha ciliata* (Hemiptera: Tingidae) in subtropical China. *PLoS One* 8, e54372.

Khuroo, A.A., Rashid, I., Reshi, Z., Dar, G.H. and Wafai, B.A. (2007) The alien flora of Kashmir Himalaya. *Biological Invasions* 9, 269–292.

Lana, X., Martinez, M.D., Serra, C. and Burgueno, A. (2004) Spatial and temporal variability of the daily rainfall regime in Catalonia (northeastern Spain), 1950–2000. *International Journal of Climatology* 24, 613–641.

Li, B., *et al.* (2009) *Spartina alterniflora* invasions in the Yangtze River estuary, China: an overview of current status and ecosystem effects. *Ecological Engineering* 35, 511–520.

Li, B., Shao, J.J., Wei, S.J., Ma, D. and Li, H. (2013) Global change and plant invasions. In: Wu, J.G., An, S.Q. and Leng, X. (eds) *Lectures in Modern Ecology (VI): Global Climate Change and Ecological Patterns and Processes.* Higher Education Press, Beijing, pp. 66–94.

Li, J.J., Peng, P.H. and He, W.M. (2012a) Physical connection decreases benefits of clonal integration in *Alternanthera philoxeroides* under three warming scenarios. *Plant Biology* 14, 265–270.

Li, K.R., Cao, M.K., Yu, L. and Wu, S.H. (2005) Assessment of vulnerability of natural ecosystems in China under the changing climate. *Geographical Research* 24, 653–663.

Li, Y.N., Li, F.R., Lou, D.F. and Zhang, Q.H. (2007) Biological invasions in Shenzhen and Hong Kong. *Plant Quarantine* 21, 29–31.

Li, Z.X., He, Y.Q., Wang, P.Y., Theakstone, W.H., An, W.L., Wang, X.F., Lu, A.G., Zhang, W. and Cao, W.H. (2012b) Changes of daily climate extremes in southwestern China during 1961–2008. *Global and Planetary Change* 80–81, 255–272.

Li, Z.Y., Liu, T.J., Xiao, N.W., Li, J.S. and Chen, F.J. (2011) Effects of elevated CO_2 on the interspecific competition between two sympatric species of *Aphis gossypii* and *Bemisia tabaci* fed on transgenic Bt cotton. *Insect Science* 18, 426–434.

Liang, X., Zhang, L.Q. and Zhao, G.Q. (2006) A comparison of photosynthetic characteristics between *Spartina alterniflora* and *Phragmites australis* under different CO_2 concentrations. *Acta Ecologica Sinica* 26, 842–848.

Liu, B.H., Xu, M., Henderson, M., Qi, Y. and Li, Y.Q. (2004) Taking China's temperature: daily range, warming trends, and regional variations, 1955–2000. *Journal of Climate* 17, 4453–4462.

Liu, H. (2009) Statistical analysis on the precipitation and temperature over China in nearly 50 years and preliminary study on the impact of ocean. MS thesis, Ocean University of China, Qingdao, China.

Liu, J.Z., Ge, Y.M., Zhou, Y.F. and Tian, G.M. (2010) Effects of elevated CO_2 on growth and nutrient uptake of *Eichhornia crassipes* under four different nutrient levels. *Water Air and Soil Pollution* 212, 387–394.

Logan, J.A. and Powell, J.A. (2001) Ghost forest, global warming, and the mountain pine beetle (Coleoptera: Scolytidae). *American Entomologist* 47, 160–173.

Lortie, C.J., Aarssen, L.W., Budden, A.E., Koricheva, J.K., Leimu, R. and Tregenza, T. (2007) Publication bias and merit in ecology. *Oikos* 116, 1247–1253.

Lu, F.P., Fu, Y.G., Huang, G.X., Huang, Y.R., Xu, X.L., Lu, H. and Chen, Q. (2011) Effect of temperature on the development and reproduction of cassava green mite *Mononychellus tanajioa. Chinese Journal of Tropical Crops* 32, 1720–1724.

Lu, P., Sang, W.G. and Ma, K.P. (2007) Activity of stress-related antioxidative enzymes in the invasive plant Crofton weed (*Eupatorium adenophorum*). *Journal of Integrative Plant Biology* 49, 1555–1564.

Lu, X.M. and Ding, J.Q. (2010) Flooding compromises compensatory capacity of an invasive plant: implications for biological control. *Biological Invasions* 12, 179–189.

Lv, S., Zhang, Y., Steinmann, P., Yang, G.J., Yang, K., Zhou, X.N. and Utzinger, J. (2011) The emergence of angiostrongyliasis in the People's Republic of China: the interplay between invasive snails, climate change and transmission dynamics. *Freshwater Biology* 56, 717–734.

Middleton, B.A. (2009) Effects of hurricane Katrina on tree regeneration in *Taxodium distichum* swamps of the gulf coast. *Wetlands* 29, 135–141.

New, M., Todd, M., Hulme, M. and Jones, P. (2001) Precipitation measurements and trends in the twentieth century. *International Journal of Climatology* 21, 1899–1922.

Ni, G.Y., Wang, C.W. and Peng, S.L. (2005) Effects of different temperature on seed germination of *Meremia biosiana. Ecology and Environment* 14, 898–900.

Pimentel, D. (2011) *Biological Invasions: Economic and Environmental Costs of Alien Plant, Animal, and Microbe Species*, 2nd edn. CRC Press, Boca Raton, Florida.

Porter, J. (1995) The effects of climate change on the agricultural environment for crop insect pests with particular reference to the European corn borer and grain maize. In: Harrington, R. and Stork, N.E. (eds) *Insects in a Changing Environment.* Academic Press, San Diego, California, pp. 94–113.

Qi, S.Y., Guo, X.H. and Zhao, M. (2009) Effects of elevated CO_2 and O_3 concentrations on leaf morphological characteristics of invasive plant *Galinsona parviflora. Journal of Northeast*

Normal University (Natural Science Edition) 41, 149–152.

Qiu, C.H., Wang, Q.Z. and Yu, Y. (2011) Predicted potential distribution of *Praxelis clematidea* in China. *Chinese Journal of Applied and Environmental Biology* 17, 774–781.

Ren, G.Y., Feng, G.L. and Yan, Z.W. (2010) Progress in observation studies of climate extremes and changes in mainland China. *Climatic and Environmental Research* 15, 337–353.

Ren, Y.P., Gu, S., Jiang, S., Wang, Y.Z. and Zheng, S.X. (2008) Influence of light, temperature and salinity on seed germination of *Flaveria bidentis* (Compositae), a new exotic plant. *Acta Botanica Yunnanica* 30, 477–484.

Sala, O.E., *et al.* (2000) Global biodiversity scenarios for the year 2100. *Science* 287, 1770–1774.

SEPA (State Environmental Protection Administration) (1998) *China's Biodiversity: A Country Study*. China Environmental Science Press, Beijing.

Shi, G.Y., Kang, H., Liang, S.C. and Li, J.M. (2009) Responses of photosynthesis and transpiration characteristics of *Spartina anglica* to different CO_2 concentration. *Guangxi Science* 16, 322–325.

Silva, L. and Smith, C.W. (2004) A characterization of the non-indigenous flora of the Azores Archipelago. *Biological Invasions* 6, 193–204.

Simberloff, D. and Von Holle, B. (1999) Positive interactions of nonindigenous species: invasional meltdown? *Biological Invasions* 1, 21–32.

Smith, S.D., Huxman, T.E., Zitzer, S.F., Charlet, T.N., Housman, D.C., Coleman, J.S., Fenstermaker, L.K., Seemann, J.R. and Nowak, R.S. (2000) Elevated CO_2 increases productivity and invasive species success in an arid ecosystem. *Nature* 408, 79–81.

Song, L.Y., Wu, J.R., Li, C.H., Li, F.R., Peng, S.L. and Chen, B.M. (2009) Different responses of invasive and native species to elevated CO_2 concentration. *Acta Oecologica – International Journal of Ecology* 35, 128–135.

Song, L.Y., Chow, W.S., Sun, L.L., Li, C.H. and Peng, C.L. (2010) Acclimation of photosystem II to high temperature in two *Wedelia* species from different geographical origins: implications for biological invasions upon global warming. *Journal of Experimental Botany* 61, 4087–4096.

Stange, G. (1997) Effects of changes in atmospheric carbon dioxide on the location of hosts by the moth, *Cactoblastis cactorum*. *Oecologia* 110, 539–545.

Sun, Y.C., Su, J.W. and Ge, F. (2010) Elevated CO_2 reduces the response of *Sitobion avenae* (Homoptera: Aphididae) to alarm pheromone. *Agriculture Ecosystems and Environment* 135, 140–147.

Thomson, A.J. and Benton, R. (2007) A 90-year sea warming trend explains outbreak patterns of western spruce budworm on Vancouver Island. *Forestry Chronicle* 83, 867–869.

Vilà, M., Corbin, J.D., Dukes, J.S., Pino, J. and Smith S.D. (2007) Linking plant invasions to global environmental change. In: Mooney, H.A. and Hobbs, R.J. (eds) *Invasive Species in a Changing World*. Island Press, Washington, DC, pp. 95–113.

Visser, M.E. and Holleman, L.J.M. (2001) Warmer springs disrupt the synchrony of oak and winter moth phenology. *Proceedings of the Royal Society of London Series B: Biological Sciences* 268, 289–294.

Walther, G.R., *et al.* (2009) Alien species in a warmer world: risks and opportunities. *Trends in Ecology and Evolution* 24, 686–693.

Wan, F.H., Guo, J.Y. and Zhang, F. (2009) *Research on Biological Invasions in China*. Science Press, Beijing.

Wang, D.H. and Fang, C. (2003) *Conserving Biodiversity and Strengthening Management of Nature Reserves*. China Environmental Press, Bejing.

Wang, F.G., Xing, F.W., Ye, H.G., Chen, X.Y., Tan, G.G. and Mai, B.L. (2004) Preliminarily study on invasive alien species in Macau. *Acta Scientiarum Naturalium Universitatis Sunyatseni* 43, 105–110.

Wang, H.B., Zhang, Z., Kong, X.B., Liu, S.C. and Shen, Z.R. (2007) Preliminary deduction of potential distribution and alternative hosts of invasive pest, *Dendroctonus valens* (Coleoptera: Scolytidae). *Scientia Silvae Sinicae* 43, 71–76.

Wang, J. (2010) *Risk Analysis and Eco-economic Losses Evaluation of Spartina alterniflora in Fujian Province*. MS thesis, Fujian Agriculture and Forestry University, Fuzhou, China.

Wang, R.L., Han, M., Liang, X.T., Song, Y.Y., Su, Y.J. and Zeng, R.S. (2011a) Responses of biomass allocation patterns and allelopathic potential of *Bidens pilosa* L. to elevated temperature. *Ecology and Environmental Sciences* 32, 126–130.

Wang, R.L., Zeng, R.S., Peng, S.L., Chen, B.M., Liang, X.T. and Xin, X.W. (2011b) Elevated temperature may accelerate invasive expansion of the liana plant *Ipomoea cairica*. *Weed Research* 51, 574–580.

Wang, S.L., Li, Z.Q., Wang, Z.X., Zhang, M., Xu, J. and Liu, R.H. (2010a) Spatial patterns of invasive animals in China and its relationship with environmental and anthropological factors. *Resources and Environment in the Yangtze Basin* 19, 1283–1289.

Wang, X.X., Wang, G.H. and Ge, F. (2011c) Effect of elevated CO_2 on the body size, enzyme activity and host selection behavior of *Bemisia tabaci* biotype B, *Acta Ecologica Sinica* 31, 0629–0637.

Wang, X.Y., Huang, X.L., Jiang, L.Y. and Qiao, G.X. (2010b) Predicting potential distribution of chestnut phylloxerid (Hemiptera: Phylloxeridae) based on GARP and Maxent ecological niche models. *Journal of Applied Entomology* 134, 45–54.

Weber, E. and Li, B. (2008) Plant invasions in China: What is to be expected in the wake of economic development? *Bioscience* 58, 437–444.

Weber, E., Sun, S.G. and Li, B. (2008) Invasive alien plants in China: diversity and ecological insights. *Biological Invasions* 10, 1411–1429.

Williams, P.H., Margules, C.R. and Hilbert, D.W. (2002) Data requirements and data sources for biodiversity priority area selection. *Journal of Biosciences* 27, 327–338.

Williamson, M. (1996) *Biological Invasions*. Chapman and Hall, London.

Wu, G., Chen, F.J. and Ge, F. (2006) Response of multiple generations of cotton bollworm *Helicoverpa armigera* Hubner, feeding on spring wheat, to elevated CO_2. *Journal of Applied Entomology* 130, 2–9.

Wu, G., Ge, F., Wan, F.H., Xiao, N.W. and Li, J.S. (2011) Responses of invasive insects to global climate change. *Chinese Journal of Applied Entomology* 48, 1170–1176.

Wu, S.H., Hsieh, C.F., Chaw, S.M. and Rejmanek, M. (2004) Plant invasions in Taiwan: insights from the flora of casual and naturalized alien species. *Diversity and Distributions* 10, 349–362.

Wu, S.H., Sun, H.T., Teng, Y.C., Rejmanek, M., Chaw, S.M., Yang, T.Y.A. and Hsieh, C.F. (2010) Patterns of plant invasions in China: taxonomic, biogeographic, climatic approaches and anthropogenic effects. *Biological Invasions* 12, 2179–2206.

Xie, M., Wan, F.H., Chen, Y.H. and Wu, G. (2011) Effects of temperature on the growth and reproduction characteristics of *Bemisia tabaci* B-biotype and *Trialeurodes vaporariorum*. *Journal of Applied Entomology* 135, 252–257.

Xie, Y., Li, Z.Y., William, P.G. and Li, D.M. (2001) Invasive species in China – an overview. *Biodiversity and Conservation* 10, 1317–1341.

Xoplaki, E., Gonzalez-Rouco, J.F., Luterbacher, J. and Wanner, H. (2004) Wet season Mediterranean precipitation variability: influence of large-scale dynamics and trends. *Climate Dynamics* 23, 63–78.

Xu, H.G. and Qiang, S. (2011) *China's Invasive Alien Species*. Science Press, Beijing.

Xu, H.G., Qiang, S., Han, Z.M., Guo, J.Y., Huang, Z.G., Sun, H.Y., He, S.P., Ding, H., Wu, H.R. and Wan, F. (2006) The status and causes of alien species invasion in China. *Biodiversity and Conservation* 15, 2893–2904.

Xue, X., Wang, T., Guo, J., Yang, Y.H., Peng, F. and Liu, L.C. (2012) Impact of global warming on drought in China. *Sciences in Cold and Arid Regions* 4, 0201–0210.

Yang, Q.H., Ye, W.H., Deng, X. and Xu, K.Y. (2002) Characteristics of exotic plant invasion and their damages in China. *Ecologic Science* 21, 269–274.

Yang, R., Wan, F.H., Guo, J.Y., Xie, M. and Zhong, G.P. (2009) Invasive alien species in China: the current status and trends of occurrence. In: Wan, F.H., Guo, J.Y. and Zhang, F. (eds) *Research on Biological Invasions in China*. Science Press, Beijing, pp. 10–24.

Yao, J., Zhang, L.W. and Yu, X.W. (2008) Advances in red turpentine bark beetle, *Dendroctonus valens* LeConte. *Journal of Anhui Agricultural University* 35, 416–420.

Yu, H. and Wan, F.H. (2009) Cloning and expression of heat shock protein genes in two whitefly species in response to thermal stress. *Journal of Applied Entomology* 133, 602–614.

Zhang, C.Y., Liu, W., Xu, Z.F., Cao, H.L. and Ye, W.H. (2006) Responses of vegetative growth and photosynthesis to temperature in the invasive species *Alternanthera philoxeroides* and its indigenous congener *A. sessilis*. *Journal of Tropical and Subtropical Botany* 14, 333–339.

Zhai, P.M., Sun, A.J., Ren, F.M., Liu, X.N., Gao, B. and Zhang, Q. (1999) Chances of climate extremes in China. *Climatic Change* 42, 203–218.

Zhai, P.M., Zhang, X.B., Wan, H. and Pan, X.H. (2005) Trends in total precipitation and frequency of daily precipitation extremes over China. *Journal of Climate* 18, 1096–1108.

Zhu, H. and Ma, R.J. (2009) Photosynthetic characteristics comparison between an invasive plant, *Lantana camara* L., and associated species. *Acta Ecologica Sinica* 29, 2701–2708.

Ziska, L.H. (2010) Global climate change and carbon dioxide: assessing weed biology and management. In: Rosenzweig, C. and Hillel, D. (eds) *Handbook of Climate Change and Agroecosystems: Impacts, Adaptation and Mitigation*. World Scientific Publishing, Hackensack, New Jersey, pp. 191–208.

Ziska, L.H., Blumenthal, D.M., Runion, G.B., Hunt, E.R. and Diaz-Soltero, H. (2011) Invasive species and climate change: an agronomic perspective. *Climatic Change* 105, 13–42.

Part III

Management: Detection and Prevention

13 Identifying Invasive Species in Real Time: Early Detection and Distribution Mapping System (EDDMapS) and Other Mapping Tools

Rebekah D. Wallace and Charles T. Bargeron

Center for Invasive Species and Ecosystem Health, University of Georgia, Tifton, Georgia, USA

Abstract

Historically, the documentation of native and non-native species distributions was limited to herbarium records, textbooks and field guides. Distribution maps were updated infrequently and the resulting information was often inaccessible. Recently, as public awareness of invasive species has increased, professional and citizen science-based monitoring and management programmes have grown, leading to the collection of large amounts of data across disparate databases. The central focus of the Early Detection and Distribution Mapping System (EDDMapS) is to foster the collection, amalgamation and sharing of these data to show a more complete map of the threat of invasive species and how this issue impacts the nation as a whole. Incorporating data from a variety of sources can raise issues with standardization, privacy and ownership, but these issues can be resolved with proper planning and policies regarding data sharing. With the advancement of technology and interest in its utilization for scientific research and public outreach, tracking invasive species is easier, and results can be accessed and distributed to both researchers and the general public. Smartphones have increased the accessibility of information, and as applications have become more central to smartphone use, invasive species programmes have secured funding for application development for the identification, reporting and management of target invasive species. This has increased the reporting of invasive species by citizen scientists and the general public, and has encouraged timely reporting, verification and management by Early Detection and Rapid Response programmes. Training citizen scientists and establishing data collection standards can encourage data collection, which can aid researchers in modelling invasive species range changes in response to a variety of factors, including climate change. As funding and public awareness increase the utilization of technology for invasive species documentation, more and better tools will be developed for invasive species management.

Introduction

Invasive species are a well-recognized threat to species diversity and ecosystem function. Common characteristics of invasive plants include low seed mass and short time to maturity, which facilitates quick reproductive rates. As a consequence, invasive

 Invasive Species and Global Climate Change (eds L.H. Ziska and J.S. Dukes)

species are recognized as having a rapid response to environmental conditions which can lead to range changes (Rejmánek and Richardson, 1996). Climate change will affect invasive and native species' distributions, leading to the geographic expansion of some species and shifting the geographic distributions of some established species (Hellmann *et al.*, 2008). As climate change shifts native species out of their adapted environment, invasive species may encounter less competition (Byers, 2002). Anecdotally, climate change will not decrease the impact of invasive species on the environment, due to the current presence of species across many ecosystems (Qian and Ricklefs, 2006). Accurate determination of the effect of climate changes on invasive species distribution is reliant on thorough documentation of the historical distribution of an invasive species population, if that distribution has changed over time and if that change can be attributed to climate change. Data which could help answer those questions can be collected fairly quickly and easily now with tools that have been developed to record and report the distribution of invasive species in various parts of the world.

Realizing a need for documenting the populations and spread of invasive species, the University of Georgia – Center for Invasive Species and Ecosystem Health (Bugwood) launched the Early Detection and Distribution Mapping System (EDDMapS) in 2005 as a tool for the Southeast Exotic Pest Plant Council. The two primary goals of EDDMapS are inherent in the name: to serve as an early detection and rapid response (EDRR) tool and to function as a distribution mapping system. Since 2005, EDDMapS has collaborated with regional groups and local and state agencies on invasive species mapping projects. In 2010, EDDMapS was granted funding by the US Forest Service (USFS), Forest Health Protection to solicit data from mapping groups nationwide. This permits better identification of the leading edges of invasive species infestations and of data gaps in existing distribution maps.

EDDMapS: An Overview

At a fundamental level, most databases require that observation records include the reporter, species, observation date and state and county of the observation. EDDMapS encourages users to supply as much additional information as the reporter knows about the locality, GPS coordinates, habitat, infestation size and distribution, as this information can be invaluable to EDRR programmes (Crall *et al.*, 2010; Meyerson, 2011). As of March 2013, EDDMapS had received 2 million reports, and over 1 million of those records had GPS coordinates attached to them. There were over 12,000 EDDMapS users, and 2000 of them had reported invasive species occurrences. EDDMapS has invasive species records for over 1900 plant species, primarily across the USA and Canada, but with a small number of records in other countries as well. In addition, EDDMapS Biocontrol, the biological control agent release mapping website of Bugwood, has documented releases of 73 agents for the control of invasive plants across the USA. Recruiting invasive species occurrence data from groups and individuals has involved the support of many state and regional projects. The ultimate goal of EDDMapS is to create a clearinghouse for all invasive species data for use by the public, to increase awareness and to document the impact of invasive species.

Comparing Efforts

EDDMapS is not the only US-based reporting and aggregation database; others exist and serve multiple objectives. The US Geological Survey (USGS) has a program to track non-indigenous aquatic species (USGS-NAS), which previously tracked all taxa but, due to funding cuts, now focuses solely on freshwater vertebrate and invertebrate introductions and freshwater and marine fish. Without aggregation databases such as EDDMapS, the aquatic plant and marine invertebrate data that were obtained by the USGS-NAS program might have been lost.

Now, though, EDDMapS houses the aquatic plant data, and other databases that have similar interests, such as the Smithsonian Environmental Research Center – National Exotic Marine and Estuarine Species Information System (NEMESIS), host the remaining information. All data are still available to the public.

Other invasive species data-aggregating websites, such as the Global Invasive Species Information Network (GISIN), focus primarily on gathering data from other databases. GISIN is run by a group of collaborators led by the USGS and is designed to share global-level invasive species data. All data sent to GISIN must be formatted to their standards prior to submission and the data are assigned globally unique identifiers (GUID), which include the name of the source/reporter. This ensures that any record is unique across databases and can be tracked to an original source. Reported data are also available for download from GISIN for use by individuals or groups. Data are viewable on maps on the GISIN website and are colour coded by native or invasive status within countries.

The US Department of Agriculture (USDA), Animal and Plant Health Inspection Service – Plant Protection and Quarantine (USDA-APHIS-PPQ) works with states to run a programme that tracks pests of agricultural concern: the Cooperative Agricultural Pest Survey (CAPS). The data from the surveys across the USA are housed by the National Agricultural Pest Information System (NAPIS) and are accessible by State Land Grant Universities, State Departments of Agriculture, state PPQ headquarters, other USDA agencies and CAPS members. The survey covers plants, invertebrates (insects, nematodes, gastropods, etc.) and diseases of agricultural significance for both native and non-native species. Maps of the survey results are available on the NAPIS Pest Tracker website (e.g. Africanized honeybee, http://pest.ceris.purdue.edu/map.php?code=ISAEAEA).

Internationally, programmes in other countries are working to survey and record the occurrence and spread of invasive species. Delivering Alien Invasive Species Inventories for Europe (DAISIE) is a cooperation of multiple programmes across dozens of countries in Europe. Its purpose is to serve as a 'one-stop shop' for information on invasive species in Europe, functioning as: an inventory for invasive species, a resource for the prevention and control of invasive species and a provider of assessments and information on the impacts of invasive species, as well as distribution data. It provides information and maps with each species of concern and lists countries, with the status of the species distribution within that country. Experts are listed within the website and can be filtered by country, species, habitat and subject area (genetics, impact, conservation, etc.).

Benefits and Challenges in Data Aggregation

EDDMapS aggregates occurrence data from a variety of sources to produce accurate maps of invasive species' ranges across the USA and specific international projects. Bugwood staff recruit data from local, state, regional and national survey programmes, as well as from volunteers and citizen scientists. Data from the different databases can reinforce each other, e.g. through separate confirmed sightings within state and county databases. Databases can also complement each other by filling in 'data gaps' to provide a more complete, uniform distribution map (Crall *et al.*, 2010). These maps serve several different functions in EDRR programmes. Distribution maps provide a visual representation of the spread of the invasive species and can be used to draw attention to the increasing problem and spread of invasive species. Kudzu (*Pueraria montana* var. *lobata*) is termed the 'vine that ate the south'; however, there are reports of kudzu in New York, Massachusetts, northern Michigan and even on the west coast of the USA, in Washington and Oregon. With long-established infestations such as kudzu, multiflora rose (*Rosa multiflora*) and purple loosestrife (*Lythrum salicaria*), locating and housing all mapped observations in one database to show the complete range of the species can be a challenge.

Aggregating data from multiple sources into a database with a specific goal (e.g. invasive species mapping) also serves to refine the data from other programmes by extracting the relevant information (e.g. invasive species occurrence and control efforts) and providing a forum for the display of those data. In addition to working with specific programmes and projects to upload data, EDDMapS' staff recruit data from a variety of historical and literature resources: existing species occurrence databases, herbariums and defunct regional mapping efforts and citizen science programmes. Many of these existing databases, such as the Biota of North America Program (BONAP) and the US Department of Agriculture – Natural Resources Conservation Service PLANTS Database (PLANTS), include very extensive databases with distribution data for introduced as well as native species. Other agencies have formal survey programmes to track native and invasive species in certain ecosystems, such as the USFS – Forest Inventory Analysis (USFS FIA) programme. Mapping of invasive species may fall under different departments or agencies within each state, and the goals of each agency may be reflected in the data that are collected. Often, these agencies or programmes will join together to achieve certain goals, and as a result will form regional initiatives that may either be focused on, or have a component for, invasive species mapping and management. Some of these programmes form Cooperative Invasive Species Management Areas (CISMAs), Soil and Water Conservation Districts (SWCD), Watershed Coalitions or other cooperative projects. EDDMapS serves to find and bring all these disparate databases. In this way, EDDMapS can serve as an intermediary by providing a national aggregation system to display distribution data and allow data access while fostering inter-agency 'co-operation'. This aids EDRR programmes by giving them access to multiple data sources and allows those programmes to use these data in modelling software to fine-tune survey programs.

Some barriers exist in recruiting data from organizations, agencies and established mapping projects. Legal, technical and other barriers can complicate data sharing between data owners and external databases. Often, there is a lack of infrastructure or standardization that can inhibit data sharing. Research councils, funding agencies and publishing companies do not always include provisos in their programmes stipulating that data should be contributed to open databases to be available for use by other researchers (Nelson, 2009).

Another important factor that impacts data recruitment is funding. Funding for open databases is necessary for a variety of reasons; it can be used to hire individuals to actively recruit and format data, provide computer software and hardware and other technology to house and display the data, and develop tools and websites to make the data available. Some funding is conditional, e.g. can only be used to collect certain kinds of data. Such restrictions may put the recruitment of data from other sources on hold (Nelson, 2009). Formatting data becomes more important when data sets are not standardized or differ in data quality. Formatting may be needed to standardize the structure of new data or to reconcile data from older technologies and software formats. Some data sets also have privacy or ownership issues; the owners, or landowners in the case of invasive species, may not wish for the public to have access to personal data (Nelson, 2009). However, this issue can be minimized by addressing privacy concerns properly and allowing reporters to retain ownership of the data.

Citizen Science

Citizen science data can benefit research programmes, because they can cover large spatial areas over long periods of time. The aggregation of large amounts of citizen science data can reveal patterns that can further research enquiry. These enquiries can then be evaluated using the same citizen scientist programmes (Cooper *et al.*, 2012).

Some concerns exist with data recruited from citizen science programmes or individuals collecting data, as they may not have been classically educated or trained in taxonomy or scientific surveying, or are not deemed 'professionals' by scientists. However, citizen scientist survey methods are more likely to bias results based on location, timing and intensity or effort expended when collecting data than based on differences in abilities and training (Engel and Voshell, 2002; Genet and Sargent, 2003; Schmeller *et al.*, 2009; Cooper *et al.*, 2012). Citizen scientists' data density in the USA and Canada tends to correlate with human population density, though, paradoxically, highly urbanized areas are underrepresented (Cooper *et al.*, 2012). This can bias results when populations are monitored, as more populated areas are surveyed more often and less populated areas are surveyed less often (McLaren and Cadman, 1999; Bray and Schramm, 2001). However, these variables can be taken into account during data analysis (Cooper *et al.*, 2012). Another concern is training citizen scientists to report instances when a species or event is surveyed for but not encountered. There are some cases in which asking the right questions can prompt the observer to report 'absence' data, i.e. situations where a site has been evaluated and the event (or species) not observed, and these data can be valuable to studies (Dhondt *et al.*, 1998). Overall, citizen scientist programmes can be very useful, cost-effective data collection tools when combined with data validation (Cooper *et al.*, 2012).

Improved Quality Control and Gap Identification

Aggregating data from multiple sources, with each source retaining ownership of the data, allows access to a broader reviewing audience to validate the data from individual, state, federal or regional databases. In a few instances, local experts have discovered and corrected misidentifications of high-priority species in data sets that have been imported from previously inaccessible databases. Without a publicly available data aggregation site to display maps of these species, those misidentifications would have gone unnoticed. This is also important in instances where those out-of-place populations have been confirmed. Such confirmations draw attention to little-known populations of invasive species. Identifying populations while they are small and isolated is a key component to many EDRR programmes (Westbrooks and Eplee, 2011).

Incorporating data from a variety of sources improves the accuracy and timeliness of reports, which facilitates the mapping of temporal changes in invasive species distributions. A variety of recent developments have made it easier to track shifts in distributions; these include advancements in technology; more widespread availability of identification materials; the formation of various invasive species awareness programmes; the wider use of volunteers and citizen scientists for surveying; and the promotion of bio-blitzes and other mass surveying events. Programmes concerned with a single or a limited number of invasive species can concentrate their surveying and potentially have better success in creating more complete distribution maps. These maps can be used by EDRR programmes to evaluate how invasive species spread and how quickly the species can cross county, regional and state lines.

Recruiting Occurrence Data

EDDMapS is an important part of many EDRR programmes and serves many functions within those programmes. While many records entered have only the reporter, date of observation, the species name and the location as state and county, much more information can be included in each record. Basic information about an observed invasive species can also include:

- Stage of life cycle (seedling, mature, flowering, fruiting, etc.)
- Gross area and infested area
- Per cent coverage of infestation

- Abundance and distribution of infestation (single plant, scattered clumps, dense monoculture, linear distribution, etc.)
- Number of individuals

In addition, information about habitat and location can be just as important as the species information listed above. Some additional environmental information that can be useful includes:

- Habitat (ruderal, forest, lake, wetlands, grasslands, urban, etc.)
- Locality (description of the location of the observation or habitat)
- Disturbance (type of disturbance or management practices which have an impact on the habitat)
- GPS coordinates and/or Public Land Survey System map information
- National and local ownership (national forest, privately owned, etc.)

Other information that can be included, if known:

- Herbarium name and record number (if a sample was sent to a herbarium)
- Identifier (if someone other than the reporter identified the sample)
- Survey information (type of survey, method of surveying, etc.)
- Precision of coordinates
- Method of documenting the location of the observation (GPS, paper maps, online map, etc.)
- Project (if the observation is part of an existing EDDMapS project or programme)

Collecting additional data for each occurrence can lead to standardization issues. Data standards are increasingly important as data sharing is becoming more prevalent between aggregation systems and databases. EDDMapS data standards are based on the North American Invasive Species Mapping Standards that was approved by the North American Invasive Species Management Association (NAISMA). The standards not only apply to data formatting but also to the surveying methods and vocabulary used in reporting. Each entry in the standards includes the field name, definition, usefulness, if it is required (core element status), coding (if there is a code related to this field), data value (the format of the data within the field) and an example of the data. Conforming to a set of standards that multiple states and programmes utilize facilitates the sharing of data across multiple platforms.

Though there may be issues with uniformity of data standards across programmes and disciplines, additional data strengthens the record. This can be invaluable when EDRR programmes validate a sighting and prepare control measures. This information can also be utilized in future survey programmes if a species develops a pattern in the habitat or situations in which observations occur. Another useful function is the ability to document any revisits to a site of a previous observation. This feature is vital to EDRR programmes to keep track of the efficacy of control efforts over time. Several reports in EDDMapS have multiple revisit entries that document changes in invasive species populations over time.

Data, in addition to contributing to EDDMapS, may also be sent to state regulatory agencies and state weed coordinators. Every reporter within EDDMapS can set up an alert when invasive species are reported based on several filtering options. Several states have designated reviewers, often state weed coordinators or invasive species programme directors, to validate observations reported in EDDMapS. Many of these reviewers have created alerts to notify them when any reports are made within their state for either all invasive species or for specific species that are of concern to their programmes. This helps to ensure accuracy within the database and those reviewers can also alert members of EDRR programmes to reports of high-priority species. When a report is made, the reviewers will determine the validity of the report, including: relationship to the programme's goals, the likelihood of the report being accurate (based on species being reported, reporter of the observation, the location of the report, etc.) and visual verification. If an observation record is in question, either the reviewer can contact the EDDMapS staff and relay his or her concerns or direct contact can be made with the reporter (i.e. if the reporter's profile is publicly available). If the report is unable to be verified or is determined to be in error, the record is removed. This system of checks allows for a rapid response time on questionable records.

Certain smartphone applications, such as the EDDMapS West application, are built such that all observational reports automatically send an alert to the state weed coordinator for validation. This is especially useful with small populations of recently introduced invasive species, as it facilitates EDRR management efforts. These alerts are also useful when documenting high-priority species or federally noxious weeds; they allow the EDRR programmes to react quickly, either to verify the sighting or to confirm a misidentification. Such quick response to newly established populations can reduce the overall cost of invasive species management (Westbrooks and Eplee, 2011).

The vast majority of EDDMapS species reported are invasive plant species. As of March 2013, Florida was the only state that was mapping all invasive species taxa. Florida's invasive species programme has led to increased public participation in the reporting of invasive animals, especially exotic invasive reptiles. There is interest in other states and agencies to map all invasive species taxa, but to do so requires the involvement and cooperation of different state and federal regulatory agencies. Nationwide animal mapping programmes already exist for aquatic species, USGS-NAS (http://nas.er.usgs.gov/) and NEMESIS (http://invasions.si.edu/nemesis/) and feral pigs, National Feral Swine Mapping System (http://128.192.20.53/nfsms/). Many other invasive animal species either are so widespread that mapping is not necessary (e.g. rats and starlings) or are restricted geographically so that local mapping efforts are more efficacious.

EDDMapS does not currently map regulatory animal and disease pests. Due to trade and regulatory issues, insects and plant pathogens of regulatory concern are tracked by regulatory agencies. As different levels of government can enact regulations, this can become particularly confusing. Also, as many regulations have not been updated in decades, many invasive plant species on regulatory lists or the Federal Noxious Weed List reflect previous agronomic concerns, particularly for eastern USA (Windle, 2011). The process of registering a species to be regulated can take up to several years, and a delay can allow that species to become a widespread problem. The process of registration and control can be slowed dramatically if the resulting control efforts impact the interests of another industry, such as shipping or the green industry (Windle, 2011). Regulatory pest reporting is currently available through some Bugwood Apps, such as Outsmart Invasives and the Great Lakes Early Detection Network; however, all reports are sent to the state regulatory agencies to facilitate EDRR programmes.

Technology and Detection Tools: Applications and More

Currently, most mapping programmes record invasive species observations using paper, which can result in missing data fields or mistakes during data recording (Crall *et al.*, 2010). From there, to share, the data must then be entered into a database, which is then made available to other databases. Organizing, transferring and formatting data from hard copies can take months and can potentially lead to errors in the data (Crall *et al.*, 2010). Utilizing smartphone and other digital formats for data entry can, simultaneously, standardize data, prevent errors introduced by transcription and upload it to an online database for use by the mapping program and potentially by other aggregation databases.

Bugwood has partnered with several programmes to customize EDDMapS websites, produce smartphone applications and/or develop mapping projects tailored to the needs of specific programmes. EDDMapS Biocontrol is designed to recruit biological control release data from state and regional programmes and display maps of biological control agents and the target invasive species' host. EDDMapS Alberta is a website developed for the Alberta Invasive Plants Council to monitor a select number of invasive species for their EDRR programmes. The Florida Invasive Species Partnership has worked with EDDMapS since 2007 to map invasive plants and animals, and the programme has been specifically successful in promoting the IveGot1 application to encourage the public to report occurrences of invasive reptiles. The Integrated Pest Management–Pest Information Platform for Extension and Education (IPM PIPE) programmes are composed of individuals from multiple agencies, universities, regionals groups and private industries whose purpose is to create a shared system that concentrates information and tools for IPM practitioners. IPM PIPE is working with Bugwood to provide real-time updates of survey information on pests of US agricultural commodities.

Bugwood has worked with several projects to supply tools for use by project partners, citizen scientists and the public. The Missouri River Watershed Coalition (MRWC), encompassing 13 western states in the USA, has an e-mail alert system to notify a state weed coordinator when a target invasive species is reported using a smartphone application built for them. In addition to working with regional projects, invasive species reporting and control program software has been developed for use by private companies and federal agencies. The USFS Southern Research Station published a book, *A Field Guide for the Identification of Invasive Plants in Southern Forests*, and then contracted Bugwood to develop an application that housed all the information within the book and allowed the reporting of those species to EDDMapS. Overall, these tools make identifying and reporting invasive species fast, easy and tailored to the needs of individual programmes to encourage public awareness of invasive species and the reporting of occurrences as they are observed. This degree of specificity facilitates EDRR programmes locally.

Smartphones have expanded the reach of citizen scientist programmes by providing a way to report field observations. The ability of smartphones to tag reports automatically with GPS coordinates and allow the user to upload biological information, location comments and images in-field, in the presence of the invasive species event, has lessened the stigma of data gathered by 'non-scientists' for research purposes (Kwok, 2009). This also provides a way for communication between citizen scientists and programme directors to give instructions and reminders (Kwok, 2009). Concerns do exist while using smartphones for reporting. To keep prices low, sensors contained within smartphones are not as precise as devices built specifically to measure data (e.g. GPS units, air-quality sensors, cameras, etc.) (Kwok, 2009).

As of August 2012, of the 1.8 million observation records that had been uploaded in EDDMapS: 76.3% were uploaded through the bulk upload service; 23.1% were uploaded via EDDMapS websites (including regional EDDMapS websites); and 0.54% were uploaded via smartphones (including iPhones, android phones and other smartphones). The majority of data, 57.7%, was entered through bulk uploading by government agencies and by regional mapping efforts, with the majority of the remainder obtained from herbariums and literature (35.3%), universities (0.96%), individual reports on the EDDMapS home page (0.3%) and private companies (0.01%). As cellphone service expands and more of the population adopts smartphones with built-in cameras, GPS technology and Internet service, the overall number of reports via smartphone applications versus website accessed via desktop computers should increase. With this in mind, Bugwood has worked with several programmes and projects to develop websites and smartphone applications tailored to specific needs.

The IveGot1 smartphone application was the first application developed by Bugwood. The application is a project between the National Park Service, Florida Fish and Wildlife Conservation Commission and the University of Florida – Center for Aquatic and Invasive Plants. The purpose of IveGot1 is to provide an integrated invasive species reporting system and outreach campaign, of which a smartphone application, a website with direct access to invasive species identification and reporting and the IveGot1 hotline (1-888-IVEGOT1) for reporting live animals are all integral components. The smartphone application was designed specifically to facilitate invasive species reporting in Florida, to be used in combination with existing EDRR programmes. The smartphone application allows the reporter to take a picture of the invasive species and make the report on the spot, rather than relying on computer access. When animal reports are made, an alert is sent out immediately to reviewers to respond to the threat as soon as possible. As of August 2012, 245,458 reports had been submitted by the website and smartphone application. The application is available in iPhone and android phone applications stores.

The Mid-Atlantic Early Detection Network (MAEDN) is a collaborative effort between the Mid-Atlantic Invasive Plant Council (MAIPC), Bugwood and the National Park Service. MAIPC is a system of naturalists, citizen scientists, gardeners and experts who are concerned with the growing problem of invasive species in the Mid-Atlantic States. To help document their efforts, MAEDN contracted with Bugwood to create a website to suit their regional needs and to develop a smartphone application. The MAEDN smartphone application can be used on iPhones and android phones to identify and report invasive species, primarily plants as well as some invasive insects and one invasive plant disease. As previously stated, invasive animal reporting is not mapped on the primary EDDMapS website: all reports are sent to the appropriate regulatory agency for that specific state. Thus, the distribution maps tab for pests may not show any reports. As of August 2012, 9777 records had been reported via the MAEDN website or through the smartphone application. Like IveGot1, this application is also available in iPhone and android phone applications stores.

In 2010, the MRWC and Bugwood launched the MWRC-EDDMapS website to encourage states within, or partners of, the MRWC to report new or potential invasive species. As the MRWC is concerned only with new or potential invasive species, the list of reportable species does not include plants that are already widespread throughout the region. This focus allows the MRWC to concentrate on identifying and treating invasive species while populations are small, before they become widespread. A smartphone application became available in 2011. When a report is made on the website or smartphone application, an alert is sent out to the state weed coordinator for the state in which the report was made. The weed coordinator reviews the report and, once the report is confirmed, the record becomes publicly available. This process allows the state weed coordinators to review all reports as they come in, to evaluate the reports for accuracy and to alert the local EDRR programmes if necessary. As of August 2012, 345 reports had been made via the MRWC website or smartphone applications. This application has also been made available in iPhone and android phone applications stores. MRWC-EDDMapS has recently been renamed EDDMapS West, to reflect the coverage of the website and the application better.

In 2010, the USFS published two guides: *A Field Guide for the Identification of Invasive Plants in Southern Forests* and *A Management Guide for Invasive Plants in Southern Forests*. The USFS contracted Bugwood to develop an application that contained the identification and management information from both guides, while also allowing for species reporting. The primary focus of the application is the identification and reporting of invasive plant species, and the management guide is included as a .pdf file. Since the user is interested in identifying a potentially unknown species, there are several ways to navigate to invasive plant species within the application; plant habit (grasses and canes, trees, shrubs, forbs or vines), by common name, by scientific name or by a search function. The application contains all the pictures and information included in the guide to aid the users in identifying an unknown species properly. The application also includes a link for each species to view the distribution maps from the EDDMapS Mobile website. This application is available on the iPhone and the iPad.

Although most of the tools developed by Bugwood have been in conjunction with state and regional programmes and are designed for use by experts and citizen scientists, Bugwood has also worked with private companies to develop tools specific to their needs. These tools allow for the importation of shape files (polygons and polylines) to aid in tracking control efforts by management teams. Information can be included for each site/polygon that can assist in developing a management history, which can inform future management plans based on previous successes and failures. All the information and data that are collected, in turn, can be used with the company's in-house database. Tools have also been developed to aid managers in creating control plans that are species- and site-specific.

As invasive species are a global issue, species tracking can be specific for the biogeographical needs of countries and/or regions. The Biological Records Centre in the UK created an application called PlantTracker to encourage the reporting of invasive plant species. The data gathered by the application is housed in the Indicia data warehouse and verified records are viewable on a map. Currently, PlantTracker recruits data for 14 species and displays point records for the countries within the UK: England, Wales, Scotland and Northern Island, as well as Ireland. The application includes photographic ID guides as well as images and metadata for verification.

Mapping and EDRR

Since EDDMapS functions as a clearinghouse for invasive species distribution data, all publicly available data that are uploaded into the system are free for download by any user for a variety of uses. Modelling programs can benefit greatly from using EDDMapS in a variety of ways. Most of the required and optional data that are entered can be used in modelling programs to predict the spread of invasive species. In this regard, records containing data with the life stage, GPS coordinates, habitat, infestation, management and disturbances, in addition to the observation date and species, are more valuable than records containing only the date of observation, species and state and county. Including habitat or environmental data in records can aid modelling programs to find patterns in occurrence reports (Jeschke, 2011).

Range changes of invasive species related to climate change have been documented: the ascidian tunicates, *Styela clava* and *Molgula manhattensis*, have moved northward along the US Pacific coast in the past 20–50 years (Carlton, 2000). Yellow starthistle (*Centaurea solstitialis*) and tamarix (*Tamarix* spp.) are likely to expand their ranges with climate change, whereas leafy spurge is predicted to contract within its known range (Bradley *et al.*, 2009). Some species will also experience range shifts, expansion into new areas and contractions in established areas. It is predicted that by 2100, cheatgrass (*Bromus tectorum*) will expand northward into Idaho, Montana and Wyoming, but will contract in some of the southern invaded areas of the Great Basin, and spotted knapweed (*Centaurea biebersteinii*) will shift to higher elevations and will contract in the lower elevations within Montana, Wyoming, Utah and Colorado (Bradley *et al.*, 2009).

A recent study suggests a ubiquitous poleward shift for many pest species (Bebber *et al.*, 2013). However, in many cases, distributions of specific invasive species will be difficult to relate to climate change, as many may have not reached the climatic limits of their new environment (Hellmann *et al.*, 2008). Including information on date, habitat and changes in abiotic parameters (e.g. seasonal temperatures, total and seasonal precipitation, etc.) when recording invasive species are important components of determining if climate change is affecting distribution.

In the south-eastern USA, cogongrass (*Imperata cylindrica*) provides an example of the value of including additional data with invasive species reporting. This species commonly invades areas after a disturbance occurs; mining operations, construction or forest operations, and naturally occurring fires and floods all provide potential habitat (MacDonald, 2004). This pattern was discovered by documenting disturbance data when making invasive species observations. It is likely that propagative materials of cogongrass are introduced along highways and interstate by construction equipment (MacDonald, 2004). Information of this kind can be combined with maps or data on human-caused disturbances (road construction projects, logging, etc.) and habitats prone to natural disturbance (wildfires, flooding, etc.) for use in modelling programs to predict areas where cogongrass is likely to invade. This can direct EDRR survey programmes to regions where there is a high probability for cogongrass invasion, increasing the chances of discovering populations as they are newly established.

EDRR programmes which encourage the documentation of biological or edaphic circumstances that promote the establishment of invasive species can lead, in some cases, to management success. The cogongrass initiatives in the southern USA have promoted awareness of cogongrass invasion and have led to an increasing number of reports by county agents, land managers and homeowners. As of July 2012 in Georgia, 666 total populations had been reported, with 111 of those sites classified as eradicated (McClure, 2012). To ensure that all populations are eradicated, sites are examined and retreated 3 years after the population is no longer visible. This allows the treatment of plants that have been missed in the survey and any sprouts that emerge after the areas are surveyed. As part of the EDRR programme, EDDMapS provides a venue for the public to report cogongrass sightings.

The Florida Invasive Species Partnership also has several examples of utilizing EDDMapS in their EDRR programme. When a report of an invasive animal is made in Florida, several reviewers are immediately e-mailed an alert. In one specific instance, a report of sacred ibis (*Threskiornis aethiopicus*) was made on 2 November 2011 on National Park Service property. An alert was immediately sent out and the National Park Service contacted USDA Wildlife Services, a unit within the USDA that controls nuisance animals, to remove the bird as part of an eradication programme. Technicians from Wildlife Services responded to the call and eliminated the sacred ibis within 3 h of the initial report. An update was made to the original record, stating the capture of the bird. A quick response time is necessary for animals, such as birds, and this occurrence is a prime instance of EDRR. Tools such as custom-designed websites and smartphone applications have made invasive species reporting quicker and easier than ever.

The Future of EDDMapS

EDDMapS continues to recruit data through the promotion of its use by mapping programs and projects, through continued solicitation of data from existing species occurrence databases, encouragement of invasive species reporting by citizen scientists, collaboration with invasive species mapping programmes and the development of project- or region-specific websites and applications. Prior to a 2010–2015 USFS cooperative agreement, most data recruitment was restricted to parameters set by other regional or state project grants. The agreement with the USFS has allowed Bugwood to devote more effort to recruiting data nationwide, especially in areas where lack of funding had previously limited data collection. At present, Bugwood anticipates that continued support by national and regional programs will allow EDDMapS to create a set of baseline national distribution maps for invasive plants in 2014.

Technology development is ongoing and will focus on working with existing and emerging invasive species mapping and control programmes to create tools, or refine existing tools, to continue improving invasive species detection and management. These efforts will include the development of customized websites and applications to target specific regional and programmatic needs. Refinement of tools, programs and websites allows for greater regional sensitivity in detecting and responding to new and existing invasive threats. Programmes that are concerned solely with identifying and mapping invasive species would find fact sheets and distribution maps needed for their tools, but would have no need for management information and control efforts. Conversely, programmes that are concerned only with tracking new invasive species may have a smaller list of target species and so would not need an exhaustive list on their website or application. EDDMapS encourages EDRR programmes to use the database to inform landowners of invasive species populations and to strategize collaboratively regarding their control. Toward this end, Bugwood staff are developing a universal system for documenting efforts to control invasive species populations. As part of EDRR, documentation aids in evaluating the efficacy of management tactics over time as revisits to populations occur. This also aids programmes in tracking the overall efforts in invasive species management.

Until recently, the only option available for invasive species mapping in the field involved carrying a notebook, GPS device, camera and, potentially, identification guides, and then later spending time on a computer inputting the information into one or more online or internal databases. With smartphone advancements and application development, a notebook, camera, GPS device and identification guides can all be on a 'computer' that fits in one hand. EDDMapS' efficacy will be directed towards facilitating mapping and control efforts through improved tools and partnership enhancement. It is anticipated that as technology advances, more features and options will become available for invasive species documentation. This translates to more information that is available to EDRR programmes and a quicker response time after detection; such rapid responses may be essential in the context of concurrent changes in climate and invasive species biology.

References

Bebber, D.P., Ramotowski, M.A.T. and Gurr, S.J. (2013) Crop pests and pathogens move polewards in a warming world. *Nature Climate Change* doi:10.1038/NCLIMATE1990.

Bradley, B.A., Oppenheimer, M. and Wilcove, D.S. (2009) Climate change and plant invasions: restoration opportunities ahead? *Global Change Biology* 15, 1511–1521.

Bray, G.S. and Schramm, H.L. (2001) Evaluation of a statewide volunteer angler diary program for use as a fishery assessment tool. *North American Journal of Fisheries Management* 21, 606–615.

Carlton, J.T. (2000) Global change and biological invasions in the oceans. In: Mooney, H.A. and Hobbs, R.J. (eds) *Invasive Species in a Changing World.* Island Press, Washington, DC, pp. 31–54.

Cooper, C.B., Wesley, M., Hochachka, C. and Dhondt, A.A. (2012) The opportunities and challenges of citizen science as a tool for ecological research. In: Dickinson, J.L. and Bonney, R. (eds) *Citizen Science: Public Participation in Environmental Research.* Comstock Publishing, Ithaca, New York, pp. 99–113.

Crall, A.W., Newman, G.J., Jarnevich, C.S., Stohlgren, T.J., Waller, D.M. and Graham, J. (2010) Improving and integrating data on invasive species collected by citizen scientists. *Biological Invasions* 12, 3419–3428.

Dhondt, A.A., Tessaglia, D.L. and Slothower, R.L. (1998) Epidemic mycoplasmal conjunctivitis in House Finches from eastern North America. *Journal of Wildlife Diseases* 34, 265–280.

Engel, S.R. and Voshell, J.R. (2002) Volunteer biological monitoring: Can it accurately assess the ecological condition of streams? *American Entomologist* 48,164–177.

Genet, K.S. and Sargent, L.G. (2003) Evaluation of methods and data quality from a volunteer-based amphibian call survey. *Wildlife Society Bulletin* 31, 703–714.

Hellmann, J.J., Byers, J.E., Bierwagen, B.G. and Dukes, J.S. (2008) Five potential consequences of climate change for invasive species. *Conservation Biology* 22, 534–543.

Jeschke, J.M.R. (2011) Modeling. In: Simberloff, D. and Rejmánek, M. (eds) *Encyclopedia of*

Biological Invasions. University of California Press, Berkeley, California, pp. 145–147.

Kwok, R. (2009) Phoning in data. *Nature* 458, 959–961.

McClure, M. (2012) Cogongrass in Georgia updated 2012. *Georgia Forestry Commission* (http://www.gatrees.org/forest-management/forest-health/cogongrass/CogongrassinGeorgiaUpdate2012.pdf, accessed 27 January 2014).

MacDonald, G.E. (2004) Cogongrass (*Imperata cylindrical*) – biology, ecology, and management. *Critical Reviews in Plant Sciences* 23, 367–380.

McLaren, A.A. and Cadman, M.D. (1999) Can novice volunteers provide credible data for bird surveys requiring song identification? *Journal of Field Ornithology* 70, 481–490.

Meyerson, L.A. (2011) Databases. In: Simberloff, D. and Rejmánek, M. (eds) *Encyclopedia of Biological Invasions*. University of California Press, Berkeley, California, pp. 145–147.

Nelson, B. (2009) Empty archives. *Nature* 461, 160–163.

Qian, H. and Ricklefs, R.E. (2006) The role of exotic species in homogenizing the North American flora. *Ecology Letters* 9, 1293–1298.

Rejmánek, M. and Richardson, D.M. (1996) What attributes make some plant species more invasive? *Ecology* 77, 1655–1661.

Schmeller, D.S., *et al.* (2009) Advantages of volunteer-based biodiversity monitoring in Europe. *Conservation Biology* 23, 307–316.

Westbrooks, R.G. and Eplee, R.E. (2011) Early detection and rapid response. In: Simberloff, D. and Rejmánek, M. (eds) *Encyclopedia of Biological Invasions*. University of California Press, Berkeley, California, pp. 145–147.

Windle, P.N. (2011) Regulation (US). In: Simberloff, D. and Rejmánek, M. (eds) *Encyclopedia of Biological Invasions*. University of California Press, Berkeley, California, pp. 145–147.

14 Global Identification of Invasive Species: The CABI Invasive Species Compendium as a Resource

Hilda Diaz-Soltero[1] and Peter R. Scott[2]

[1]*USDA, Office of the Secretary, Washington, DC, USA;* [2]*CAB International, Wallingford, UK*

Abstract

The number, spread and impact of invasive species in the latter half of the 20th century has been without historical precedent. Now, as human activity causes a precipitous rise in greenhouse gas emissions (e.g. CO_2), there is growing concern that climate change may also be a significant, long-term driver enhancing invasive species introduction and spread. New and more powerful tools that facilitate linking invasive species and climate change are required to identify and manage these consequences. One such tool, which exploits a wide range of traditional and social media, is the Invasive Species Compendium, or ISC. The ISC is a scientific, web-based encyclopedia that compiles the latest information on the invasive species that have the most negative impacts on the environment, the economy and/or animal or human health. The information in the ISC, updated weekly with the latest scientific findings, can be used; to infer future climate change impacts on an invasive species; to understand the potential environmental and/or economic impacts of the species, and to identify ways to control and manage the species in question. This chapter discusses the value and efficacy of the ISC, with a particular emphasis on its application in a globally warmed future.

Introduction

As climate change and rising CO_2 levels impose additional risks regarding the introduction and spread of invasive species (Mooney and Hobbs, 2000), new and effective management methods are needed. Chief among these is the strategy of detection and prevention, i.e. to keep track of invasives through visual recognition and assess their potential for environmental or economic harm. The introduction and demography of invasive species is tied to global trade (Hulme, 2009), with consequences for ecology and agriculture (Pimentel *et al.*, 2005; Ziska *et al.*, 2011). To manage invasive species, a comprehensive means to identify, track and, if possible, eliminate them is needed. It is from this perspective that the global Invasive Species Compendium (ISC) was developed. Previously, knowledge about a given invasive species was often sporadic, specific to a given region, and not integrated spatially or temporally. Increasing international trade, in combination with changes in climate, lend urgency to the need for an integrated, publically accessible, continuously updated invasive species resource at no cost to the user. Such a resource may be especially relevant to developing countries that may lack technical information on invasive species management or in countries/regions where trade and climate are undergoing rapid changes. To develop such a resource, the global ISC was initiated as a means to provide scientific information on invasive species that was widely accessible, region specific and temporally accurate.

The Invasive Species Compendium: An Overview

The ISC is a multimedia encyclopedia that includes information regarding several thousand invasive species (available at http://www.cabi.org/isc/). The ISC is, at present, the most extensive and authoritative compilation on all taxa of invasive species. It includes, globally, known invasive organisms across all ecosystems – aquatic, marine or terrestrial – that cause the most harm. Since its release in 2012, the tool has been updated weekly and is available on an open-access basis.

The ISC was developed by a consortium of organizations from 12 countries that contributed financially to its development. The effort to develop the ISC was led by CAB International (CABI) in association with the US Department of Agriculture (USDA). They are partnered by expert organizations including the European and Mediterranean Plant Protection Organization (EPPO) and the World Organisation for Animal Health (OIE). CABI, an international, non-profit, science-based, knowledge organization, has a mission to provide information and apply scientific expertise to solve problems in agriculture and the environment (CABI, 2012).

The objective of the ISC effort was to provide an international resource with respect to invasive species. In that regard, the ISC covers the identification, biology, distribution, impact and management of thousands of invasive species. Invasive species of all taxa are covered, including plants, fungi, bacteria, viruses and animals. Subcategories of animals are further distinguished into insects, nematodes, molluscs and vertebrates.

The impacts of these organisms on natural ecosystems and on biodiversity are considered, as well as their impacts on systems managed for agriculture, agroforestry, aquaculture and animal health. Economic impacts are documented, as well as potential adaptive uses (see Chapter 20, this volume). While an underlying goal is to ingrate global knowledge, a geographic information system (GIS) is also used to display distributions of invasive species at the country, state and province levels. The ISC also includes information the invasive species' pathways and vectors, as well as available methods regarding control and management. Extensive reference materials are included, including images, library documents (including peer-reviewed journal articles), conference papers and reports, bibliographic abstracts and a glossary of technical terms, as well as comprehensive links to relevant identification keys and image libraries. As of June 2013, there were over 1682 full invasive species data sheets and over 7291 abbreviated data sheets in the ISC.

Data Sources for the ISC

The content for the ISC data sheets is derived globally from thousands of expert scientists. The ISC's veracity derives from their expertise, from corroboration through peer review and through rigorous editing for quality control.

Additional content is derived from existing compilations of knowledge on invasive species, including expert organizations such as the Inter-American Biodiversity Information Network, the currently inactive Global Invasive Species Programme and the Smithsonian National Museum of Natural History. Content is presented dynamically in the ISC through an advanced web-based platform combining databases, text and images. GIS, taxonomic relationships and data suitable for decision-support tools, such as risk analysis data, are also provided by the ISC in an interactive framework.

Any user accessing the ISC can also link to information both within the ISC and with other local or remote knowledge bases. The content is updated weekly, and the ISC will be maintained, enhanced and updated regularly into the foreseeable future.

Management and Financing of the ISC

The ISC Consortium, a group of supporting organizations (Table 14.1), provides the strategic direction and scientific leadership for the ISC project, as well as financial resources. The Consortium's goal is to provide a ubiquitous resource that: (i) is easily accessed; (ii) is relevant at different biogeographical scales; (iii) is biologically accurate; and (iv) provides control and management options for invasive species. Because such a resource must be temporally relevant, the ISC continues to be updated regularly as new information becomes available.

A primary goal of the ISC is accessibility. It should be widely available to an extensive group of stakeholders. To make the ISC free to its users, development costs are shared among Consortium Members. At present, the budget for the development and

Table 14.1. Consortium members of the Invasive Species Compendium.

Australia, Group Membership
Cooperative Research Centre for National Plant Biodiversity, CRCNPB
Grains Research and Development Corporation, GRDC
Horticulture Australia Limited, HAL
Invasive Animals Cooperative Research Centre, IACRC
Australian Centre for International Agricultural Research (ACIAR)
Canadian Food Inspection Agency; Forest Service; and International Development Agency
Caribbean Island Countries (sponsored by European Union DGDEV)
France, Ministry of the Environment
Forum for Agricultural Research in Africa (sponsored by DFID)
India, Ministry of Agriculture
Malaysian Agricultural Research and Development Institute
Mexico
National Health, Safety and Quality Service for Agri-Food, SENASICA
Comisión Nacional para el Conocimiento y Uso de la Biodiversidad, CONABIO
Monsanto Corporation
Netherlands Ministry of Economic Affairs, Agriculture and Innovation
Secretariat of the Pacific Community (SPC)
Swiss Agency for Development and Cooperation
Syngenta Crop Protection
UK
Department for Environment Food and Rural Affairs, DEFRA
Department for International Development (DFID)/CABI/FARA
US Agency for International Development (USAID)
US Department of Agriculture:
Agricultural Research Service (ARS);
Animal and Plant Health Inspection Service (APHIS);
Foreign Agricultural Service (FAS);
Forest Service (USFS);
Invasive Species Coordination Program;
Natural Resources Conservation Service (NRCS);
Rural Development (RD)
US Department of Commerce, NOAA's National Ocean Service (NOS)
US Department of the Interior, U.S. Fish and Wildlife Service (FWS)

Note: New Members are expected from Latin America, Asia, Australasia, Europe and North America. Any country, non-governmental organization, private company or industry can become a member; most members are in the environment or agriculture arena, and work on invasive species issues. Participation is voluntary and by invitation.

maintenance of the ISC is US$4.75 million over an 8-year period until 2016. Contributions from more than 29 Members of the Development Consortium, based in 12 countries, have amounted to more than US$4.75 million to date. Of this, US$3 million covered the phase of development and release of the definitive ISC. The remaining US$1.75 million covers the task of updating, maintenance and enhancement. Additional funds are sought continuously, to include emerging invasive species and to enhance the current information base. Such funds are also critical in extending the period of updating, maintenance and enhancement beyond 2016.

New Consortium Members are asked to make a one-time contribution of US$175,000 (US$130,000 for developing country organizations), paid over 1 or 2 years. Each new Member makes an essential contribution towards the development target. No distinction is made regarding when a member joins the Consortium.

Organizations that join the Consortium benefit in several ways. Participants have the opportunity to leverage the contributions of all the other Members and to receive the benefits of a multi-million dollar development based on global expertise. Membership allows participants to ensure that their own priorities are duly considered in the ISC's development, ongoing maintenance and enhancement. Membership presents a visible token of participation and leadership in a project that is positioned to reduce the economic and environmental impact of invasive species globally. Members have an opportunity to discuss and influence common interests in an international forum of like-minded representatives of public and private sector organizations and development assistance agencies.

Development of the ISC

The development of the ISC has taken place in phases. The first phase was completed after 18 months in 2008 and delivered an alpha version covering work to document 1000 invasive species. This was made available to the development Consortium Members for use and feedback. During the second phase, information on 500 more species was added, with enhanced development of the IT platform and with the development of advanced search tools. The beta version, with extended content and enhanced functionality, was released in October 2010. A revised beta version was released to the public in June 2011, with the final, definitive version in April of 2012 (see Table 14.2; CABI, 2012).

Table 14.2. Features of the Invasive Species Compendium.

The April 2012 version of the ISC includes:

- Full data sheets on more than 1500 invasive species and related topics prepared by 1000 specialists from throughout the world, peer-reviewed and edited by CABI. A single data sheet for an invasive species includes:
  - Identity: including names, taxonomy, description
  - Images
  - Distribution: including global and regional maps
  - Biology and Ecology: including habitat, host species, genetics, physiology, environmental requirements, epidemiology, vectors, natural enemies
  - Impact: including economic, environmental, cultural
  - Management: including diagnosis, detection and control (regulatory, genetic, cultural, chemical, biological)
  - Gaps in knowledge/research needs
  - References: linked to abstracts and, where available, to full text
- Additional data sheets on a further 7000 species
- 4000 images
- More than 75,000 abstracts of the published literature on invasive species (added to weekly)
- Full text of more than 1085 relevant published documents (added to weekly)
- Powerful search and browse facilities

The third phase of the ISC relates to sustainability. Specifically, the ISC will be updated continually, both in regard to information and also to accessibility and use. This process will involve the input and participation of Consortium Members, as well as feedback from users. At present, the business plan covers these additional costs until 2016, with the intention of indefinite continuation.

How can the ISC be improved for the future? Proposed actions for Stage 4 of the ISC include: better GIS tools for assessing the impact of climate change on invasive species biology; training people to use the ISC; linking population decline or the extinction of native, threatened species to specific invasive species and documenting it in the ISC; expanding information on biological mechanisms that enhance invasive species establishment; developing new scientific tools for extinction vulnerability (population and habitat viability analysis and meta-model analysis); establishing and extending the capacity of the ISC to communicate and exchange data with national invasive species databanks/inventories; enhancing/updating information on the biological control of invasives (updated global database on biological control); and augmenting the number of invasive species with full data sheets.

Use and Access

Open access, free at the point of use, was proposed and agreed at the ISC Inception Workshop in 2006. The costs of ISC maintenance are covered by the ISC Consortium contributions.

The ISC is freely available for use as a global knowledge base to address practical issues concerning the impact of thousands of invasive species – on the environment, on biodiversity and on agriculture, forestry and fisheries, and on animal health. In June 2012, the Compendium content was licensed under a Creative Commons Attribution-NonCommercial-NoDerivs 2.0 UK: England and Wales Licence. It spans the boundary between invasive species' impacts on food security (through the productive use of land and water) and concurrent impacts on the environment and biodiversity.

The ISC has intrinsic value for preventing the introduction of invasive species (including providing data for tools with predictive capability) by providing timely access to information on potential introductory pathways, as well as the management and containment of established invasive species. In addition, it has utility in extension, training, public awareness and research and development, and may provide a scientific tool for policy decisions. The ISC is especially valuable in the local production of educational materials – for example, public information notices and training techniques specific for a given biogeographical region. All information, text and photographs in the ISC can be downloaded to prepare educational materials specific to local needs.

Application of the ISC for the Detection and Prevention of Invasive Species with Climate Change

Anthropogenic increases in atmospheric carbon dioxide, and the resultant increases in climatic variation, are among the most challenging issues in global biology (IPCC, 2007).

These changes have a number of implications for the biology of invasive species. For example, increased carbon dioxide could favour weedy invasive species such as cheatgrass, exacerbating the environmental damage from large, cheatgrass-fuelled fires in some regions (Ziska *et al.*, 2005). Although additional data are needed, initial studies have suggested that, on average, invasive species may show a stronger response to both recent and projected changes in carbon dioxide relative to native plant species (Song *et al.*, 2009). Climate change, particularly warmer temperatures, could also result in shifts of invasive species' ranges, sometimes leading to net expansions or contractions (Parmesan, 1996; Bradley *et al.*, 2009).

Clearly, there is a need to have tools that can assist in predicting the behaviour of invasive species under different climate change scenarios. The ISC can provide a first assessment. Here are some examples on how the ISC can be used for purposes related to climate change.

Assessing the future range of an invasive species

The ISC contains information on the preferred climate for a species: temperature, rainfall, soils (of a plant species), etc. These data are contained in tables in the Biology and Ecology section of the ISC's Invasive Species data sheets. The optimal climate data relative to optimal growth and fecundity of an invasive species can be exported via text or table formats, enabling them to be used for specific and sophisticated climate models. The ISC can also export distribution data to a Google map using a KML (Keyhole Markup Language; a map file format that can display distribution data) download tool. Layering freely available present-day climate data under the distribution of a species could be used to explore, using maximum entropy or other techniques, the range change of the species predicted by future-climate-change models.

What are the invasives in my country?

The ISC can make a list of the invasive species present (those that have a full data sheet) for any country. For some countries, the information can be obtained at the state or province level. This can serve as a starting point for land managers, scientists and others to examine the regional threats posed by invasive species for their area. Such an assessment may also identify adjoining invasives that could potentially migrate in response to climate change (McDonald *et al.*, 2009).

Assessing the threat of new invasives

A list of known invasive species can be requested from other geographically close countries; or, alternatively, from countries with similar climate and/or soils. Cross-comparisons can be made in order to elucidate new threats or to assess the economic or environmental vulnerability of a given country or region to proximate invasive species. The ISC can then provide a list of invasive species that may present a risk. The GIS capability of the ISC can be used to do this task. As a result, any country can prepare prevention actions and early detection and rapid response (EDRR) programmes as management tools to cope with new invasive threats.

This assessment can be a two-step process. The ISC search capability can be deployed to compile a list of invasive species in country X that are not in country Y at present. Extrapolating the ISC data using the ISC search capability can be used with other resources such as WorldClim (www.worldclim.org) climate projection models to investigate the invasive species of a country (country X) with a climate that is the same as or similar to that predicted for another country (Y) some time in the future. Those species identified as present in X but not in Y could then be prioritized in national invasive species prevention and management programmes in country Y; programmes which could also be informed by other data in the ISC such as those relating to known pathways of introduction.

Analysing pathways for invasive species' introductions with climate change

The ISC will have a data sheet for over 50 pathways used by invasive species. For a particular proposed climate change scenario, pathways that might be used by an invasive species to enter a country can be identified, and as a result, management strategies to mitigate or prevent introductions can be developed and implemented.

Analysing invasive species by habitat under climate change

In the near future, over 45 habitat data sheets will be included in the ISC, providing lists of associated invasive species and collating information on the specific risks, threats and management challenges invasive species pose to each of those habitats. The name of the habitats in the ISC is contained in the 'Advanced Search Help'. Some of the habitats included are agricultural lands, forests, grasslands, roadsides, mangroves, salt marshes, lakes, streams, coral reefs, etc. For example, if we are interested in knowing the invasive species that are found in the habitat called 'terrestrial-managed, forests', we can query the ISC to list those invasive species associated with such a habitat.

If there is particular concern about the range expansion of an invasive species habitat under climate change, the ISC can also help. In our example, we might wish to pursue how invasive species within the habitat, 'terrestrial-managed, forests', might be impacted for a given IPCC scenario. The ISC can be queried for a list of invasives in that specific habitat throughout the world. The list can then be refined by the specific climate parameters (e.g. change in minimum temperature) that are projected for the climate model in question. Identification of the invasive species associated with that habitat will also generate a list of species in neighbouring countries or regions with strong trade links. That way, in conjunction with other techniques described above, an estimated list of all of the potential invaders within that habitat and the range that habitat might occupy between regions or countries, can be examined.

Summary

Anthropogenic carbon dioxide and associated changes in global climate will influence the distribution and biology of invasive species. Given the economic and environmental damage inflicted by invasives, it is imperative that management tools that can track, discern and help to prevent the establishment of invasive species become a global priority. In this regard, the Invasive Species Compendium (ISC), by initiating and maintaining a comprehensive scientific database of current and potential threats, associated movements and biological and ecological characteristics, will provide a freely available, unique, easily accessible, worldwide resource to assist with the detection and management of invasive species on a global basis.

Acknowledgements

The authors acknowledge the collaboration and information on the ISC provided by Lucinda Charles, Gareth Richards and Elizabeth Dodsworth of CABI.

References

Bradley, B.A., Wilcove, D.S. and Oppenheimer, M. (2009) Climate change and plant invasions: restoration opportunities ahead? *Global Change Biology* 15, 1511–1521.

CABI (2012) Invasive Species Compendium. CAB International, Wallingford, UK (http://www.cabi.org/isc/, accessed 27 January 2014).

Hulme, P.E. (2009) Trade, transport and trouble: managing invasive species pathways in an era of globalization. *Journal of Applied Ecology* 46, 10–18.

IPCC (Intergovernmental Panel on Climate Change) (2007) *Climate Change 2007: Impacts, Adaptation and Vulnerability.* IPCC Secretariat, Geneva, Switzerland.

McDonald, A., Riha, S., DiTommaso, A. and DeGaetano, A. (2009) Climate change and the geography of weed damage: analysis of US maize systems suggests the potential for significant range transformations. *Agriculture, Ecosystems and Environment* 130, 131–140.

Mooney, H.A. and Hobbs, R.J. (2000) *Invasive Species in a Changing World.* Island Press, Washington, DC, 457 pp.

Parmesan, C. (1996) Climate and species range. *Nature* 382, 765–766.

Pimentel, D., Zuniga, R. and Morrison, D. (2005) Update on the environmental and economic costs associated with alien-invasive species in the United States. *Ecological Economics* 52, 273–288.

Song, L., Wu, J., Li, C., Li, F., Peng, S. and Chen, B. (2009) Different responses of invasive and native species to elevated CO_2 concentration. *Acta Oecologia* 35, 128–133.

Ziska, L.H., Reeves, J.B. III and Blank, R. (2005) The impact of recent increases in atmospheric

CO_2 on biomass production and vegetative retention of cheatgrass (*Bromus tectorum*): implications for fire disturbance. *Global Change Biology* 11, 1325–1332.

Ziska, L.H., Blumenthal, D.M., Runion, G.B., Hunt, E.R. Jr and Diaz-Soltero, H. (2011) Invasive species and climate change: an agronomic perspective. *Climatic Change* 105, 13–42.

15 The Biogeography of Invasive Plants – Projecting Range Shifts with Climate Change

Bethany A. Bradley

Environmental Conservation, University of Massachusetts, Amherst, Massachusetts, USA

Abstract

Correlative models between species occurrences and climate (here referred to as 'habitat suitability models') have become increasingly popular for forecasting risk from invasive plants under current and future climate scenarios. These models have the potential to inform management and monitoring efforts by prioritizing landscapes considered at highest risk under a changing climate. However, a wide range of choices regarding climatic predictor variables, modelling approaches and even distributional data sets influences the resulting projections. The effects of these choices are seldom defined explicitly, which reduces their utility for scientists and managers alike. This chapter reviews common practices of habitat suitability modelling as they apply to invasive plants. The chapter also reviews major findings of recent projections of range shifts in invasive plants. In both cases, the aim is to explore how different choices of predictors, models and input data can influence conclusions in a habitat suitability modelling framework and develop recommendations for best practices.

Introduction

Identifying and eradicating early infestations is the most effective strategy for preventing invasions (Moody and Mack, 1988; Strayer, 2009). Projections of potential range shifts with climate change could provide a valuable tool for informing ecosystem monitoring and targeting high-risk species (Stohlgren and Schnase, 2006). In response to this need, spatial models of species habitat under current and future climate conditions have become increasingly prevalent in recent years, with many investigators focusing on invasive plants.

Conceptually, the approach is relatively straightforward. Habitat suitability modelling uses empirical relationships between species occurrence and environmental variables (e.g. climate, topography, land use) to define habitat spatially across landscapes or regions (Fig. 15.1). Typically, this process involves collecting locations where an invasive plant is present along with spatial maps or models of the environmental conditions (Fig. 15.1a). Next, presence locations are used to specify environmental conditions in which the invasive plant can establish successfully, defining its potential habitat in environmental space (Fig. 15.1b). Finally, environmental habitat is projected back on to geographic space to determine the areas where the species could potentially establish (Fig. 15.1c). For invasive species, the potential range usually includes substantial land area that has not yet been invaded.

There are numerous synonymous or near-synonymous terms used to describe this

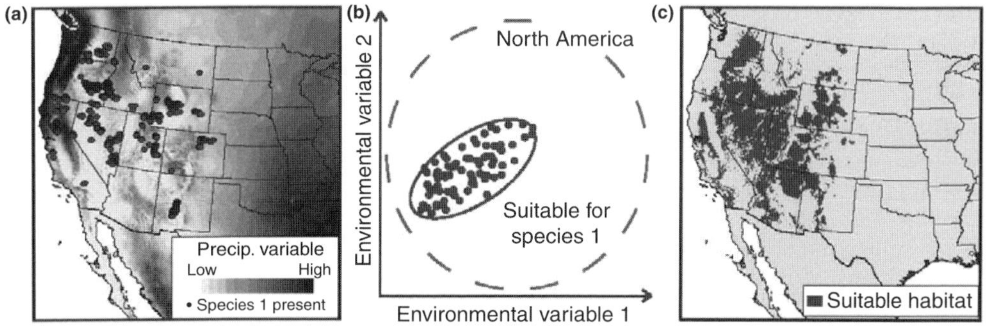

Fig. 15.1. Schematic representation of habitat suitability modelling. (a) Geographic occurrences of an invasive species are compared to climatic variables to calculate (b) climatic suitability for that species. (c) Climatic suitability is mapped back on to geographic space to create a model of suitable habitat, or potential invasive species range.

modelling process (Franklin, 2009). This chapter refers to it as 'habitat suitability modelling'. Other common terms include 'species distribution modelling', 'ecological niche modelling' and 'bioclimatic envelope modelling'. All of these models aim to predict the potential distribution of species, although the latter term typically implies that the environmental predictor variables are related to climate only.

Habitat suitability models (HSMs) do not contain any physiological information about invasive plant species, nor do they include any mechanistic information about spread or competition. They are purely empirical. As a result, they have been criticized, because correlation between climate and distribution does not necessarily imply a causal relationship between climate and distribution (Dormann, 2007). Environmental variables often have a high degree of correlation – for example, locations that have high annual precipitation also tend to be wet in the summer. So, the environmental variables that are empirically the best predictors of species distribution may not actually be the ones influencing plant growth. For a hypothetical plant, annual precipitation might provide the best fit empirically, even though summer precipitation actually affects how quickly the plant can grow and how extensive the invasion could ultimately be.

Model Limitations

How important is climate for modelling plant invasion risk? At a regional scale, climate can limit plant growth and competitive ability, defining broad boundaries on where species can establish, persist and potentially become problematic. However, at landscape and local scales, which are much more management relevant, other environmental conditions such as topography, soils and disturbance may be far more important (Pearson and Dawson, 2003). Moreover, the locations of initial introduction and dispersal ability influence how quickly non-native species are able to become an important component of regional flora (Stohlgren *et al.*, 2011; Stohlgren *et al.*, Part II, Chapter 10, this volume). All of these components play important roles in understanding invasion risk. Here, the focus is only on the approaches to modelling regional invasion risk in response to climate and climate change.

Predictor Variables

Under current climate conditions, switching one climatic predictor for another correlated predictor will have little effect on the overall projection. Studies have shown that empirical models based on current climate have similar levels of accuracy to mechanistic models (Hijmans and Graham, 2006; Estes *et al.*, 2013). However, there is no way to measure accuracy under future climate conditions. And, if physiologically important variables are neglected in HSMs in favour of correlated but physiologically unimportant variables, this choice could have dramatic effects on projections of future range under climate change. To follow our hypothetical example above, if our plant's growth is influenced primarily by summer precipitation, then a model of future habitat based on annual average precipitation could skew projections of range shifts with climate change. Skewed projections are particularly likely if seasonal climate predictors respond differently; for example, if summer precipitation decreases but annual precipitation remains the same.

The best way to work around this problem is to select climatic predictor variables that are known to be physiologically important to the target species. This information can sometimes be found from experimental and observational data. In the absence of physiological information, climatic variables that provide the best empirical fit can be compared to expert knowledge of the species or ecosystem in order to select appropriate predictors (Bradley *et al.*, 2010a). In all cases, models of current and future distribution should be treated as hypotheses (due not only to climatic predictor uncertainty but also to uncertainty in climate projections and changes to species–environment interactions). If the climatic predictors are later found to represent physiology poorly, then any projections of future habitat should be treated with caution.

Invasive Species Equilibrium

One underlying assumption of HSMs is that the target species is at equilibrium (or near equilibrium) with current environmental conditions (Guisan and Zimmermann, 2000). If a species is not in equilibrium, then distribution points will be missing from locations where climate is actually suitable (where the species could establish if it were introduced) and spatial models will under-predict potential establishment. Invasive species, by definition, are not at equilibrium (Václavík and Meentemeyer, 2012). Their continued spread through landscapes and regions fails the equilibrium assumption.

However, for the purposes of modelling, equilibrium does not require that the species occur everywhere that it could potentially establish *geographically*. Rather, equilibrium requires that the species encompass the *environmental* space where it could potentially establish. Environmental equilibrium is achieved much earlier than geographic equilibrium. For example, Welk (2004) showed that predictions of HSMs for purple loosestrife (*Lythrum salicaria*) invasion in North America stopped changing after distribution points from the first 120–150 years of invasion were added. Occurrence points from the past 50 years of invasion did not change the overall prediction, even though they doubled the available data. Thus, it is a reasonable assumption that long-established and/or widely introduced invasive species approximate climatic equilibrium and are appropriate for use in habitat suitability modelling. If a species is in the early stages of invasion and distribution points are only available in a localized area, then it is safe to assume that habitat models will underestimate its potential range vastly. The amount of time that it takes a species to approximate climatic equilibrium likely varies with the introduction and dispersal mechanism. For example, plants introduced as ornamentals may be widely distributed early on and approximate equilibrium sooner.

Modelling Considerations

Choice of model

A variety of different methodological approaches have been applied to habitat suitability modelling. These methods can be broken down roughly into presence/absence, presence/pseudo-absence and presence-only models. Presence/absence models require distribution not only for the location of species presence but also for locations of species absence. Absence data for invasive plants can be problematic, because of the previously mentioned concerns with equilibrium. An absence might mean that the species cannot establish at that location. But, it could also mean that the species can establish, but has not yet been introduced. As a result, presence/absence models are typically used only at landscape or local scales to ask questions such as how local disturbance influences invasion (e.g. Bradley and Mustard, 2006). At regional scales, the likelihood of a 'false' absence skewing model results is much greater and there are few data sets that include both presence and absence. Hence, presence/absence models are rarely used to predict climatic suitability or to forecast species range shifts in response to climate change.

Presence/pseudo-absence models use species occurrences and a set of pseudo-absence points chosen at random from the available environmental space, or 'background'. Suitable climate conditions are identified by comparing the occurrences to the available climate conditions. If more occurrence points are found than expected relative to the available climate space, then those climate conditions are considered more suitable for invasive plant establishment. Many examples of presence/pseudo-absence models have been developed (Elith *et al.*, 2006), but the most widely used approach currently is based on a maximum entropy (or MaxEnt) approach developed by Phillips *et al.* (2006).

There has been considerable debate within the modelling community on how to define 'background' from which to sample pseudo-absence points (VanDerWal *et al.*, 2009a; Lobo *et al.*, 2010; Acevedo *et al.*, 2012). If defined too broadly, the model for current climate conditions might be very precise, but the predictor variables and probability relationships will be minimally informative. For example, a broad background of North America relative to an invasion in the Sonoran Desert is more likely to describe the unique climatic conditions of that desert (relative to North America) rather than the species itself. Too broad a background will underpredict future range with climate change and may bias the climatic predictor variables if they are not defined physiologically (Acevedo *et al.*, 2012). If defined too narrowly, pseudo-absence points are more likely to be placed in locations that are actually climatically suitable, thereby biasing the projection. To date, there is no standard approach for defining background in order to minimize bias in presence/pseudo-absence models. For invasive plants, which are not in equilibrium geographically, the likelihood of creating biased models when selecting pseudo-absence points is high (Lobo *et al.*, 2010).

Presence-only models do not assume any information about absence locations or background environmental space (see Tsoar *et al.*, 2007, for examples). One example of a presence-only model is Mahalanobis Distance, which is a multivariate technique that defines perpendicular major and minor axes within the data and calculates distance from the data centroid relative to the covariance of axes lengths (Farber and Kadmon, 2003). This model measures only distance from the presence data; absences are unnecessary. Presence-only models will almost always predict a larger area of suitable habitat than presence/pseudo-absence models, because they have no direct mechanism for excluding some subsets of environmental space. Jiménez-Valverde *et al.* (2008) note that this larger predicted area has often been interpreted as an indication of overprediction, and therefore poor model performance. However, accuracy assessments are all performed relative to current species distribution, and may be unreliable measures of the accuracy of potential future range. It is equally probable that presence/pseudo-absence models underpredict future range due to biased definition of the background extents (Jiménez-Valverde *et al.*, 2008). A presence-only approach to modelling future habitat suitability under climate change is the least likely to underpredict potential range or be biased in unknown ways.

Although presence-only models are less prone to underprediction (false negatives), presence/pseudo-absence models may be less prone to overprediction (false positives). A balance between minimizing both of these errors may ultimately be most useful for guiding regional monitoring and management. Hence, an ensemble approach (Araujo and New, 2007) that combines the projections of multiple models and climate projections could identify priority management areas.

No-analogue climate

The Earth's surface is not currently experiencing all the possible combinations of climate conditions that could exist. In some locations, rising temperatures and altered precipitation will change local climate to conditions that historically have not been experienced. These 'no-analogue' climate conditions (Williams and Jackson, 2007) may pose a problem for habitat modelling. It is possible to imagine climate conditions that do not currently exist on Earth, but in which a plant could still establish. Maybe the conditions are slightly warmer than current temperatures, or maybe they reflect an annual monsoon arriving a month later. In either case, our lack of invasive plant establishment data does not signify that the species cannot tolerate the environmental conditions, just that the location does not exist.

When projecting into these no-analogue climate conditions, models must extrapolate beyond the available data. The more complex the model, the greater the chances it will produce a problematic relationship beyond the current environmental conditions on which the model is trained. This problem is more likely in presence/absence or presence/pseudo-absence models, and provides further rationale for including presence-only models when modelling range shifts with climate change. Some presence/pseudo-absence models have developed methods for dealing with no-analogue conditions. For example, 'clamping' in MaxEnt (Phillips *et al.*, 2006) sets the value of any location outside the training range as equal to the edge of the training range. However, as a general rule, there is much higher uncertainty in projecting range shifts into climate conditions with no current analogue (Elith *et al.*, 2010). Identifying and highlighting no-analogue climate conditions (Williams and Jackson, 2007) within habitat projections would help to convey uncertainty in risk assessments.

Range shifts with climate change

Although rising levels of CO_2 increase the growth of invasive plants relative to native plants (Blumenthal and Kray, Part I, Chapter 5, this volume), higher temperatures and altered precipitation do not necessarily provide a similar advantage. At the cold edge of invasive plant ranges, warming temperatures might enable plants to expand. For example, kudzu's (*Pueraria lobata*) range is likely limited at its northern edge in the USA by its intolerance of frost. With higher temperatures, this range margin is likely to move northwards (Bradley *et al.*, 2010b). However, at the warm edge, higher temperatures could increase evapotranspiration and reduce invasive plant competitiveness. Observations of native species have shown a consistent shift in species distribution towards the colder edge and away from the warmer edge with warming temperatures (Parmesan and Yohe, 2003; Root *et al.*, 2003). Similar range shifts are likely for invasive plants (Bradley *et al.*, 2009).

A number of projections of invasive plant range shifts with climate change have shown that both expanded and contracted range is likely (Fig. 15.2). In North America, of five problematic invasive plants, two showed primarily range expansion, two showed primarily range contraction and one showed roughly equal expansion and contraction (Bradley *et al.*, 2009). In South Africa, 30 grass species showed primarily contraction of potential range due to climate warming, with up to 50% loss of potential range by mid-century (Parker-Allie *et al.*, 2009). In Australia, three hawkweed species showed a loss of potential range of 20% by 2030 (Beaumont *et al.*, 2009). In North America, projections of range shifts by 2035 for 12

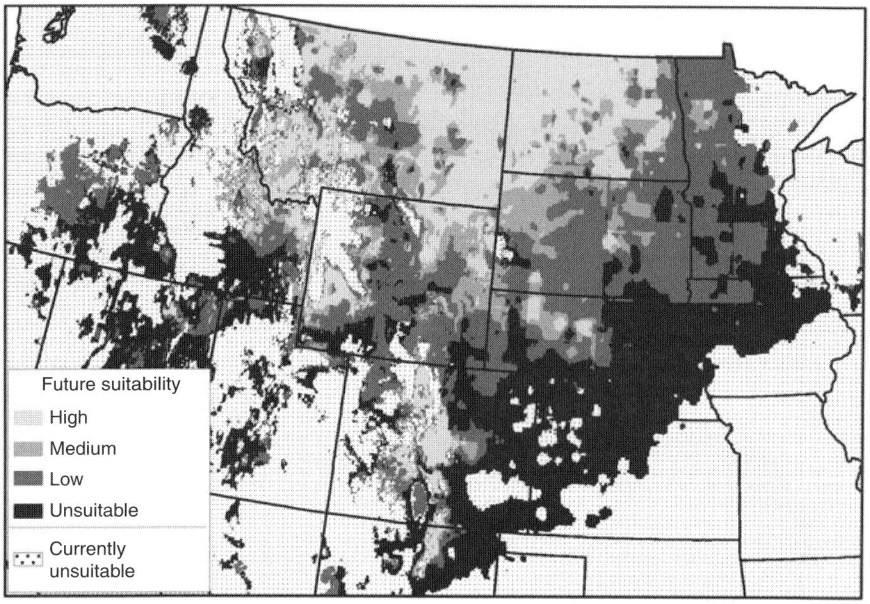

Fig. 15.2. Future climate conditions are likely to lead to substantial habitat loss for leafy spurge (*Euphorbia esula*) in the USA by the end of the century (adapted from Bradley *et al.*, 2009). Black areas on the map are currently climatically suitable, but lose suitability under climate change.

grasses and forbs showed more contraction of potential land area than expansion (Holcombe *et al.*, 2010). In Australia, models for 72 weeds of national significance projected that range will contract for the vast majority of species, with losses averaging 40% of potential range by mid-century (O'Donnell *et al.*, 2012). Losses of potential range suggest that establishment of these species might not be quite so widespread in the future.

These studies have focused primarily on shifts in potential range. However, loss of habitat within the currently invaded range is also likely. For example, of the three species with reduced potential range in North America, 30% of currently invaded areas were likely to become climatically unsuitable by the end of the century (Bradley *et al.*, 2009). This loss of habitat suggests that climate change may cause some invasive plants to become less competitive in, or even retreat from, areas that they have already invaded. Invasive species retreat highlights the intriguing possibility of widespread restoration opportunities due to climate change (Bradley *et al.*, 2009). However, the term 'restoration' is poor for this type of opportunity, because altered climate conditions make it counterproductive to restore the native vegetation that existed prior to invasion. Rather, a type of transformative restoration may be needed to revegetate landscapes with non-invasive, regionally native species that can survive the novel climate conditions and will serve the desired function of the ecosystem (e.g. soil stability, carbon storage, animal habitat) (Bradley and Wilcove, 2009). The development of ecological goals and acceptable risks for assisted migration are needed if these opportunities are to be realized (Harris *et al.*, 2006; McLachlan *et al.*, 2007; Richardson *et al.*, 2009).

Although a number of studies now suggest that range contractions will be just as important, if not more so, than range expansion for invasive plants, the total sample size still remains small. There are thousands of non-native species that have become invasive, but the potential ranges of only a handful have been modelled. For the majority of invasive plants that have not been modelled, potential range shifts are likely to mimic those observed in native species. Warmer temperatures are causing an upward elevational and a poleward directional shift in distributions, and a similar response is likely for invasive plants. For regional managers attempting to catalogue future risk, species that have invaded neighbouring regions with slightly warmer climates likely pose the greatest risk under climate change.

Vulnerable ecosystems

The pool of invasive species with climate change projections is small relative to all global invaders and tends to be focused on the most problematic examples. This small sample size limits the potential for identifying regions or biomes that might be more susceptible under future climate conditions. To date, only one study in Australia has attempted to assess whether some biomes have higher risk than others. O'Donnell *et al.* (2012) modelled 72 weeds of national significance in Australia under current and future climate conditions. They projected two main invasion 'hotspots', one in the south-east and one in the south-west part of the country. Both hotspots were primarily temperate ecosystems. However, this finding might reflect the unique geography of Australia rather than temperate ecosystems as higher risk. Temperate ecosystems are the poleward destination of all tropical invasive plants in Australia, and the country's central desert reduces the available range. Interestingly, the authors note that the invasion hotspots also occur in areas with intensive land use, so disturbance is likely to create more opportunities for the establishment of novel species in these particular areas.

Identifying the most vulnerable ecoregions in the near term may be linked more closely to trade, particularly imports and the availability of ornamental plants, than to climatic suitability. The horticulture industry is the primary source of non-native plant introductions (Reichard and White, 2001; Mack and Erneberg, 2002), and trade is on the rise (Hulme, 2009; Bradley *et al.*, 2012). Ecoregions with higher rates of trade (Vila and Pujadas, 2001) and/or wider distribution of aliens versus natives (Stohlgren *et al.*, 2011) ultimately may see the most dramatic shifts in invasive species in response to climate change.

Invasion Risk: Establishment Versus Impact

Invasive plants are unique in that they have multiple different ranges: the range in which they are native and all of the locations where they have been introduced and successfully established. Often, the native range and the invaded ranges encompass different climatic space (Broennimann *et al.*, 2007), which greatly expands the potential geographic locations of invasive species establishment. Models based on the invaded range alone are likely to underestimate the potential for establishment, and therefore underestimate the potential range under climate change.

However, using all possible distribution data in a suitability model assumes that establishment is the most important stage of invasion risk to predict. For native species, trying to figure out where a plant could establish and persist under climate change is important for informing conservation (Thomas *et al.*, 2004; Kremen *et al.*, 2008), restoration and even assisted migration (Richardson *et al.*, 2009). But, for invasive species, the management applications of understanding potential establishment are much less direct. Many invasive plants have wide climatic tolerances (Pysek and Richardson, 2007), and could establish under a variety of conditions. For example, cheatgrass (*Bromus tectorum*) is established in all 50 states and most Canadian provinces (Fig. 15.3) (USDA-NRCS, 2012). But, it is abundant and has a high impact only in western states, where its early season growth helps it outcompete native species for limited water resources (Mack, 1981; Knapp, 1996) and where its fine fuel biomass

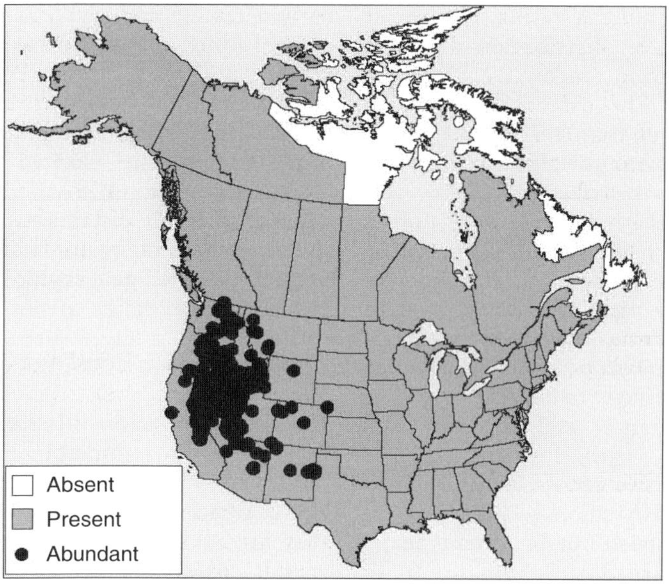

Fig. 15.3. Comparison of establishment range to abundance/impact range for cheatgrass (*Bromus tectorum*) in North America. State-/province-level establishment information is available on the USDA PLANTS website (USDA-NRCS, 2012). Point locations of high abundance come from USGS GAP analysis data for the north-west and south-west and from California.

increases fire frequency (D'Antonio and Vitousek, 1992; Brooks *et al.*, 2004; Balch *et al.*, 2013) (Fig. 15.3). These types of impacts are not a concern in eastern states, even though the species is well established.

The climatic and geographical space where an invasive species has an impact is a smaller subset of its total establishment. This space has been termed 'damage niche' (McDonald *et al.*, 2009; DiTommaso *et al.*, Part III, Chapter 16, this volume) or 'impact niche' (Bradley, 2013). Modelling the current and future geographical extents of the impact niche is much more management relevant than establishment or range alone (Hulme, 2006; McDonald *et al.*, 2009). Yet, the two invasion stages of establishment and impact (Lockwood *et al.*, 2007) are often confusingly lumped together in model projections under the single term 'invasion risk'.

If the goal of a model is to identify the current and potential range of species establishment, then all available distribution data should be used (including data from the native range and any other invaded ranges). Not using all available establishment data will result in an underestimate of potential establishment. But, what if the goal is a more management-relevant assessment of the current and future geographical extents of impact? The choice of distribution data in this case is not straightforward. Very few distribution data sets contain any sort of measure of impact, particularly because impact itself is notoriously difficult to quantify, changes through time as species interact and could be measured for a range of different native competitors and/or ecosystem processes. An easier-to-measure proxy for impact in the case of invasive plants is abundance. Some plant species have meaningful impacts at low abundance, but high abundance always leads to greater impacts (Parker *et al.*, 1999). Unfortunately, even abundance or cover is rarely available in distribution data sets, particularly at the coarse (>1 km) spatial resolutions typically used to model climatic habitat.

One thing is clear, distribution data, even if limited to the invaded range, are a poor proxy for impact. As a result, models of suitability based on distribution data will overestimate impact (Bradley, 2013). This is partially due to the establishment niche being much larger than the impact niche. But, this phenomenon is also a function of invasive plant distribution data sets themselves. Herbarium collections, often available online from sources such as the Global Biodiversity Information Facility (http://www.gbif.org/), tend to be focused on collecting individuals where the species is rare. Hence, invasive plant herbarium records are more likely to be located in sites with low abundance, and therefore low impact. Indeed, herbarium records for problematic invasive plants in the western USA were located in regions where managers considered the species to be absent nearly 60% of the time (Bradley, 2013).

Invasive plant distribution data sets collected by managers and compiled regionally, such as the Early Detection and Distribution Mapping System (http://EDDMaps.org/; Wallace and Bargeron, Part III, Chapter 13, this volume) or the invasive plant atlas of New England (http://www.eddmaps.org/ipane/), tend to be focused on early detection and rapid response. Because control is more effective for early infestations (Moody and Mack, 1988), treatment tends to be focused on areas with low abundance. Hence, distribution data compiled regionally also tend to be skewed towards low abundance (Marvin *et al.*, 2009; Bradley, 2013). As a result, distribution data available in the invaded range are likely to be a poor proxy for the geographic and climatic conditions that define the impact or damage niche.

Establishment Models Do Not Predict Impact

HSMs create a surface of suitability values that are related to the probability that a species could establish at a given location (Fig. 15.4a). MaxEnt, for example, produces a 'gain' value ranging from zero to one, where high numbers indicate higher probability of establishment. It seems intuitive that higher probability of establishment should also be related to higher abundance, and therefore impact. If true, distribution data alone could be used to

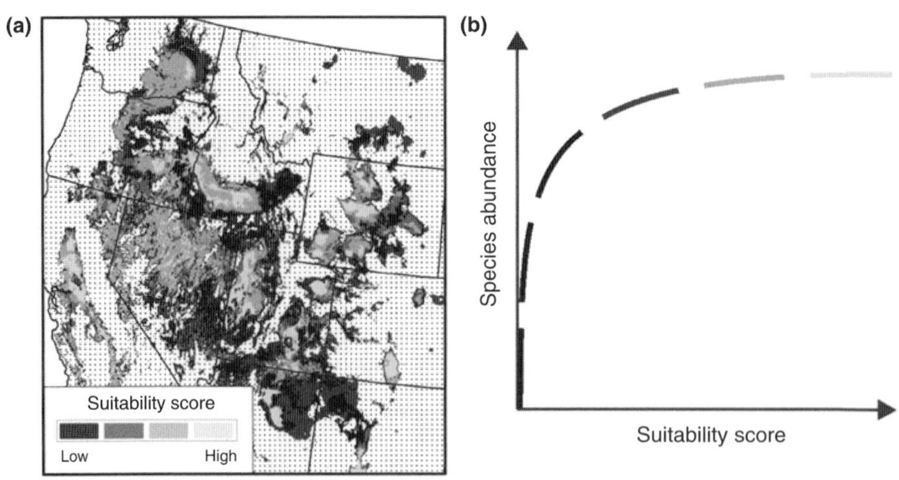

Fig. 15.4. Habitat suitability scores based on distribution data are linked poorly to abundance. (a) Probability of establishment for an invasive plant in the western USA based on MaxEnt gain values (Phillips *et al.*, 2006). White is unsuitable, lighter shades have higher suitability. (b) Typical relationship between suitability scores and species abundance.

predict impact by identifying locations with the highest suitability for establishment.

Relationships between modelled suitability based on distribution data and abundance have been tested for a number of species, including birds (VanDerWal *et al.*, 2009b), mammals (Tôrres *et al.*, 2012), arthropods (Jiménez-Valverde *et al.*, 2009) and plants (Pearce and Ferrier, 2001; Nielsen *et al.*, 2005; Estes *et al.*, 2013). Rather than finding a strong relationship between suitability for establishment and measured abundance, these studies have found low or insignificant correlations. In several cases, the relationship is represented best as a step function, where low suitability corresponds to absence but high suitability corresponds to any value of abundance from low to high (Fig. 15.4b).

How Can We Predict Impact?

If distribution data are a poor proxy for impact niche, and suitability for establishment is a poor predictor of impact, is there any way to predict impact niche under current and future climate conditions? One positive finding is that models appear to be able to predict abundance (a proxy for impact) using continuous abundance data. Kulhanek *et al.* (2011) showed that observed abundance of invasive carp in Minnesota lakes (USA) could be used to predict the abundance of carp in South Dakota lakes. Unfortunately, continuous abundance data are extremely rare, particularly at the regional scales needed to link abundance to climate. A few regional abundance data sets have been collected in the USA, including the forest inventory and analysis (FIA) data set (http://www.fia.fs.fed.us/) used by the US Forest Service to inventory forest resources and the national gap analysis program (GAP) data sets used to validate state and regional land cover classifications. However, these data are focused on native species and are likely to overlook all but the most problematic invasive plants.

One innovative approach to modelling the impact or damage niche is to create a subset of distribution data based on locations known to have a high impact. McDonald *et al.* (2009) modelled current and future potential damage to crop species by identifying the states with high economic losses to two invasive species (Bridges, 1992), velvetleaf (*Abutilon theophrasti*) and Johnsongrass (*Sorghum halepense*), and using those state locations as distribution data for suitability models. This approach could be applied more widely to invasive plants based on expert knowledge of where the species has the highest impact, with the results less likely to overestimate impact than distribution data alone.

Finally, although suitability models based on distribution data are correlated poorly to abundance (Fig. 15.3), a recent study suggests that suitability models based on high-abundance locations are well correlated with abundance. Estes *et al.* (2013) showed that suitability values derived from a MaxEnt (Phillips *et al.*, 2006) model based on locations of high-yield maize in South Africa were correlated linearly ($R^2 = 0.6$) to continuous measurements of maize yield. This finding suggests that models based on point locations of known high abundance, such as the ones created by McDonald *et al.* (2009), could do better than just modelling impact/non-impact by predicting levels of potential abundance. The study is based on a single plant species and needs wider testing, but the approach of modelling damage or impact niche is very promising for invasive plant risk assessments.

Conclusions

Although habitat suitability modelling is conceptually straightforward, the case of forecasting plant invasions due to climate change presents a unique set of challenges. While models of establishment risk under current climate are likely robust, choices of modelling approach, distribution data and climatic predictors can have substantial effects on projections of range shifts with climate change. The limitations and likely biases of model and data choices are communicated poorly to managers, and better efforts towards transparency are needed.

The sample size of predictive models for invasive plants remains small relative to the total pool of global invaders. However, models have consistently shown both potential for expansion and likely contraction of invasive plant range. It is currently unknown whether the impact niche is projected to follow similar expansion and contraction patterns as the total range. Likely invasive plant expansion suggests that managers should be including new species in monitoring efforts, particularly the problematic species from warmer-climate neighbours. Invasive plant contraction due to climate change may open a window of unique opportunity for the targeted restoration of high-priority invaded ecosystems.

References

Acevedo, P., Jiménez-Valverde, A., Lobo, J.M. and Real, R. (2012) Delimiting the geographical background in species distribution modelling. *Journal of Biogeography* 39, 1383–1390.

Araujo, M.B. and New, M. (2007) Ensemble forecasting of species distributions. *Trends in Ecology and Evolution* 22, 42–47.

Balch, J.K., Bradley, B.A., D'Antonio, C.M. and Gomez-Dans, J. (2013) Introduced annual grass increases regional fire activity across the arid western USA (1980–2009). *Global Change Biology* 19, 173–183.

Beaumont, L.J., Gallagher, R.V., Downey, P.O., Thuiller, W., Leishman, M.R. and Hughes, L. (2009) Modelling the impact of *Hieracium* spp. on protected areas in Australia under future climates. *Ecography* 32, 757–764.

Bradley, B.A. (2013) Distribution models of invasive plants over-estimate potential impact. *Biological Invasions* 15, 1417–1429.

Bradley, B.A. and Mustard, J.F. (2006) Characterizing the landscape dynamics of an invasive plant and risk of invasion using remote sensing. *Ecological Applications* 16, 1132–1147.

Bradley, B.A. and Wilcove, D.S. (2009) When invasive plants disappear: transformative restoration possibilities in the western United States resulting from climate change. *Restoration Ecology* 17, 715–721.

Bradley, B.A., Oppenheimer, M. and Wilcove, D.S. (2009) Climate change and plant invasion: restoration opportunities ahead? *Global Change Biology* 15, 1511–1521.

Bradley, B.A., Blumenthal, D.M., Wilcove, D.S. and Ziska, L.H. (2010a) Predicting plant invasions in an era of global change. *Trends in Ecology and Evolution* 25, 310–318.

Bradley, B.A., Wilcove, D.S. and Oppenheimer, M. (2010b) Climate change increases risk of plant invasion in the Eastern United States. *Biological Invasions* 12, 1855–1872.

Bradley, B.A., Blumenthal, D.M., Early, R., Grosholz, E.D., Lawler, J.J., Miller, L.P., Sorte, C.J.B., D'Antonio, C.M., Diez, J.M., Dukes, J.S., Ibanez, I. and Olden, J.D. (2012) Global change, global trade, and the next wave of plant invasions. *Frontiers in Ecology and the Environment* 10, 20–28.

Bridges, D.C. (1992) *Crop Losses Due to Weeds in the United States, 1992.* Weed Science Society of America, Champaign, Illinois.

Broennimann, O., Treier, U.A., Muller-Scharer, H., Thuiller, W., Peterson, A.T. and Guisan, A. (2007) Evidence of climatic niche shift during biological invasion. *Ecology Letters* 10, 701–709.

Brooks, M.L., D'Antonio, C.M., Richardson, D.M., Grace, J.B., Keeley, J.E., Ditomaso, J.M., Hobbs, R.J., Pellant, M. and Pyke, D. (2004) Effects of invasive alien plants on fire regimes. *Bioscience* 54, 677–688.

D'Antonio, C.M. and Vitousek, P.M. (1992) Biological invasions by exotic grasses, the grass fire cycle, and global change. *Annual Review of Ecology and Systematics* 23, 63–87.

Dormann, C.F. (2007) Promising the future? Global change projections of species distributions. *Basic and Applied Ecology* 8, 387–397.

Elith, J., *et al.* (2006) Novel methods improve prediction of species' distributions from occurrence data. *Ecography* 29, 129–151.

Elith, J., Kearney, M. and Phillips, S. (2010) The art of modelling range-shifting species. *Methods in Ecology and Evolution* 1, 330–342.

Estes, L.D., Bradley, B.A., Beukes, H., Hole, D.G., Lau, M., Oppenheimer, M., Schulze, R., Tadross, M.A. and Turner, W.R. (2013) Comparing mechanistic and empirical model projections of crop suitability and productivity: implications for ecological forecasting. *Global Ecology and Biogeography* 22, 1007–1018.

Farber, O. and Kadmon, R. (2003) Assessment of alternative approaches for bioclimatic modeling with special emphasis on the Mahalanobis distance. *Ecological Modelling* 160, 115–130.

Franklin, J. (2009) *Mapping Species Distributions: Spatial Inference and Prediction.* Cambridge University Press, Cambridge, UK.

Guisan, A. and Zimmermann, N.E. (2000) Predictive habitat distribution models in ecology. *Ecological Modelling* 135, 147–186.

Harris, J.A., Hobbs, R.J., Higgs, E. and Aronson, J. (2006) Ecological restoration and global climate change. *Restoration Ecology* 14, 170–176.

Hijmans, R.J. and Graham, C.H. (2006) The ability of climate envelope models to predict the effect of climate change on species distributions. *Global Change Biology* 12, 2272–2281.

Holcombe, T.R., Stohlgren, T.J. and Jarnevich, C.S. (2010) From points to forecasts: predicting invasive species habitat suitability in the near term. *Diversity* 2, 738–767.

Hulme, P.E. (2006) Beyond control: wider implications for the management of biological invasions. *Journal of Applied Ecology* 43, 835–847.

Hulme, P.E. (2009) Trade, transport and trouble: managing invasive species pathways in an era of globalization. *Journal of Applied Ecology* 46, 10–18.

Jiménez-Valverde, A., Lobo, J.M. and Hortal, J. (2008) Not as good as they seem: the importance of concepts in species distribution modelling. *Diversity and Distributions* 14, 885–890.

Jiménez-Valverde, A., Diniz, F., De Azevedo, E.B. and Borges, P.A.V. (2009) Species distribution models do not account for abundance: the case of arthropods on Terceira Island. *Annales Zoologici Fennici* 46, 451–464.

Knapp, P.A. (1996) Cheatgrass (*Bromus tectorum* L.) dominance in the Great Basin Desert – history, persistence, and influences to human activities. *Global Environmental Change: Human and Policy Dimensions* 6, 37–52.

Kremen, C., *et al.* (2008) Aligning conservation priorities across taxa in Madagascar with high-resolution planning tools. *Science* 320, 222–226.

Kulhanek, S.A., Leung, B. and Ricciardi, A. (2011) Using ecological niche models to predict the abundance and impact of invasive species: application to the common carp. *Ecological Applications* 21, 203–213.

Lobo, J.M., Jiménez-Valverde, A. and Hortal, J. (2010) The uncertain nature of absences and their importance in species distribution modelling. *Ecography* 33, 103–114.

Lockwood, J., Hoopes, M. and Marchetti, M. (2007) *Invasion Ecology*. Blackwell Publishing, Malden, Massachusetts, USA.

McDonald, A., Riha, S., Ditommaso, A. and Degaetano, A. (2009) Climate change and the geography of weed damage: analysis of US maize systems suggests the potential for significant range transformations. *Agriculture Ecosystems and Environment* 130, 131–140.

Mack, R.N. (1981) Invasions of *Bromus tectorum* L. into Western North America: an ecological chronicle. *Agro-Ecosystems* 7, 145–165.

Mack, R.N. and Erneberg, M. (2002) The United States naturalized flora: largely the product of deliberate introductions. *Annals of the Missouri Botanical Garden* 89, 176–189.

McLachlan, J.S., Hellmann, J.J. and Schwartz, M.W. (2007) A framework for debate of assisted migration in an era of climate change. *Conservation Biology* 21, 297–302.

Marvin, D.C., Bradley, B.A. and Wilcove, D.S. (2009) A novel, web-based ecosystem mapping tool using expert opinion. *Natural Areas Journal* 29, 281–292.

Moody, M.E. and Mack, R.N. (1988) Controlling the spread of plant invasions – the importance of nascent foci. *Journal of Applied Ecology* 25, 1009–1021.

Nielsen, S.E., Johnson, C.J., Heard, D.C. and Boyce, M.S. (2005) Can models of presence–absence be used to scale abundance? – Two case studies considering extremes in life history. *Ecography* 28, 197–208.

O'Donnell, J., Gallagher, R.V., Wilson, P.D., Downey, P.O., Hughes, L. and Leishman, M.R. (2012) Invasion hotspots for non-native plants in Australia under current and future climates. *Global Change Biology* 18, 617–629.

Parker, I.M., Simberloff, D., Lonsdale, W.M., Goodell, K., Wonham, M., Kareiva, P.M., Williamson, M.H., Von Holle, B., Moyle, P.B., Byers, J.E. and Goldwasser, L. (1999) Impact: toward a framework for understanding the ecological effects of invaders. *Biological Invasions* 1, 3–19.

Parker-Allie, F., Musil, C.F. and Thuiller, W. (2009) Effects of climate warming on the distributions of invasive Eurasian annual grasses: a South African perspective. *Climatic Change* 94, 87–103.

Parmesan, C. and Yohe, G. (2003) A globally coherent fingerprint of climate change impacts across natural systems. *Nature* 421, 37–42.

Pearce, J. and Ferrier, S. (2001) The practical value of modelling relative abundance of species for regional conservation planning: a case study. *Biological Conservation* 98, 33–43.

Pearson, R.G. and Dawson, T.P. (2003) Predicting the impacts of climate change on the distribution of species: are bioclimate envelope models useful? *Global Ecology and Biogeography* 12, 361–371.

Phillips, S.J., Anderson, R.P. and Schapire, R.E. (2006) Maximum entropy modeling of species geographic distributions. *Ecological Modelling* 190, 231–259.

Pysek, P. and Richardson, D.M. (2007) Traits associated with invasiveness in alien plants: Where do we stand? In: Nentwig, W. (ed.)

Biological Invasion, Ecological Studies. Springer-Verlag, Berlin and Heidelberg, pp. 97–126.

Reichard, S.H. and White, P. (2001) Horticulture as a pathway of invasive plant introductions in the United States. *Bioscience* 51, 103–113.

Richardson, D.M., *et al.* (2009) Multidimensional evaluation of managed relocation. *Proceedings of the National Academy of Sciences* 106, 9721–9724.

Root, T.L., Price, J.T., Hall, K.R., Schneider, S.H., Rosenzweig, C. and Pounds, J.A. (2003) Fingerprints of global warming on wild animals and plants. *Nature* 421, 57–60.

Stohlgren, T.J. and Schnase, J.L. (2006) Risk analysis for biological hazards: What we need to know about invasive species. *Risk Analysis* 26, 163–173.

Stohlgren, T., *et al.* (2011) Widespread plant species: natives versus aliens in our changing world. *Biological Invasions* 13, 1931–1944.

Strayer, D.L. (2009) Twenty years of zebra mussels: lessons from the mollusk that made headlines. *Frontiers in Ecology and the Environment* 7, 135–141.

Thomas, C.D., *et al.* (2004) Extinction risk from climate change. *Nature* 427, 145–148.

Tôrres, N.M., De Marco, P., Santos, T., Silveira, L., De Almeida Jácomo, A.T. and Diniz-Filho, J.A.F. (2012) Can species distribution modelling provide estimates of population densities? A case study with jaguars in the Neotropics. *Diversity and Distributions* 18, 615–627.

Tsoar, A., Allouche, O., Steinitz, O., Rotem, D. and Kadmon, R. (2007) A comparative evaluation of presence-only methods for modelling species distribution. *Diversity and Distributions* 13, 397–405.

USDA-NRCS (2012) The PLANTS Database (http://plants.usda.gov). National Plant Data Center, Baton Rouge, Lousiana.

Václavík, T. and Meentemeyer, R.K. (2012) Equilibrium or not? Modelling potential distribution of invasive species in different stages of invasion. *Diversity and Distributions* 18, 73–83.

VanDerWal, J., Shoo, L.P., Graham, C. and Williams, S.E. (2009a) Selecting pseudo-absence data for presence-only distribution modeling: How far should you stray from what you know? *Ecological Modelling* 220, 589–594.

VanDerWal, J., Shoo, L.P., Johnson, C.N. and Williams, S.E. (2009b) Abundance and the environmental niche: environmental suitability estimated from niche models predicts the upper limit of local abundance. *American Naturalist* 174, 282–291.

Vila, M. and Pujadas, J. (2001) Land-use and socio-economic correlates of plant invasions in European and North African countries. *Biological Conservation* 100, 397–401.

Welk, E. (2004) Constraints in range predictions of invasive plant species due to non-equilibrium distribution patterns: purple loosestrife (*Lythrum salicaria*) in North America. *Ecological Modelling* 179, 551–567.

Williams, J.W. and Jackson, S.T. (2007) Novel climates, no-analog communities, and ecological surprises. *Frontiers in Ecology and the Environment* 5, 475–482.

16 Identifying Climate Change as a Factor in the Establishment and Persistence of Invasive Weeds in Agricultural Crops

Antonio DiTommaso,[1] Qin Zhong[2] and David R. Clements[3]

[1]*Department of Crop and Soil Sciences, Cornell University, Ithaca, New York, USA;* [2]*Department of Ecology, College of Agriculture, South China Agricultural University, Guangzhou, China;* [3]*Biology Department, Trinity Western University, Langley, British Columbia, Canada*

Abstract

Climate change may lead to extensive shifts in invasive weed distributions by impacting their colonization and persistence in new habitats. These alterations will require an enhanced capacity to predict the future range of these species and the implementation of effective management strategies. This is particularly critical for agro-ecosystems where climate-induced changes in the composition and impact of weedy species are likely to add great uncertainty to the biodiversity, stability and productivity of these systems. In recent years, various modelling approaches have been employed to examine the current and potential distributions of invasive weeds at a range of scales in response to projected climate change. This chapter details the efforts under way to map the distributions of invasive weed species based on climate change projections for the USA. The chapter also identifies specific agricultural systems that are most vulnerable to invasion in a changing climate. It concludes with a brief discussion of the uncertainties, future directions and challenges of current modelling approaches and provides suggestions for improvement. Although bioclimatic envelope modelling has provided a useful approach for projecting potential weed species distribution changes resulting from climatic change (see Chapter 15), there is a need to incorporate more details of weed physiology and genetics, as well as habitat complexity, in these models. Given rapidly increasing anthropogenic impacts, land-use changes and climate variability, such refinements to the modelling approach could be critical to managing plant invasions under climate change.

Introduction

Invasive plants, or invasive weeds in the context of agriculture, have become a serious threat and an issue of great concern in North America and throughout the world. In cropping systems, including forage crops, an estimated 500 introduced plant species have become invasive weed pests in the USA. A quarter of the US agricultural gross national product is lost each year to invasive species. These threats may be exacerbated by climate change (Dukes and Mooney, 1999; Weltzin *et al.*, 2003; Thuiller *et al.*, 2007).

Because introduced species often become invasive due to a lack of predators in their new location, biogeographical expansion may be limited primarily by climatic factors. At regional scales, therefore, climate change may lead to extensive shifts in invasive weed distributions. Such shifts, in turn, will require new or more powerful management strategies to control their colonization and impacts on agricultural productivity (Bradley *et al.*, 2010). Climate change is also likely to create novel climate conditions that currently do not occur and which could afford new opportunities for invasive colonization. Overall, the interaction of climate with agricultural invasives adds great uncertainty to biodiversity, agro-ecosystem stability and crop productivity (Williams and Jackson, 2007; Ziska *et al.*, 2011).

This chapter describes the various efforts and modelling approaches that are being used to map current and future distributions of invasive weed species based on climate projections for the USA. The chapter also identifies specific agricultural systems that are most vulnerable to invasion with climate change. Finally, suggestions are provided to improve current modelling approaches that could help to meet the additional challenges of tracking potential range expansions and the accompanying impacts of invasive weeds under climate change.

Mapping Current Invasive Weed Distributions

A fundamental approach to understanding and managing invasive weeds is to determine their distribution and abundance. There has been an increasing effort to map and model the distribution of invasive weeds with the aim of: (i) identifying habitat suitability for their establishment and colonization; (ii) determining regions at risk of invasions; and (iii) providing insight into the abiotic factors that limit their distribution.

Regional invasive weed distribution maps include demographic information that is useful for risk assessments and also for early detection and eradication efforts. However, such maps may often be incomplete, with regional variability. Given the lack of higher-resolution weed distribution data, Bradley and Marvin (2011) initiated new regional distribution maps for ten problematic invasive plants in the western USA, created from expert knowledge obtained from weed managers in over 300 counties. Invasive plant experts identified infestations on paper, and the results were digitized into a regional geographic information system (GIS). Over 40% of requests were returned, resulting in maps with uniform spatial coverage and distribution data suitable for assessing invasive plant abundance across the western USA.

Bromus tectorum and *Cirsium arvense* were the most abundant and widespread of the species surveyed; however, the high concentrations and broad spatial extent of other invasive plants, such as *Cynoglossum officinale*, *Lepidium draba* and *Linaria dalmatica*, highlight the ongoing problems invasive species pose for western ecosystems, rangelands and croplands. These results reinforce the critical role that regional mapping efforts can play in assessing and communicating invasion risk.

With the development of GIS and the growing availability of digital landscape data, a common approach to the study of the geographic distributions of invasive species has been to describe quantitatively a number of physical conditions at the known distribution points of a species, such as climate, elevation or soil type (Austin *et al.*, 1990; Kearney and Porter, 2004). These combinations of conditions are referred to variously as the 'fundamental niche' of a species, consistent with Hutchinson's definition of the fundamental niche as the potential set of resources utilized by an organism (Hutchinson, 1957). Such information can serve as the quantitative basis for niche modelling, which, in turn, can be used to develop many types of predictive distribution models or programs. For instance, Pande *et al.* (2007) identified the environmental characteristics associated with *Celastrus orbiculatus*, a deciduous liana introduced into North America in the late 19th century, then employed map algebra combined with a logistic regression model to construct an accurate map of the current occurrence of this invasive plant in southern Illinois. However, the predictive capacity of such an approach may decline if the model is applied to novel situations; for example, under scenarios of climatic change or species introductions. Such extrapolations should be based on a mechanistic, rather than a descriptive, understanding of the niche if it is to provide maximal predictive power (Kearney and Porter, 2004).

Climate Change Projections

Climate change has had significant impacts on the distribution, abundance, phenology and physiology of a wide range of species over the past several decades (Walther *et al.*, 2002; Parmesan, 2006; Jarvis *et al.*, 2008). Overall, the negative impact of weedy species on plant community function is expected to increase with climate change (Thuiller *et al.*, 2006), including increases in invasive species distributions (Kriticos *et al.*, 2003b), shifts in the distribution ranges of invasives and rapid adaptive potential and microevolutionary change. Recent research has focused on several key questions, including the potential range expansion of weedy agricultural species, the role of phenotypic plasticity, evolutionary adaptation and genetic change in response to climate change and other influential factors. The resulting information allows a better estimation of the biogeographical areas where invasive weeds may spread in the future, thus providing assistance for planning and prioritizing those regions that should receive increased monitoring efforts or additional management resources.

Models have been adopted to predict how the distribution of invasive weeds might be altered with changes in climate. Among these, bioclimatic models are commonly used for projecting the potential range of invasive species for risk assessment, and more generally for species range shifts under the influence of climate change (Guisan and Zimmermann, 2000; Peterson, 2003). This approach defines the potential limits of species distributions using various combinations of the species' known range, physiological tolerances, biotic interactions and dispersal potential (Elith and Leathwick, 2009; Kearney and Porter, 2009; Soberón and Nakamura, 2009). The models are then used to predict the geographic locations that the species could potentially occupy at some point in the future, and risk assessments are based on assumptions about the species' ability to shift its range to these locations.

To forecast future habitats, climate variables obtained from different atmosphere–ocean general circulation models (AOGCMs) (Parry, 2007) are used to assess the target species' responses to novel environmental scenarios such as introduction into a new region, climate change or to inter-annual variation in weather patterns. Due to inconsistencies in AOGCM projections (Macadam *et al.*, 2010) and differences among bioclimatic model methodologies (Elith *et al.*, 2006; Tsoar *et al.*, 2007), some studies have developed comparisons among different methodologies and climate change projections to predict climatic habitats and then combined these results to quantify the consistency among the models.

Climate Change and Invasive Weed Ranges (Geographical Distribution)

Climate change has contributed to a large-scale shift in the geographic distribution of species over the past several decades (Walther *et al.*, 2002; Root *et al.*, 2003; Parmesan, 2006). Similar forcings of temperature and precipitation can also exert major influences on weed distribution, with negative consequences for crop production. For example, many tropical C_4 grass weeds cause serious agronomic problems in the southern USA but do not occur at troublesome levels in the US maize belt or in Canadian maize growing areas (Patterson, 1995). Global warming also could allow the northward range expansion of other exotic weeds such as *Rottboellia cochinchinensis* and *Centaurea maculosa* in North America. Potential synergistic effects between rising atmospheric CO_2 and climate change are likely to have a significant impact on the geographical distribution of weeds and on the severity of weed infestations (Cobb and Reade, 2010). Additional recorded range expansions of weeds or weed genotypes, as well as associated adaptive traits, were recently presented by Clements and DiTommaso (2011).

Several niche-based models, such as the bioclimatic prediction system (BIOCLIM; Busby, 1991), domain model (DOMAIN; Carpenter *et al.*, 1993), ecological niche factor analysis (ENFA; Hirzel *et al.*, 2002) and the genetic algorithm for the rule-set prediction modelling system (GARP; Stockwell, 1999; Stockwell *et al.*, 2006) have been developed and widely used to predict the geographic distribution range of exotic species. Most of these models have also been used as effective approaches to predict weed distribution due to warming temperatures and other climatic alterations.

CLIMEX is a process-oriented climatic niche model for simulating species biology to climate variability (Sutherst and Maywald, 1985). It has been used successfully to project the potential distribution of invasive weeds, such as *Acacia nilotica* (Kriticos *et al.*, 2003b), *Cryptostegia grandiflora* (Kriticos *et al.*, 2003a) and *Senna obtusifolia* (Dunlop *et al.*, 2006), as well as their risk assessment. For instance, the potential spread of *Panicum virgatum* was assessed with CLIMEX (Barney and DiTomaso, 2010). Model results show that >8.7 million km^2 of North America has suitable to very favourable habitat, most of which occurs east of the Rocky Mountains. The non-native range of western North America is largely unsuitable to *P. virgatum* as a crop or potential weed, unless irrigation or permanent water is available. An overall increase in suitable habitat is predicted over the coming century.

A broad range of logistic regression methods such as boosted regression trees (BRT; Leathwick *et al.*, 2006; De'ath, 2007), maximum entropy (MaxEnt; Phillips *et al.*, 2006), multivariate adaptive regression splines (MARS; Leathwick *et al.*, 2005) and generalized additive models (GAM; Yee and Mitchell, 1991; Ferrier *et al.*, 2002) have also been adopted to predict species distributions.

Among these approaches, the MaxEnt model, a general-purpose machine learning method for characterizing probability distributions from limited information (Phillips *et al.*, 2006), is commonly used. MaxEnt was recently employed to assess the current invasion risk for three of the most dominant and aggressive invasive plants in the south-eastern USA: *Pueraria lobata*, *Ligustrum sinense* and *Imperata cylindrica* (Bradley *et al.*, 2010). Results show that the current invasion risk in this region of the USA is widespread for all three species, although the invasion risk from *I. cylindrica* is more restricted to the Gulf Coast region. Climate change is likely to enable all three species to expand their ranges greatly. The ranges of *P. lobata* (Fig. 16.1) and *L. sinense* (Fig. 16.2) are projected to expand north into Ohio, Pennsylvania, New York and the New England states by 2100. Range expansion for *I. cylindrica* is expected to be as far north as Kentucky and Virginia.

Climate is one of the major factors determining the distribution of plant species at broad spatial scales. However, climate is not the only force guiding the geographic spread of a species. Studies indicate that human disturbance and environmental stress affect alien invasions by altering resource availability and making disturbed habitats more susceptible to invasion (Davis *et al.*, 2000). A model that combined climatic and human influences would likely increase the accuracy of predictions made from climate information alone. A typical example is the study of *Lonicera japonica*, a common invasive horticultural plant in the USA (Beans *et al.*, 2012). Using MaxEnt, the current distribution and the future range expansion of this species has been predicted for the USA. Findings show that *L. japonica* has the potential to spread into additional areas with suitable climates and that human influence may increase the likelihood of establishment.

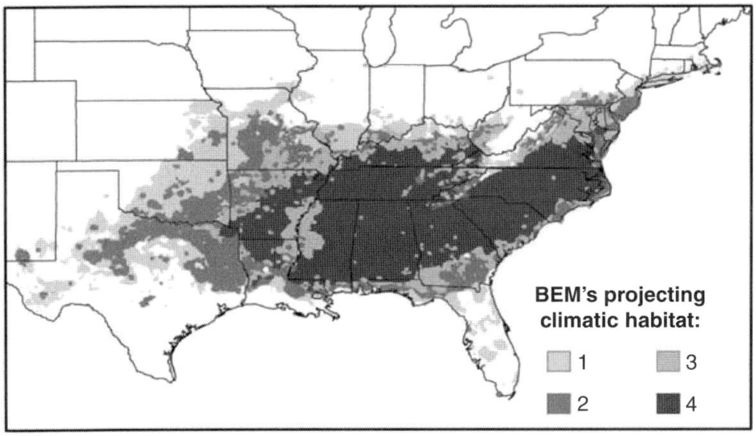

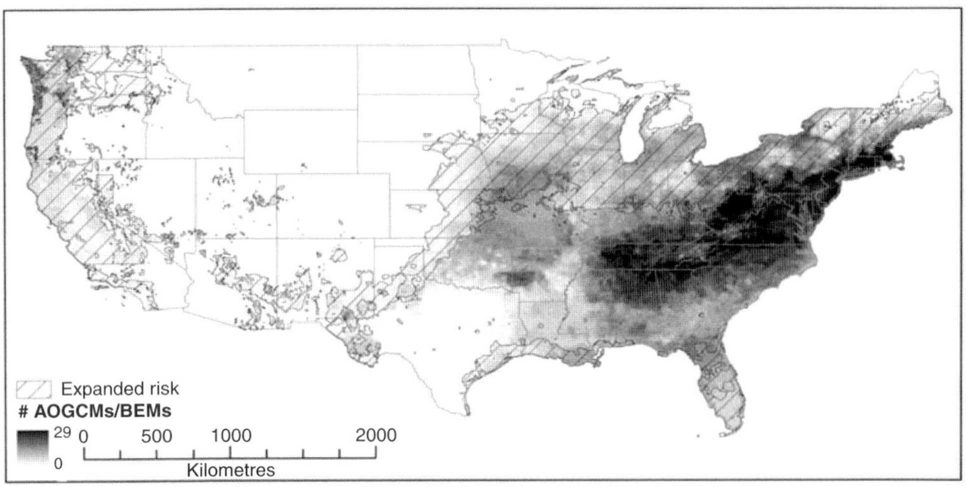

Fig. 16.1. *Pueraria lobata* climatic habitat under current climate conditions. Areas identified as climatic habitat by all four bioclimatic envelope models (BEMs) (both the constraint and the release sets of variables were used to construct bioclimatic envelopes using the MaxEnt and Mahalanobis distance methodologies) are more likely at risk from *P. lobata* invasion. (Source: Bradley, B.A., Wilcove, D.S. and Oppenheimer, M. (2010) Climate change increases risk of plant invasion in the Eastern United States. *Biological Invasions* 12, 1855–1872, Figs 1F and 6E. Reproduced with kind permission from Springer Science and Business Media.)

Climate Change and Weed Risk Assessment

Weed risk assessment and weed risk management rely in large part on predictions of habitat suitability for determining invasive threats (Crossman and Bass, 2008). The more suitable the receiving habitat is for an exotic species, the greater the likelihood that the species will be invasive.

Generally, assessment of the likely impact of climate change on the distribution of a species involves three steps: (i) modelling the effect of climate on the distribution of the species under current climatic conditions; (ii) generating future climate

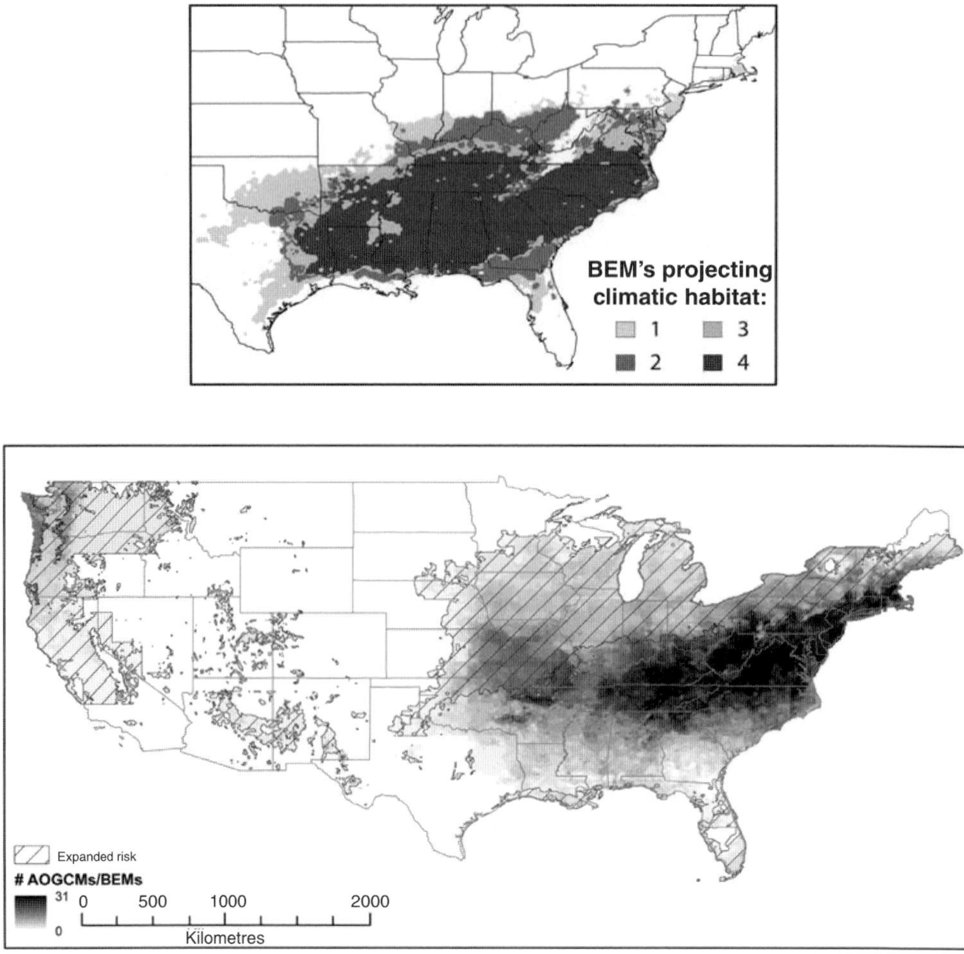

Fig. 16.2. *Ligustrum sinense* climatic habitat under current climate conditions. Areas identified as climatic habitat by all four BEMs (both the constraint and the release sets of variables were used to construct bioclimatic envelopes using the MaxEnt and Mahalanobis distance methodologies) are more likely at risk from *L. sinense* invasion. (Source: Bradley, B.A., Wilcove, D.S. and Oppenheimer, M. (2010) Climate change increases risk of plant invasion in the Eastern United States. *Biological Invasions* 12, 1855–1872, Figs 2F and 7E. Reproduced with kind permission from Springer Science and Business Media.)

scenarios; and (iii) using the future climate scenarios in the distribution model. A number of pattern-based studies have predicted potential shifts and/or expansion of invasive plant habitats using this approach (Rouget *et al.*, 2001; Thuiller *et al.*, 2005; Schussman *et al.*, 2006). For example, Kriticos *et al.* (2006) used CLIMEX models to explore the effects of climate change on three weed species in Australia with contrasting climatic requirements: *Acacia nilotica*, a dry tropical species, *Chromolaena odorata*, a wet tropical species, and *Buddleja davidii*, a temperate species. The projected future climate suitability of these target weeds was then interpreted in a GIS map.

The integration of environmental pattern and ecological process for better conservation management and planning has received much attention recently (Rouget *et al.*, 2006; Thuiller *et al.*, 2008). Crossman *et al.* (2011) estimated the impacts of climate change on exotic plant habitats and incorporated elements of dispersal to develop a management index for identifying invasive plant threats under climate change. In this study, the spatial distribution of habitat suitability was modelled at the landscape scale for multiple exotic plant species under current climate and a climate change scenario for 2030. Expert opinions of the dominant dispersal mechanisms and weed status were used to model the relative dispersal threat of each exotic plant species. These pattern and process outputs were integrated to create a multispecies management priority index in an effort to synthesize the inherently complex outputs from multiple models of multiple species. This multispecies management index combined pattern and process to identify geographic locations at greatest threat from invasion under climate change.

Effect of Climate Change on the 'Damage Niche' of Agricultural Weeds

Arable agricultural land represents a man-made habitat with a high proportion of exotic plant species (weeds) and specific disturbance regimes (Pysek *et al.*, 2005). The primary negative impact of weeds on crop growth and yield is through competition for the limited environmental resources for plant growth. McDonald *et al.* (2009) introduced the term 'damage niche' to refer to the suite of factors under which specific weed species were judged troublesome to the production of a given crop. This term integrates environment, production and competition factors to identify geographic areas where specific weed species are damaging to crop yield.

Bioclimatic niche models may provide a robust means to identify the damage niche for agricultural weeds. For instance, to explore the implications of climate change for the distribution of damaging agricultural weeds in the USA, McDonald *et al.* (2009) coupled ensemble climate change projections of annual temperature and precipitation with survey data of two troublesome weed species: *Abutilon theophrasti* in the northern USA (Fig. 16.3) and *Sorghum halepense* in the southern USA (Fig. 16.4). These results indicated that the composition of damaging weed communities might be fundamentally altered by climate change. Regions such as the north-eastern USA may prove particularly vulnerable, with emerging climate conditions favouring few weed species of present-day significance. In contrast, regions like the mid-south are likely to experience fewer shifts, even with a similar magnitude in climate change. Overall, by the end of the century in the maize belt of the USA, *S. halepense*, a predominantly southern weed species at present, may become common and damaging to maize production, with its damage niche advancing 200–600 km north of its present-day distribution.

Stratonovitch *et al.* (2012) modelled the impact of climate change on the distribution of agricultural weeds and consequent crop yield loss (or damage niche) using *Alopecurus myosuroides* in winter wheat (*Triticum aestivum*) for the UK. Due to rising CO_2 concentration and its effect on the radiation-use efficiency of wheat, weed-free wheat yields were predicted to increase. The distribution of *A. myosuroides* was predicted to remain broadly similar, with a possible northward shift in range. Overall, the competitive balance was shifted in favour of the deeper-rooted crop under climate change, particularly on sites with lighter soils, owing to more frequent and severe drought stress events. Although the damage niche of *A. myosuroides* was predicted to decrease under climate change, weeds with contrasting physiology, such as C_4 species, may be better adapted to future conditions and pose a more serious threat (Stratonovitch *et al.*, 2012).

Influences of Climate and Land-use Changes

Studies have demonstrated that climate and land use affect the abundance and

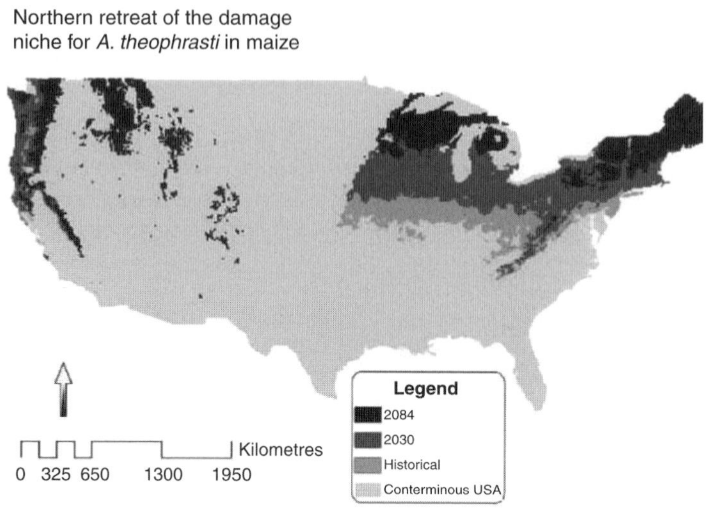

Fig. 16.3. Historical and projected distribution of the damage niche for *Abutilon theophrasti* in US maize cropping systems. Projections are for climatology centred on 2030 and 2084 under a 'business-as-usual' GHG emission scenario. Towards the end of the century, the damage niche for *A. theophrasti* may experience a poleward retreat of approximately 200–650 km north of present-day boundaries. (Source: McDonald, A., Riha, S., DiTommaso, A. and DeGaetano, A. (2009) Climate change and the geography of weed damage: analysis of US maize systems suggests the potential for significant range transformation. *Agriculture, Ecosystems and Environment* 130, 131–140, Fig. 6. Reproduced with kind permission from Elsevier Publishing.)

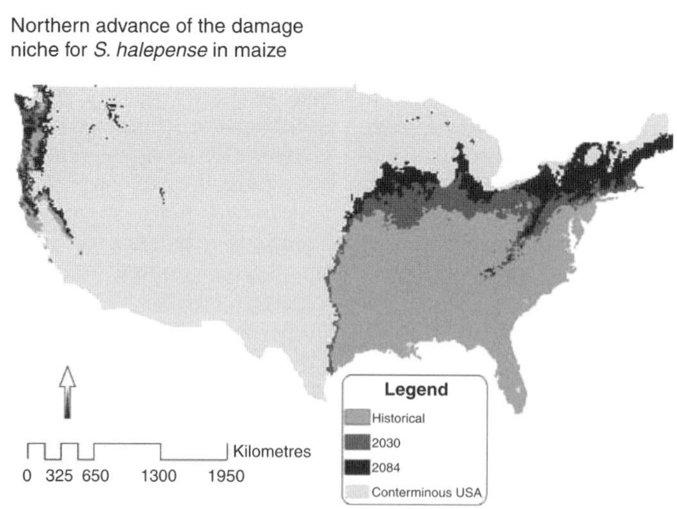

Fig. 16.4. Historical and projected distribution of the damage niche for *Sorghum halepense* in US maize cropping systems. Projections are for climatology centred on 2030 and 2084 under a 'business-as-usual' GHG emission scenario. Towards the end of the century, the damage niche for *S. halepense* may experience a poleward advance of approximately 200–600 km north of present-day boundaries. (Source: McDonald, A., Riha, S., DiTommaso, A. and DeGaetano, A. (2009) Climate change and the geography of weed damage: analysis of US maize systems suggests the potential for significant range transformation. *Agriculture, Ecosystems and Environment* 130, 131–140, Fig. 7. Reproduced with kind permission from Elsevier Publishing.)

distribution of arable invasive weed species. Climate change will alter the land use of arable habitats by encouraging the cultivation of new crop species and cultivars. The introduction of new crop species, in turn, could promote invasion by new weed species (Pysek *et al.*, 2005). These changes are expected to be most marked in regions where climatic factors constrain agricultural production currently (Mela, 1996).

We have already witnessed the northward movement of a number of weed species to Canada from the USA during the 20th century, with the increased cultivation of crops in Canada associated with the weeds in question (Clements *et al.*, 2004). Invasibility of habitats is affected by land use, which in arable habitats comprises, among other things, crop rotations and the application of herbicides and fertilizers (e.g. Haas and Streibig, 1982). Land-use practices with lower chemical inputs, such as with organic farming or integrated weed management, support richer weed species communities than conventional practices (Clements *et al.*, 1994), but may promote greater invasion risk (Török *et al.*, 2003; Kiss and Béres, 2006).

Responses to land use differ among weed species with specific traits, e.g. monocots and dicots, native and alien species. Grasses and dicots differ in herbicide tolerance and in importance as weeds in different parts of Europe (Schroeder *et al.*, 1993). Neophyte (i.e. alien species introduced after AD 1500) weed species have been found to be more numerous in maize and canola than in cereals (Pysek *et al.*, 2005). Hyvönen *et al.* (2011) compared the species richness and abundance of all species and three functional groups (dicotyledons and grasses, C_3 and C_4 species, species alien and native to Europe), as well as the occurrence of the 13 most common individual species between three land-use types (fallow, low input and conventionally cultivated) and six regions. The findings showed that fallowing and low-input cultivation, as well as warmer climate, supported greater weed species richness and abundance. Non-cultivated fallows had more weeds than cultivated fields, in most cases. No difference was found in the species richness and abundance of grasses or in the occurrence of individual species between low-input and conventionally cultivated fields. Likewise, weed species diversity was promoted by reduced tillage and/or increased crop rotation in Canadian cropping systems (Murphy *et al.*, 2006).

Uncertainties

Agricultural weeds face powerful and fluctuating selection pressures that result from changes in cropping systems, agricultural technology and pest management strategies. The success of weeds has often been attributed to a 'general-purpose genotype' (Parker *et al.*, 2003), implying a high level of phenotypic plasticity and the ability to infest a broad range of cropping systems over a range of abiotic conditions. The adaptive variation, including both phenotypic and genetic variation, of weed species undergoing geographic expansion has been observed in croplands (Clements *et al.*, 2004).

Some weed species (e.g. *Datura stramonium*) were observed to have an exceptionally high degree of phenotypic plasticity despite genetic uniformity (Núñez-Farfán, 2000). Through selection for traits that favour rapid development, including heavier seeds, longer cotyledons and earlier anthesis (Weaver *et al.*, 1985), *D. stramonium* has extended its range northward in the eastern USA and Canada. *Poa annua* plants from disturbed environments tend to produce many seeds in their first growing season, but suffer high rates of mortality, whereas plants from less disturbed but more competitive environments (e.g. pastures) behave more like biennials (Law *et al.*, 1977; Law, 1979). Besides adaptive variation in seed germination, life history and physiological traits, weedy species have been reported to be capable of relatively rapid genetic change in response to climate fluctuation (e.g. Franks *et al.*, 2007), enhancing their ability to invade new areas in response to anthropogenic ecosystem modification (Clements *et al.*, 2004). Some weeds, *Capsella bursa-pastoris* for example, exhibited greater genetic heterogeneity in more disturbed, cultivated agricultural sites than populations found growing in lawns and other habitats that were not cultivated (Bosbach and Hurka, 1981).

It should be emphasized that agriculture is continuously changing, with the introduction of new crop species and genotypes and new types of management practices. These changes represent a heightened risk of invasion through the potential for evolutionary adaptation within invading plant species. However, all these changes are largely overlooked in bioclimatic models (Clements and DiTommaso, 2011). In a changing climate, evolutionary change of weedy species does happen, and sometimes even relatively rapidly. These facts fail to meet a key assumption of bioclimatic models in which the genetic and phenotypic composition of species are assumed to be constant over space and time, thereby making it challenging to predict future weed geographic ranges. The possibility that agricultural weed populations will evolve new traits in response to emerging climate and non-climate selection pressures cannot be discounted.

Agriculture is a highly managed ecosystem strongly perturbed by human activities. Consequently, many factors other than climate substantially influence actual weedy species distributions, including crop–weed competitive interactions, seed dispersal limitations and patterns of disturbance (Guisan and Thuiller, 2005). In general, any aspect of climatic change (e.g. CO_2, temperature) that affects the growth or fitness of weeds and crops differentially will alter crop–weed competitive interactions (Patterson, 1995; Furness et *al*., 2005).

For example, Furness *et al.* (2005) demonstrated that increased UV-B radiation could alter weed–crop interactions through species-specific impacts on plant morphology. Cobb and Reade (2010) reported that the evolutionary rate in the development of herbicide resistance was affected by temperature and soil water availability.

The innate ability of weeds to adjust phenotypes quickly in response to rapid environmental change may provide a significant competitive advantage in agroecosystems. Any environmental change that differentially affects the morphology, growth or reproduction of interacting plant communities has the potential to modify the spatial extent of the damage niche (Patterson, 1995). Consequently, it will be necessary to assess the importance of key aspects of weed function, biotic interactions and disturbance regimes (such as farming practices and field management) for a more robust understanding of a species' current and extended distribution of a given invasive weed range. Some ecophysiological models incorporating complex interactions of changes in physiological constraints and crop–weed competition in the context of spatially heterogeneous soil properties were recently developed to predict the impact of climate change on the damage niche of an agricultural weed (Stratonovitch *et al.*, 2012).

Biotic interactions generally have been thought to be unimportant in determining species distributions, especially at large scales. However, Suttle *et al.* (2007) demonstrated experimentally that biotic interactions could be more important than direct climate effects for species occurrences. Overall, biotic interactions in arable cropping systems involving pathogens, pests and mutualists such as soil mycorrhizae and rhizosphere bacteria are known to affect the distribution and spatial patterns of invasive weeds.

These interactions can be considered in the context of a changing climate in complex ways (Pearson and Dawson, 2003; Guisan and Thuiller, 2005; Kearney and Porter, 2009). Although some models have integrated data on population dynamics, life-history traits and dispersal (e.g. Keith *et al.*, 2008), or have incorporated the dynamic nature of plant phenology in species distribution models (e.g. Morin and Chuine, 2005), the development of tools that explicitly and comprehensively account for these biotic interactions remains a challenge.

Although cropped fields are managed extensively, management practices are not static in time and space. New classes of herbicides, cultivars, tillage innovations, use of irrigation and seed-cleaning practices can all influence the geographic distribution of and crop damage caused by agricultural weeds. For example, Chancellor (1985) studied changes in the weed flora in the UK during 20 years of arable cropping and found that some species (i.e. *Fumaria officinalis* and *Papaver rhoeas*) moved little, whereas others (i.e. several annual *Polygonum* species) showed sudden range expansions which were interpreted as resulting from sowing contaminated crop seed. Similarly, Thomas and Leeson (2007) tracked changes in the weed flora over the 20th century in western Canada and found that certain species developed higher populations over larger areas (e.g. *Galium aparine*, *Kochia scoparia* and *Echinochloa crus-galli*), while other weed species declined. These changes were related to new agronomic practices and the invasion of alien species. Evidence also suggests that the recent introduction of glyphosate-resistant crops can alter weed community composition significantly (Harker *et al.*, 2005). Overall, bioclimatic envelope model predictions may be problematic, since these models do not account for barriers to seed dispersal, local soil or management factors, biotic interactions and the capacity of species to adapt to new environments (Davis *et al.*, 2005; Thuiller *et al.*, 2008).

Some studies have confirmed that climate change can cause extreme beneficial impacts on the spread and growth of invasive weeds. These facts are consistent with bioclimatic model predictions that assume species can disperse to any climatically suitable location (e.g. Thuiller *et al.*, 2005). However, recent studies demonstrate that not all suitable habitat patches will be occupied by an invasive species (Ehrlén and Eriksson, 2000). Species distribution depends on the availability of suitable habitats, the capacity to disperse to these habitats and the capacity of populations to persist after establishment. Ehrlén and Eriksson (2000) examined occupancy patterns in seven perennial forest herbs and found that species distribution could be characterized as the result of processes operating both among and within patches (e.g. seed size). As a result, bioclimatic models may give an unrealistically optimistic impression of species' capacities to adapt to climatic change, because dispersal and colonization rates will limit the realization of potential range shifts (Huntley *et al.*, 1995; Midgley *et al.*, 2006), as may barriers to dispersal. Conversely, assuming that species occur only at locations with favourable climate cannot provide insight into the actual range dynamics that occur as a species shifts its range. In reality, climate change is likely to be highly dynamic and species range expansions would be aided if populations could survive short periods when climate is unfavourable for them. When conditions improve, populations that survived at a range margin would produce dispersing individuals that could further extend the species range (Jackson *et al.*, 2010).

Although many different methodological approaches have been used to determine the geographic distribution of invasive species with climate change, there are several issues that may contribute to uncertainties in the models. These include (i) the quality and quantity of the data: e.g. finer-scale survey data for troublesome weeds in cropping systems are limited; (ii) data scale uncertainty: species distributional data may not match environmental data; (iii) difficulty in evaluating predicted distributions of different climate projections by means of independent validation, rather than re-substitution or data splitting (but it is difficult to locate such independent data sets); and (iv) species distribution patterns and ecological characteristics, i.e. ecological properties of the target species might affect model performance.

Future Directions and Challenges

Agricultural weeds, especially invasive weed species, are currently receiving considerable attention in connection with global climate change. There is mounting evidence of increases in their distribution, expansion of ranges, changes in herbicide efficacy, increasing economic impacts from crop–weed competition and decreases in agro-ecosystem biodiversity.

Minimizing the negative impacts on cropping systems from these consequences requires effective management strategies that address system vulnerability and potential responses. There are many different methodological approaches currently available for assessing the potential effects of climate change on invasive weed biology through the use of quantitative scenarios and visualizations. These methods and techniques have been widely used and have provided valuable insights into ecological and evolutionary processes associated with current and future anthropogenic climate change.

Among these available tools, effective demographic models of species distribution and impact are essential for the management of invasive weeds. Recently, there have been numerous examples and cases in which these models, ranging from spatially explicit mechanistic models (e.g. Hill *et al.*, 2001) to physiologically based (e.g. Walther *et al.*, 2002) and correlative bioclimatic envelope models (e.g. Lawler *et al.*, 2009) have been applied in weedy species climate modelling at varying geographic and taxonomic scales (Heikkinen *et al.*, 2007).

An increasing number of studies have sought to improve the robustness or reliability of the models used, whereas other studies have aimed to develop more realistic projections of species' responses to climatic change. For instance, Huntley *et al.* (2010) envisaged a unified framework for an integrated model that would examine species' spatial responses and abundance patterns in the context of climatic change. This integrated model is comprised of four modules: (i) climatic suitability; (ii) habitat availability/suitability; (iii) population dynamics; and (iv) dispersal. Unfortunately for most species, insufficient data are available and various simplifications are necessary. However, even simple integrated models (e.g. Anderson *et al.*, 2009) may represent an important advance in predictive capacity.

Some researchers have proposed more comprehensive frameworks (Barney and Whitlow, 2008; Catford *et al.*, 2009; Foxcroft *et al.*, 2011). Perkins *et al.* (2011) presented an invasion triangle that embodied the multifactor process of invasion, in which three major processes that resulted in invasion: (i) attributes of the potential invader; (ii) biotic characteristics of a potentially invaded site; and (iii) environmental conditions of the site, were incorporated as the three sides of a triangle (Fig. 16.5). While obviously incomplete, these conceptual frameworks represent the developmental trend of modelling methodology for weedy species and hold great potential to be applied in both research and management contexts.

Future research on invasive weeds should focus on improving model capacity in simulating important processes such as seed dispersal, soil characteristics, weed traits associated with invasion, crop–weed competitive dynamics, agronomic management, evolutionary potential and capacity to adapt to new environments. A major challenge is integrating these associated variables into process-based crop growth and development models. Model improvement will facilitate the prediction of damage niche changes, but could also be used to test key hypotheses regarding species' interactions and physiological constraints (Heikkinen *et al.*, 2007; Preston *et al.*, 2008) and the probability of establishment of invasive agricultural weeds under different climate change scenarios. Model improvements could also contribute to better predictive capacity of future invasive weed ranges and provide more contextualized information to policy makers and crop managers.

To establish these integrated models, additional knowledge relative to their limitations and uncertainties is required. In addition, abiotic information (e.g. processes associated with niche shifts), as well as biological inputs (e.g. detailed data sets of individual species), will be essential. In this regard, the development of integrated models requires careful balancing of model complexity with data availability (Stratonovitch *et al.*, 2012).

There is also a clear and immediate need to address the adaptive capacity of invasive agricultural weeds in the context of climate change (Clements and DiTommaso, 2011). As suggested by Stratonovitch *et al.* (2012),

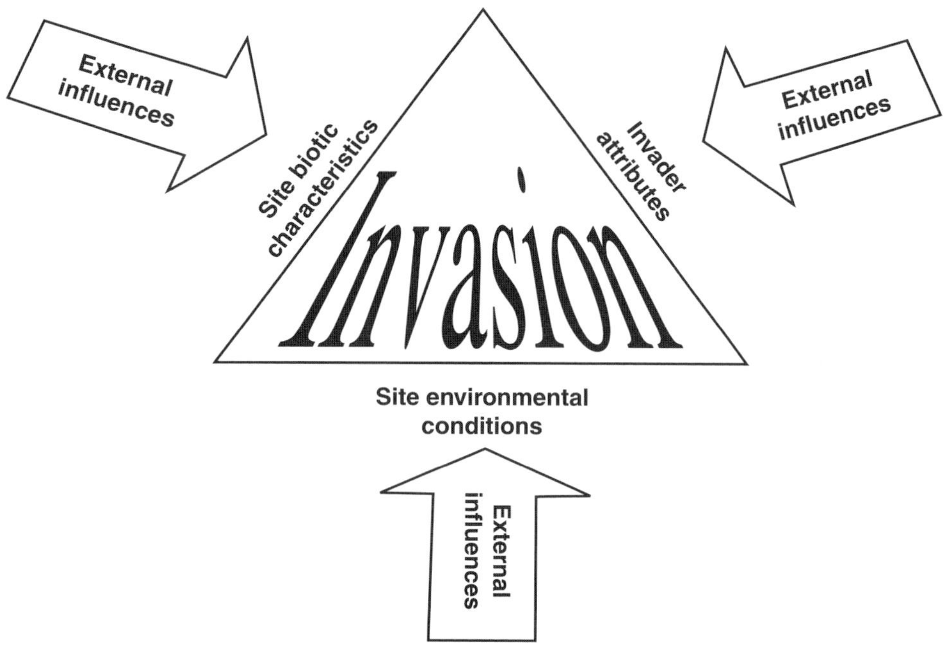

Fig. 16.5. Conceptual diagram of the invasion triangle. Invader attributes refer to inherent characteristics of the introduced species (including competitive ability, novel weapons, evolution of invasiveness and ecosystem engineering). Site biotic characteristics refer to the intrinsic biological characteristics of a site that influence its vulnerability to invasion (including diversity and the presence of potential enemies and mutualists). Site environmental conditions refer to the environmental or physical conditions of a site that influence its invasibility (including the amount of unused resources and habitat suitability). Invader attributes, site biotic characteristics and site environmental conditions all interact to determine the potential for invasion. The external influences arrows indicate that all sides can be changed by outside factors, such as introduction effort, global environmental change or changes in land use and disturbance regime. (Source: Perkins, L.B., Leger, E.A. and Nowak, R.S. (2011) Invasion triangle: an organizational framework for species invasion. *Ecology and Evolution* 1, 610–625, Fig. 1. Reproduced with kind permission from the authors.)

the propensity of weeds to adapt to shifting environmental circumstances should be accounted for in models to avoid underestimating possible shifts in their distribution and impact. The incorporation of data on intraspecific variability within weed species (particularly at the leading edge of its distribution) will, therefore, need to be an important component of future models.

To predict the potential spread of invasive weeds in arable systems, it is crucial to recognize the significance of human adaptive responses to climate change and incorporate them with climate variables. Agricultural adaptation to climate change, including changes in the distribution of lands under cultivation and the alteration of crop species and field management practices, may also contribute to range expansion of agricultural weeds. Identifying the potential magnitude of these human adaptive responses and developing combined predictive models may increase significantly our predictive ability in areas with high human impacts.

Weed evolution is an important consideration in developing predictions of range expansion under climate change (Clements and DiTommaso, 2011). To assess the value of weed evolution in the development of predictive models of invasive species spread in a changing climate, specific traits (phenotypically or genetically based) which would likely make such species successful in novel environments must be evaluated carefully (Clements and DiTommaso, 2012) and be incorporated into the models. The other important consideration is how the range expansion of a species' distribution edge influences selection pressures in a changing climate (Clements and DiTommaso, 2012). Although the potential mechanisms of weed evolution involve many different traits, thereby making predicting the exact course of evolution difficult, an ensemble approach to modelling species' distribution changes should be performed to improve estimates of future weed species range expansion and invasiveness. This ensemble may involve experiments examining the characteristics and population dynamics of a given weed and how they would be affected by climate change. Concomitantly, studies predicting the spread of invasive weeds using CLIMEX-type models that do not incorporate evolutionary change but can provide a relatively rapid analysis of the general trend in range expansions of these weeds could also be pursued.

Summary

Bioclimatic envelope models have been the main approach used to project potential weed species' distribution changes resulting from climatic change in the past two decades. Advances in these models have enhanced our ability to determine which agro-ecosystems may be most vulnerable to future changes in climate at regional scales. Despite ongoing debate about the underlying assumptions, inherent simplifications, critical limitations and reliability of these modelling techniques, an increasing number of studies have demonstrated the general robustness of these models. Importantly, multiple efforts are under way to address these deficiencies and increase confidence in model results. These efforts include clarification and further illustration of the niche concept, the proposing of comprehensive frameworks and integrated models to outline strategies for developing predictive models for various invasion scenarios and to evaluate the predictive accuracy of the models. Efforts are also underway to resolve practical issues by relating collected data sets or field observations to environmental predictor variables. To make more reliable and robust predictions of weed species distribution and associated changes in abundance patterns, important parameters, such as dispersal, competition and other biotic interactions involving pests and pathogens, and cropping management practices are being considered for incorporation in the models. Given the rapidly increasing anthropogenic impacts, land-use changes and climate variability, agroecosystem responses to these effects are more complex and difficult to predict with high confidence. As a way forward, it is critical that weed species distribution models incorporate known physiological and ecological processes in a broader systems context. Equally important will be the development of ensemble analytical approaches for parameterization, model selection and evaluation.

References

Anderson, B.J., Akcakaya, H.R., Araujo, M.B., Fordham, D.A., Martinez-Meyer, E., Thuiller, W. and Brook, B.W. (2009) Dynamics of range margins for metapopulations under climate change. *Proceedings of the Royal Society B: Biological Sciences* 276, 1415–1420.

Austin, M.P., Nicholls, A.O. and Margules, C.R. (1990) Measurement of the realized qualitative niche: environmental niches of five eucalyptus species. *Ecological Monographs* 60, 161–177.

Barney, J. and Whitlow, T. (2008) A unifying framework for biological invasions: the state factor model. *Biological Invasions* 10, 259–272.

Barney, J.N. and DiTomaso, J.M. (2010) Bioclimatic predictions of habitat suitability for the biofuel switchgrass in North America under current and future climate scenarios. *Biomass and Bioenergy* 34, 124–133.

Beans, C., Kilkenny, F. and Galloway, L. (2012) Climate suitability and human influences combined explain the range expansion of an invasive horticultural plant. *Biological Invasions* 14, 2067–2078.

Bosbach, K. and Hurka, H. (1981) Biosystematic studies on *Capsella bursa-pastoris* (*Brassicaceae*): enzyme polymorphism in natural populations. *Plant Systematics and Evolution* 137, 73–94.

Bradley, B., Wilcove, D. and Oppenheimer, M. (2010) Climate change increases risk of plant invasion in the Eastern United States. *Biological Invasions* 12, 1855–1872.

Bradley, B.A. and Marvin, D.C. (2011) Using expert knowledge to satisfy data needs: mapping invasive plant distributions in the Western United States. *Western North American Naturalist* 71, 302–315.

Broennimann, O., Treier, U.A., Müller-Schärer, H., Thuiller, W., Peterson, A.T. and Guisan, A. (2007) Evidence of climatic niche shift during biological invasion. *Ecology Letters* 10, 701-709.

Busby, J.R. (1991) Bioclim – a bioclimatic analysis and prediction system. In: Margules, C.R. and Austin, M.P. (eds) *Nature Conservation: Cost Effective Biology Survey and Data Analysis.* CSIRO, Australia, pp. 64–68.

Carpenter, G., Gillison, A.N. and Winter, J. (1993) DOMAIN: a flexible modelling procedure for mapping potential distributions of plants and animals. *Biodiversity and Conservation* 2, 667–680.

Catford, J.A., Jansson, R. and Nilsson, C. (2009) Reducing redundancy in invasion ecology by integrating hypotheses into a single theoretical framework. *Diversity and Distributions* 15, 22–40.

Chancellor, R.J. (1985) Changes in the weed flora of an arable field cultivated for 20 years. *Journal of Applied Ecology* 22, 491–501.

Clements, D.R. and DiTommaso, A. (2011) Climate change and weed adaptation: can evolution of invasive plants lead to greater range expansion than forecasted? *Weed Research* 51, 227–240.

Clements, D.R. and DiTommaso, A. (2012) Predicting weed invasion in Canada under climate change: evaluating evolutionary potential. *Canadian Journal of Plant Science* 92, 1013–1020.

Clements, D.R., Weise, S.F. and Swanton, C.J. (1994) Integrated weed management and weed species diversity. *Phytoprotection* 75, 1–18.

Clements, D.R., DiTommaso, A., Jordan, N., Booth, B.D., Cardina, J., Doohan, D., Mohler, C.L., Murphy, S.D. and Swanton, C.J. (2004) Adaptability of plants invading North American cropland. *Agriculture, Ecosystems and Environment* 104, 379–398.

Cobb, A.H. and Reade, J.P.H. (2010) *Herbicides and Plant Physiology*, 2nd edn. Wiley-Blackwell, Oxford, UK.

Crossman, N.D. and Bass, D.A. (2008) Application of common predictive habitat techniques for post-border weed risk management. *Diversity and Distributions* 14, 213–224.

Crossman, N.D., Bryan, B.A. and Cooke, D.A. (2011) An invasive plant and climate change threat index for weed risk management: integrating habitat distribution pattern and dispersal process. *Ecological Indicators* 11, 183–198.

Davis, M.A., Grime, J.P. and Thompson, K. (2000) Fluctuating resources in plant communities: a general theory of invasibility. *Journal of Ecology* 88, 528–534.

Davis, M.B., Shaw, R.G. and Etterson, J.R. (2005) Evolutionary responses to changing climate. *Ecology* 86, 1704–1714.

De'ath, G. (2007) Boosted trees for ecological modeling and prediction. *Ecology* 88, 243–251.

Dukes, J.S. and Mooney, H.A. (1999) Does global change increase the success of biological invaders? *Trends in Ecology and Evolution* 14, 135–139.

Dunlop, E.A., Wilson, J.C. and Mackey, A.P. (2006) The potential geographic distribution of the invasive weed *Senna obtusifolia* in Australia. *Weed Research* 46, 404–413.

Ehrlén, J. and Eriksson, O. (2000) Dispersal limitation and patch occupancy in forest herbs. *Ecology* 81, 1667–1674.

Elith, J. and Leathwick, J.R. (2009) Species distribution models: ecological explanation and prediction across space and time. *Annual Review of Ecology, Evolution, and Systematics* 40, 677–697.

Elith, J., *et al.* (2006) Novel methods improve prediction of species' distributions from occurrence data. *Ecography* 29, 129–151.

Ferrier, S., Watson, G., Pearce, J. and Drielsma, M. (2002) Extended statistical approaches to modelling spatial pattern in biodiversity in northeast New South Wales. I. Species-level modelling. *Biodiversity and Conservation* 11, 2275–2307.

Foxcroft, L.C., Pickett, S.T.A. and Cadenasso, M.L. (2011) Expanding the conceptual frameworks of plant invasion ecology. *Perspectives in Plant Ecology, Evolution and Systematics* 13, 89–100.

Franks, S.J., Sim, S. and Weis, A.E. (2007) Rapid evolution of flowering time by an annual plant in response to a climate fluctuation. *Proceedings of the National Academy of Sciences* 104, 1278–1282.

Furness, N.H., Jolliffe, P.A. and Upadhyaya, M.K. (2005) Ultraviolet-B radiation and plant competition: experimental approaches and underlying mechanisms. *Photochemistry and Photobiology* 81, 1026–1037.

Guisan, A. and Thuiller, W. (2005) Predicting species distribution: offering more than simple habitat models. *Ecology Letters* 8, 993–1009.

Guisan, A. and Zimmermann, N.E. (2000) Predictive habitat distribution models in ecology. *Ecological Modelling* 135, 147–186.

Haas, H. and Streibig, J.C. (1982) Changing patterns of weed distribution as a result of herbicide use and other agronomic factors. In: Le Baron, M.H. and Gressel, J. (eds) *Herbicide Resistance in Plants*. Wiley, New York.

Harker, K.N., Clayton, G.W., Blackshaw, R.E., O'Donovan, J.T., Lupwayi, N.Z., Johnson, E.N., Gan, Y., Zentner, R.P., Lafond, G.P. and Irvine, R.B. (2005) Glyphosate-resistant spring wheat production system effects on weed communities. *Weed Science* 53, 451–464.

Heikkinen, R.K., Luoto, M., Virkkala, R., Pearson, R.G. and Körber, J.-H. (2007) Biotic interactions improve prediction of boreal bird distributions at macro-scales. *Global Ecology and Biogeography* 16, 754–763.

Hill, J.K., Collingham, Y.C., Thomas, C.D., Blakeley, D.S., Fox, R., Moss, D. and Huntley, B. (2001) Impacts of landscape structure on butterfly range expansion. *Ecology Letters* 4, 313–321.

Hirzel, A.H., Hausser, J., Chessel, D. and Perrin, N. (2002) Ecological-niche factor analysis: how to compute habitat-suitability maps without absence data? *Ecology* 83, 2027–2036.

Huntley, B., Berry, P.M., Cramer, W. and McDonald, A.P. (1995) Modelling present and potential future ranges of some European higher plants using climate response surfaces. *Journal of Biogeography* 22, 967–1001.

Huntley, B., Barnard, P., Altwegg, R., Chambers, L., Coetzee, B.W.T., Gibson, L., Hockey, P.A.R., Hole, D.G., Midgley, G.F., Underhill, L.G. and Willis, S.G. (2010) Beyond bioclimatic envelopes: dynamic species' range and abundance modelling in the context of climatic change. *Ecography* 33, 621–626.

Hutchinson, G.E. (1957) Concluding remarks. *Cold Spring Harbor Symposia on Quantitative Biology* 22, 415–427.

Hyvönen, T., Glemnitz, M., Radics, L. and Hoffmann, J. (2011) Impact of climate and land use type on the distribution of Finnish casual arable weeds in Europe. *Weed Research* 51, 201–208.

Jackson, S.T., Betancourt, J.L., Booth, R.K. and Gray, S.T. (2010) Ecology and the ratchet of events: climate variability, niche dimensions, and species distributions. *Proceedings of the National Academy of Sciences* 106, 19685–19692.

Jarvis, A., Lane, A. and Hijmans, R.J. (2008) The effect of climate change on crop wild relatives. *Agriculture, Ecosystems and Environment* 126, 13–23.

Kearney, M. and Porter, W.P. (2004) Mapping the fundamental niche: physiology, climate, and the distribution of a nocturnal lizard. *Ecology* 85, 3119–3131.

Kearney, M. and Porter, W. (2009) Mechanistic niche modelling: combining physiological and spatial data to predict species' ranges. *Ecology Letters* 12, 334–350.

Keith, D.A., Akçakaya, H.R., Thuiller, W., Midgley, G.F., Pearson, R.G., Phillips, S.J., Regan, H.M., Araújo, M.B. and Rebelo, T.G. (2008) Predicting extinction risks under climate change: coupling stochastic population models with dynamic bioclimatic habitat models. *Biology Letters* 4, 560–563.

Kiss, L. and Béres, I. (2006) Anthropogenic factors behind the recent population expansion of common ragweed (*Ambrosia artemisiifolia* L.) in Eastern Europe: is there a correlation with political transitions? *Journal of Biogeography* 33, 2156–2157.

Kriticos, D.J., Sutherst, R.W., Brown, J.R., Adkins, S.W. and Maywald, G.F. (2003a) Climate change and biotic invasions: a case history of a tropical woody vine. *Biological Invasions* 5, 147–165.

Kriticos, D.J., Sutherst, R.W., Brown, J.R., Adkins, S.W. and Maywald, G.F. (2003b) Climate change and the potential distribution of an invasive alien plant: *Acacia nilotica* ssp. *indica* in Australia. *Journal of Applied Ecology* 40, 111–124.

Kriticos, D.J., Alexander, N.S. and Kolomeitz, S.M. (2006) Predicting the potential geographic distribution of weeds in 2080. In: Preston, C., Watts, J.H. and Crossman, N.D. (eds) *Fifteenth Australian Weeds Conference*. Weed Science Society of Victoria, Melbourne, Australia, pp. 27–34.

Law, R. (1979) The cost of reproduction in annual meadow grass. *American Naturalist* 113, 3–16.

Law, R., Bradshaw, A.D. and Putwain, P.D. (1977) Life-history variation in *Poa annua*. *Evolution* 31, 233–246.

Lawler, J.J., Shafer, S.L., White, D., Kareiva, P., Maurer, E.P., Blaustein, A.R. and Bartlein, P.J. (2009) Projected climate-induced faunal change in the Western Hemisphere. *Ecology* 90, 588–597.

Leathwick, J.R., Rowe, D., Richardson, J., Elith, J. and Hastie, T. (2005) Using multivariate adaptive regression splines to predict the distributions of New Zealand's freshwater diadromous fish. *Freshwater Biology* 50, 2034–2052.

Leathwick, J.R., Elith, J., Francis, M.P., Hastie, T. and Taylor, P. (2006) Variation in demersal fish species richness in the oceans surrounding New Zealand: an analysis using boosted regression trees. *Marine Ecology Progress Series* 321, 267–281.

Macadam, I., Pitman, A.J., Whetton, P.H. and Abramowitz, G. (2010) Ranking climate models by performance using actual values and anomalies: implications for climate change impact assessments. *Geophysical Research Letters* 37, L16704.

McDonald, A., Riha, S., DiTommaso, A. and DeGaetano, A. (2009) Climate change and the geography of weed damage: analysis of U.S. maize systems suggests the potential for significant range transformations. *Agriculture, Ecosystems and Environment* 130, 131–140.

Mela, T.J.N. (1996) Northern agriculture: constraints and responses to global climate change. *Agricultural and Food Science in Finland* 5, 229–234.

Midgley, G.F., Hughes, G.O., Thuiller, W. and Rebelo, A.G. (2006) Migration rate limitations on climate change-induced range shifts in Cape Proteaceae. *Diversity and Distributions* 12, 555–562.

Morin, X. and Chuine, I. (2005) Sensitivity analysis of the tree distribution model Phenofit to climatic input characteristics: implications for climate impact assessment. *Global Change Biology* 11, 1493–1503.

Murphy, S.D., Clements, D.R., Belaoussoff, S., Kevan, P.G. and Swanton, C.J. (2006) Promotion of weed species diversity and reduction of weed seedbanks with conservation tillage and crop rotation. *Weed Science* 54, 69–77.

Núñez-Farfán, J.F.J. (2000) Evolutionary ecology of *Datura stramonium*: genetic variation and costs for tolerance to defoliation. *Evolution* 54, 789–797.

Pande, A., Williams, C., Lant, C. and Gibson, D. (2007) Using map algebra to determine the mesoscale distribution of invasive plants: the case of *Celastrus orbiculatus* in Southern Illinois, USA. *Biological Invasions* 9, 419–431.

Parker, I.M., Rodriguez, J. and Loik, M.E. (2003) An evolutionary approach to understanding the biology of invasions: local adaptation and general-purpose genotypes in the weed *Verbascum thapsus*. *Conservation Biology* 17, 59–72.

Parmesan, C. (2006) Ecological and evolutionary responses to recent climate change. *Annual Review of Ecology, Evolution, and Systematics* 37, 637–669.

Parry, M.L. (2007) *Climate Change 2007: Impacts, Adaptation and Vulnerability: Contribution of Working Group II to the Fourth Assessment Report of the Intergovernmental Panel on Climate Change*. Cambridge University Press, Cambridge, UK, and New York.

Patterson, D.T. (1995) Weeds in a changing climate. *Weed Science* 43, 685–701.

Pearson, R.G. and Dawson, T.P. (2003) Predicting the impacts of climate change on the distribution of species: are bioclimate envelope models useful? *Global Ecology and Biogeography* 12, 361–371.

Perkins, L.B., Leger, E.A. and Nowak, R.S. (2011) Invasion triangle: an organizational framework for species invasion. *Ecology and Evolution* 1, 610–625.

Peterson, A.T. (2003) Predicting the geography of species' invasions via ecological niche modeling. *Quarterly Review of Biology* 78, 419–433.

Phillips, S.J., Anderson, R.P. and Schapire, R.E. (2006) Maximum entropy modeling of species geographic distributions. *Ecological Modelling* 190, 231–259.

Preston, K.L., Rotenberry, J.T., Redak, R.A. and Allen, M.F. (2008) Habitat shifts of endangered species under altered climate conditions: importance of biotic interactions. *Global Change Biology* 14, 2501–2515.

Pyšek, P., Jarošík, V., Chytrý, M., Kropáč, Z., Tichý, L. and Wild, J. (2005) Alien plants in temperate weed communities: prehistoric and recent invaders occupy different habitats. *Ecology* 86, 772–785.

Root, T.L., Price, J.T., Hall, K.R., Schneider, S.H., Rosenzweig, C. and Pounds, J.A. (2003) Fingerprints of global warming on wild animals and plants. *Nature* 421, 57–60.

Rouget, M., Richardson, D.M., Milton, S.J. and Polakow, D. (2001) Predicting invasion dynamics of four alien *Pinus* species in a highly fragmented semi-arid shrubland in South Africa. *Plant Ecology* 152, 79–92.

Rouget, M., Cowling, R.M., Lombard, A.T., Knight, A.T. and Kerley, G.I.H. (2006) Designing large-scale conservation corridors for pattern and process. *Conservation Biology* 20, 549–561.

Schroeder, D., Schaerer, H.M. and Stinson, C.S.A. (1993) A European weed survey in 10 major crop systems to identify targets for biological control. *Weed Research* 33, 449–458.

Schussman, H., Geiger, E., Mau-Crimmins, T. and Ward, J. (2006) Spread and current potential distribution of an alien grass, *Eragrostis lehmanniana* Nees, in the southwestern USA: comparing historical data and ecological niche models. *Diversity and Distributions* 12, 582–592.

Soberón, J. and Nakamura, M. (2009) Niches and distributional areas: concepts, methods, and assumptions. *Proceedings of the National Academy of Sciences* 106, 19644–19650.

Stockwell, D. (1999) The GARP modelling system: problems and solutions to automated spatial prediction. *International Journal of Geographical Information Science* 13, 143–158.

Stockwell, D.R.B., Beach, J.H., Stewart, A., Vorontsov, G., Vieglais, D. and Pereira, R.S. (2006) The use of the GARP genetic algorithm and internet grid computing in the Lifemapper world atlas of species biodiversity. *Ecological Modelling* 195, 139–145.

Stratonovitch, P., Semenov, M.A. and Storkey, J. (2012) A process-based approach to modelling impacts of climate change on the damage niche of an agricultural weed. *Global Change Biology* 18, 2071–2080.

Sutherst, R.W. and Maywald, G.F. (1985) A computerised system for matching climates in ecology. *Agriculture, Ecosystems and Environment* 13, 281–299.

Suttle, K.B., Thomsen, M.A. and Power, M.E. (2007) Species interactions reverse grassland responses to changing climate. *Science* 315, 640–642.

Thomas, A.G. and Leeson, J.Y. (2007) Tracking long-term changes in the arable weed flora of Canada. In: Clements, D.R. and Darbyshire, S.J. (eds) *Invasive Plants: Inventories, Strategies and Action.* Canadian Weed Science Society, Victoria, Canada, pp. 43–69.

Thuiller, W., Richardson, D.M., Pyšek, P., Midgley, G.F., Hughes, G.O. and Rouget, M. (2005) Niche-based modelling as a tool for predicting the risk of alien plant invasions at a global scale. *Global Change Biology* 11, 2234–2250.

Thuiller, W., Lavorel, S., Sykes, M.T. and Araújo, M.B. (2006) Using niche-based modelling to assess the impact of climate change on tree functional diversity in Europe. *Diversity and Distributions* 12, 49–60.

Thuiller, W., Richardson, D.M. and Midgley, G.F. (2007) Will climate change promote alien plant invasions? *Biological Invasions* 193, 197–211.

Thuiller, W., Albert, C., Araujo, M.B., Berry, P.M., Paterson, J., Cabeza, M., Guisan, A., Hickler, T., Sykes, M.T., Midgley, G.F., Schurr, F.M. and Zimmermann, N.E. (2008) Predicting global change impacts on plant species' distributions: future challenges. *Perspectives in Plant Ecology, Evolution and Systematics* 9, 137–152.

Török, K., Botta-Dukát, Z., Dancza, I., Németh, I., Kiss, J., Mihály, B. and Magyar, D. (2003) Invasion gateways and corridors in the Carpathian basin: biological invasions in Hungary. *Biological Invasions* 5, 349–356.

Tsoar, A., Allouche, O., Steinitz, O., Rotem, D. and Kadmon, R. (2007) A comparative evaluation of presence-only methods for modelling species distribution. *Diversity and Distributions* 13, 397–405.

Walther, G.R., Post, E., Convey, P., Menzel, A., Parmesan, C., Beebee, T.J., Fromentin, J.M., Hoegh-Guldberg, O. and Bairlein, F. (2002) Ecological responses to recent climate change. *Nature* 416, 389–395.

Weaver, S.E., Dirks, V.A. and Warwick, S.I. (1985) Variation and climatic adaptation in northern populations of *Datura stramonium*. *Canadian Journal of Botany* 63, 1303–1308.

Weltzin, J.F., Belote, R.T. and Sanders, N.J. (2003) Biological invaders in a greenhouse world: will elevated CO_2 fuel plant invasions? *Frontiers in Ecology and the Environment* 1, 146–153.

Williams, J.W. and Jackson, S.T. (2007) Novel climates, no-analog communities, and ecological surprises. *Frontiers in Ecology and the Environment* 5, 475–482.

Yee, T.W. and Mitchell, N.D. (1991) Generalized additive models in plant ecology. *Journal of Vegetation Science* 2, 587–602.

Ziska, L., Blumenthal, D., Runion, G., Hunt, E. and Diaz-Soltero, H. (2011) Invasive species and climate change: an agronomic perspective. *Climatic Change* 105, 13–42.

17 Assessing and Managing the Impact of Climate Change on Invasive Species: The PBDM Approach

Andrew Paul Gutierrez[1,2] and Luigi Ponti[1,3]

[1]*Center for the Analysis of Sustainable Agricultural Systems (CASAS Global NGO), Kensington, California, USA;* [2]*Division of Ecosystem Science, College of Natural Resources, University of California, Berkeley, California, USA;* [3]*Laboratorio Gestione Sostenibile degli Agro-Ecosistemi (UTAGRI-ECO), Agenzia Nazionale per le Nuove Tecnologie, l'Energia e lo Sviluppo Economico Sostenibile (ENEA), Centro Ricerche Casaccia, Rome, Italy*

Abstract

Assessing the geographic distribution and abundance of invasive species is critical for developing sound management and/or eradication policies. Ecological niche modelling approaches (ENMs) that make implicit assumptions about biology and mathematics are commonly used to predict the potential distribution of invasive species based on their recorded distribution. An alternative approach is physiologically based demographic modelling (PBDM), which explicitly incorporates the mathematics and the observed biology, including trophic interactions, to predict the temporal phenology and dynamics of a species across wide geographic areas. The invasive weed, yellow starthistle (YST) (*Centaurea solstitialis*), and its interactions with annual grasses and herbivorous biological control agents is used to demonstrate the utility of the PBDM approach for analysing complex invasive species problems. The PBDM predicts the distribution and relative abundance of YST accurately across the western USA, and the results are used to assess the effects of temperature, rainfall, competition from grasses and the efficacy of biocontrol efforts. Such an effort could also be used to include the direct effects of rising carbon dioxide on YST biology. A bioeconomic model could be developed to show how the YST PBDM analysis can also be used to assess the biological and economic effects of climate change on YST infestation levels regionally. Finally, this chapter discusses the need for a unified system for assessing invasive species problems at the field, regional and global levels, with the goal of enhancing the development of efficacious policy and management decisions.

Introduction

A critical step to assessing biological invasions is to predict their potential geographic distribution and dynamics in time and space. This capacity is a critical component in developing strategies for their management, including eradication. To do this, we must recognize that species are affected not only by climatic change but also by trophic relationships (Davis *et al.*, 1998). However, assessing the impact of these factors on invasive species biology has proven difficult (Gallien *et al.*, 2010).

The most commonly used methods to characterize the ecological niche of invasive species fall under the ambit of *ecological niche modelling* (ENMs) approaches that attempt to characterize the distribution of species using aggregate weather and other factors in areas where the species has been recorded. ENMs are often used to predict the potential range of invasive species following introduction (see Chapter 15 this volume.). ENMs may be statistical models (e.g. generalized linear models, generalized additive models, generalized boosted models, genetic algorithm for rule-set prediction or principal components analysis; see Estrada-Peña, 2008), physiological index models (e.g. CLIMEX; Sutherst and Maywald, 1985; Sutherst *et al.*, 2007), models based on information theory (MaxEnt; Phillips *et al.*, 2006; Phillips and Dudík, 2008) and other methods (see Elith and Leathwick, 2009). Elith *et al.* (2011) provide a useful exposé of the MaxEnt approach.

ENMs assume that the current geographic distribution of a species is the best indicator of its climatic requirements, that the distribution is in equilibrium with current climate and that climate niche conservatism will be maintained in both space and time (Beaumont *et al.*, 2009). But there are several deficiencies: incorporating trophic interactions in ENMs is difficult (Davis *et al.*, 1998); the assumed native range may be in error (e.g. the species may have been misidentified); a single species may be composed of different ecotypes; aggregate weather data used to delineate distribution may miss important short-term weather effects. Different ENM approaches may also give different results, as the region considered and sample size affect the demographic projection (Wisz *et al.*, 2008) and other factors (see Lozier *et al.*, 2009, for other cautions). An additional deficiency is that ENMs make implicit ecological and mathematical assumptions having no mechanistic basis (Soberón and Nakamura, 2009). For these reasons, the results and transferability (validation) of ENMs are often conjectural, with the predictive power being potentially lower than spatial interpolation (Bahn and McGill, 2007). As a result, ENM predictions of invasive species ranges in response to climate change may also be unsound, as climate change effects not captured by the ENM may influence species abundance and subsequent trophic interactions (van der Putten *et al.*, 2010; Wardle *et al.*, 2011). Despite these acknowledged shortcomings, ENMs have been used to assess the potential distribution of invasive species under climate change (Jeschke and Strayer, 2008). However, because ENMs are unable to predict the dynamics of invasive species, we suggest they are marginally useful for developing management strategies.

The main advantage of ENM approaches is their relative simplicity to implement and low biological data requirements. Thuiller *et al.* (2005) has suggested ENMs may provide an unbiased first step for screening the potential range of invasive plants in new areas; but as Hickler *et al.* (2009) caution, process-based representations of water balance in terrestrial ecosystems need to be included. The use of ENMs for heterothermic animal species such as insects has also provided useful insights when based on species biology (see Lozier and Mills, 2011).

A more direct approach, however, is to model explicitly the mechanistic weather-driven, age-structured biology and dynamics of a species and of important species in its food chain or web (i.e. the system) using weather-driven PBDMs. This approach is independent of data concerning the assumed native range of the species. PBDMs estimate prospectively the phenology, dynamics and distribution of species in time and space as driven by observed or climate model scenarios. The PBDM approach has early roots in the work of Fitzpatrick and Nix (1968), Gutierrez *et al.* (1974) and de Wit and Goudriaan (1978). The underpinning concepts were summarized in Gutierrez and Baumgärtner (1984) and Gutierrez (1992, 1996).

PBDMs may be developed for an individual, a single species population or a tritrophic food web system that includes the bottom-up effects of plant growth and development on herbivores and the top-down action of natural enemies (Gutierrez *et al.*, 2010b). In their most complete form, PBDMs capture the daily age-structured dynamics of dry matter acquisition and allocation (see Chapter 4). The same model may be used to model the growth dynamics of plants or animals having very different resource acquisition and allocation biology (Gutierrez *et al.*, 1987). As such, we believe that PBDMs may be a more effective method to capture biogeographical and physiologically relevant parameters likely to be influenced with climate change.

Estimating the Physiological Effects of Weather

The biological functioning of a species is highly dependent on local weather, and if climate changes, the species dynamics will respond to the new weather regime. For example, if temperature increases, the ontogenetic development and growth rates of plants and other heterothermic species may change in positive or negative ways. Additional factors associated with climatic change such as moisture and CO_2 concentration [CO_2] are likely to change globally, as well. For example, the highly invasive weed, yellow starthistle (*C. solstitialis* L.), grew more than six times faster in response to elevated CO_2 in community settings (Dukes *et al.*, 2011). In some plants, increases in the potential growth rates may be slowed by extant physical, morphological and physiological constraints, and in some cases photoperiod and other factors may act as developmental switches (e.g. Morison and Lawlor, 1999). These are important considerations that call for physiological-based modelling approaches to capture the expression of the biology as it varies with changes in weather, including climate change.

Physiological Index Model

The origin of the PBDM approach is the concept of plant growth indices proposed by Fitzpatrick and Nix (1968) for Australian pastures, which Gutierrez *et al.* (1974) extended to the study of the effects of weather on the distribution of insect populations. The Fitzpatrick–Nix physiological indices were also the basis for the development of the widely used CLIMEX ENM algorithm (Sutherst and Maywald, 1985). The growth index approach was an important bridge to the development of PBDMs, and hence is reviewed in greater detail here.

Fitzpatrick and Nix (1968) proposed that plants had a concave growth response to various factors that, when normalized, yielded growth indices having minimum and maximum limits and optimal values (Fig. 17.1b). Assuming current [CO_2], the effects of temperature T on the photosynthetic ($f(T)$) and respiration ($r(T)$) rates (i.e. the Q_{10} rule) in a hypothetical plant given constant light is illustrated in Fig. 17.1a. The difference, $f(T) - r(T)$, is the potential assimilation rate ($G(T)$) (Fig. 17.1b). The normalized growth response (an index) is computed by dividing $G(T) = f(T) - r(T)$ by the maximum potential growth rate $G_{max}(T)$ (Eqn 17.1).

$$0 \leq GI(t) = (f(T) - r(T))/G_{max} < 1 \tag{17.1}$$

Normalized growth responses to other factors may also be computed in a similar fashion, with the limiting effects of all factors estimated as the product of the various indices (e.g. light ($LI(t)$), temperature ($TI(t)$), moisture ($MI(t)$), nitrogen ($NI(t)$), etc.) (Eqn 17.2; see Fitzpatrick and Nix, 1968).

$$0 \leq \phi = GI(t) = LI(t) \times TI(t) \times MI(t) \times NI(t) \times \ldots < 1 \tag{17.2}$$

This model assumes that concentrations of any factor or combination of factors away from the optimum slows plant growth rates, and may be limiting in the extreme. This is

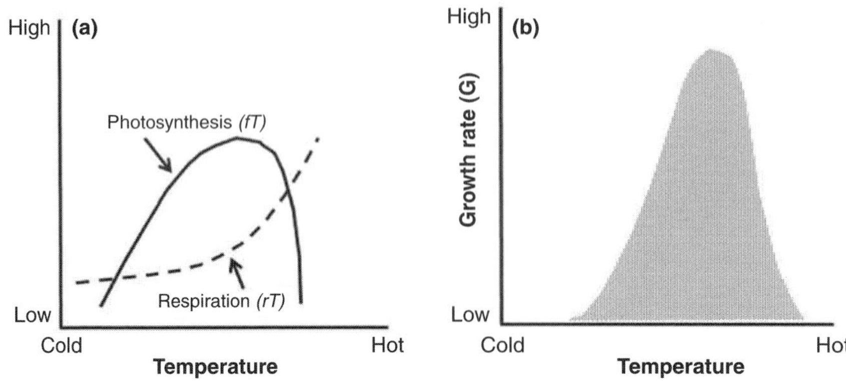

Fig. 17.1. The effects of temperature on growth rates: (a) photosynthesis and respiration and (b) plant growth rate (i.e. $f(T) - r(T)$) at given light and [CO_2].

not a new idea (see von Liebig, 1840; Shelford, 1931; Andrewartha and Birch, 1954). The same approach can be applied to other organisms (say arthropods), with limiting factors being weather variables, resource levels, availability of mates, oviposition sites, density, etc. (see below). One enduring example of this approach is Haefner (1970), who mapped the effects of temperature, salinity and dissolved oxygen on the survival of sand shrimp. A major difficulty, however, is determining, a priori, the limits and optimum for each index.

The physiological index niches of three hypothetical species in three weather dimensions are depicted in a stylized manner in Fig. 17.2a. In practice, the physiological indices can be computed using weather data during periods when the species are active in the field, and used to define their ecological niches. For example, the average *MI* index for the plant and *TI* indices for three aphids (*Rhopalosiphum padi* Koch, *Rhopalosiphum maidis* Fitch and *Aphis craccivora* Koch) were estimated during periods when winged aphids (alate) were trapped in some 40 pasture sites across south-east Australia during 1968–1970 (Fig. 17.2b; Gutierrez *et al*., 1974; see also Gutierrez and Yaninek, 1983, for other species). During periods when the aphids were active, *MI* was computed using a soil water balance model based on pan evaporation data, and *TI* was computed using average daily temperatures. The weather varied greatly across south-east Australia during this period, and only a few locations were favourable. A bivariate normal model was used to average *MI* and *TI* values for the periods of activity (i.e. the *MI* × *TI* space is the 95% tolerance region), with highest favourability occurring in the middle of the bivariate distribution space.

As observed in the field, the model suggested that *R. padi* was active during the cooler, wetter periods (autumn–mild winters), *A. craccivora* was active during periods with mild temperatures and adequate soil moisture for pasture growth (autumn or spring) and *R. maidis* winged forms were active during warmer and drier periods (late spring–early summer). Interpreting the model for future time and other locations suggests that populations of each species could develop only when the computed indices fall within the species' ecological niche tolerance region. Further, we note that the geographic distribution of these species is not static, and the migration of alate forms plays a key role in the life cycle of these aphids. Hence, whether populations of these species develop in an area also depends on the arrival of migrants during the favourable period.

Different species may have different responses to weather, and this problem is illustrated in a tritrophic setting where the trace of weekly *MI* and *TI* values at one location during a year is the line

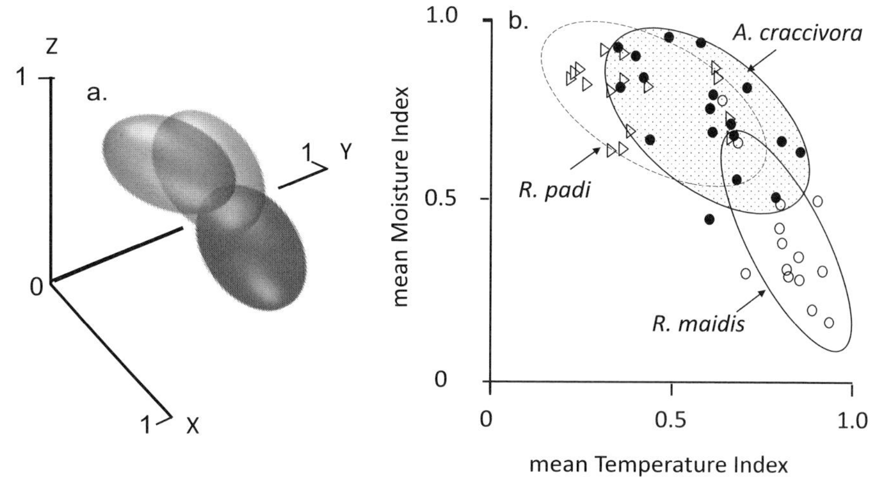

Fig. 17.2. The ecological niche space: (a) three hypothetical species and (b) three aphids in Australian pastures, with the data being the average *MI* and *TI* values during the period when aphids were present (Gutierrez *et al.*, 1974). The ellipses are bivariate normal fits to the data.

circumscribing the grey area in Fig. 17.3a. Other years and other locations would have different *MI* and *TI* traces, indicating different weather patterns and hence levels of favourability. The ecological niche tolerance regions for the plant, a herbivorous species, and its natural enemy are superimposed in the same *MI* × *TI* space in Fig. 17.3a. The relationships suggest that the different populations may grow during periods when the *MI* × *TI* trace falls within their ecological niche boundary, provided its resource (species) is also present. Some species may enter dormancy or survive in microclimates during unfavourable periods or, as in our cowpea aphid examples, may survive in the larger region via migration from areas of permanence.

Figure 17.3a shows a good climatic match for the three species to current climate for relatively long periods. In contrast, the weather scenario in Fig. 17.3b suggests marginal favourability.

The weather scenarios, say from Fig. 17.3b, could be viewed as climate change. We note that in the short run, the physiological niches of the species do not change, though in some species selection may occur on a shorter timescale (see Meineke *et al.*, 2013), but predicting the rate of that change remains vexing. Climate change could lead to trophic and spatial mismatches (Schweiger *et al.*, 2008), disruption of pollination (Schweiger *et al.*, 2010) and, potentially, new communities composed of more adapted species (i.e. succession) (George *et al.*, 2009).

Although Fig. 17.3 relates weather to trophic interactions, how to extrapolate this link to project demographic change, particularly in a rapidly shifting climate, is the crux of this chapter. Sakai *et al.* (2001) proposes that population biology holds promise for understanding and managing invasive species demography, but we argue that, in a climate change context, the models must be weather driven (e.g. Aurambout *et al.*, 2009). In doing so, greater realism can be added to the PBDM approach, with subsequent improvements in our understanding of climate-driven invasive species dynamics.

Age-structured Physiologically Based Demographic Model (PBDM)

We first make the links to the physiological index concepts using a single species model

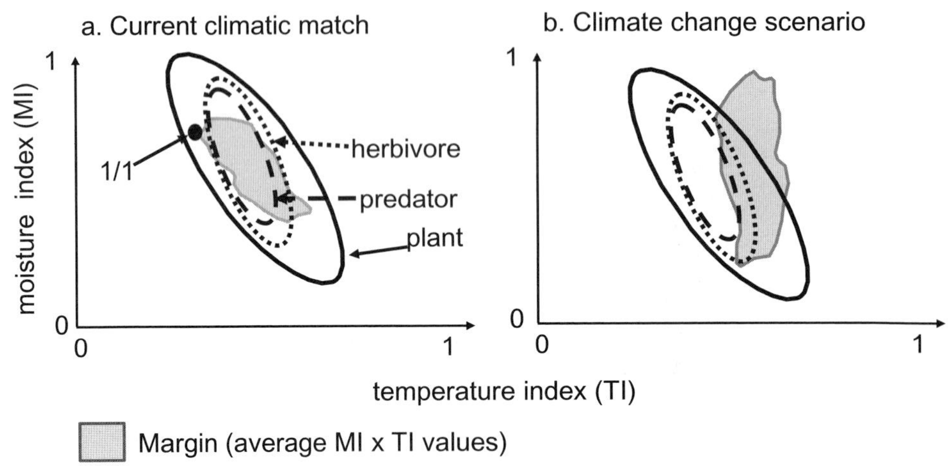

Fig. 17.3. Hypothetical ecological space on temperature (*TI*) and moisture (*MI*): (a) a well-adapted tritrophic system, and (b) an alternative scenario of decreasing favourability. The margins of the grey areas are the average weekly *TI* × *MI* values starting 1 January.

lacking age structure, where N is species density and R is its resource density. In dynamics model Eqn 17.3, $f(R, N, T)$ is the functional response of N to R, T is temperature, θ is the efficiency of converting resource R to N (number or mass), $r(T)$ is the respiration rate (i.e. the Q_{10} rule in a time-varying environment) (Fig. 17.1), $\mu(T)$ is the temperature-dependent net death rates and ϕ is the product of all limiting factors affecting the growth rate of N (Eqn 17.2) (see Gutierrez and Baumgartner, 1984; Gutierrez, 1992; Gutierrez *et al.*, 1994).

$$\frac{dN}{dt} = \phi\theta[f(R,N,T) - r(T)]N - \mu(T)N \qquad (17.3)$$

The single-species model may be recast with parallel structure to include mass structure (not shown), and one or more higher-trophic level consumers (C) may be added where $f_c(N, C, T)$, θ_c, r_c, μ_c and ϕ_c have analogous definitions as given above (see Gutierrez *et al.*, 1994; Eqn 17.4).

$$\frac{dN}{dt} = \phi\{\theta[f(R,N,T) - r(T)]N - \mu(T)N - g(N,C,T,)C\} \qquad (17.4)$$

$$\frac{dC}{dt} = \phi_c\{\theta_c[g(N,C,T) - r_c(T)]C - \mu_c(T)C\}$$

Adding age structure to Eqn 17.4 (Gutierrez, 1996) enables more nuanced modelling of the interactions of R, N and C. A variety of age/mass structured models may be used to imbed the time-varying dynamics in an age-specific manner (see Di Cola *et al.*, 1999). Such models may be used at the individual or population level, and apply equally well to plant and animal species (see Chapter 4). For example, plants pass through various life stages, and each plant has subunit populations (e.g. leaves, stem root, fruit), with fruit having distinct developmental stages and potential growth rates (i.e. bud stage, growing buds, flowers, maturing growing fruit and ripe fruit). Similarly, a holometabolous insect species has distinct stages: egg (i.e. e = eggs), larval (l), pupal (p) and adults (a), but the model can be structured for hemimetabolous life cycles as well. Individuals enter the first age class, say as an egg, and age through the life stages at temperature-dependent rates, exiting from the final age class, or via mortality (the double arrows) from any age or stage (Fig. 17.4). One may also add developmental variability using models such as those developed by Manetsch (1976) and Vansickle (1977) (see Di Cola *et al.*, 1999, for a review) wherein the distribution of

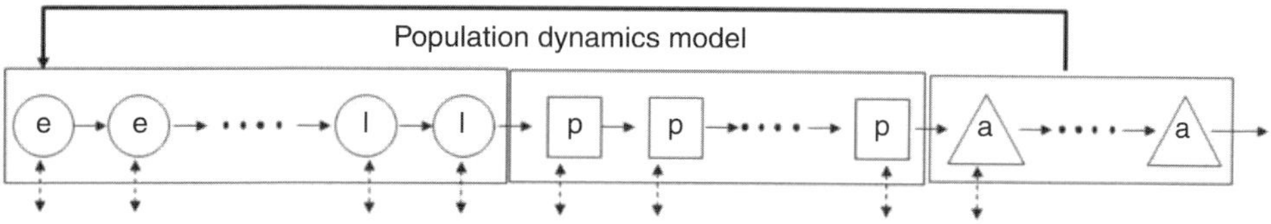

Fig. 17.4. Population dynamics: an age structure model for the dynamics for the egg (symbol e), larval (l), pupal (p) and adult (a) stages, with flows (ageing) between age classes and stages, with the double arrows indicating net age-specific mortality.

maturation times depends on the number of age classes in each stage. The numerical solution for the time-varying form of this model is given in the Appendix to this chapter (see Severini *et al.*, 2005; Gutierrez and Ponti, 2013a).

Parameterizing PBDMs

PBDMs are time-varying life tables (*sensu* Gilbert *et al.*, 1976), and the parameters may be estimated from laboratory and field age-specific life-table studies conducted across gradients of ecological factors (e.g. temperature). Optimally, studies on dry matter allocation and growth would also be performed at the same time. The parameterization process is simplified by the fact that models for all species have similar subcomponent models that may vary with temperature and other factors (e.g. density, nutrition) and can be captured using simple functions. PBDMs may include aspects of behaviour such as host preferences, physiology, dormancy, and other factors. At a minimum, developmental times of life stages, and age-specific resource acquisition, reproduction and mortality rates and dormancy as applicable must be estimated (see review in Gutierrez and Ponti, 2013a; see the Appendix to this chapter). While this may appear to be a difficult task, in reality it is quite simple and straightforward.

The effect of temperature on the developmental rate is illustrated in Fig. 17.5a (Brière *et al.*, 1999). Similar demand- and search-driven resource acquisition processes (functional response) occur in all trophic levels (e.g. photosynthesis, water and nitrogen) acquisition by plants, and resource acquisition by higher trophic levels; Gutierrez, 1992, 1996; Gutierrez and Baumgärtner, 1984, 2007) and at the economic level (Regev *et al.*, 1998), and all can be modelled using the same functional form (Fig. 17.5b). Depending on the biology of search, the functional response model (e.g. $f(R, N, T)$ may be type II if the search rate is a constant or type III if it is an increasing function of consumer density C. The choice of predator (single attacks) or parasitoid (multiple attacks per host) form of the model depends on the acquisition biology being modelled. For example, a predator model is used for photosynthesis, because a quantum of light can be captured only once, while a parasitoid form might be used for herbivores (or parasitoids) when multiple attacks are possible (e.g. on fruit).

Per capita age-specific fecundity at the optimum temperature (i.e. T_{opt} in Fig. 17.5d,e) may be captured by a simple function (Fig. 17.5c; Bieri *et al.*, 1983) and scaled for the effects of temperature by the function $\phi(T)$ (Fig. 17.5d) and other factors (Eqn 17.2). Mortality varies with temperature and may be captured by a convex function (e.g. Fig. 17.5e). Of course, in trophic interactions, we must include predation by higher trophic levels (e.g. $g(N, C, T)$, see Eqn 17.4). Dormancy in some species may allow them to survive unfavourable periods (see Hahn and Denlinger, 2011). We note, however, that dormancy may be absent in some species, may be transient in others, may be induced by low host density and/or high temperatures, or day length (Fig. 17.5f), or it may depend on

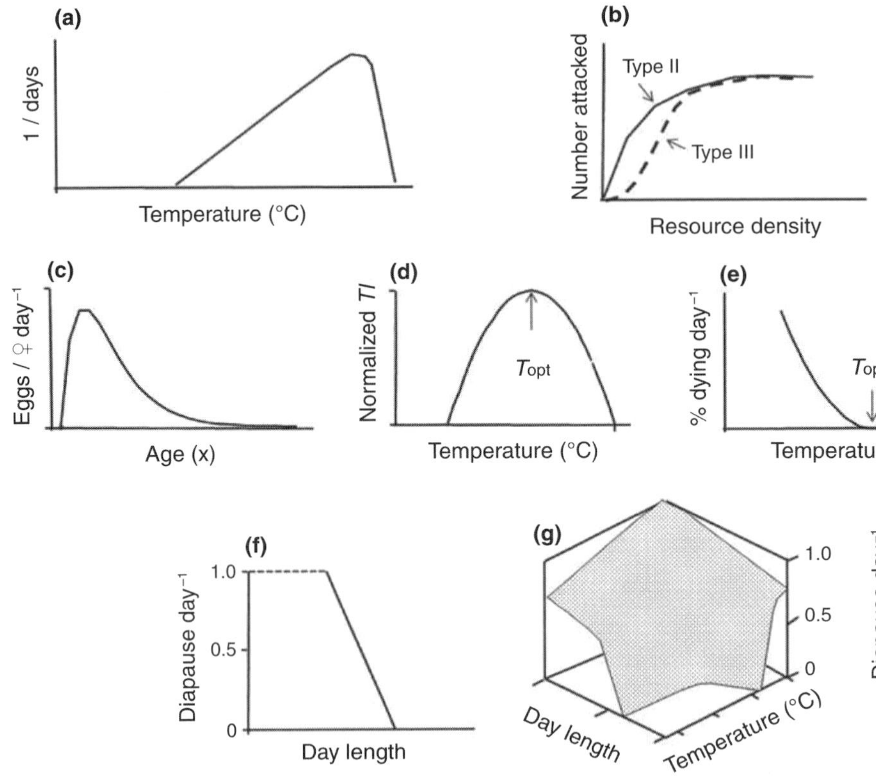

Fig. 17.5. Biological submodels: (a) the rate of development on temperature (Brière *et al.*, 1999); (b) type II and type III functional responses; (c) the per capita fecundity profile on female age in days at the optimum temperature (T_{opt} in Fig. 17.5d,e) (Bieri *et al.*, 1983); (d) the normalized effects of temperature on longevity (and gross fecundity) (Gutierrez, 1996); (e) the effects of temperature on mortality and two examples of diapause induction; (f) the proportion of diapause induction as a function of day length (e.g. grapevine moth, Gutierrez *et al.*, 2012) and (g) as a function of day length and temperature (e.g. pink bollworm, Gutierrez *et al.*, 1981). (Figure modified from Gutierrez and Ponti, 2013a.)

multiple factors (e.g. temperature, day length (Fig. 17.5g) and nutrition). As appropriate, the time of dormancy initiation and termination, and the daily rates of dormancy induction must be computed in the model.

Energy (dry matter) flow in an age-structured tritrophic system model consisting of a plant–herbivore–pathogen–parasitoid–predator system) is illustrated in Fig. 17.6a–c (e.g. Gutierrez *et al.*, 2008, 2011). The plant provides bottom-up effects via resource availability to herbivores, while higher trophic levels provide top-down regulation.

Regional Analysis

Several systems have been modelled using this basic structure in a GIS context (Table 17.1). The analysis path is as follows: (i) run the model using daily weather data (observed or climate model data) across all grid points in the landscape; (ii) map the georeferenced model data using GIS; and (iii) perform marginal analysis of the output data (i.e. $\partial y/\partial x_i$) to assess the relative contribution of various factors in the regulation of the pest. The model could also be used as the objective function in a bioeconomic analysis of the invasive species

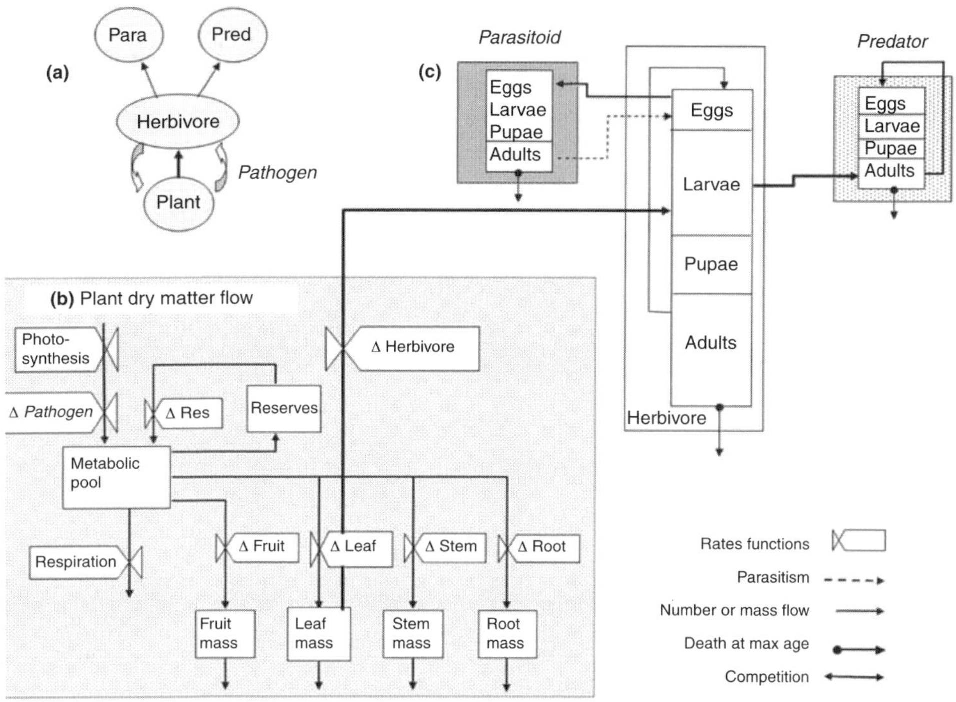

Fig. 17.6. A plant–herbivore–pathogen–parasitoid–predator system: (a) the basic tritrophic relationships; (b) dry matter flow in the plant as modified by the pathogen; and (c) to the herbivore and a parasitoid and predator showing stage-specific attack preferences. (Modified from Gutierrez *et al.*, 2008, 2011.)

problem. To illustrate these phases, we review the biological control of the invasive Mediterranean annual yellow starthistle (YST) in California (Gutierrez *et al.*, 2005).

The yellow starthistle system

YST is an important noxious weed that heavily infests pastures in the western USA and has been the focus of an ongoing biological control effort for more than 25 years. Growth of the thistle is affected by competition for light and water from exotic Mediterranean annual grasses (*G*) and attack from exotic seed head (= capitula)-feeding insect herbivores introduced for biological control (Fig. 17.7a). Among the species introduced are two weevils (*Bangasternus orientalis* (*BO*) and *Eustenopus villosus* (*EV*)) and two picture-winged flies (*Chaetorellia succinea* (*CS*) and *Urophora sirunaseva* (*US*)). Gutierrez *et al.* (2005) analysed the system using the PBDM approach to estimate the distribution and abundance of YST and to estimate the impact of the interacting herbivores on YST and each other across the varied ecological zones of California. Here, the analysis is extended to the western USA.

Germination of YST (and Mediterranean grasses) in California normally begins during autumn, with the intensity and patterns determined by seed-bank density and the timing and quantity of autumn–spring rains and favourable temperatures. The germination patterns vary greatly across years, regions and specific locations. Germination of an annual cohort of seed occurs over several years, providing an important stabilizing mechanism for coping with variable rainfall. Winter dormancy in the insect herbivores provides a bridging mechanism between seasons. Season length

Table 17.1. Summary of PBDM systems evaluated in a GIS context.

Host plant	Herbivores	Parasitoids	Predators	Pathogens	Regions	Reference	Origins
Lucerne	Spotted alfalfa aphid	*Aphelinus semiflavus*, *Praon palitans*, *Trioxys complanatus*	*Hippodamia convergens*	*Erynia neoaphidis* (insect)	California	Gutierrez and Ponti, 2013a,b	Temperate
Cotton	Pink bollworm				USA/Mexico	Gutierrez *et al.*, 2006; Gutierrez and Ponti, 2013a	Tropical
Citrus	Asian citrus psyllid	*Tamarixia radiata*	Coccinellid beetles		USA/Mexico	Gutierrez and Ponti, 2013c	Tropical
Grapevine	European grapevine moth				USA/Mexico	Gutierrez *et al.*, 2012	Temperate
	Glassy-winged sharpshooter	*Gonatocerus ashmeadi*, *Gonatocerus triguttatus*		*Xylella fastidiosa* (plant)	USA/Mexico	Gutierrez *et al.*, 2011; Gutierrez and Ponti, 2013a	Subtropical
	Grapevine mealybug	*Anagyrus pseudococci*, *Leptomastidea abnormis*	*Cryptolaemus montrouzieri*		California	Gutierrez *et al.*, 2008	Temperate
Olive	Olive fly				USA/Mexico, Mediterranean Basin	Gutierrez and Ponti, 2013a; Gutierrez *et al.*, 2009; Ponti *et al.*, 2009a,b, unpublished	Subtropical
Perennial plant	Light brown apple moth				USA/Mexico	Gutierrez *et al.*, 2010a	Temperate
Fruit tree hosts	Medfly				USA/Mexico/Italy	Gutierrez and Ponti, 2011	Tropical
California laurel	Oleander scale	*Aphytis chilensis*	*Rhyzobius lophanthae*		California	Gutierrez and Pizzamiglio, 2007	Temperate
Vertebrate host	Screwworm				USA/Mexico	Gutierrez and Ponti, 2013a	Tropical
Yellow starthistle	*Bangasternus orientalis*, *Eustenopus villosus*, *Urophora sirunaseva*, *Chaetorellia succinea*				California	Gutierrez *et al.*, 2005	Temperate

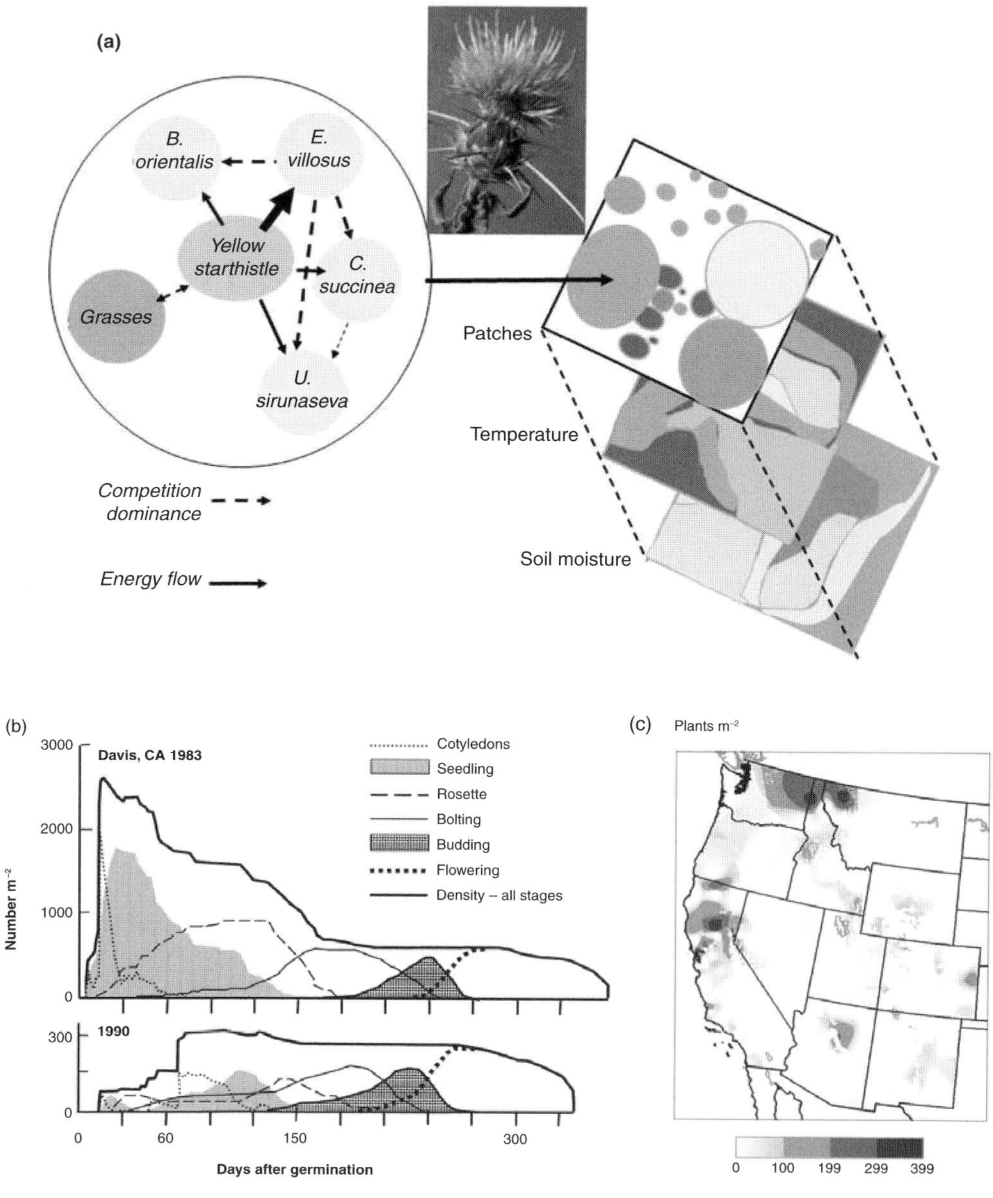

Fig. 17.7. The yellow starthistle PBDM system: (a) trophic interactions (competition between yellow starthistle and exotic annual grasses and herbivory by seed-feeding insects as affected by temperature and soil moisture (see text)); (b) the simulated phenology of YST from germination through senescence during a wet (1983) and a dry (1990) year (see Gutierrez *et al.*, 2005); and (c) the prospective average distribution of flowering plants m^{-2} during the period September 1982 to September 2003 in the western USA.

for YST is determined mostly by temperature and the availability of soil moisture, estimated using a mechanistic soil water balance model that predicts the evapotranspiration rate in response to weather and developing leaf area index (Ritchie, 1972).

The phenology and population dynamics of plant stages at Davis, California, from germination in autumn to senescence and death during a wet (1983) and a dry (1990) year are illustrated in Fig. 17.7b (see Gutierrez *et al.*, 2005). The number of germinating seeds (cotyledon stage) during the wet 1983 period was tenfold that in 1990, with the steep pattern of germination in 1983 starting after the first heavy rains. In contrast, germination during dry 1990 occurred over a protracted 110-day period due to low, infrequent rainfall. Figure 17.7c shows the predicted average distribution of flowering YST plants across the varied ecological zones of the western USA. In California, the average distribution of YST is restricted to areas with more than 350 mm of annual rainfall. Similar maps can be drawn for each plant stage, and for the herbivorous insect species, but aside from illustrating the predicted distribution, the maps are not highly informative.

Marginal analysis of simulated yellow starthistle data

Using marginal analysis, we review the average impact of each herbivore species and grass in suppressing YST, their competitive interactions and the effect of competition from grasses on YST populations. The simulated action of the different combinations of herbivores and grass (abbreviations *Bo*, *Ev*, *Us*, *Cs* and *G*) at the 72 sites across the varied ecological zones of California were summarized using linear multiple regression wherein species' absence or presence (i.e. 0 or 1) and the interactions were used as the independent variables. Only independent variables having slopes significantly different from zero were retained in the multiple regression models (see Gutierrez *et al.*, 2005, for full details). The goal of the regression analysis was not prediction (explained variance) but rather to estimate the direction and average impact of the various factors. (We note, however, that prediction increases with decreasing geographic scale.)

Regressing mature flowering plant density on season length (*dd*), total rainfall (*mm*) and the presence/absence of the four herbivores, grass and their interactions yielded multiple regression Eqn 17.5. The weevil, *B. orientalis*, and the fly, *U. sirunaseva*, had no significant effect in reducing plant densities and were not included in the final model.

$$plants\ m^{-2} = 219.4 - 0.012dd - 0.01mm - 12.0Ev - 30.0Cs + 8.6EvCs - 7.9G \quad (17.5)$$
$$R = 0.22, F = 27.6, df = 3{,}233$$

Plant density declines with factors that increase season length (*dd*, *mm*) and with *E. villosus*, *C. succinea* and competition from annual grasses. A longer season enables more generations of *Ev* and *Cs*. The interaction *EvCs* decreases control and is a measure of interspecific competition. Using marginal analysis, average *dd* (2656) and *mm* (466) and herbivore presence, the predicted high average plant density is 142 m^{-2}, indicating, as observed, poor control.

A regression of $\log_{10}$ seed density on the same independent variables yielded Eqn 17.6.

$$\log_{10} seed\ density = 3.30 + 0.00007dd + 0.0002mm - 0.18Ev - 0.36Cs + 0.16Ev \times Cs \quad (17.6)$$
$$R = 0.42, F = 141.4, df = 3{,}234$$

Again, *B. orientalis* and *U. sirunaseva*, on average, had no significant effect in reducing seed densities. Taking the antilog of Eqn 17.6 and substituting mean values for *dd* and *mm* across sites shows that yellow starthistle seed densities increase with season length and total rainfall but decrease with *Ev* and *Cs* presence. The action of *E. villosus* and *C. succinea*, on average, reduce seed production by 58% across California, with *C. succinea* having the greatest impact. The impact of *C. succinea* was reduced by its interaction with *E. villosus* because it kills fly larvae when they co-occur in capitula. The interaction of *EvCs* increases seed survival by 12.8%, offsetting much of *Ev*'s contribution to seed reduction. Most important, the proportion of seeds that survive in attacked capitula is far higher than required for plant replacement.

Interspecific competition among YST herbivores

The interactions of the four herbivore species (and grass) were also examined using $\log_{10}$ cumulative larval days during the season (i.e. abundance) as the dependent variable and the presence/absence and interaction terms of the other species and grass as the independent variables. The results of the marginal analyses suggest (see Gutierrez *et al.*, 2005):

B. orientalis

Only the presence of the weevil, *E. villosus*, has a significant large negative effect on the abundance of the weevil, *B. orientalis*. Without *E. villosus*, average cumulative *B. orientalis* abundance would be 1096 m^{-2} larval days, but with *E. villosus*, the average of *B. orientalis* drops to 3.5 m^{-2}, because *E. villosus* is a superior competitor.

E. villosus

Competition from *B. orientalis*, *C. succinea* and *U. sirunaseva* on log *E. villosus* is not significant, but the indirect effects of grass competition on YST growth, and hence capitula production, reduced average *E. villosus* larval days by 23%.

C. succinea

E. villosus reduces the average abundance of the fly, *C. succinea*, by an average of 62%.

U. sirunaseva

The combined action of *E. villosus* (weevil) and *C. succinea* (fly) decreased the average larval days of the fly *U. sirunaseva*, with *C. succinea* having the larger effect. In the absence of competitors, *U. sirunaseva*'s average larval days would be 10,715 m^{-2}, but with competition it is reduced to 19.5 m^{-2}, because both *E. villosus* and *C. succinea* are superior competitors in cases of multiple occurrences in capitula.

In summary, lack of control of yellow starthistle is due to the high survivorship of seed and to plant compensation that increase per plant seed production at lower plant densities. The analysis suggests that herbivory that reduces the plant's ability to compensate and/or kill whole plants before seed maturity would be most likely to lead to the control of yellow starthistle. This is what occurred in the control of the noxious Klamath weed in California by the root-feeding Klamath weed beetle (Huffaker and Kennett, 1959). The results of our YST model are consistent with field observations and illustrate the utility of a well-parameterized PBDM.

However, increases in CO_2 levels have been shown to increase YST growth rates greatly, and this will likely complicate the biological control and management of this weed (Dukes *et al.*, 2011). The effects of increasing [CO_2] can be incorporated in the current PBDM via the demand parameter (see Appendix Eqn A3), though a more physiologically nuanced submodel could also be incorporated. In addition, the modular structure of the PBDM allows integration of the effects of additional biological control species and is used to assess the evolving temporal and spatial changes in YST density with climate change and as an aid in guiding ongoing management policy. It could also be used as a template to assess other weed (or crop) species.

Furthermore, PBDMs can be used as the production function in economic analyses of invasive species problems, by replacing the static production and damage control function approaches commonly used in agricultural economic analyses (e.g. Pemsl *et al.*, 2008). This approach would enable assessment of the effects of climate change on the system.

Management of Invasive Insects: A Policy Perspective

Effective management of invasive species requires approaches that go beyond the existing environmental policy (Pyšek and Richardson, 2010). Some invasive species may disrupt extant food webs, while others may be of minor consequence (see Tylianakis *et al.*, 2008). It is to our advantage to differentiate the threat posed once an invasive pest is introduced. Currently, the presence of invasive species does not include differentiation of actual risk, and as a result, invasives are often handled in a crisis mode by government agencies (Larson, 2005). Yet, such agencies often lack the capacity to predict the potential geographic range and relative abundance of the exotic species (i.e. invasiveness) under current weather, much less under future climate change (Gutierrez and Ponti, 2013a). McNeely *et al.* (2001) have suggested that the eradication of an invasive species is but one component of invasion policy, which must include a combination of prevention measures and early detection, with effective management (including eradication) being the last resort (see Simberloff *et al.*, 2011). The capacity to predict the occurrence and resulting impact of an invasive species is pivotal to developing science-based policy, including environmental, social and economic decisions (Larson *et al.*, 2010). Policy decisions should also extend to invasive species management, including eradication, control and, in some cases, acceptance (Myers *et al.*, 2000). Decisions made for the management of invasive species based on a ENM approach are likely suboptimal, as discussed herein, and hence there is a greater need to be able to assess the effects of species' interactions and other factors including the effects of climate change. To that end, we propose that knowledge gaps can be bridged effectively by developing biologically comprehensive, weather-driven models such as PBDMs that can utilize sound data in a GIS setting with fine space/time resolution, including data from satellite remote sensing (e.g. Roiz *et al.*, 2011; Table 17.1) and climate models (e.g. Ponti *et al.*, 2009b).

Venette *et al.* (2010) have argued that pest risk models that incorporate a high degree of biological complexity (e.g. PBDMs) are less utilized because substantially more biological data at temporal and spatial scales are required and that decision makers with political and legislative constraints cannot wait for research to fill the data gaps. Policy decisions based on simple assessments may, over time, prove to result in greater environmental and economic damage, as illustrated by some of the cases reviewed in Chapter 4 and Table 17.1.

It is widely acknowledged that management/eradication efforts against some invasive pests have been ongoing for decades (Gutierrez and Ponti, 2013a). In the absence of basic biological information, large sums of public monies are often spent in control efforts, efforts that could be strengthened if a fraction of the funds could be applied to the research needed to develop the requisite biological data and models that could improve policy and management decisions (Gutierrez and Ponti, 2013a). The recent termination of eradication efforts against the light brown apple moth (*Epiphyas postvittana*) in California is a poignant case in point (see Gutierrez *et al.*, 2010a; Lozier and Mills, 2011).

Overall, the propensity to dismiss biologically intense approaches in favour of short-term 'solutions' can be counterproductive and often leads to inadequate and/or misdirected research funding and continuing gaps in scientific knowledge regarding the long-term failures or success of invasive species management programmes (e.g. the YST example above; see Simberloff *et al.*, 2005). The field of invasion biology has matured sufficiently so that a synthesis and global policy framework for research and implementation for dealing with diverse invasive species can be developed. This synthesis would harness extant tools and develop new ones such as PBDMs to enhance scientifically based decision making at the policy level (e.g. Yemshanov *et al.*, 2009).

References

Andrewartha, H.G. and Birch, L.C. (1954) *The Distribution and Abundance of Animals*. University of Chicago Press, Chicago, Illinois.

Aurambout, J.P., Finlay, K.J., Luck, J. and Beattie, G.A.C. (2009) A concept model to estimate the potential distribution of the Asiatic citrus psyllid (*Diaphorina citri* Kuwayama) in Australia under climate change – a means for assessing biosecurity risk. *Ecological Modelling* 220, 2512–2524.

Bahn, V. and McGill, B.J. (2007) Can niche-based distribution models outperform spatial interpolation? *Global Ecology and Biogeography* 16, 733–742.

Beaumont, L.J., Gallagher, R.V., Thuiller, W., Downey, P.O., Leishman, M.R. and Hughes, L. (2009) Different climatic envelopes among invasive populations may lead to under estimations of current and future biological invasions. *Diversity Distribution* 15, 409–420.

Bieri, M., Baumgärtner, J., Bianchi, G., Delucchi,V. and von Arx, R. (1983) Development and fecundity of pea aphid (*Acyrthosiphon pisum* Harris) as affected by constant temperatures and by pea varieties. *Mitteilungen der Schweizerischen entomologischen Gesellschaft* 56, 163–171.

Brière, J.F., Pracros, P., Le Roux, A.Y. and Pierre, J.S. (1999) A novel rate model of temperature-dependent development for arthropods. *Environmental Entomology* 28, 22–29.

Davis, A.J., Jenkinson, L.S., Lawton, J.H., Shorrocks, B. and Wood, S. (1998) Making mistakes when predicting shifts in species range in response to global warming. *Nature* 391, 783–786.

de Wit, C.T. and Goudriaan, J. (1978) *Simulation of Ecological Processes*, 2nd edn. Pudoc Publishers, Wageningen, Netherlands.

Di Cola, G., Gilioli, G. and Baumgärtner, J. (1999) Mathematical models for age-structured population dynamics. In: Huffaker, C.B. and Gutierrez, A.P. (eds) *Ecological Entomology*, 2nd edn. John Wiley and Sons, New York, pp. 503–534.

Dukes, J., Chiariello, N.R., Loarie, S.R. and Field, C.B. (2011) Strong response of an invasive plant species (*Centaurea solstitialis* L.) to global environmental changes. *Ecological Applications* 21(6), 1887–1894.

Elith, J. and Leathwick, J.R. (2009) Species distribution models: ecological explanation and prediction across space and time. *Annual Review of Ecology Evolution and Systematics* 40, 677–697.

Elith, J., Phillips, S.J., Hastie, T., Dudík, M., Chee, Y.E. and Yates, C.J. (2011) A statistical explanation of MaxEnt for ecologists. *Diversity and Distribution* 17, 43–57.

Estrada-Peña, A. (2008) Climate, niche, ticks, and models: what they are and how we should interpret them. *Parasitology Research* 103, 87–95.

Fitzpatrick, E.A. and Nix, H.A. (1968) The climatic factor in Australian grasslands ecology. In: Moore, R.M. (ed.) *Australian Grasslands*. Australian National University Press, Canberra, pp. 3–26.

Frazer, B.D. and Gilbert, N.E. (1976) Coccinellids and aphids: a qualitative study of the impact of adult lady birds (Coleoptera: Coccinellidae) preying on field populations of pea aphids (Homoptera: Aphididae). *Journal of the Entomological Society of British Colombia* 73, 33–56.

Gallien, L., Münkemüller, T., Albert, C.H., Boulangeat, I. and Thuiller, W. (2010) Predicting potential distributions of invasive species: where to go from here? *Diversity and Distribution* 16, 331–342.

George, K., Ziska, L.H., Bunce, J.A., Quebedeaux, B., Hom, J.L., Wolf, J. and Teasdale, J.R. (2009) Macroclimate associated with urbanization increases the rate of secondary succession from fallow soil. *Oecologia* 159, 637–647.

Gilbert, N., Gutierrez, A.P., Frazer, B.D. and Jones, R.E. (1976) *Ecological Relationships*. Freeman and Co, New York.

Gutierrez, A.P. (1992) The physiological basis of ratio-dependent predator–prey theory: the metabolic pool model as a paradigm. *Ecology* 73, 1552–1563.

Gutierrez, A.P. (1996) *Applied Population Ecology: A Supply–Demand Approach*. John Wiley and Sons, New York.

Gutierrez, A.P. and Baumgärtner, J.U. (1984) Multitrophic level models of predator–prey energetics: II. A realistic model of plant–herbivore–parasitoid–predator interactions. *Canadian Entomologist* 116, 933–949.

Gutierrez, A.P. and Baumgärtner, J.U. (2007) Modeling the dynamics of tritrophic population interactions. In: Kogan, M. and Jepson, P. (eds) *Perspectives in Ecological Theory and Integrated Pest Management*. Cambridge University Press, Cambridge, UK, pp. 301–360.

Gutierrez, A.P. and Pizzamiglio, M.A. (2007) A regional analysis of weather mediated competition between a parasitoid and a coccinellid predator of oleander scale. *Neotropical Entomology* 36(1),70–83.

Gutierrez, A.P. and Ponti, L. (2011) Assessing the

invasive potential of the Mediterranean fruit fly in California and Italy. *Biological Invasion* 13, 2661–2676.

Gutierrez, A.P. and Ponti, L. (2013a) Eradication of invasive species: why the biology matters. *Environmental Entomology* 42, 395–411.

Gutierrez, A.P. and Ponti, L. (2013b) Deconstructing the control of the spotted alfalfa aphid (*Therioaphis maculata*). *Agricultural and Forest Entomology* 15, 272–284.

Gutierrez, A.P. and Ponti, L. (2013c) Prospective analysis of the geographic distribution and relative abundance of asian citrus psyllid (Hemiptera: Liviidae) and citrus greening disease in North America and the Mediterranean Basin. *Florida Entomologist* 96, 1375–1391.

Gutierrez, A.P. and Yaninek, J.S. (1983) Responses to weather of eight aphid species commonly found in pastures in southeastern Australia. *Canadian Entomologist* 115, 1359–1364.

Gutierrez, A.P., Havenstein, D.E., Nix, H.A. and Moore, P.A. (1974) The ecology of *Aphis craccivora* Koch and subterranean clover stunt virus. III. A regional perspective of the phenology and migration of the cowpea aphid. *Journal of Applied Ecology* 11, 21–35.

Gutierrez, A.P., Butler, G.D. Jr and Ellis, C.K. (1981) Pink bollworm: diapause induction and termination in relation to fluctuating temperatures and decreasing photophases. *Environmental Entomology* 10, 936–942.

Gutierrez, A.P., Schulthess, F., Wilson, L.T., Villacorta, A.M., Ellis, C.K. and Baumgärtner, J.U. (1987) Energy acquisition and allocation in plants and insects: a hypothesis for the possible role of hormones in insect feeding patterns. *Canadian Entomologist* 119, 109–129.

Gutierrez, A.P., Mills, N.J., Schreiber, S.J. and Ellis, C.K. (1994) A physiologically based tritrophic perspective on bottom-up-top-down regulation of populations. *Ecology* 75, 2227–2242.

Gutierrez, A.P., Pitcairn, M.J., Ellis, C.K., Carruthers, N. and Ghezelbash, R. (2005) Evaluating biological control of yellow starthistle (*Centaurea solstitialis*) in California: a GIS based supply–demand demographic model. *Biological Control* 34, 115–131.

Gutierrez, A.P., d'Oultremont, T., Ellis, C.K. and Ponti, L. (2006) Climatic limits of pink bollworm in Arizona and California: effects of climate warming. *Acta Oecologica* 30, 353–364.

Gutierrez, A.P., Daane, K.M., Ponti, L., Walton, V.M. and Ellis, C.K. (2008) Prospective evaluation of the biological control of vine mealybug: refuge effects and climate. *Journal of Applied Ecology* 45, 524–536.

Gutierrez, A.P., Ponti, L. and Cossu, Q.A. (2009) Effects of climate warming on olive and olive fly (*Bactrocera oleae* (Gmelin)) in California and Italy. *Climatic Change* 95, 195–217.

Gutierrez, A.P., Mills, N.J. and Ponti, L. (2010a) Limits to the potential distribution of light brown apple moth in Arizona–California based on climate suitability and host plant availability. *Biological Invasions* 12, 3319–3331.

Gutierrez, A.P., Ponti, L. and Gilioli, G. (2010b) Climate change effects on plant–pest–natural enemy interactions. In: Hillel, D. and Rosenzweig, C. (eds) *Handbook of Climate Change and Agroecosystems: Impact, Adaptation and Mitigation*. Imperial College Press, UK, pp. 209–237.

Gutierrez, A.P., Ponti, L., Hoddle, M., Almeida, R.P.P. and Irvin, N.A. (2011) Geographic distribution and relative abundance of the invasive glassy-winged sharpshooter: effects of temperature and egg parasitoids. *Environmental Entomology* 40, 755–769.

Gutierrez, A.P., Ponti, L., Cooper, M.L., Gilioli, G., Baumgärtner, J. and Duso, C. (2012) Prospective analysis of the invasive potential of the European grapevine moth (*Lobesia botrana* (Den. & Schiff.)) in California. *Agricultural and Forest Entomology* 14, 225–238, doi:10.1111/j.1461-9563.2011.00566.x.

Haefner, P.A. (1970) The effects of low dissolved oxygen concentrations on temperature–salinity tolerance of the sand shrimp, *Crangon septemspinosa* Say. *Physiological Zoology* 43, 30–37.

Hahn, D.A. and Denlinger, D.L. (2011) Energetics of insect diapause. *Annual Review of Entomology* 56, 103–121.

Hickler, T., Fronzek, S., Araújo, M.B., Schweiger, O., Thuiller, W. and Sykes, M.T. (2009) An ecosystem model-based estimate of changes in water availability differs from water proxies that are commonly used in species distribution models. *Global Ecology and Biogeography* 18, 304–313.

Huffaker, C.B. and Kennett, C.E. (1959) A ten year study of the vegetational changes associated with biological control of Klamath weed. *Journal of Range Management* 12, 69–82.

Jeschke, J.M. and Strayer, D.L. (2008) Usefulness of bioclimatic models for studying climate change and invasive species. *Annals of the New York Academy of Science* 1134, 1–24.

Larson, B.M.H. (2005) The war of the roses: demilitarizing invasion biology. *Frontiers in Ecology and the Environment* 3, 495–500.

Larson, D.L., Phillips-Mao, L., Quiram, G., Sharpe, L., Stark, R., Sugita, S. and Weiler, A. (2010) A framework for sustainable invasive species management: environmental, social, and economic objectives. *Journal of Environmental Management* 92, 14–22.

Lozier, J.D. and Mills, N.J. (2011) Predicting the potential invasive range of the light brown apple moth (*Epiphyas postvittana*) using biologically informed and correlative species models. *Biological Invasions* 13, 2409–2421.

Lozier, J.D., Aniello, P. and Hickerson, M.J. (2009) Predicting the distribution of Sasquatch in western North America: anything goes with ecological niche modeling. *Journal of Biogeography* 36, 1623–1627.

McNeely, J.A., Mooney, H.A., Neville, L.E., Schei, P.J. and Waage, J.K. (eds) (2001) *A Global Strategy on Invasive Alien Species*. IUCN, Gland, Switzerland, 50 pp.

Manetsch, T.J. (1976) Time-varying distributed delays and their use in aggregate models of large systems. *IEEE Transactions on Systems, Man, and Cybernetics* 6, 547–553.

Meineke, E.K., Dunn, R.R., Sexton, J.O. and Frank, S.D. (2013) Urban warming drives insect pest abundance on street trees. *PLoS One* 8(3), e59687, doi:10.1371/ journal.pone.0059687.

Monteith, J.L. (1965) Light distribution and photosynthesis in field crops. *Annals of Botany* (London) [N.S.] 29, 17–38.

Morison, J.I.L. and Lawlor, D.W. (1999) Interactions between increasing CO_2 concentration and temperature on plant growth *Plant, Cell and Environment* 22, 659–682.

Myers, J.H., Simberloff, D., Kuris, A.M. and Carey, J.R. (2000) Eradication revisited: dealing with exotic species. *Trends in Ecology and Evolution* 15, 316–320.

Nicholson, A.J. and Bailey, V.A. (1935) The balance of animal populations. Part 1. *Proceedings of the Zoological Society of London* 3, 551–598.

Pemsl, D.E., Gutierrez, A.P. and Waibel, H. (2008) The economics of biotechnology under ecosystem disruption. *Ecological Economics* 66, 177–183.

Phillips, S.J. and Dudík, M. (2008) Modelling of species distributions with Maxent: new extensions and a comprehensive evaluation. *Ecography* 31, 161–175.

Phillips, S.J., Anderson, R.P. and Schapire, R.E. (2006) Maximum entropy modelling of species geographic distributions. *Ecological Modelling* 190, 231–259.

Ponti, L., Cossu, Q.A. and Gutierrez, A.P. (2009a) Climate warming effects on the *Olea europaea–Bactrocera oleae* system in Mediterranean islands: Sardinia as an example. *Global Change Biology* 15, 2874–2884.

Ponti, L., Gutierrez, A.P. and Ruti, P.M. (2009b) The olive–*Bactrocera oleae* (Diptera Tephritidae) system in the Mediterranean Basin: a physiologically based analysis driven by the ERA-40 climate data. *Notiziario sulla Protezione delle Piante – III Serie* 1, 113–128.

Pyšek, P. and Richardson, D.M. (2010) Invasive species, environmental change and management, and health. *Annual Review of Environment and Resources* 35, 25–55.

Regev, U., Gutierrez, A.P., Schreiber, S.J. and Zilberman, D. (1998) Biological and economic foundations of renewable resource exploitation. *Ecological Economics* 26, 227–242.

Ritchie, J.T. (1972) Model for predicting evaporation from a row crop with incomplete cover. *Water Resources Research* 8, 1204–1213.

Roiz, D., Neteler, M., Castellani, C., Arnoldi, D. and Rizzoli, A. (2011) Climatic factors driving invasion of the tiger mosquito (*Aedes albopictus*) into new areas of Trentino, northern Italy. *PLoS ONE* 6, e14800.

Sakai, A.K., *et al.* (2001) The population biology of invasive species. *Annual Review of Ecology and Systematics* 32, 305–332.

Schweiger, O., Settele, J., Kudrna, O., Klotz, S. and Kühn, I. (2008) Climate change can cause spatial mismatch of trophically interacting species. *Ecology* 89, 3472–3479.

Schweiger, O., Biesmeijer, J.C., Bommarco, R., Hickler, T., Hulme, P.E., Klotz, S., Kühn, I., Moora, M., Nielsen, A. and Ohlemüller, R. (2010) Multiple stressors on biotic interactions: how climate change and alien species interact to affect pollination. *Biological Reviews* 85, 777–795.

Severini, M., Alilla, R., Pesolillo, S. and Baumgärtner, J. (2005) Grapevine and *Lobesia botrana* (lep. Tortricidae) phenology in the Castelli Romani area. *Rivista Italiana di Agrometeorologia* 3, 34–39.

Shelford, V.E. (1931) Some concepts of bioecology. *Ecology* 12, 455–467.

Simberloff, D., Parker, I.M. and Windle, P.N. (2005) Introduced species policy, management, and future research needs. *Frontiers in Ecology and the Environment* 3, 12–20.

Simberloff, D., Genovesi, P., Pyšek, P. and Campbell, K. (2011) Recognizing conservation success. *Science* 332, 419.

Soberón, J. and Nakamura, M. (2009) Niches and distributional areas: concepts, methods, and assumptions. *Proceedings of the National Academy of Sciences* 106, 19644–19650.

Sutherst, R.W. and Maywald, G.F. (1985) A computerized system for matching climates in ecology. *Agriculture, Ecosystems and Environment* 13, 281–299.

Sutherst, R.W., Maywald, G.F. and Bourne, A.S. (2007) Including species interactions in risk assessments for global change. *Global Change Biology* 13, 1843–1859.

Thuiller, W., Richardson, D.M., Pysek, P., Midgley, G.F., Hughes G.O. and Rouget, M. (2005) Niche-based modelling as a tool for predicting the risk of alien plant invasions at a global scale. *Global Change Biology* 11, 2234–2250.

Tylianakis, J.M., Didham, R.K., Bascompte, J. and Wardle, D.A. (2008) Global change and species interactions in terrestrial ecosystems. *Ecology Letters* 11, 1351–1363.

van der Putten, W.H., Macel, M. and Visser, M.E. (2010) Predicting species distribution and abundance responses to climate change: why it is essential to include biotic interactions across trophic levels. *Philosophical Transactions of the Royal Society* B 365, 2025–2034.

Vansickle, J., (1977) Attrition in distributed delay models. *IEEE Transactions on Systems, Man, and Cybernetics* 7, 635–638.

Venette, R.C., Kriticos, D.J., Magarey, R.D., Koch, F.H., Baker, R.H.A., Worner, S.P., Gómez Raboteaux, N.N., Dobesberger, E.J., Yemshanov, D. and De Barro, P.J. (2010) Pest risk maps for invasive alien species: a roadmap for improvement. *Bioscience* 60, 349–362.

von Liebig, J. (1840) *Chemistry and its Applications to Agriculture and Physiology* (4th edn 1847). Taylor and Walton, London.

Wardle, D.A., Bardgett, R.D., Callaway, R.M. and van der Putten, W.H. (2011) Terrestrial ecosystem responses to species gains and losses. *Science* 332, 1273–1277.

Watt, K.E.F. (1959) A mathematical model for the effects of densities of attacked and attacking species on the number attacked. *Canadian Entomologist* 91, 129–144.

Wisz, M.S., Hijmans, R.J., Li, J., Peterson, A.T., Graham, C.H. and Guisan, A. (2008) Effects of sample size on the performance of species distribution models. *Diversity and Distributions* 14, 763–773.

Yemshanov, D., McKenney, D.W., Pedlar, J.H., Koch, F.H. and Cook, D. (2009) Towards an integrated approach to modeling the risks and impact of invasive forest species. *Environmental Reviews* 17, 163–178.

Appendix

Distributed Maturation Time Model

The numerical solution of this model for a life stage i with i =1, ... , K age classes is Eqn A1 (text Fig. 17.4; see Manetsch, 1976; Vansickle, 1977; Di Cola *et al.*, 1999; Severini *et al.*, 2005). The state variable is density (as a rate, $n_i(t)$) in units of number (or mass). Births flow into the first age class (i =1), survivors exit at the maximum age ($i = k$). Note that this is the time-varying form (Vansickle, 1977; Severini *et al.*, 2005). The model can be adapted for flows in other dimensions as well (mass, etc.).

$$n_i(t+1) = n_i(t) + \frac{V}{\Delta(t)}\left\{ n_{i-1}(t) - \left(1 + \frac{\Delta(t) + [\mu_i - 1]\cdot\Delta(t-1)}{V}\right)\cdot n_i(t)\right\} \quad \text{(A1)}$$

The forcing variable is temperature (T) and time (t) is a day that from the perspective of the heterotherm organisms is of variable length in physiological time units above its lower thermal threshold (see below). The variable V is the proportional change in development (ageing), $\Delta(T)$ is the mean developmental time. The function variable $\mu_i(t)$ is the proportional net loss rate that includes the rich biology affecting the age-specific births, deaths, growth, predation, net immigration and other factors included in Eqn A1.

The density of cohort i is $N_i(t) = n_i(t) \cdot \Delta(T(t))/k$, the total density in the life stage is $N(t) = \sum_1^k N_i(t)$ and $k = \Delta(T_{opt})^2/\text{var}$. The model allows multiple thermal thresholds and multiple age and timescales for the different species and their substages. The definitions of variables in Eqn A1 are expanded below.

Parameterizing the Dynamics Model

All of the submodels below are parameterized by fits to data.

The developmental rates and times

In PBDMs, the time step is a day, while the change in age is in physiological time units above the lower thermal threshold for development. The effect of temperature on the developmental rate (Fig. 17.5a) is captured by Eqn A2 (Brière *et al.*, 1999).

$$R(T(t)) = {}^1/_{days(T)} = \frac{a(T(t) - T_L)}{1 + b^{T-T_U}} \quad \text{(A2)}$$

The constants a and b and the lower and upper thresholds T_L and T_U are estimated from data. The average development of a cohort initiated at some time t_0 is completed when the daily rates sum to 1 (i.e. $\int_{t_0}^{t} R(T(t))\,dt = 1$. Development can also be affected by nutrition and other factors.

Resource acquisition (*f*(*R*,*N*))

Similar demand- and search-driven resource acquisition processes occur in all trophic levels (e.g. photosynthesis, water and nitrogen acquisition by plants and resource acquisition by higher trophic levels; Gutierrez, 1992, 1996; Gutierrez and Baumgärtner, 1984, 2007) including the economic one (Regev *et al.*, 1998). The per capita resource acquisition rate S (= supply) is computed using the ratio-dependent functional response model (Eqn A3; Fig. 17.5b, solid line) where D is the per capita consumer demand that varies with temperature and other factors (e.g. [CO_2], see also Gutierrez, 1992).

$$S = f(N,C) = D(1 - e^{\frac{\alpha N}{DC}}) \quad \text{(A3)}$$

Depending on the biology of search, Eqn A3 is type II if α is constant but is type III if α is an increasing function of C (i.e. $\alpha(C)$). In plant physiology, the function $\alpha(C = LAI)$ would be the well-known Monteith (1965) light extinction model, and in animal ecology the equally well-known Nicholson and Bailey model (1935) (see Gutierrez, 1996). The type II model is related to the model proposed by Watt (1959) and is the predator form of the Frazer and Gilbert (1976) parasitoid model. The choice of the predator (single attacks) or parasitoid (multiple attacks) form of the model depends on the attack biology being modelled. Dividing both sides of Eqn A3 by D yields the supply–demand ratio ($0 \leq S/D < 1$) (i.e. search is always imperfect).

Per capita age-specific fecundity

($F(T, x)$, Fig. 17.5c) may be captured by a simple per capita reproduction function $f(x)$ (Bieri *et al.*, 1983) scaled by the effects of temperature ($\phi(T)$, Fig. 17.5d).

$$F(T, x) = \phi(T)h(x), \qquad \text{(A4)}$$

where

$$h(x) = \frac{cx}{d^x}$$

Temperature-dependent mortality

Similarly, mortality varies with temperature and the rate per unit time may be captured by upward convex functions such as Fig. 17.5e with constants d and e.

$$0 \leq \mu(T) = d \times (T - T_{opt})^2 + e \leq 1 \qquad \text{(A5)}$$

Of course, in trophic interactions, we must include the predation effects of higher trophic levels ($g(N, C, T)$, see Eqns A3 and A4).

Dormancy

Various forms of dormancy allow different species to survive unfavourable periods. Dormancy is absent in many species: it may be transient in some; may be induced by low host density and/or high temperatures; it may be induced by day length (Fig. 17.5f) or it may depend on temperature and day length (Fig. 17.5g); but also may be influenced by nutrition and other factors (see, for example, pink bollworm, Gutierrez *et al.*, 1981). As appropriate, the time of dormancy initiation, the daily rates of dormancy induction and dormancy termination must be computed in the model (see Gutierrez *et al.*, 2006, 2011).

Part IV

Management: Control and Adaptation

18 Climate, CO_2 and Invasive Weed Management

Lewis H. Ziska

Crop Systems and Global Change Laboratory, USDA-ARS, Beltsville, Maryland, USA

Abstract

Given the economic and environmental harm caused by invasive weeds, one of the fundamental objectives of weed biologists is to manage invasive populations in order to minimize their impact following introduction. At present, in most developed countries, chemical application remains the principal means of management regarding spread and impact. Yet, it is becoming increasingly clear that there are a number of direct and indirect effects between climate change and rising carbon dioxide levels [CO_2] that are likely to alter herbicide efficacy. For those data that suggest a reduction in effectiveness, primarily with rising CO_2 levels, there are several (as yet untested) possible physiological and/or physical mechanisms. At present, it appears that a single ubiquitous explanation for reduced efficacy in the context of climate change is unlikely. A number of alternative non-chemical weed control methods are available, and a combination of available management methods under the rubric of integrated pest management may provide a robust strategy to minimize climate change/[CO_2] impacts on weed management. Overall, looking forward, it is highly likely that current chemical control practices will need to change to cope with emerging challenges related to increasing herbicide resistance and extreme variation in climate parameters (e.g. drought). It is also clear that research to adapt herbicide management to these challenges deserves more time and attention than it has received to date.

Introduction

As illustrated in previous chapters, the best means to keep invasive weeds and other invasive species at bay is to develop global prevention and detection strategies that identify and track these species so that they can be excluded or prevented from establishing themselves into new regions.

However, even with detection and prevention, it is likely that new invasive introductions will continue to occur (e.g. Bebber *et al.*, 2013). Consequently, it is worthwhile to focus on those practices that are currently being utilized to keep the populations of invasive weeds at acceptable levels, and how those practices, in turn, are likely to be affected by a changing climate, or increases in atmospheric carbon dioxide (CO_2) levels.

In management vernacular, 'acceptable levels' has a dual meaning: first, to keep any new introduction from multiplying beyond their point of establishment, or, if possible, to eliminate them altogether; second, if detection occurs after initial introduction and proliferation (i.e. eradication is not possible), then to minimize the economic or environmental impact of the invasive species while preventing further spread.

Although exceptions exist (e.g. witchweed, Shaw *et al.*, 1962), the eradication of newly introduced invasive species is rare (Timmons, 2005). Consequently, the emphasis following the introduction and recognition of a new invasive threat is to employ available management methods to keep the invasive population at

 Invasive Species and Global Climate Change (eds L.H. Ziska and J.S. Dukes)

acceptable levels within a given biogeographical region.

Globally, these methods vary greatly and may include the identification and implementation of cultural, mechanical, chemical and biological options. However, among developed countries, such as Australia and the USA, the application of chemicals, i.e. herbicides, represents the most widely used alternative for invasive weed control. In other locations, cultural weed control, such as the physical removal of weeds by hand, may be more widespread.

Among invasive pests, weedy species are recognized as eliciting the highest costs to both agriculture and ecosystems (Pimentel *et al.*, 2005; Oerke, 2006). But what is an invasive weed? Often, a weed, if new to an area, will be labelled as invasive; yet other weeds, while non-native, have been around sufficiently long enough to have become accepted as part of the 'naturalized' weed community. Ragweed (*Ambrosia artemisiifolia*), for example, is native to North America but is considered invasive in Eastern Europe, where it has only occurred in recent decades. Lambsquarters (*Chenopodium album*) is from Eurasia but has a long history in North America and, consequently, is usually not considered invasive.

For this chapter, rather than defining an invasive weed *per se*, the emphasis will be on how climate change and/or CO_2 level is likely to alter the control of weed species within developed countries; i.e. how chemical management is likely to be affected by anthropogenically induced climatic change and/or rising CO_2 levels. At present, the interaction between climate, CO_2 and chemical management represents the greatest amount of empirical data available. However, an examination of how climate and/or CO_2 may impact alternative weed management strategies will also be included. While such an assessment is primarily theoretical at present, it may provide a starting point in identifying and pursuing future objectives for non-chemical approaches and integrated pest control in a changing climate.

Herbicides: History, Application and Efficacy

History

Although some attempts were made in the early 20th century to control weeds by chemical applications (most notably, the use of sodium arsenite to control aquatic plants by the US Army Corps of Engineers), labour shortages during the Second World War greatly intensified the pressure to develop chemicals to control weed infestations. By the 1940s, 2,4-D (2,5 dicholorophenoxyacetic acid), a plant growth regulator, was introduced and became widely utilized by the agricultural community as a means to control broadleaved weeds (Timmons, 2005). Increasing mechanization and chemical utilization at the farm level, concurrent with post-war demographic shifts to suburbs with less farm labour, increased herbicide use significantly. By the mid-1960s, herbicides, as the primary means of weed control, were being applied on 55% of US farms (www.nass.usda.gov).

By the 1980s and 1990s, the introduction of new herbicides with amino acid inhibitors led to commercial ventures to increase their use by genetically modifying crops to introduce herbicide resistance (e.g. 'Round-up Ready' soybean). Adoption and wide application of this class of herbicides has, in turn, led to application intensification. For example, as of 2007, approximately 94% of the soybeans grown in North America had been genetically engineered for herbicide resistance (USDA National Agricultural Statistics Service, 2007). At present, the acreage treated with herbicides in the USA is about twice that of fungicide and insecticide use combined (see Figure 8.1 in Ziska and Dukes, 2011).

Application

The effectiveness of herbicidal weed control depends on the timing of herbicide application relative to crop development.

While there are a number of timing options, the two broad categories are pre-plant or pre-emergence, where the herbicide is applied directly to the soil (Fryer and Evans, 1968), and post-emergence, where the herbicide is applied directly to emerged plants. However, depending on the stage of growth of the weed relative to the crop, herbicides can be applied pre-plant, pre-emergence, at emergence, post-emergent, or some combination thereof.

Post-emergent applications are made before canopy closure. Herbicides can be applied by spraying directly over the rows, or between rows. They can be incorporated into the soil, or applied to the soil surface. If applied in spray form, the spray can be directed to the tops of leaves, or under the plant canopy. Sprays that include the active chemical ingredient needed to kill the weed are modified by surfactants or wetting agents in order to increase the transfer and uptake of the active ingredient. Herbicides can be selective or non-selective, depending on the mode of action and specific physiology of the plant to be controlled.

There are numerous sources that explain in much greater depth the chemistry and application of herbicides. Interested readers are referred to the *Herbicide Handbook*, published by the Weed Science Society of America (WSSA, 2007).

Efficacy

The effectiveness of any herbicide for weed control is a function of the amount applied, when it is applied relative to the biology of the weed, and the environmental conditions during and after application.

Abiotic or physical factors that influence herbicide efficacy include soil moisture and temperature, wind speed, rainfall, relative humidity and air temperature. These factors may influence both plant uptake of the herbicide and/or its degradation following application. For example, wind can interfere directly with surface application, lead to herbicide 'drift' and accelerate the drying of herbicide solutions, with a subsequent reduction in plant absorption. Similarly, relative humidity and air temperature influence evaporation and affect droplet size and herbicide drift. Light may alter herbicide absorption, particularly with high temperature, as it may increase vaporization and metabolic degradation (Coupland, 1986).

There are also a number of plant-based biological aspects that can affect herbicide efficacy. For example, herbicides are dependent on metabolic function; hence, younger, actively growing plants are, in general, more susceptible to herbicides. In addition, meristem morphology (i.e. grasses versus broadleaf weeds), leaf characteristics (e.g. cuticular thickness, pubescence, stomatal number) and root-to-shoot ratio can all influence herbicide uptake and translocation (WSSA, 2007).

Such characteristics are often species specific; consequently, there are a number of species by environmental interactions that affect herbicide effectiveness. For example, depending on species, temperature and humidity will affect water absorption and transpiration, with subsequent effects on chemical absorption and movement. Water availability may also affect leaf characteristics, such as thickness and orientation, and alter herbicide absorption. Overall, any environmental condition that reduces metabolic plant function is likely to affect the herbicide's mode of action adversely.

Herbicide Efficacy, CO_2 and Climate Change: Potential Interactions

Climatic interactions

There are a number of means by which climate change, particularly abiotic variables, can alter herbicide efficacy. For example, an increase in the number of extreme weather events (e.g. heavy rainfall) may limit opportunities to apply herbicides in the field. Similarly, greater variability in rain, wind, humidity or soil/air temperatures may reduce the effectiveness of application (e.g. coverage) and herbicide retention times. Windier conditions would reduce suitable opportunities for herbicide application by increasing the risk of drift. Increasing soil temperatures and/or drought would increase herbicide volatilization. Overall, increased variation in environmental factors associated with climatic change may reduce application opportunities, alter herbicide persistence following application and increase herbicide injury of crops and other non-target organisms.

Biotic interactions

Rising carbon dioxide and temperature could increase photosynthetic rates, with a subsequent stimulation of enzymatic activity and pigment production. Consequently, chemical inhibitors of these metabolic functions (e.g. atrazine) could increase in efficacy. Conversely, protein content per gramme of leaf tissue has been shown to decline with rising CO_2 levels (Ehleringer *et al.*, 2002); such a decline would reduce demand for aromatic and branch chain amino acids, with a reduction in the efficacy of amino acid inhibitors such as glyphosate. Leaf development and morphology would also be affected by CO_2 and/or climate in ways that could reduce herbicide effectiveness. For example, if rising CO_2 or temperature increased leaf area or leaf number, this could increase the interception of post-emergent herbicides. Alternatively, increasing CO_2 and/or water stress could result in changes in leaf anatomy and surface characteristics (e.g. a reduction in stomatal number) that could reduce chemical absorption. Accumulation of leaf starch, which can occur in some plants grown at elevated CO_2 (e.g. Arp, 1991), could also interfere with herbicide mode of action. At the whole plant level, warmer temperatures and more CO_2 could increase metabolism, but also decrease the time spent in the seedling stage; i.e. the period of greatest herbicide sensitivity. Temperature and CO_2 increases could also alter the allocation of assimilate to different plant organs, with subsequent changes in plant development, e.g. timing of flowering, leaf initiation, pollen production, root-to-shoot ratios, etc. (Ziska and Bunce, 2006). Such changes, in turn, could affect herbicide uptake, translocation and efficacy.

At the community or ecosystem level, CO_2 and climatic extremes could affect flowering times and pollen production differentially between related species. Such a differential response could, in turn, alter gene flow, with changes in the occurrence of herbicide resistance among weedy populations, especially if genetically modified plant species are present (Snow, 2002; Ellstrand, 2003; Ziska *et al.*, 2012). The impact of long-term evolutionary changes being altered by CO_2 and/or climate is less certain; only a single study has demonstrated that a weed species (field mustard, *Brassica nigra*) could undergo rapid genetic changes to adapt to the sudden onset of drought (Franks *et al.*, 2007). Clearly, additional information is necessary to understand gene flow in the context of CO_2/climate.

Carbon Dioxide, Climate Change and Chemical Management: A Review of Existing Studies

There are a number of reports of rising CO_2 having a negative effect on herbicide efficacy independent of any impact of climate change. In a greenhouse study, Ziska *et al.* (1999) examined whether elevated CO_2 altered the glyphosate sensitivity of redroot pigweed and common lambsquarters (a C_4 and a C_3 species, respectively). Elevated CO_2 did not affect the response of redroot pigweed, but reduced the overall effectiveness of glyphosate for lambsquarters (Ziska *et al.*, 1999). In a separate study, a C_3 perennial grass, quackgrass (*Elytrigia repens*), when grown at the higher (770 ppm) CO_2 concentration and treated with the recommended dosage of glyphosate, demonstrated increased tolerance to glyphosate for two different cohorts (Ziska and Teasdale, 2000). Field trials for Canada thistle (*Cirsium arvense*) showed reduced herbicide efficacy with both glyphosate and glufosinate when grown at elevated carbon dioxide (Ziska *et al.*, 2004; Ziska and Goins,

2006). Archambault *et al.* (2001) reported that herbicide efficacy was often reduced with elevated CO_2, but that the effects were dependent on the mode of action, weed species and competition. Archambault (2007) showed that the efficacy of glufosinate-ammonium declined for wild oat control with increased temperatures and/or CO_2 concentration. Manea *et al.* (2011) showed that three of four C_4 grass species displayed increased tolerance to glyphosate under elevated CO_2 (Table 18.1).

Potential Mechanisms

Elevated CO_2

Given these results, are there commonalities in the biological response of these plant species to elevated CO_2 levels that could account for their greater herbicide tolerance? If CO_2 stimulates plant growth, then reduced efficacy may simply be related to plant size and an increased dilution effect at the time of herbicide application, as has been suggested by Manea *et al.* (2011). For Canada thistle, applications of glufosinate and glyphosate at ambient and elevated CO_2 levels resulted in complete plant desiccation within 2 weeks (Ziska *et al.*, 2004). However, regrowth of the elevated CO_2 plants occurred within a month after spraying. Because aboveground biomass was killed, absorption and mode of action did not appear to be affected by CO_2 concentration. The greater extent of regrowth for the elevated CO_2 grown plants was associated with a shift in root over shoot biomass at the time of spraying (Ziska *et al.*, 2004). This suggests that CO_2 stimulation of belowground biomass could dilute the effect of any systemic herbicides (e.g. glyphosate), provided belowground structures are not killed. This is consistent with Patterson and Flint (1990), who hypothesized that chemical control could be affected adversely if additional CO_2 resulted in greater root, rhizome or tuber growth; in part, because each belowground structure could give rise asexually to a new plant. In other studies, glyphosate tolerance did increase with plant size for lambsquarters and quackgrass; however, cohort age at the time of spraying was not a factor in enhanced tolerance to the herbicide relative to the ambient CO_2

Table 18.1. Documentation of statistically significant changes in chemical (herbicide) efficacy as a function of elevated relative to ambient carbon dioxide concentrations. Herbicides were applied at the manufacturers' recommended dosage.

Weed species	Photosynthetic pathway	Herbicide (active ingredient)	Change in efficacy?	Reference
Canada thistle (*Cirsium arvense*)	C_3	Glyphosate	Declined	Ziska *et al.*, 2004
Canada thistle (*Cirsium arvense*)	C_3	Glufosinate	Declined	Ziska, unpublished
Dallisgrass (*Paspalum dilatatum*)	C_4	Glyphosate	Declined	Manea *et al.*, 2011
Lambsquarters (*Chenopodium album*)	C_3	Glyphosate	Declined	Ziska *et al.*, 1999
Lovegrass (*Eragrostis curvula*)	C_4	Glyphosate	Declined	Manea *et al.*, 2011
Quackgrass (*Elytrigia repens*)	C_3	Glyphosate	Declined	Ziska and Teasdale, 2000
Redroot pigweed (*Amaranthus retroflexus*)	C_4	glyphosate	None	Ziska *et al.*, 1999
Rhodes grass (*Chloris gayana*)	C_4	Glyphosate	Declined	Manea *et al.*, 2011
Smut grass (*Sporobolus indicus*)	C_4	Glyphosate	None	Manea *et al.*, 2011

condition (Ziska *et al.*, 1999; Ziska and Teasdale, 2000). While dilution may be a common factor, it would be premature to assume that this is the only mechanism for CO_2-induced changes in herbicide tolerance. Given the diversity of herbicidal modes of action, the range of genetic and phenotypic variability among weeds and the potential means by which CO_2 can alter plant biology, it seems unlikely that a single common biological explanation is valid.

Climate change and CO_2 interactions

As previously discussed, there are a number of potential means whereby climate could also affect herbicide efficacy; however, there are only a handful of experimental data that address this issue. Ziska and Goins (2006) examined possible changes in weed species demographics associated with field-grown Round-up Ready™ soybeans with and without applications of glyphosate over a 2-year period. In this study, there was a large precipitation shift between years, with a record amount of rainfall in 2003 followed by a near-normal year in 2004. Using this contrast as a portent of extreme climate events, Ziska and Goins observed that precipitation itself was a strong selection factor for weed populations, with only shallow-rooted weedy grasses able to grow in the anoxic environment created by the large amount of rainfall. These grasses were C_4 species, and did not respond to the elevated CO_2 treatment. As a result, CO_2 concentration had no effect on herbicide efficacy in a very wet year. In contrast, a variety of weeds, including C_4 and C_3 broadleaf species, were observed during the year with near-normal precipitation levels. Elevated CO_2 resulted in a significant growth stimulation of C_3 broadleaf weeds, with a subsequent decline in glyphosate efficacy (Table 18.2). In a study by Valerio *et al.* (2011), tomato, a C_3 crop species, was grown from emergence to anthesis using replacement series mixtures with *Amaranthus retroflexus*, a C_4 weed species, at three different CO_2 concentrations; 400, 600 and 800 μmol mol^{-1}, with and without water stress. Under well-watered conditions, leaf photosynthetic rates and plant height, leaf area and biomass all increased with elevated CO_2 concentration for tomato relative to *A. retroflexus*, consistent with the kinetics of C_3 photosynthesis. However, if water was limiting, as is expected if climate change increases the occurrence of drought, a significant positive effect of CO_2 concentration was noted for plant height and biomass of *A. retroflexus* with increased competitive ability (Valerio *et al.*, 2011). Similarly, a study by Alberto *et al.* (1996) examined competition between rice (*Oryza*

Table 18.2. Averages and statistical significance for aboveground biomass (g m^{-2}) as a function of weed species grown at two CO_2 concentrations (ambient and elevated, ambient +250 ppm) with and without the application of glyphosate (+GLY, –GLY) in stands of Round-Up Ready™ soybean. There was a record amount of rainfall in 2003, and selection of shallow-rooted C_4 grasses, whereas 2004 was a near-normal precipitation year during which a greater variety of weeds was observed. Application of glyphosate controlled 100% of the C_4 grasses in 2003. Conversely, a significant GLY × CO_2 interaction was observed for C_3 broadleaf weeds in 2004. Data are from Ziska and Goins, 2006. (*)P <0.10; *P <0.05; **P <0.01; ***P <0.001.

	Averages				Level of significance		
	Ambient	Elevated	+GLY	–GLY	CO_2	GLY	GLY × CO_2
2003							
C_4 grasses	138.1	128.3	0.0	266.1		***	
2004							
C_3 broadleaf	48.3	303.9	18.5	333.7	***	***	***
C_4 broadleaf	166.1	257.1	0.0	423.2		***	
C_4 grasses	157.1	23.1	0.5	179.8	(*)	*	

sativa, C_3) and a weedy competitor, smooth barnyard grass (*Echinocholoa glabrescens*, C_4) at two different carbon dioxide concentrations (ambient and ambient +200 ppm) and two different temperatures (day/night of 27/21 and 37/29°C). This study confirmed that at 27/21°C, increased carbon dioxide favoured the crop (i.e. the C_3 species) over the weed; however, with concomitant changes in both CO_2 and temperature, the C_4 weed was favoured, primarily because higher temperatures resulted in floral sterility, reduced photosynthetic stimulation and increased seed yield loss for rice relative to the C_4 weed (Alberto *et al.*, 1996). Overall, these data suggest that CO_2 and climate (e.g. temperature, water availability) may interact to alter chemical weed management in ways that are not predicted, based simply on the photosynthetic pathway of a given weed species.

Future Chemical Management

There are a number of studies to date indicating that increasing CO_2 concentration per se, or in combination with some aspect of climate extremes (e.g. drought), is likely to alter the overall efficacy of chemical weed control. The overall basis for CO_2 impacts on efficacy may be related to biomass partitioning and/or dilution; however, this is by no means a ubiquitous explanation. The interaction of climate and CO_2 may be more subtle, as evidenced by the effect of extreme weather events on weed demographics (e.g. more C_4 grasses in an extremely wet year), or the interaction of drought on C_3:C_4 species competition. Clearly, there is an urgent need for additional experimentation in this area.

It seems likely that there will be no single strategy that will compensate for reduced chemical efficacy with increasing CO_2 levels and/or climatic change. Potentially, different surfactants, or other chemical carriers, could increase the absorption of active herbicide ingredients even under high wind or extreme rainfall (Froud-Williams, 1996). The incorporation of more extreme scenarios in climate modelling could also help optimize herbicide application times. Another promising strategy may be to adapt existing environmental performance profiles for a given weed (e.g. quackgrass and glyphosate, Coupland and Caseley, 1981; Coupland, 1983) by the inclusion of CO_2 and temperature response profiles for herbicide efficacy. Such profiles could also be used for a number of invasive weeds or crop–weed associations (Archambault *et al.*, 2001).

There is a further consideration in adapting herbicide management to increasing CO_2 levels and/or climate change; the recognition that the ease and simplicity of herbicide use, particularly glyphosate, carries a long-term cost. That cost is becoming more evident as greater usage of a single chemical over large geographic areas increases selection pressure and selects for weedy resistance to that chemical.

One recent example is that of Palmer amaranth (*Amaranthus palmeri*), one of the few amaranths considered invasive (www.plants.usda.gov/java/profile?symbol=AMPA). An initial report documented glyphosate-resistant Palmer amaranth in a few counties in Georgia in 2005 and 2006 (Culpepper *et al.*, 2008). As of 2012, the resistant line of Palmer amaranth is present in several million acres of farmland throughout the lower Corn Belt. Unfortunately, Palmer amaranth is not the exception. As Heap (2011) has documented (Fig. 18.1), the increase in herbicide resistance is increasing logarithmically. Given this increase, it would be short-sighted to assume that any CO_2 or climatic impact that reduced herbicide efficacy could be compensated by more frequent sprayings or more concentrated herbicide.

Alternatives to Chemical Management

Although chemical control is the primary means for the management of pest species in developed countries, non-chemical control remains the predominant strategy for much of the world. Unfortunately, at present, there are virtually no experimental data indicating how those methods will be impacted by anthropogenic climate change.

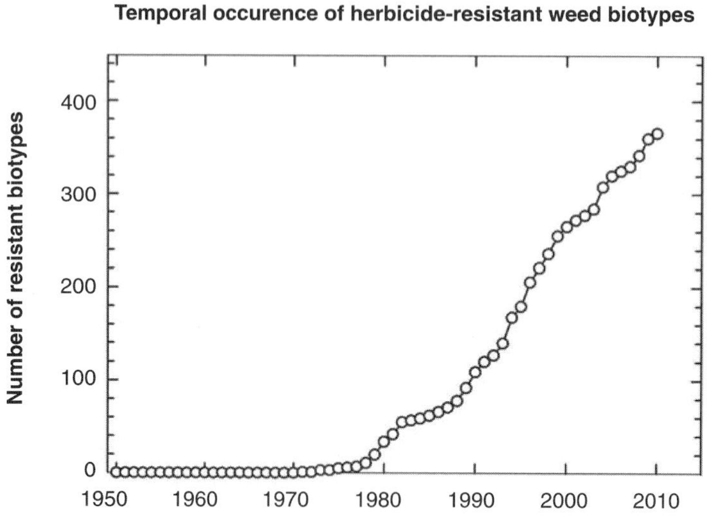

Fig. 18.1. Temporal occurrence since 1950 in the number of herbicide-resistant weed biotypes. Data are from Ian Heap, *www.weedscience.com*.

However, an analysis and description of these non-chemical means for agronomic and invasive weed management will help to understand the basis for their effectiveness and, hopefully, provide a starting point in developing a theoretical construct of how, potentially, these methods could be altered by CO_2 and/or climate.

Physical control

It is estimated that most farmers (250 out of 350 million) still employ hand pulling or hoeing as the primary means of weed control (Hill, 1982; Melander *et al.*, 2005). Weeding by hand still accounts for a considerable portion of preharvest labour input in the developing world (~60%, Webb and Conroy, 1995). However, hand pulling may be particularly efficacious when a known invasive plant first appears, or if the area is small or if the crop is of high cash value. For example, for glyphosate-resistant Palmer amaranth, small populations should be removed by hand or hoe in the seedling stage to prevent spread (Sosnoskie *et al.*, 2011). However, hand hoeing may not be economically viable for large farms, or if the degree of infestation is high. Furthermore, perennial weeds, particularly those that grow from underground structures, will require multiple visits. If farm mechanization is available, then tillage, or cultivation to bury weed populations or tear shoots from roots, may act as an effective means of physical weed control. The kind of tillage can vary, from primary (initial, usually deep, soil break-up at the beginning of the growing season) to secondary tillage (usually lighter passes of less than 10 cm). Overall, tillage, or physical removal, is a very effective means of controlling weeds between rows, but may be less effective for weeds within rows, particularly if the weed resembles the crop.

Cultural control

This type of weed control frequently reflects management choices made at the farm level, and is determined by local resources. For example, flooding is often employed as a means of physical weed control, since a number of agronomic and invasive weeds do not survive anaerobic conditions. On a global basis, it is one of the most effective non-chemical means culturally to control weed populations in rice (paddy) systems. The imposition of flooded conditions in rice systems to eliminate weed competition is one factor that allows paddy or irrigated rice to achieve the highest yields among rice systems (e.g. upland, rainfed; Watson, 2002). Mulching, or the deposition of surface material, is another means of cultural weed control that can prevent seed germination and manage weeds effectively. However, because of cost, mulches are usually used in small areas (e.g. gardens) or for high-value crops (e.g. strawberries). Examples of effective mulches include straw, sawdust, wood chips and grass clippings, black plastic and shredded paper. Living mulches such as cover crops or green manure can also be an effective weed management technique (Teasdale and Mohler, 2000). Crop rotation can also influence weed–crop competition. This is because many of the most troublesome weeds are those that simulate the life cycle of the crop or desired vegetation; in turn, alternating crops can prevent an accumulation of those weeds within the seed bank.

Biological control

The introduction of natural enemies to lower the population level or competitive ability of an agronomic or invasive weed (or any invasive species) is a means to keep the level of that pest below an economic or environmental threshold. Biological control can represent a low-cost, permanent strategy, because an effective biocontrol agent is self-sustaining. Although initial costs may be high, they are non-recurring, and if established successfully, no additional costs may be needed. In addition, biocontrol agents can control a pest over large regional areas without the use of pesticides.

However, there are also limits to biological control. Biocontrol agents must be highly selective, since such agents may be unable to distinguish between related species. A thorough understanding of the biology is necessary to ensure that a control agent rather than a new pest is selected (e.g. the introduction of cane toads as a biocontrol agent in Australia). Specificity of the control agent is critical, but such specificity may require a great deal of research; unfortunately, in agriculture, a number of weed species are usually present, so that a single, specific biocontrol agent may be inadequate.

Potential Interactions between Alternative Pest Control, CO_2 and Climate

There are a number of means by which climate and CO_2 concentration could affect physical weed control. For example, crop plants could be stimulated to grow faster with increasing carbon dioxide, resulting in less time for mechanical control. Climate extremes (e.g. precipitation) could limit opportunities for field operations. Rising levels of carbon dioxide could also result in further belowground carbon sequestration, particularly in roots with subsequent increases in asexual reproduction (e.g. Rogers *et al.*, 1994). Data for some invasive plants are consistent with this hypothesis, with four out of five invasive plants showing an increase in root-to-shoot ratio for the increase in atmospheric CO_2 that occurred during the 20th century (Ziska, 2003). Overall, a greater enhancement of roots to shoot would have potentially negative effects on tillage; for example, the chopping up of a larger root system induced by rising CO_2 levels would provide greater opportunity for asexual recolonization of perennial weeds like Canada thistle (Ziska and Goins, 2006). Declining water for irrigation could adversely affect the use of flooding as a weed control measure. Rice cultivation can adjust to less water, for example direct-seeded rice, but greater emphasis must be given to post-emergent weed control in order to keep rice yields high. Rising CO_2 and/or climate change could also alter the efficacy of mulching. For example, mulches are less effective against perennial than annual weeds, because food reserves in belowground structures of perennials can permit these species to grow through thick mulches. Hence, if elevated CO_2 results in an increase in root-to-shoot ratios, this may make perennials (e.g. Canada thistle) even more resistant to control by mulch application. Similarly, increasing climatic extremes could accelerate mulch degradation. However, mulches could also help by conserving soil moisture during drought, or helping to modify temperature extremes, thereby aiding the crop, and presumably increasing crop competition with weedy species. For crop rotations, crop type will be chosen based on climatic variables (e.g. earlier springs, milder winters), as well as fertilizer supply, water availability, etc. As climate changes, a number of management decisions related to weed demographics, field access, etc., will necessitate new management directions. Biological agents and weeds may respond differentially to climate change. The effectiveness of biocontrol agents (e.g. insects) is dependent on synchrony with the plant community and environment. Rising temperatures have already altered a number of insect species as well as plant–insect interactions (Singer and Parmesan, 1993; Parmesan, 1996, *inter alia*). Such climate induced de-synchrony would reduce the efficacy of biocontrol agents.

Integrated Pest Control

Different management strategies for keeping population(s) of undesirable weedy species in check can, on occasion, conflict (Zimdahl, 1993). Integrated pest management or control (IPM, IPC) is a strategy designed to prevent conflicts by wielding all management tools into a series of evaluations that begin with monitoring, threat assessment, application of control measures and follow-up. Such a strategy recognizes that, first, not all weeds (or other pest species) require control. Consequently, monitoring and identification are critical to distinguish significant from routine threats. Second, an action threshold must be determined beforehand, i.e. how high can the population get before action (control) is required. Third, prevention is emphasized, so that cultural methods, such as rotation, weed-free seed, cleaning of farm equipment to prevent pest dispersion, etc., are practised routinely. If monitoring, identification and action thresholds indicate that a pest species population is a threat and control is required, then a management decision is made. Such decisions should reflect an understanding of the potential cost, environmental damage and efficacy. Control strategies may include a combination of methods to reduce pest populations to acceptable levels. (Broadcast, non-discriminatory chemical spraying in these circumstances is usually considered as a last resort.) Following the application of control method(s), evaluations of the efficacy of the technique should be incorporated into future management decisions.

Such an approach can provide flexibility and continuity to pest management. It may therefore assist land managers to cope with potential changes in invasive species management resulting from rising carbon dioxide and climate change. For example, monitoring and recording will be an especially critical aspect of assessing new weed introductions. Changes in IPM/IPC to account for potential declines in herbicide or cultural weed control, and to identify how efficacy and/or timing could be improved, may also serve as a means of adaptation to climate change and/or rising CO_2 levels. Land managers could also exploit climate extremes (e.g. drought) that could facilitate weed or pest control. Overall, an integrated strategy may represent an evolving system that could, with proper feedback, provide a useful and effective management tool for invasive pest control, even in the context of climatic change.

Summary

There are, globally, a diverse set of management practices that are used to keep weeds (and other pests, native or invasive) at bay. Often, these practices reflect the technological and cultural capacity of a given region, with developed and developing countries having a greater reliance on chemical and cultural weed control, respectively. Although there are numerous potential interactions between pest management and human-induced increases in atmospheric carbon dioxide and/or climatic change, most empirical evidence to date points to a reduction in herbicide efficacy; however, it seems likely that all management strategies will be affected to some degree and must adjust accordingly. Overall, with respect to invasive species, there is a greater need for global surveillance to determine distribution and establishment, a reappraisal of the effectiveness of non-chemical and chemical strategies, research to determine potential adaptation methods that could exploit environmental extremes as a means of pest control, and a greater emphasis on integrated management rather than non-selective chemical applications.

References

Alberto, A.M., Ziska, L.H., Cervancia, C.R. and Manalo, P.A. (1996) The influence of increasing carbon dioxide and temperature on competitive interactions between a C_3 crop, rice (*Oryza sativa*), and a C_4 weed (*Echinochloa glabrescens*). *Australian Journal of Plant Physiology* 23, 795–802.

Archambault, D.J. (2007) Efficacy of herbicides under elevated temperature and CO_2. In: Newton, P.C.D., Carran, A., Edwards, G.R. and Niklaus, P.A. (eds) *Agroecosystems in a Changing Climate*. CRC Press, Boston, Massachusetts, pp. 262–279.

Archambault, D.J., Li, X., Robinson, D., O'Donovan, J.R. and Klein, K.K. (2001) The effects of elevated CO_2 and temperature on herbicide efficacy and weed/crop competition. *Report to the Prairie Adaptation Research Collaborative No 29*, Ottawa, Canada.

Arp, W.J. (1991) Effects of source–sink relations on photosynthetic acclimation to elevated CO_2. *Plant, Cell and Environment* 14, 869–875.

Bebber, D.P., Ramotowski, M.A.T. and Gurr, S.J. (2013) Crop pests and pathogens move polewards in a warming world. *Nature Climate Change* doi:10.1038/NCLIMATE1990.

Coupland, D. (1983) Influence of light, temperature and humidity on the translocation and activity of glyphosate in *Elymus repens*. *Weed Research* 23, 347–355.

Coupland, D. (1986) The effect of environmental factors on the performance of fluazifop-butyl against *Elymus repens*. *Annals of Applied Biology* 108, 353–363.

Coupland, D. and Caseley, J.C. (1981) Environmental influences on the effects of glyphosate on *Agropyron repens*. In: *Conference on Grass Weeds in Cereals*, 6–7 January 1981, University of Reading, Berkshire, UK. Association of Applied Biologists, UK, pp. 177–186.

Culpepper, A.S., Whitaker, J.R., MacRae, A.W. and York, A.C. (2008) Distribution of glyphosate resistant Palmer amaranth (*Amaranthus palmerii*) in Georgia and North Carolina during 2005 and 2006. *Journal of Cotton Science* 12, 306–310.

Ehleringer, J.R., Cerling, T.E. and Dearing, M.D. (2002) Atmospheric CO_2 as a global change driver of plant–animal interactions. *Integrative and Comparitive Biology* 42, 424–430.

Ellstrand, N.C. (2003) Current knowledge of gene flow in plants: implications for transgene flow. *Philosophical Transactions of the Royal Society London* (B) 358, 1163–1170.

Franks, S.J., Sim, S., Weis, A.E. (2007) Rapid evolution of flowering time by an annual plant in response to a climate fluctuation. *Proceedings of the National Academy of Sciences* 104, 1278–1282.

Froud-Williams, R.J. (1996) Weeds and climate change: implications for their ecology and control. *Aspects of Applied Biology* 45, 187–196.

Fryer, J.D. and Evans, S.A. (1968) *Weed Control Handbook*. Blackwell Press, Oxford, UK, 402 pages.

Heap, I. (2011) The International Survey of Herbicide Resistant Weeds (http://www.weedscience.com/summary/home.aspx, accessed 3 February 2014).

Hill, G.D. (1982) Impact of weed science and agricultural chemicals on farm productivity in the 1980s. *Weed Science* 30, 426–429.

Manea, A., Leishman, M.R. and Downey, P.O. (2011) Exotic C_4 grasses have increased tolerance to glyphosate under elevated carbon dioxide. *Weed Science* 59, 28–36.

Melander, B., Rasmussen, I.A., Barberi, P. (2005) Integrating physical and cultural methods of weed control: examples from European research. *Weed Science* 53, 369–381.

Oerke, E.C. (2006) Crop losses to pests. *Journal of Agricultural Science* 144, 31–43.

Parmesan, C. (1996) Climate and species range. *Nature* 382, 765–766.

Patterson, D.T. and Flint, E.P. (1990) Implications of increasing carbon dioxide and climate change for plant communities and competition in natural and managed ecosystems. In: Kimball, B.A., Rosenburg, N.J. and Allen, L.H. Jr (eds) *Impact of Carbon Dioxide, Trace Gases and Climate Change on Global Agriculture*, ASA Special Publication No 53. American Society of Agronomy, Madison, Wisconsin, pp. 83–110.

Pimentel, D., Zuniga, R. and Morrison, D. (2005) Update on the environmental and economic costs associated with alien-invasive species in the United States. *Ecological Economics* 52, 273–288.

Rogers, H.H., Runion, G.B. and Krupa, S.V. (1994) Plant responses to CO_2 enrichment with emphasis on roots and the rhizosphere. *Environmental Pollution* 83, 155–189.

Shaw, W.C., Shepherd, D.R., Robinson, E.L. and Sand, P.F. (1962) Advances in witchweed control. *Weeds* 10, 182–192.

Singer, M.C. and Parmesan, C. (1993) Sources of variation in patterns of plant–insect association. *Nature* 361, 251–253.

Snow, A.A. (2002) Transgenic crops – why gene flow matters. *Nature Biotechnology* 20, 542.

Sosnoskie, L.M., Webster, T.M., Culpepper, A.S. and Kichle, J. (2011) The biology and ecology of Palmer amaranth: implications for control. *Circular No 1000.* Cooperation Service of the State of Georgia, Tifton, Georgia.

Teasdale, J.R. and Mohler, C.L. (2000) The quantitative relationship between weed emergence and the physical properties of mulches. *Weed Science* 48, 385–392.

Timmons, F.L. (2005) A history of weed control in the United States and Canada. Weed Science 53, 748–761.

USDA, National Agricultural Statistics Service (2007) http://www.nass.usda.gov/, accessed 3 February 2014.

Valerio, M., Tomecek, M.B., Lovelli, S. and Ziska, L.H. (2011) Quantifying the effect of drought on carbon dioxide-induced changes in competition between a C_3 crop (tomato) and a C_4 weed (*Amaranthus retroflexus*). *Weed Research* 51, 591–600.

Watson, A.K. (2002) Flooding: weed control in rice. In: Pimentel, D. (ed.) *Encyclopedia of Pest Management.* Marcel Dekker, New York, pp. 283–286.

Webb, M. and Conroy, C. (1995) The socio-economics of weed control on small-holder farms in Uganda. In: *Proceedings, Brighton Crop Protection Conference – Weeds.* British Crop Protection Council, Farnham, UK, pp. 157–162.

WSSA (Weed Science Society of America) (2007) *Herbicide Handbook,* 9th edn. WSSA, Champaign, Illinois, 240 pages.

Zimdahl, R.L. (1993) *Fundamentals of Weed Science.* Academic Press, New York, 454 pages.

Ziska, L.H. (2003) Evaluation of the growth response of six invasive species to past, present and future carbon dioxide concentrations. *Journal of Experimental Botany* 54, 395–404.

Ziska, L.H. and Bunce, J.A. (2006) Plant responses to rising carbon dioxide. In: Morison, J.I.L. and Morecroft, M.D. (eds) *Plant Growth and Climate Change.* Blackwell Publishing, Oxford, UK, pp. 17–47.

Ziska, L.H. and Dukes, J.S. (2011) *Weed Biology and Climate Change.* Wiley-Blackwell, Ames, Iowa, 248 pages.

Ziska, L.H. and Goins, E.W. (2006) Elevated atmospheric carbon dioxide and weed populations in glyphosate treated soybean. *Crop Science* 46, 1354–1359.

Ziska, L.H. and Teasdale, J.R. (2000) Sustained growth and increased tolerance to glyphosate observed in a C_3 perennial weed, quackgrass (*Elytrigia repens* (L.) Nevski), grown at elevated carbon dioxide. *Australian Journal of Plant Physiology* 27, 159–164.

Ziska, L.H., Teasdale, J.R. and Bunce, J.A. (1999) Future atmospheric carbon dioxide concentrations may increase tolerance to glyphosate. *Weed Science* 47, 608–615.

Ziska, L.H., Faulkner, S.S. and Lydon, J. (2004) Changes in biomass and root:shoot ratio of field-grown Canada thistle (*Cirsium arvense*), a noxious, invasive weed, with elevated CO_2: implications for control with glyphosate. *Weed Science* 52, 584–588.

Ziska, L.H., Gealy, D.R., Tomecek, M.B., Jackson, A.K. and Black, H.L. (2012) Recent and projected increases in atmospheric CO_2 concentration can enhance gene flow between wild and genetically altered rice (*Oryza sativa*). *PLoS ONE* 7, doi:10.1371/journal.pone.0037522.

19 Early Detection and Rapid Response: A Cost-effective Strategy for Minimizing the Establishment and Spread of New and Emerging Invasive Plants by Global Trade, Travel and Climate Change

Randy G. Westbrooks,[1] Steven T. Manning[2] and John D. Waugh[3]

[1]*Invasive Plant Control, Inc, Whiteville, North Carolina, USA;* [2]*Invasive Plant Control, Inc, Nashville, Tennessee, USA;* [3]*Integra LLC, Washington, DC, USA*

Abstract

Over the past 50 years, considerable effort has been made by state and national agencies as well as other partners to minimize the establishment and spread of newly introduced and/or emerging invasive plants through single agency-led programmes, inter-agency councils and task forces and, most recently, the landscape approach to early detection and rapid response (EDRR). Examples of single agency-led programmes include the USDA-Carolinas Witchweed Eradication Program in the USA and the Kochia Eradication Project in Western Australia (EDRR 1.0). In recent years, state inter-agency councils and task forces have been formed to address all types of new invasive species – particularly newly introduced species that are not already regulated by federal or state agencies. The Delaware Invasive Species Council, the Ontario Invasive Plant Council and the Beach Vitex Task Force are good examples of this new trend in inter-agency partnering (EDRR 2.0). The landscape approach to EDRR involves the development of EDRR capacity at all levels of the landscape – local to national. It includes individual public and private land units, geographic land units (watersheds, biomes, corridors, etc.) and political land units (towns, counties, states/provinces and nations) (EDRR 3.0). From a societal standpoint, due to global climate change and increased global trade and travel, it is important to emphasize that the impacts of invasive species on food security, human health, and biodiversity will continue to increase unless steps are taken now to minimize their introduction, establishment and spread. Development of EDRR capacity at all levels of the landscape is a proven strategy for achieving those goals.

Introduction

Early detection and rapid response (EDRR) is a coordinated framework for the management of new invasive species introductions. Elements include early detection, identification and vouchering, verification and

 Invasive Species and Global Climate Change (eds L.H. Ziska and J.S. Dukes)

archiving, rapid assessment and rapid response (Westbrooks, 2004). Introduced species typically show a lag time before exponential population growth (Crooks and Soulé, 1999). This period before rapid expansion is the 'golden hour' for EDRR; once invasive populations cross a critical threshold, total elimination is often expensive and, in many cases, impractical. The goal of EDRR is the total elimination of an introduced invasive species from a defined area before it can become well established – while eradication is still cost-effective and practical (Waugh, 2009).

According to biogeographer John Kartesz, over 5000 species of introduced plants have established free-living populations outside of cultivation in the USA since European colonization began in the 1500s (Kartesz, 2012). Of this total, Kartesz has documented over 1300 species that are invasive, resulting in losses and control costs amounting to at least US$19.184 billion year^{-1} – as described below. This represents less than 5% of the 33,739+ invasive plant species that have been documented in the world (to date) by weed geographer Rod Randall (Randall, 2007, 2012), and points to the threat we face from this global issue. The problem has expanded, in large part because of a lack of attention to newly introduced and emerging species; landowners and managers generally focus, understandably, on widespread invaders that are known to impact land productivity and ecosystem balance seriously.

Introduced weed species cause serious economic problems. In the USA, for example, weeds are estimated to reduce crop yields by about 12% year^{-1} (Monaco *et al.*, 2002), which is currently at least US$20.567 billion year^{-1} in lost revenues (based on an estimated value of US$171,393,620,000 of field and miscellaneous crops in 2011; USDA, 2012). Over the past two decades, various estimates have been given for the cost of weed control across the USA. In 1999, it was estimated that over US$13.6 billion was being spent to control weeds in various settings. This includes US$4 billion for chemical weed control and US$3 billion for cultural and other methods of weed control in crops; US$1.5 billion in lawns, gardens and golf courses; and US$100 million for aquatic weeds control (Pimentel *et al.*, 1999, 2002; Monaco *et al.*, 2002). In 2011, the US Environmental Protection Agency estimated that over US$5.8 billion was spent on chemical weed control in 2007. This includes US$4.2 billion on herbicides and growth regulators in crops, US$896 million by industry, communities and government agencies and US$749 million on lawns and gardens (Grube *et al.*, 2011).

Assuming a conservative figure of approximately US$8.5 billion spent on weed control, this brings total estimated losses and control costs of weeds in the USA to at least US$29.067 billion year^{-1}. Since introduced species account for about two-thirds of the naturalized flora in the USA (as deliberate introductions or contaminants of crop seeds) (Mack and Erneberg, 2002), and assuming an equal distribution of weed species in agriculture, the total economic impact on the US economy from non-native invasive weed species would now equal or exceed US$19.184 billion year^{-1}. Additional research is required to measure more accurately the economic impacts of weeds in general, and of non-native weeds in particular. We present these figures to give the reader an indication of how little attention has been given to this important economic facet of food security.

On a global scale, it was estimated in 2001 that invasive species cost as much as US$1.4 trillion annually, or 5% of planetary GDP (Pimentel *et al.*, 2001). This is comparable to climate change costs, which have been projected in the range of 1–20% of GDP (Stern, 2006).

The annual costs to the US economy from weeds alone are similar to the impacts of a major hurricane. This does not take into account the costs of other invasive species such as invasive insects, diseases and aquatic nuisance species. Yet, because the impacts are distributed over time and space, and the costs are incremental, they do not *feel* like disasters and are not treated as such. In a similar fashion, the non-linear effects of changing climate are bound to have more surprises in store, which may include more costly biological invasions.

Over the past 58 years, since the discovery of the parasitic witchweed (*Striga asiatica* (L.) O. Kuntze) in the eastern Carolinas, three distinct approaches have evolved to minimize the establishment and spread of new invasive plants in the USA. These includes single agency-led weed eradication programmes (e.g. the USDA-Carolinas Witchweed Eradication Program), which were established to address regulated federal/state noxious weeds (EDRR 1.0), state inter-agency councils (e.g. the Delaware Invasive Species Council) and task forces (e.g. the Beach Vitex Task Force) to address both regulated and non-regulated invasive plants (EDRR 2.0), and most recently, a growing effort to develop landscape-level EDRR systems for all types of land units (individual, geographic and political) at all levels of the landscape, from local to national (EDRR 3.0). The common goal of all these efforts is to develop new capacity to detect and eradicate new invasive plants early – to minimize costs, from an economic and an ecological standpoint. *Further development of EDRR capacity at all levels of the landscape would be a practical and cost-effective approach for minimizing this economic toll on the global economy – especially since climate change could permit new and established invaders to spread even more.*

The 'Philosophy' of EDRR

Preventing the introduction of invasive species into a country or state to begin with is the first line of defence against new invasions. However, even the best prevention efforts by plant regulatory agencies will not stop all invasive species introductions. Next to prevention, the most cost-effective way to manage the potential negative impacts of new invasive plants is through early detection and rapid response (EDRR). Such efforts include detecting new invaders when they first arrive in a given area and then mobilizing rapidly to control and eradicate them before they become well established and start to spread again. The environmental and economic costs associated with detecting and eradicating new invasive species before they become established in a given area are much less than the costs of long-term management of widespread species.

According to Dr Paul Sand, who served as National Weed Specialist with the US Department of Agriculture, Agriculture Research Service/Animal and Plant Health Inspection Service (USDA ARS/APHIS) from 1959 to 1985, EDRR as a management strategy for invasive plants is not a new idea (personal communication). Before the introduction of organic pesticides during World War II, it was standard practice for farmers to seek out and remove the weeds in their fields before they became well established. However, with the advent of chemical weed control, farmers were able to 'suppress' weed populations to ensure maximum productivity of their crops. This new dependency on chemicals ultimately led to herbicide resistance in some weed populations and gave farmers and land managers a false sense of security about low-level weed populations, including new, 'troublesome', weed species that were not (yet) affecting production systems. This tendency to ignore existing species – to permit them to spread continually – along with ongoing introductions of new species (both intentional and unintentional) has led to many continental-scale problems, including kudzu (*Pueraria montana* (Lour.) Merr.) and leafy spurge (*Euphorbia esula* L.). As such, it highlights the need to re-examine EDRR as a practical management strategy for preventing the establishment and spread of new invasive species – especially if climate becomes less of a limiting factor in keeping current invasive populations from expanding.

The track record of EDRR applications in the context of limiting the damage associated with invasive species makes it a ubiquitous and valuable prevention strategy. As such, it is reasonable to expect it to grow in importance as an effective response mechanism in the context of elevated CO_2 and rapid climate change – but this assumption should be tested. The need to invest in science to monitor and evaluate the impacts of invasive species, as well as the efficacy of EDRR and various control measures, can only grow in urgency as new species are introduced in the years ahead.

That climate could alter the introduction and spread of invasives has been discussed previously in this book (e.g. Chapters 3–5); however, there are two of the general principles previously discussed that are related directly to the efficacy of EDRR. First, because of the lack of natural predators, climate (temperature and precipitation) is one of the most important environmental factors in determining the geographic range of an invasive species. Second, climate change may alter trade routes (e.g. opening of the Northwest Passage), with subsequent rapid introduction of new invasive species in new geographic areas. Considering that climate change could open up new areas for invasion, and that increased trade and travel will facilitate those invasions, it seems only logical to consider the usefulness of EDRR as an adaptive management strategy in addressing these emerging problems.

To appreciate fully the preventative value of EDRR from a climate change perspective, it is worthwhile to look at its implementation history and successes. To that end, lessons learned in the programmes and projects described here should provide guidance in addressing more effectively confirmed new invasive species as well as historically limited species (limited by climate) as they begin to appear and spread to new regions under new climate conditions.

Overview of Notable Single Agency-led Federal/State Weed Eradication Programme – EDRR 1.0: Issues and Lessons Learned

The establishment of single agency-led weed eradication programmes (EDRR 1.0) was the first effort in the modern weed management era that utilized EDRR as a primary strategy in responding to a newly introduced weed. Single agency-led weed eradication programmes are typically established by federal, state and/or local agencies to address specific federal/state/local regulated noxious weeds that theoretically can be eradicated from a predefined land area (e.g. watershed, county or state). Three notable examples over the past 50 years in the USA include the USDA-Carolinas Witchweed Eradication Program, the USDA-Utah Goatsrue Eradication Program and the USDA-Idaho Common Crupina Eradication Feasibility Study.

Witchweed (*Striga asiatica* (L.) O. Kuntze)

Witchweed was a parasitic weed of maize and other grass crops from Africa and Asia that was first detected in Columbus County, in south-eastern North Carolina, in July 1956. Once it was realized that witchweed posed a serious threat to maize production in the USA, surveys were conducted to determine the total extent of the infestation (ultimately over 174,824 ha (432,000 acres) in 38 counties in eastern North Carolina and South Carolina), a federal/state quarantine was established in the infested area to prevent its spread (6 September 1957) and a federal/state programme was funded to eradicate it from the infested area (initially US$3 million year^{-1} starting in FY 1958) (Iverson *et al.*, 2011). In 1976, it was formally listed as a Federal Noxious Weed by the USDA APHIS. To date, over 99.7% of the total infestation has been eradicated through federal/state efforts, at a cost of about US$250 million – or about US$1430.00 ha^{-1} (US$576.00 acre^{-1}).

One major factor in the success of this programme was the discovery that ethylene gas could be used at certain times of the year to induce suicidal germination of 99% of witchweed seeds in a field. Without ethylene, it is doubtful that the programme could have been sustained long enough to achieve the goal of eradication by merely denying the further reproduction of emergent witchweed plants. In terms of dollars saved, the witchweed programme has been a resounding success. Left unattended, witchweed could cause a 10% loss of the national maize crop per year – a loss 1.2 billion bushels of the average 12 billion bushel national crop (US$6 billion year^{-1} at US$5.00 bushel^{-1}) (Iverson *et al.*, 2011).

Issues and lessons learned

Farmers and landowners with infested fields have never played an active role in the witchweed eradication programme. This has almost certainly extended the time required for this programme to achieve its unparalleled results – 58 years. In the future, impacted and potential stakeholders who stand to gain the most from such programmes should have some meaningful role in the overall effort. Otherwise, there is no incentive for such government-sponsored programmes to ever come to an end.

Goatsrue (*Galega officinalis* L.)

Goatsrue (also called professor weed) is a legume from Eastern Europe that was first introduced by a professor from Utah State University in 1891 as an experimental livestock forage plant on western rangelands. Once the plant was determined to be poisonous to livestock, the research plots were abandoned and the plant began to spread. By 1970, it had spread to over 15,782 ha (39,000 acres) in Cache County, Utah. In 1976, it was formally listed as a Federal Noxious Weed by the USDA APHIS and efforts were made to organize a federal/state programme to address it. Between 1981 and 1996, over US$2 million of federal funding was provided to Utah State University under a congressional earmark to develop and implement eradication methods for goatsrue. During that time, Dr Jack Evans, a weed scientist from Utah State, led the effort and made significant progress in developing and implementing methods for reducing the infestation. However, when the congressional earmark was withdrawn in 1996, the programme was terminated, due to a lack of alternative funding, and has never been resumed (Westbrooks and Eplee, 2011).

Issues and lessons learned

The goatsrue eradication programme was a single agency-led programme that made significant progress as long as federal funding was provided to support it. However, as with the witchweed eradication programme, stakeholders (including landowners and land management agencies such as the Bureau of Land Management, the USDA Forest Service and US Fish and Wildlife) were never involved with the effort to any significant degree. Generally, a better approach for addressing new invaders like this is to establish an inter-agency task force of responsible agencies, as well as other impacted and potential stakeholders such as private landowners. Such a group, as typified by the Beach Vitex Task Force in the Carolinas and Virginia (www.beachvitex.org), can be very effective in helping to generate the support and momentum that is needed to sustain such initiatives – and in identifying impediments to achieving long-term programme goals (e.g. total eradication within a reasonable time frame). This would have been most helpful in the case of goatsrue, which is known to have a seed life of 50+ years in the soil.

Common crupina (*Crupina vulgaris* Cass.)

Common crupina is a rangeland composite from the Mediterranean region that was introduced into Idaho unintentionally some time after 1958. According to Miller and Thill (1983), it was not found in 1958 plant inventories of areas in Idaho that subsequently were found to be infested by 1980. The plant was first noticed on a ranch near Grangeville (Idaho County), Idaho, in 1968 and was collected from the 16 ha (40 acre) infestation in July 1969 (Stickney, 1972). By 1981, when common crupina was listed as a Federal Noxious Weed and the University of Idaho undertook a federally funded eradication feasibility study, the infestation had increased to 9308 ha (23,000 acres) in west central Idaho. The US$1.4 million study, which was completed in 1988, concluded that complete eradication of common crupina from the USA was still feasible. In September 1991, after US$5 million of federal funding had been appropriated for the effort, federal, state and local agencies met in Lewiston, Idaho, to discuss the establishment of a federal/state crupina eradication programme in the north-west. By that time, it had spread to 22,257 ha (55,000 acres) in Idaho, 3237 ha (8000 acres) in Oregon, 162 ha (400 acres) in Washington state and 8 ha (20 acres) in California. Moreover, at that meeting, due to environmental concerns about the impact of pesticides on spawning sockeye salmon in the Salmon River, no consensus was reached by programme partners and the crupina eradication effort was abandoned (Westbrooks *et al.*, 2000). Since that time, crupina has continued to spread and efforts to find a suitable/effective biological control agent have been unsuccessful to date.

Issues and lessons learned

If an eradication effort had been launched soon after the original 16 ha (40 acres) infestation of crupina was detected near Grangeville, Idaho, in 1968, it was quite possible that this newly introduced invasive plant could have been totally eliminated in a short period of time. In the future, eradication feasibility studies such as this should focus on developing a suite of management strategies for addressing the invader in all types of habitats – including biologically sound control methods for sensitive areas where native and endangered species are at particular risk. In addition, while the study is being conducted to develop effective control methods, programme partners should do everything possible to prevent the invader from continually reproducing and spreading – anything to slow the spread. The common crupina infestation more than doubled during that 10-year study.

One shortcoming of EDRR 1.0 approaches is that they are highly reactive and short change the early detection process by focusing primarily on a limited number of regulated species. As will be demonstrated in subsequent iterations, the 'golden hour' has often passed by the time that many high-risk species are identified. Therein lies a paradox – the authority, and the resources, to act decisively in managing other new invaders (e.g. beach vitex (*Vitex rotundifolia* L. f.) in the Carolinas) is often absent unless impacted and potential stakeholders can pull together to address them.

New Approaches for Inter-agency Partnering to Address New and Emerging Invasive Plants: State Early Detection and Rapid Response Systems – EDRR 2.0

In recent years, primarily due to the limitations and sustainability of single agency-led federal/state weed eradication programmes, new inter-agency approaches for managing new and emerging invasive plants have evolved. Since 1998, the effort has been focused on the development of state EDRR systems for invasive plants (EDRR 2.0) – and ultimately, a national EDRR system for invasive plants. The elements of a state EDRR system include: (i) a state EDRR coordinating committee to lead the development of the system; (ii) a state EDRR work plan; (iii) a state EDRR target list; and (iv) a state early detection network of trained volunteers and agency personnel. Coordinated activities by system partners include: (i) identification and vouchering of submitted specimens by cooperating botanists; archiving of new invasive plant field records in state herbaria and online databases (e.g. the University of Georgia's Early Detection Distribution and Mapping System (EDDMapS); see Chapter 13); (ii) rapid assessment of confirmed new records by cooperating biologists; and (iii) rapid response to confirmed new invaders through single agency-led weed eradication programmes and/or inter-agency initiatives such as cooperative weed management areas with a geographic focus (e.g. the South Fork Weed Management Area in Park County, Wyoming; Parsons, 2011) and invasive plant task forces with a single species focus (e.g. the Beach Vitex Task Force; Westbrooks and Brabson, 2011).

Development of a national EDRR system, as described above, was first identified as a long-term programmatic goal by the US Federal Interagency Committee for the Management of Noxious and Exotic Weeds (FICMNEW) in 1998, and was subsequently endorsed for development by the North American Weed Management Association. From 1998 to 2003, FICMNEW undertook a national effort to encourage the establishment of state invasive species councils and committees (e.g. the Wyoming Weed Team and the Delaware Invasive Species Council). The development of state EDRR systems was a primary reason for organizing such groups. In 2003, a conceptual design plan to provide guidance in the development of system elements was published by FICMNEW (FICMNEW, 2003).

Since 2003, the effort has been focused largely on the development of state EDRR systems in cooperation with established state invasive species councils, as well as single species focus task forces (e.g. the Delaware Invasive Species Council, State Exotic Pest Plant Councils and the Beach Vitex Task Force) (Westbrooks, 2011; Westbrooks and Eplee, 2011). One good example in this area is the effort to eradicate beach vitex from beaches and communities along the South Carolina coast. Since 2004, over 220 known sites of beach vitex have been treated, cleared and replanted through the work of the Beach Vitex Task Force in South Carolina. Task force crews have searched diligently and have no reason to expect to find more infested sites – especially on uninhabited barrier islands – since 99+% of the seed production is controlled and contained (Beach Vitex Task Force, 2013).

Issues and lessons learned

The constant turnover of volunteer personnel involved with state invasive species councils and task forces, as well as periodic changes in agency priorities, make it difficult to sustain long-term interest and participation in state invasive species councils, task forces and the development of state-level EDRR systems. In order for the inter-agency approach to developing EDRR capacity to work, it is important to continually train the next generation of invasive species field specialists and managers to ensure continuity and leadership as programmes and priorities change over the years.

Development of Landscape-level EDRR Systems for Individual, Geographic and Political Land Units: Local to National – EDRR 3.0

Currently, due to reduced funding for federal/state/local weed eradication efforts, and a notable shift in national priorities toward energy and security issues, a new effort is being made to develop EDRR capacity at all levels of the landscape – local to national (EDRR 3.0). The primary objective of this new approach to EDRR is to encourage individual landowners and managers, and private citizens everywhere, to 'take steps to detect and eradicate all new invasive plants on lands that they own or manage'. The landscape approach to EDRR, which is simply an expansion of the effort to develop state-level EDRR systems, includes the development of EDRR capacity on all types of land units at all levels of the landscape, including: (i) individual public and private land units (e.g. parks, forests, refuges, preserves, farms, ranches and other private lands); (ii) geographic land units (watersheds, biomes and cooperative weed management areas); and (iii) political land units (municipalities, counties, states/provinces and countries).

The long-term goal of this new effort is a renewed focus on EDRR as the preferred management strategy for addressing *all* new and emerging invasive plants on a designated land unit – *not just regulated species*. Under this scenario, any invasive plant that is new to a land unit (even species that are widespread elsewhere, such as kudzu and leafy spurge) would be targeted immediately for control and eradication. The main point with this approach is that landowners and managers need to take steps to 'identify, control and eradicate new invaders while it is still practical'. Ultimately, considering the many thousands of potential invasive species that have yet to be spread around the world, our long-term success depends on landowners and managers adopting a comprehensive approach for managing *all* new and emerging invasive species.

An example of an EDRR 3.0 effort that is now being organized is the hydrilla control and eradication project at Lake Waccamaw (Lake Waccamaw State Park) in south-eastern North Carolina. The goal of the project, which is being launched by cooperating state and local agencies and organizations, as well as local citizens, is to eradicate a recently discovered 246 ha (608 acre) infestation of hydrilla (*Hydrilla verticillata* (L.) f. Royle) before it has a chance to spread across all of this 3636 ha (8935 acre) natural Carolina Bay (which has a maximum depth of 3.35 m (11 feet)). Project organizers have decided to use aquatic herbicides (e.g. fluridone) to eliminate hydrilla from the lake, because mechanical removal methods as well as biological control with the triploid white amur grass carp (*Ctenopharyngodon idella* Val.) pose a serious threat to endemic species in the lake, including the federally listed threatened Waccamaw silverside minnow (*Menidia extensa* Hubbs & Rainey). The programme, which is being funded by state and local agencies, is expected to cost US$450,000.00 year^{-1} and will take 7–10 years to complete. To ensure that the lake is not invaded like this again in the future, project partners are also planning to develop an early detection and survey programme that will utilize agency field personnel and trained volunteers to monitor the lake for the appearance of high profile invaders such as crested floating heart (*Nymphoides cristata* (Roxb.) O. Kuntze) and zebra mussel (*Dreissena polymorpha* Pallas).

By nature, EDRR 3.0 projects on individual land units (e.g. the Lake Waccamaw hydrilla eradication project) are independent, self-sustaining efforts. However, EDRR 3.0 landscape-level projects that include multiple land units within defined geographic areas (e.g. cooperative weed management areas) or designated political land units (cities, counties, states or nations) require, by necessity, common data resources and close coordination of field activities between the various partners.

High Profile New and Emerging Invasive Plants in the USA and Canada – EDRR Opportunities

There is no dispute that global trade and travel are continually spreading all types of plants and animals far beyond their native biogeographical realms to distant parts of the world. Here are four notable examples of emerging introduced invasive plants in the eastern USA and Canada that are now being spread by natural vectors (wind, water and animals) and by people. These species can be managed locally – even regionally – utilizing the principles and practices of EDRR.

Cogongrass – a threat to pine savannah forest ecosystems

Cogongrass (*Imperata cylindrica* (L.) Beauv.) is a perennial, colony-forming grass up to 1.8 m (6 ft) tall – with no apparent aboveground stem – that is native to South-east Asia (Fig. 19.1). It thrives in a wide variety of habitats and is tolerant of shade, full sun, high salinity, drought, flooding, mowing and fire. It currently affects forests, pastures, roadways and wetlands across most of the south-eastern USA.

Cogongrass was first introduced into the USA as a packing material in imported shipping crates in Mobile, Alabama, around 1911. Later, it was introduced into Alabama intentionally as a pasture grass and for erosion control. Cogongrass reproduces and spreads via wind-borne seeds (up to 3000 plant^{-1}) and rhizome fragments. The feather-light seeds can be carried as far as 24 km (15 miles) by wind. The rhizome fragments are spread unwittingly by people through contaminated hay and soil. They are also spread by machinery, equipment and vehicles. Bloodgrass, which is a cultivated variety of cogongrass, has also been sold as an ornamental grass in some parts of the country. Cogongrass is now widespread in Mississippi, Alabama, Georgia and Florida. Small infestations of the plant also occur in

Fig. 19.1. Cogongrass in a pine savannah. (Source: Chris Evans, U-GA, Bugwood Image Gallery: http://www.invasive.org/browse/detail.cfm?imgnum=1380040)

Louisiana, Texas, South Carolina, North Carolina, Virginia and Oregon.

Cogongrass has been called the perfect weed (U-GA Center for Ecosystem Health, 2012). It affects 35 crops and infests over 416 million ha (Mha) (1.25 billion acres) in 73 countries around the world. However, it is also valued in some countries for use as fodder and in construction. It is an extremely aggressive invader that forms dense, circular infestations with thick ground-level mats that act as a physical barrier to the establishment of other types of plants. In the process, it outcompetes native plants for nutrients, rooting space, moisture and sunlight. It is also allelopathic – it produces chemicals that suppress the growth of other plants. By overwhelming native plants, it drives out ground-nesting wildlife such as turkey and bobwhite quail, as well as the endangered gopher tortoise, which depends on native grasses and legumes. While cogongrass roots are very fire tolerant, the plant is extremely flammable and can cause fast-moving wildfires (to more than 448°C (840°F)) that consume native plants and animals and disrupt the natural fire regime.

The first step in controlling cogongrass is to remove ground-level thatch and older leaves by a controlled burn during the summer. This reduces biomass, stimulates regrowth from the rhizomes and allows herbicides to be applied to actively growing leaves only. If tillage of the infested area is possible, disking soon after a controlled burn will cut down on rhizome mass. To prevent spreading, equipment and vehicles should be cleaned thoroughly before leaving an infested area. After 1–4 months of regrowth, a herbicide such as Arsenal® (imazapyr) plus glyphosate should be applied. Follow-up treatments in the second and possibly the third year will be needed to eliminate the population.

Regulatory status

Cogongrass is listed as a US Federal Noxious Weed. As a result, it is automatically listed as a State Noxious Weed in a number of states, including Alabama, California, Florida, Massachusetts, Minnesota, North Carolina, Oregon, South Carolina and Vermont.

Note: while cogongrass cannot be eradicated from Mississippi, Alabama, Florida and, Georgia, it occurs at only a few locations in Oregon, Texas, Louisiana, South Carolina, North Carolina and Virginia. However, in every case where possible, it should be eradicated from individual land units such as farms, parks, forests and refuges (EDRR 3.0).

Crested floating heart – a threat to waterways in the eastern USA

Crested floating heart (*Nymphoides cristata* (Roxb.) O. Kuntze) is a herbaceous aquatic plant, native to Asia, with floating stems from a buried rootstock that grows in ditches, canals, ponds and lakes (Fig. 19.2). Crested floating heart was first introduced to the USA as a water garden plant. Once

Fig. 19.2. Crested floating heart in an 8 ha (20 acre) cove on the south shore of Lake Marion, South Carolina. (Source: images courtesy of the Santee Cooper Authority, Moncks Corner, South Carolina.)

established in a waterway, fragments of the plant are spread by wind, flowing water, boats and trailers.

Free-living populations of crested floating heart were first observed in cypress swamps and water management canals in south-east Florida in the late 1990s. Within a few years, large canals and suburban lakes in the central and eastern parts of the state had become infested. There is serious concern about this plant because herbicides that control other floating and emergent invasive plants have had little effect on it.

Crested floating heart was first detected in an 8 ha (20 acres) cove near Eutaw Creek at the south-east end of Lake Marion (Orangeburg County), South Carolina, in August 2006. This was the first time that free-living populations of the plant had been found outside of Florida. The plant was most likely introduced into the Santee Cooper Lake System (Lake Marion and Lake Moultrie) from backyard water gardens or by recreational boaters from Florida.

Initial applications of herbicides that control other aquatic weeds were largely ineffective – as already seen in Florida. Low water levels experienced during 2007–2008, as well as freezing winter temperatures, also had little effect on the plant. By mid-2007, about half of the cove at Eutaw Creek was covered by the plant. By late 2008, the plant had spread to other areas of the lake, both downstream and upstream from the original infestation. In 2009, it was estimated that about 809 ha (2000 acres) of Lake Marion's surface was infested with the plant. It has also been found in nearby Lake Moultrie, as well as the Santee River below the Lake Marion dam. Crested floating heart has also been detected in a golf course pond on Pawley's Island, South Carolina (Jack Whetstone, Clemson University, Georgetown, South Carolina, personal communication, 19 May 2010).

Control strategies

In October 2009, Santee Cooper aerially sprayed 142 surface ha (350 acres) of Lake Marion that had become infested with the plant. Currently, Santee Cooper is using a tank mix of glyphosate (Touchdown Pro®) plus imazamox (Clearcast®), combined with TopFilm® (a pesticide adjuvant/sticker), or the surfactant AquaBupH®, in efforts to control crested floating heart. Santee Cooper has also found that endothall (Aquathol K®) provides short-term control of the plant in enclosed water bodies with still water. Early results from the 2009 treatment effort showed a small reduction in total surface acres infested in the Santee Cooper lakes. Treatments have continued since then with similar results. Efforts continue toward the discovery of more effective control agents throughout the plant's expanding range. Chemical applications by airboat and helicopter have shown positive results; however, the plant continues to demonstrate the ability to regrow from root crowns that survive the initial treatment. Crested floating heart has continued to spread throughout the lake system, with some 2428 ha (6000 acres) infested as of October 2012.

The role of volunteers in the detection and reporting of crested floating heart

Based on suitable habitat for the plant, biologists estimate that crested floating heart could ultimately infest as much as 40% of the 65,000 ha (160,000 acres) Santee Cooper Lake system if it is not controlled. Therefore, Santee Cooper is conducting annual early detection and delimiting surveys across the lake system, and will conduct additional treatments as possible. At this point, it is critical that boaters and water garden enthusiasts assist in minimizing the spread of this new invader. This can be done by reporting suspected new infestations to state officials, to permit the eradication of satellite infestations, which would prevent further spread. This is a good example of the fieldwork that EDRR volunteers can do to assist agencies such as Santee Cooper and the South Carolina Department of Natural Resources in their efforts to detect and manage new invasive species such as crested floating heart.

Regulatory status

Crested floating heart is currently regulated as a State Noxious Weed in North Carolina by the North Carolina Department of Agriculture and Consumer Services. It is not being regulated in South Carolina at the present time.

Wavyleaf basketgrass (WLBG) (*Oplismenus hirtellus* (L.) P. Beauv. subspecies *undulatifolius* (Ard.) Roem. & Schult.)

WLBG is a shade-tolerant, low-growing, stoloniferous perennial grass, from 20 to 30 cm (8–12 inches) tall that is native to Europe and Asia (Fig. 19.3). WLBG is an exotic subspecies of bristle basketgrass (*Oplismenus hirtellus* (L.) P. Beauv.) that is native to the south-eastern USA – from Virginia to Texas. It occurs in temperate environments in coastal plain, piedmont and montane regions; full canopy hardwood forests, forest margins and shady riparian zones; does not tolerate direct sunlight; and appears adaptable to a wide range of pH.

It is unclear how WLBG was first introduced to the USA. It may have been a contaminant of variegated types of native bristle basketgrass that are often sold by plant nurseries in the Mid-Atlantic Region. However, WLBG is not known to be sold in the horticultural trade. WLBG was first discovered in the USA in Patapsco Valley State Park, Howard County, Maryland, by botanist, Edward Uebel, in 1996. In 2000, it was found growing along a woody stream at the Hernwood Sanitary Landfill, in Baltimore County, Maryland. It is known to occur only in Maryland and Virginia. Although limited in scope, it is probably not realistic to eradicate WLBG totally from the USA. However, if possible, it should be eradicated from local land units before it becomes well established and starts to spread even further.

Ecological and economic impacts

Once it becomes established, WLBG spreads rapidly through wooded natural areas – crowding out native herbaceous plants and preventing regeneration of native hardwood tree species. Unless it is soon eradicated completely, it will continue to spread and could have a devastating impact on the deciduous forests of eastern North America for many decades to come.

Prevention measures

Since 2006, volunteers with the Anacostia Watershed Society (AWS) have been working to detect and eradicate WLBG from Maryland and Virginia. In 2006, AWS volunteers discovered and began eradicating a 3 acre infestation in Little Paint Branch Park, in Prince George's County, Maryland. In 2007, several small colonies and patches of the plant were detected and treated in the Beltsville Area Research Center-West, also in Prince George's County, Maryland. In 2008, AWS led the effort to establish the Wavyleaf Basketgrass Task Force. The task force is focused on the mapping and eradication of known infestations and a public awareness campaign to help detect and report other infestations before they become widespread.

Physical control

WLBG is hand pulled easily at the roots and nodes. Hand pulling is thought to be more effective than herbicide treatment when the population is small enough for weeding (Westbrooks and Imlay, 2010; Wavyleaf Basketgrass Task Force, March, 2009).

Chemical control

WLBG may be treated effectively with a 1–2% solution of Roundup® (glyphosate). Current populations in Maryland and Virginia are still relatively small and eradication is still possible. General weed

(a)

(b)

Fig. 19.3. Wavyleaf basketgrass (a) and close-up of leaves (b) in Patapso Valley Park, Baltimore County, Maryland. (Source: Forestry Images, U-GA (a) http://www.forestryimages.org/browse/detail.cfm?imgnum=5389747; (b) http://www.forestryimages.org/browse/detail.cfm?imgnum=5389748)

killers such as Roundup® work best late in the year, while grass inhibitors such as Envoy Plus® (clethodim) are more effective early in the season.

Regulatory status

Wavyleaf basketgrass is currently not regulated at the local, state or federal level in the USA.

Kudzu (*Pueraria montana* (Lour.) Merr.) – a new invader in Canada

Kudzu is a climbing, deciduous vine with lobed leaflets that grows up to 30 m (100 ft) in length (Fig. 19.4). The roots and vines grow from a woody crown that may be 18 cm (7 inches) or more in diameter, 1.82 m (6 ft) or more in length and up to 181 kg (400 lbs) in weight. The preferred habitat includes open, disturbed areas such as roadsides, right-of-ways, forest edges and old fields.

Kudzu, which was native to Asia, was first introduced into the USA as an ornamental porch vine at the Philadelphia Centennial Exposition in 1876. From 1935 until the early 1950s, farmers were encouraged to plant kudzu to reduce soil erosion throughout the south. During the Great Depression, Franklin D. Roosevelt's Civilian Conservation Corps planted it widely for this reason. In 1953, it was finally recognized as a weed by the US Department of Agriculture and was removed from its list of permissible cover plants for soil conservation projects. Kudzu occurs throughout the eastern USA, as well as Oregon and Washington.

For the past 50 years, kudzu has been widely recognized as one of the most invasive plants in the USA. Kudzu's ability to dominate infested areas to the exclusion of all other plant life has led it to be termed 'the plant that ate the south' and a 'vegetal form of cancer', and its impacts on native biodiversity, regional nitrogen cycles, watershed

Fig. 19.4. Kudzu growing over a road sign along US Highway 74 Business in Chadbourn, North Carolina. (Source: Randy G. Westbrooks, 17 June 2009.)

nitrogen loads, eutrophication of freshwater and air quality are well documented (Sharkey and Loreto, 1993; Forseth and Innis, 2004). In addition, kudzu is a recognized host of soybean rust, a major threat to soybean production in North America (Lynch *et al*., 2006).

Kudzu and climate change

One of the most notable features of kudzu is its ongoing range expansion. It currently occupies about 3 Mha and is migrating northward at the rate of approximately 50,000 ha $year^{-1}$ (Pappert *et al*., 2000). As such, it is of obvious interest to examine whether rising temperatures associated with climate change are contributing to the migration of this invasive species. One of the most interesting forecasts in this regard was made more than two decades ago by Tom Sasek and Boyd Strain at Duke University. They observed that the latitudinal distribution of kudzu at that time was limited to regions south of the Mason–Dixon line by low winter temperatures of <−15°C (Fig. 7 in Sasek and Strain, 1990). As winter temperatures have warmed and hardiness zones have shifted northward (www.usna.usda.gov/Hardzone/ushzmap.html), ranges have shifted concurrently (www.eddmaps.org/ipane/distribution/necounties.cfm?sub=2425). How much of this latitudinal migration is attributable solely to warming winter temperatures is unclear, but the northward spread of kudzu is consistent with the Sasek and Strain hypothesis.

This hypothesis was further validated with the discovery in July 2009 of a kudzu population in Essex County, Ontario, Canada (Fig. 19.5). The patch, which measured 50 m × 150 m (164 ft × 492 ft), was found on a bluff overlooking Lake Erie near Leamington, Ontario – a farming community about 70 km (45 miles) south-east of Detroit, Michigan. This is much further north than kudzu was ever expected to grow in North America.

Fig. 19.5. Kudzu on the shore of Lake Erie near Leamington, Ontario, Canada. (Source: image by Rachel Gagnon, Ontario Invasive Plant Council, Toronto, Ontario, Canada, http://www.ontarioinvasiveplants.ca/files/unwantedlettersKudzuLOWRES1.pdf)

The origin of the infestation – whether it was planted or simply floated across Lake Erie from Ohio – is unknown. However, according to Rachel Gagnon, Director of the Ontario Invasive Plant Council, the infestation is well established but more or less contained – and is not known to occur elsewhere in Canada. Rowan Sage, a biology professor at the University of Toronto, has recommended that the infestation be destroyed immediately or 'the plant could wind up costing millions of dollars to eradicate' (CBC News, 2009).

Even though kudzu is currently well contained, the discovery of kudzu in Canada argues strongly in favour of early detection and rapid response measures in areas likely to be impacted by climate change (e.g. rapid warming above winter thresholds).

Ecological and economic impacts

Kudzu often grows over, smothers and eventually kills all other vegetation that it covers, including trees. In the process, it girdles tree trunks, breaks branches and even uproots entire trees through the sheer force of its weight.

Control strategies

Work by the South Carolina Kudzu Coalition in Spartanburg, South Carolina, since 2004 has shown that the only effective way to control mature kudzu plants is by killing the woody root crowns. Any crowns that are left in place following a single treatment (manual or chemical) will simply regrow. If the soil cannot be disturbed by digging (as on a landscaped site), the vines should be treated repeatedly with herbicides or cut close to the soil until the energy reserves in the root crown are exhausted. Late season cutting or mowing should be followed up, if possible, with immediate application of a systemic herbicide (e.g. glyphosate) to the cut stems, to permit translocation of the chemical into the root system. Kudzu can be completely eradicated from any site if the root crowns are totally killed or removed.

Regulatory status

Kudzu is regulated as a noxious weed or otherwise banned in a number of US states, including Connecticut, Florida, Illinois, Kansas, Kentucky, Mississippi, Missouri, Oregon, Pennsylvania, Texas, Washington and West Virginia. It is not currently regulated in any of the Canadian provinces.

EDRR on the International Stage – Kochia in Western Australia

EDRR is also being employed for addressing new invasive species in other parts of the world, particularly in Australia and New Zealand. The introduction and subsequent eradication of kochia (*Bassia scoparia* (L.) A.J. Scott) in Western Australia is a good example. Kochia, which is native to Eastern Europe and Western Asia, is an annual weed of cereal crops and pastures in warmer regions of the world. Kochia was first imported into the wheat belt of Western Australia in 1990 as a salt-tolerant pasture forage plant. Although palatable to livestock, the plant is poisonous in large quantities. It also alters fire regimes and reduces the diversity and abundance of native plant species.

By 1992, the Western Australia Department of Agriculture recognized the weedy potential of kochia and targeted it for containment and eradication in 1993. Eradication strategies for kochia included burning, grazing, chemical control and mechanical removal. By 1996, 46 of the 52 known infestations had been eliminated from the state. The last sightings of kochia in Western Australia were in March 2000.

The programme's success was due in large part to: (i) a detailed knowledge of where kochia had been planted; (ii) the conspicuous nature of the plant; (iii) limited seed dormancy and a long vegetative period prior to seeding; (iv) fences around most of the planted sites that helped to contain it (under most conditions, mature plants break away from the root system and roll across the landscape in high winds – dropping seeds along the way); (v) rapid response within 2 years of introduction; and (vi) a well-organized and -funded eradication programme and cooperation from impacted farmers.

The total cost of the kochia eradication effort in Western Australia was AUS$200,000 in combined federal and state funding, with an additional AUS$300,000 of in-kind control and monitoring activities provided by affected farmers. In retrospect, it is obvious that the introduction of kochia was a short-sighted idea that cost at least half a million dollars to correct. However, this pales in comparison to future losses and control costs that would have been incurred by Australian landowners if the plant had been allowed to continue spreading (Westbrooks and Eplee, 2011).

Evolution of EDRR with Climate Change

Globally, invasive species have been estimated to cost as much as 5% of planetary GDP (Pimentel *et al.*, 2001). This puts it on a par with climate change costs, which have been projected in the range of 1–20% of GDP (Stern, 2006). Yet, our experience with invasive species is that although the costs to the economy can be as high as the costs of other natural disasters such as fires and storms, people do not often react to biological invasions as natural disasters (unless they are directly affected), because biological invasions are 'slow onset' hazards (as contrasted with 'fast onset' catastrophes like earthquakes and floods). Therefore, funds for early detection and rapid response measures for invasions are more vulnerable than funds for more acute disaster prevention and response, such as for wildfires – in spite of the established relationships between biological invasions and catastrophic fire. Creeping, slow onset changes are often ignored as decision makers are forced to deal with more urgent contingencies (UNEP, 2012a).

The links between climate change, rising CO_2 and invasive species biology have been covered elsewhere in this book (Chapters 2–5). It is clear that climate change and introduced invasive species unleash a cascade of impacts that require management input. When that happens, healthy, functioning ecosystems can be a bulwark against climate impacts and invasive species. Degraded ecosystems increase the vulnerability of human populations to climate change (UNEP, 2012b), and there is growing evidence that the threat of invasive species is compounded by climate change (Burgiel and Muir, 2010).

Climate change, landscapes and biological invasions

It has long been known that the protection of the integrity of natural communities is best provided at the landscape scale (Noss, 1983, 1990). Landscape ecology approaches integrate complex interactions between biodiversity, climate, socio-economic and cultural factors that shape a landscape and give it its character. They incorporate nature and human dimensions in a living environment.

Landscape ecology integrates knowledge about biology with geophysical processes such as climate zones and socio-economic processes such as trade patterns, transportation corridors, patterns of ownership and legal jurisdictions, to understand the forces that shape ecosystems. A landscape approach to biological invasion would use knowledge of such patterns to predict change and manage risk, within the context of the overall management framework for an ecosystem. One advantage of a landscape approach to biological invasions, then, is that it affords opportunities to integrate invasive species management into a broader suite of ecosystem management objectives.

Ecosystem-based adaptation (EbA) is winning acceptance as a logical, comprehensive approach to climate adaptation. EbA involves the sustainable management, conservation and restoration of ecosystems to provide services that help people adapt to the adverse effects of climate change. It combines the concepts of landscape ecology with the imperatives of climate adaptation and uses the natural resilience of ecosystems as a buffer against climate change. Such an integrated management approach to a complex system is a logical way to unify the mitigation of an adaptation to climate impacts that manifest in a variety of ways (Burgiel and Muir, 2010).

Incorporating EDRR 3.0 into a comprehensive ecosystem management programme will be a more cost-effective and efficient approach than trying to deal with species in isolation. It can take advantage of economies of scale, both in terms of the resources to be deployed and in terms of the management of information to measure impacts and model trends as we prepare for the uncertainties to come. Rather than competing for funding and duplicating functions, under an EbA approach, disaster risk reduction and EDRR could be combined in much the same way that wildfire emergency response is managed in the USA – through cross-cutting budgets and an integrated incident command system.

Summary – The Utility of EDRR

Everyone agrees that prevention – the first line of defence – is the preferred strategy for managing invasive species. However, in an era when global trade has increased exponentially, people are continually introducing exotic species, both intentionally and accidentally. As a result, there is no practical way to prevent the introduction of all documented invasive species, or to know with total certainty what introduced species may become invasive within a given region (e.g. beach vitex had no history of invasiveness before it was introduced from Korea to coastal communities in the Carolinas in the mid-1980s).

As is evident from the history of EDRR in the modern weed management era, early detection and rapid response is a key strategy that can be employed to avoid the considerable economic and environmental costs associated with the widespread occurrence of invasive species. As such, it has been a successful prevention strategy for invasive species in a wide range of circumstances.

Those circumstances can also be extended to the issue of climate change. As climate change permits native and introduced species to survive and thrive in new areas, there can be no doubt about the value of EDRR in managing this man-made problem.

Indeed, one can argue that, more than ever, now is the time to refocus and develop EDRR capacity at all levels of the landscape, from local weed warriors to NGOs, to state and federal agencies; from one's backyard to the Rocky Mountains. Cogongrass, crested floating heart and wavyleaf basketgrass in the eastern USA, and kudzu in Canada, are examples of new and emerging species that could be addressed in this way.

Even without the additional threat of climate as a driver of invasive species introductions and spread, there already exists considerable potential for increased exposure to new flora and fauna. Of the 33,759 species of weed around the world that have been documented by weed geographer, Rod Randall, less than 5% are known to occur in the USA. So, it is inevitable that many additional species will become introduced here in the coming years unless stronger foreign pest exclusion rules and regulations are implemented.

Given the increased number of species that could be introduced and spread through trade and travel, and then spread further because of climate change, it is only prudent that steps be taken to build increased capacity to address them – *now*. Such an effort would involve an intensified application of the basic building blocks of good EDRR practices, including early detection, reporting, rapid assessment and rapid response.

Clearly, EDRR is not the only strategy that may be employed to address future invasive species threats. But, our history of success in addressing new invaders such as witchweed and beach vitex demonstrates that EDRR, if applied properly, could help to minimize the establishment and spread of new invaders at all levels of the landscape – from local to national. Overall, as nations rise to the challenge of climate adaptation, the opportunity to incorporate a key mechanism to protect biodiversity, food production and ecosystem services – *by employing EDRR* – should not be missed.

References

Beach Vitex Task Force (2013) http://www.beachvitex.org/, accessed 4 February 2014.

Burgiel, S. and Muir, A. (2010) *Invasive Species, Climate Change and Ecosystem-Based Adaptation: Addressing Multiple Drivers of Global Change*. Global Invasive Species Programme (GISP), Washington, DC, and Nairobi, Kenya.

CBC News (2009) Invasive plant species takes root in Canada. Canadian Broadcasting Corporation News Report about Kudzu in Ontario, Canada, 23 September 2009 (http://www.cbc.ca/news/technology/story/2009/09/23/leamington-kudzu-092309.html, accessed 4 February 2014).

Crooks, J. and Soulé, M. (1999) Lag times in population explosions of invasive species: causes and implications. In: Sandlund, O.T., Schei, P.J. and Viken, A.S. (eds) *Invasive Species and Biodiversity Management.* Kluwer Press, Dordrecht, Netherlands, pp. 103–125.

FICMNEW (2003) *A National Early Detection and Rapid Response System for Invasive Plants In the United States.* Federal Interagency Committee for the Management of Noxious and Exotic Weeds (FICMNEW), Washington, DC, 27 pp (http://www.fws.gov/ficmnew/FICMNEW_EDRR_FINAL.pdf, accessed 4 February 2014).

Forseth, I. and Innis, A. (2004) Kudzu (*Pueraria Montana*): history, physiology, and ecology combine to make a major ecosystem threat. *Critical Reviews in Plant Sciences* 23, 401–413.

Grube, A., Donaldson, D., Kiely, T. and Wu, L. (2011) *Pesticide Industry Sales and Usage – 2006 and 2007 Market Estimates.* Biological and Economic Analysis Division Office of Pesticide Programs, Office of Chemical Safety and Pollution Prevention US Environmental Protection Agency, Washington, DC, 33 pp.

Iverson, R., Westbrooks, R., Eplee, R. and Tasker, A. (2011) Overview and status of the witchweed (*Striga asiatica*) eradication program in the Carolinas. In: Leslie, A. and Westbrooks, R. (eds) *Invasive Plant Management Issues and Challenges in the United States – 2011 Overview.* American Chemical Society, Washington, DC, pp. 51–68.

Kartesz, J. (2012) The Biota of North America Program. Online Database. Chapel Hill, NC (http://www.bonap.org/, accessed 4 February 2014).

Lynch, T., Marois, J., Wright, D., Harmon, P., Harmon, C., Miles, M. and Hartman, G. (2006) First report of soybean rust caused by *Phakopsora pachyrhizi* on *Phaseolus* spp. in the United States. *Plant Disease* 90, 970.

Mack, R.N. and Erneberg, M. (2002) The United States naturalized flora: largely the product of deliberate introductions. *Annals of the Missouri Botanical Garden* 89, 176–189.

Miller, T. and Thill, D. (1983) Today's weeds: common crupina. *Weeds Today* 14, 10–11.

Monaco, T., Weller, S. and Ashton, F. (2002) *Weed Science – Principles and Practices*, Fourth Edn. John Wiley and Sons, Inc, New York, 671 pp.

Noss, R. (1983) A regional landscape approach to maintain diversity. *BioScience* 33, 700–706.

Noss, R. (1990) Indicators for monitoring biodiversity: a hierarchical approach. *Conservation Biology* 4, 1–4.

Pappert, R., Hamrick, J. and Donovan, L. (2000) Genetic variation in *Pueraria lobata* (Fabaceae), an introduced, clonal, invasive plant of the southeastern United States. *American Journal of Botany* 87, 1240–1245.

Parsons, B. (2011) New approaches for invasive plant management through cooperative weed management areas – the South Fork Weed Management Area in Park County, Wyoming. In: Leslie, A. and Westbrooks, R. (eds) *Invasive Plant Management Issues and Challenges in the United States – 2011 Overview.* American Chemical Society, Washington, DC, pp. 169–187.

Pimentel, D., Lach, L., Zuniga, R. and Morrison, D. (1999) Environmental and economic costs associated with non-indigenous species in the United States. *Bioscience* 50, 53–65.

Pimentel, D., McNair, S., Janecka, J., Wightman, J., Simmonds, C., O'Connell, C., Wong, E., Russel, L., Zern, J., Aquino, T. and Tsomondoa, T. (2001) Economic and environmental threats of alien plant, animal, and microbe invasions. *Agriculture, Ecosystems and Environment* 84, 1–20.

Pimentel, D., Lach, L., Zuniga, R. and Morrison, D. (2002) Environmental and economic costs associated with non-indigenous species in the United States. In: Pimentel, D. (ed.) *Biological Invasions – Economic and Environmental Costs of Alien Plant, Animal, and Microbe Species.* CRC Press, Boca Raton, Florida, pp. 285–303.

Randall, R. (2007) Global Compendium of Weeds. Online Database. Department of Agriculture and Food – Western Australia (http://www.agric.wa.gov.au/, accessed 4 February 2014) – in cooperation with the Hawaiian Ecosystems at Risk Project (http://www.hear.org/gcw/, accessed 4 February 2014).

Randall, R. (2012) *Global Compendium of Weeds*, Second Edn. Department of Agriculture and Food – Perth, Western Australia (http://www.agric.wa.gov.au/objtwr/imported_assets/content/pw/weed/global-compendium-weeds.pdf, accessed 4 February 2014).

Sasek, T. and Strain, B. (1990) Implications of atmospheric CO_2 enrichment and climatic change for the geographical distribution of two introduced vines in the USA. *Climatic Change* 16, 31–51.

Sharkey, T. and Loreto, F. (1993) Water-stress, temperature, and light effects on the capacity for isoprene emission and photosynthesis of kudzu leaves. *Oecologia* 95, 328–333.

Stern, N. (2006) *Stern Review: The Economics of Climate Change*. HM Treasury, London, 700 pp.

Stickney, P. (1972) *Crupina vulgaris* (Compositae: Cynareae), new to Idaho and North America. *Madrono* 21, 402.

UNEP (2012a) *Early Warning Systems: A State of the Art Analysis and Future Directions*. United Nations Environment Programme, Nairobi, 70 pp.

UNEP (2012b) Ecosystem Based Adaptation Guidance: Moving From Principles to Practice – Working Document April 2012 (http://www.unep.org/climatechange/adaptation/Portals/133/documents/Ecosystem-Based%20Adaptation/Decision%20Support%20Framework/EBA%20Guidance_WORKING%20DOCUMENT%2030032012.pdf, accessed 4 February 2014).

University of Georgia, Center for Ecosystem Health (2012) Cogongrass – The Perfect Weed. Video (http://www.cogongrass.org/mississippivideo.cfm, accessed 4 February 2014).

USDA (2012) Crop Values – 2011 Summary. USDA National Agricultural Statistics Service. Washington, DC, ISSN 1949-0372, 48 pp (http://www.nass.usda.gov/Publications/Todays_Reports/reports/cpvl0212.pdf, accessed 13 February 2014).

Waugh, J. (2009) *Neighborhood Watch: Early Detection and Rapid Response to Biological Invasion Along US Trade Pathways*. IUCN, Gland, Switzerland, 92 pp.

Westbrooks, R. (2004) New approaches for early detection and rapid response to invasive plants in the United States. *Weed Technology* 18, 1468–1471.

Westbrooks, R. (2011) Interagency partnering for weed prevention – progress on development of a national early detection and rapid response system for invasive plants in the United States. In: Leslie, A. and Westbrooks, R. (eds) *Invasive Plant Management Issues and Challenges in the United States – 2011 Overview*. American Chemical Society, Washington, DC, pp. 135–142.

Westbrooks, R. and Brabson, E. (2011) Overview and history of the Beach Vitex Task Force – an interagency partnership in action. In: Leslie, A. and Westbrooks, R. (eds) *Invasive Plant Management Issues and Challenges in the United States – 2011 Overview*. American Chemical Society, Washington, DC, pp. 159–168.

Westbrooks, R. and Eplee, R. (2011) Early detection and rapid response. In: Simberloff, D. and Rejmanek, M. (eds) *Encyclopedia of Biological Invasions*. University of California Press, Berkeley, California, pp. 169–177.

Westbrooks, R., Hayes, D. and Gregg, W. (2000) Proposed Strategies for Early Detection, Reporting, Rapid Assessment, and Rapid Response to New Invasive Plants in the United States of America. Proceedings of the First National Early Warning Workshop, Fort Collins, Colorado, June, 2000 (http://www.fws.gov/ficmnew/proceedings%20on%20the%20web/concept_paper.htm, accessed 4 February 2014).

Resources

Cogongrass:

Cogongrass Fact Sheet – BASF (http://www.cogongrass.org/basfarsenal.pdf, accessed 4 February 2014).

Cogongrass Fact Sheet – Plant Conservation Alliance (http://www.nps.gov/plants/alien/fact/imcy1.htm, accessed 4 February 2014).

Cogongrass Control Fact Sheet – BASF (http://www.cogongrass.org/basffactsheet.pdf, accessed 4 February 2014).

Cogongrass Images – U-GA, Bugwood Image Gallery (http://www.invasive.org/species/subject.cfm?sub=2433, accessed 4 February 2014).

Cogongrass Profile – USDA Plants Database (http://plants.usda.gov/java/profile?symbol=IMCY, accessed 4 February 2014).

Cogongrass Website (http://www.cogongrass.org/, accessed 4 February 2014).

Crested Floating Heart:

Crested Floating Heart in Lake Marion, South Carolina. SEEDN – EDDMapS (http://www.eddmaps.org/southeast/distribution/point.cfm?id=633409, accessed 4 February 2014).

Crested Floating Heart Profile – U-FL Center for Aquatic and Invasive Plants (http://plants.ifas.ufl.edu/node/291, accessed 4 February 2014).

Crested Floating Heart Supplier – The Water Garden, Chattanooga, TN (http://www.watergarden.org/Pond-Supplies/Lily-like-Aquatics/White-Snowflake).

Floating Heart Fact Sheet – Comparison of the Four Floating Heart Species that Occur in Florida. Colette Jacono, US Geological Survey, Gainesville, Florida (http://plants.ifas.ufl.edu/misc/pdfs/nymphoides.pdf, accessed 4 February 2014).

New nymphoides in Lake Marion (2007) *SCAPMS Newsletter* 28(1), 3 (http://www.scapms.org/images/may2007news.pdf).

Santee Cooper plans aerial spraying to fight invasive lilies. DigitelCharleston. 1 October 2009 (http://charleston.thedigitel.com/green/santee-cooper-plans-aerial-spraying-fight-invasive-6353-1001, accessed 4 February 2014).

Kudzu:

Kudzu Coalition of Spartanburg, South Carolina. Facebook page (http://www.facebook.com/pages/Kudzu-Coalition/106855309378957?sk=info, accessed 4 February 2014).

Kudzu Fact Sheet, Oregon Department of Agriculture (http://cms.oregon.egov.com/ODA/PLANT/WEEDS/Pages/profile_kudzu.aspx, accessed 4 February 2014).

Kudzu Fact Sheet, Pennsylvania Department of Conservation and Natural Resources (http://www.dcnr.state.pa.us/cs/groups/public/documents/document/dcnr_010253.pdf, accessed 13 February 2014).

Kudzu Fact Sheet, Plant Conservation Alliance (http://www.nps.gov/Plants/alien/fact/pumo1.htm, accessed 4 February 2014).

Dianne Fergusson, Swampland.com (http://swampland.com/articles/view/title:rooting_out_kudzu_the_spartanburg_way, accessed 4 February 2014).

Kudzu Profile – University of Florida, Center for Aquatic and Invasive Plants (http://plants.ifas.ufl.edu/node/354, accessed 4 February 2014).

Kudzu profile (including list of states where regulated) – USDA Plants Database (https://plants.usda.gov/core/profile?symbol=PUMOL, accessed 13 February 2014).

Kudzu Vine – One of Ontario's Most Un-wanted Invasive Plant Species. Fact Sheet. Rachel Gagnon, Ontario Invasive Plant Council (http://www.ontarioinvasiveplants.ca/files/unwantedlettersKudzuLOWRES1.pdf, accessed 4 February 2014).

Wavyleaf Basketgrass:

Westbrooks, R. and Imlay, M. (2010) Wavyleaf Basketgrass Fact Sheet. North Carolina Exotic Pest Plant Council (http://nceppc.weebly.com/uploads/6/8/4/6/6846349/wavyleaf_basketgrass.pdf, accessed 13 February 2014).

Wavyleaf Basketgrass – Maryland Invader of the Month. August, 2007 (http://www.mdinvasivesp.org/archived_invaders/archived_invaders_2007_08.html, accessed 4 February 2014).

Wavyleaf Basketgrass Profile – Global Invasive Species Database – ISSG (http://www.issg.org/database/species/ecology.asp?si=1557&fr=1&sts=&%20ang=EN&ver=print&prtflag=false, accessed 4 February 2014).

Wavyleaf Basketgrass Task Force – Meeting Minutes, 31 March 2009 (http://www.fs.fed.us/ficmnew/documents/notices/WLBG_TaskForceMinutes_033109.pdf, accessed 4 February 2014).

Wavyleaf Basketgrass Images – U-GA Bugwood Image Gallery (http://www.invasive.org/species/subject.cfm?sub=21294, accessed 4 February 2014).

Wavyleaf Basketgrass Profile – USDA Plants Database (http://www.plants.usda.gov/java/profile?symbol=OPHIU2, accessed 4 February 2014).

Wavyleaf Basketgrass in Virginia – Virginia Department of Forestry (http://virginiaforests.blogspot.com/2008/11/wavyleaf-basket-grass-invasive-exotic.html, accessed 4 February 2014).

20 Adapting to Invasions in a Changing World: Invasive Species as an Economic Resource

Matthew A. Barnes, Andrew M. Deines, Rachel M. Gentile and Laura E. Grieneisen

Department of Biological Sciences, University of Notre Dame, Notre Dame, Indiana, USA

Abstract

Management and policy decisions regarding biological invasions and other aspects of global change such as the changing climate have the potential to influence one another. Recently, interest in harvest as an invasive species management strategy has surged. Researchers have speculated about the potential positive (e.g. population and impact reduction, economic losses recouped through new profits) and negative (e.g. generation of economic incentives to encourage further introductions) consequences of such management strategies, but few experimental analyses exist despite the fact that real-world 'experiments' in harvesting invasive species continue to accrue. This chapter provides the first comprehensive list of the many ways humankind has adapted to the presence of invasive species through harvest and considers trends that emerge from this list. The spectrum ranges from collection of feral invasive populations to cultivation of potential invaders as crops, and harvested invasive species serve a variety of purposes ranging from biofuels to foods. The costs of invasion and benefit from harvest can be difficult to quantify, and this difficulty is compounded by the fact that costs and benefits are often incurred by different stakeholders. Climate and other aspects of global change accompanying invasive species harvest have rarely been taken into account, and the climate effects of invasive species harvest should become an explicit consideration of harvest strategies moving forward. Ecologists, economists, environmental managers and stakeholders must work together to ensure the successful and responsible application of ongoing harvests and to inform future efforts.

Introduction

Global environmental change continues to reshape the ways humankind accesses food and other ecosystem services (Godfray *et al.*, 2010; Cardinale *et al.*, 2012). The spread of non-indigenous invasive species represents a major driver of ecosystem change, and invasions are occurring with increasing tolls to biodiversity (Sala *et al.*, 2000), economic costs (Keller *et al.*, 2009) and threats to human and wildlife health (Daszak *et al.*, 2000). Moreover, mounting evidence suggests that biological invasions and other elements of global change such as elevating atmospheric carbon dioxide concentrations interact with one another, potentially exacerbating negative impacts (Dukes and Mooney, 1999). Accordingly, management and policy decisions regarding climate change and biological invasions also influence one another (Pyke *et al.*, 2008).

Management and policy responses to the growing burden of biological invasions include preventing introductions, early detection and rapid response to new arrivals, slowing the spread of established invasions and, as a final resort, adaptation to the persistent presence of an invader (Lodge *et al.*, 2006). As with other forms of global environmental change where management efforts continue to fall short of goals, such as with climate change and biodiversity loss (United Nations Environment Programme, 2009; Butchart *et al.*, 2010), adapting to biological invasions (i.e. changing human behaviour or simply bearing new costs) increasingly represents the only available option.

In an increasing number of cases, harvest has emerged as an adaptive response to biological invasions. As a management strategy, harvest potentially reduces invasive population sizes, produces goods or services to offset at least partially the costs of the invasion and presents opportunities for education to increase invasive species awareness and environmental stewardship (Nuñez *et al.*, 2012). However, elimination of the negative impacts associated with invasive populations (i.e. species eradication) will likely be undersupplied by markets aimed at profiting from those populations. Moreover, not all populations or their impacts will respond to harvest (i.e. population growth rates or ecological and economic damages may be independent of population density). Unintended consequences, such as the generation of economic incentives that encourage the spread of species into new habitats, also represent a risk associated with creating markets for invasive populations (Nuñez *et al.*, 2012). Furthermore, the profit motive could be abused to forgo prevention and eradication efforts during the early stages of an invasion when they might otherwise succeed. However, where invasive populations have been securely established, adaptation strategies should consider the harvesting and utilization of those invasive populations. Given the connection between biological invasions and other trends in global change (e.g. Pyke *et al.*, 2008), the impacts of invasive species management may be far-reaching.

The values associated with introduced species are difficult to quantify, because the ecological functions of species are translated into human well-being via a complex network of ecosystem services including supporting, provisioning, regulating and cultural services (Millenium Ecosystem Assessment, 2005; Cardinale *et al.*, 2012). Furthermore, translating values into monetary units is technically challenging (Chee, 2004). Accordingly, only rough estimates for a small sample of ecosystem services provided by a few invasive species have been attempted (Pimentel *et al.*, 2005), and the potential positive contributions to ecosystem services from invasive species are only beginning to be recognized (Pejchar and Mooney, 2009; Davis *et al.*, 2011; Schlaepfer *et al.*, 2011). A comprehensive list of benefits has not yet been compiled, despite being a prerequisite to evaluating the net effect of invasive species harvest and its contribution to conservation goals. Notwithstanding the difficulties of describing the monetary values of ecosystem services, the total value, composed of both market and non-market goods and services, should be considered when comparing the benefits and costs of harvesting invasive species, and approaching accurate estimates of those values represents a critical management goal. There is still much to learn about making such estimates, but we can begin by providing concrete examples of the costs and benefits of invasive species harvest estimated so far.

Examples of Harvest

Coverage of invasive species harvest in the popular media (e.g. Gorman, 2011) as well as the scientific literature (e.g. Nuñez *et al.*, 2012) has focused primarily on eating the invasive species. However, harvest occurs for a variety of other consumptive uses, including biofuel production, animal feed and fertilizer, furs, recreational hunting and fishing, timber and pulp production etc. Here, we provide the first comprehensive list of the many ways humankind has adapted to the presence of invasive species through harvest (Table 20.1), and we outline several case studies in more detail below. The list compiles the results of Web of Science and Google searches for cases meeting two criteria: (i) each example represents a situation in which harvest is recommended as a strategy for controlling feral invasive populations; (ii) we considered only cases in which the target population was not introduced initially for the purpose of being harvested. Where information is available, we provide estimates of the relative costs of species invasion and the benefits derived from their harvest.

Biofuels

Amid rising energy concerns and political pressure, the global market for biofuels was worth US$82.7 billion in 2011 and is expected to double in the next 10 years (Navigant Research, 2011). For some invasive plants with large established populations, harvest and conversion into biofuel may represent an option for the management of both biological invasions through population reduction and climate change by providing an alternative to fossil fuels (Sage *et al.*, 2009). However, the climate impacts of more deliberate agricultural production of biofuels are not well understood and could actually lead to an increase in greenhouse gas emissions (Delucchi, 2010). Traditional biofuels such as maize and sugarcane can be costly in terms of cultivation, requiring fertilizer, irrigation, pesticides and other management, which all decrease the net energy benefit derived from such crops. Increasingly, biofuel developers are considering invasive plant species, because many of the same traits that contribute to invasive success (i.e. fast growth rates, competitive growth habits, pest resistance and wide climate tolerances) also make biofuel production more efficient and economical, contributing to offsetting carbon emissions (Raghu *et al.*, 2006; Barney and DiTomaso, 2008; Glasser and Glick, 2012).

Invasive species fuel feedstock arises from either intentionally cultivated or established wild populations, and we address both below. It is important to acknowledge that, regardless of species, a population of intentionally grown plants (i.e. crops) does not represent an established invasive population. However, historically cultivation has facilitated invasions (Mack, 2000). Even species under apparently close cultivation warrant careful scrutiny, because if they escape cultivation or are abandoned in a field, former crops could begin to spread. Many current and proposed biofuel sources, although not appearing on invasive species lists in the areas in which they are being considered for cultivation, are listed as noxious weeds elsewhere in the world. Given the increasing media attention afforded to the use of potentially invasive species as cultivated sources of biofuels, as well as the particular relevance of biofuel production to the topic of global climate change, we deviate from our search criteria in the following discussion and consider not only feral populations but also species proposed for cultivation.

For example, Australian producers are encouraged to grow *Jatropha curcas*, which has been praised as an excellent biofuel crop due to its ability to grow in dry climates and on marginal land (Odeh and Tan, 2007), despite invasive status in Australia, Brazil, India and numerous Caribbean islands. Giant reed (*Arundo donax*), a riparian invader which appears on the IUCN list of 100 of the World's Most Invasive Species, is being promoted in Europe as a source of energy and pulp (Lewandowski *et al.*, 2003). Other notable invaders being promoted for biofuel cultivation include Chinese tallow (*Triadica sebifera*), under development in Texas (Low and Booth, 2008), reed canarygrass (*Phalaris arundinacea*), which is being considered for cultivation in Michigan's Upper Peninsula, despite the fact that it is a formidable invader of wetlands, rivers and lakes elsewhere in the

Table 20.1. Current exploitative uses of feral invasive populations reported in refereed and non-refereed sources. Use column indicates general categorization of species use, where: A = animal feed and fertilizer; B = biofuel production; E = education and awareness; F = food; H = hides and fur; R = recreational use; T = timber and pulp; and $ = bounty programmes. Where species use remains under development (i.e. proposed biofuels), no harvest location exists yet. Precise economic value (US$) could not be identified for all species.

Common name	Taxa	Use	Harvest location	Value	Citation
Norway maple	*Acer platanoides*	T	Washington, DC		Clark, 2009
Tree of heaven	*Ailanthus altissima*	B, E, T	USA	US$100–300 $cord^{-1}$	Virginia Department of Forestry, 2009
Persian silk tree	*Albizia julibrissin*	A	USA		Meyer, 2010
Garlic mustard	*Alliaria petiolata*	F, E, T	Washington, DC		Clark, 2009
Porcelainberry	*Ampelopsis brevipedunculata*	E	Washington, DC		Clark, 2009
Lesser burdock	*Arctium minus*	F	USA		Wanjek, 2012
Axis deer	*Axis axis*	R	Hawaii		Hawaii Department of Land and Natural Resources, 2013
Chinese mysterysnail	*Bellamya chinensis*	F	USA		Parry, 2011
Banteng	*Bos javanicus*	E, R	Australia		Bradshaw and Brook, 2007
Canadian goose	*Branta canadensis*	F	USA		Klein, 2012a
Paper mulberry	*Broussonetia papyrifera*	E	Washington, DC		Clark, 2009
Asian water buffalo	*Bubalus bubalis*	A, F, H, R	Australia	US$47 $steak^{-1}$	Jesser *et al.*, 2008
Camel	*Camelus* spp.	A, F, R	Australia		Borrel, 2009; Australian Government Department of Sustainability, Environment, Water, Population and Communities, 2010; Mahr, 2011
Goat	*Capra hircus*	F, H, E	Australia, New Zealand	US$15–30 million $year^{-1}$ export	Parkes, 1993; Parkes *et al.*, 1996; Forsyth *et al.*, 2009; The State of Queensland, Department of Employment, Economic Development and Innovation, 2010
Green crab	*Carcinus maenas*	F, $	New England, Washington	US$5 bounty	Holmes, 2001; Skonberg and Perkins, 2002
Oriental bittersweet	*Celastrus orbiculatus*	E	Washington, DC		Clark, 2009
Spotted knapweed	*Centaurea maculosa*	A	Michigan	US$4 million $year^{-1}$	Jarrell, 2011

Continued

Table 20.1. Continued

Common name	Taxa	Use	Harvest location	Value	Citation
Peacock grouper	*Cephalopholis argus*	A, R	Hawaii		Osher, 2011
Northern snakehead	*Channa argus*	F, R, $	New England	US$200–1500 prizes	Maryland Department of Natural Resources, 2012; Potomac Snakehead Tournament, 2012
Dead man's fingers	*Codium fragile*	F	Connecticut		HeadCount, 2011
Asian clam	*Corbicula fluminea*	F	North America		Klein, 2012b
Common carp	*Cyprinus carpio*	A, F, R	USA, Australia	US$26 l^{-1}[a]	Carpbusters, 2011; Charlie Carp, 2012
Queen Anne's lace	*Daucus carota*	F	Temperate North America		Wanjek, 2012
Didymo	*Didymosphenia geminata*	B			Glasser and Glick, 2012
Water hyacinth	*Eichhornia crassipes*	A, B, E	Lake Victoria		Woomer *et al.*, 2000; ZanaAfrica, 2009; Glasser and Glick, 2012
Autumn olive	*Elaeagnus umbellata*	F	North America		Invasive.org, 2011
Horse	*Equus ferus caballus*	E, R	Australia, Virginia, USA	US$1500 $tour^{-1}$, US$1442 $pony^{-1}$	Australian Government Department of Sustainability, Environment, Water, Population and Communities, 2011; Hampson and Zabek, 2011; Chincoteague, 2012
Chinese mitten crab	*Eriocheir sinensis*	F	UK		Owen, 2003
Japanese knotweed	*Fallopia japonica*	R	North America		Breyer, 2012; Eattheinvaders.org, 2012
English ivy	*Hedera helix*	E	Washington, DC		Weeks, 2011
Irish ivy	*Hedera hibernica*	E	Washington, DC		Clark, 2009
Asian shore crab	*Hemigrapsus sanguineus*	F	Connecticut		HeadCount, 2011
Rio Grande cichlid	*Herichthys cyanoguttatus*	R	New Orleans		Mardon, 2012
Rose of Sharon	*Hibiscus syriacus*	E, T	Washington, DC		Clark, 2009
Sea buckthorn	*Hippophae* spp.	F	UK		Wright, 2011
Hydrilla	*Hydrilla verticillata*	B			Glasser and Glick, 2012

Asian carp	*Hypophthalmichthys molitrix*, *Hypophthalmichthys nobilis*	A, F, R	Midwest USA	US$20 million export	Asian Carp Regional Coordinating Committee, 2010; Charlebois *et al.*, 2010; The Original Redneck Fishing Tournament, 2012
Blue catfish	*Ictalurus furcatus*	R	Eastern USA		Blankenship, 2011
Green iguana	*Iguana iguana*	F	Florida, Puerto Rico		Cera, 2009; Associated Press, 2012a
Nile perch	*Lates niloticus*	F, R	Lake Victoria		Spinney, 2010
Periwinkle	*Littorina littorea*	F	New England		Eattheinvaders.org, 2012
Japanese honeysuckle	*Lonicera japonica*	E, T	USA		Clark, 2009
Blueline snapper	*Lutjanus coeruleolineatus*	A, R	Hawaii		Osher, 2011
Blacktail snapper	*Lutjanus fulvus*	A, R	Hawaii		Osher, 2011
Macaque monkeys	*Macaca* spp.	E, R	Mauritius, Florida, Brazil, Cook Islands, Mexico, Puerto Rico		Gumert *et al.*, 2011; Gonzalez, 2012
Leatherleaf mahonia	*Mahonia bealei*	F, E	Washington, DC		Clark, 2009
White mulberry	*Morus alba*	B, F, E, T	Washington DC		Global Invasive Species Program, 2008; Clark, 2009
Nutria	*Myocastor coypus*	A, H	Southern USA		US Geological Survey, 2000; Carter and Leonard, 2002; Buskey, 2011
Eurasian watermilfoil	*Myriophyllum spicatum*	B			Glasser and Glick, 2012
Purple varnish clam	*Nuttallia obscurata*	F	BC Canada, Washington, Oregon		Fisheries and Oceans Canada, 2009; Meacham, n.d.
Salmonids	*Oncorhynchus* and *Salmo* spps.	F, R	Global		McKinnell *et al.*, 1997; Crawford and Muir, 2007; Strong, 2011
Rusty crayfish	*Orconectes rusticus*	F	Michigan, Wisconsin		Minnesota Sea Grant, 2009; Peters, 2010; Food and Water Watch, 2011; Lodge *et al.*, 2012
Tilapia	*Oreochromis* spp.	F	Global		Canonico *et al.*, 2005; Food and Water Watch, 2011
Oysters	*Ostreidae*	F	Global		Ruesink *et al.*, 2005
Signal crayfish	*Pacifastacus leniusculus*	F	UK, USA, Lake Tahoe		de Lange, 2011; Lodge *et al.*, 2012; Onishi, 2012
Asian tiger prawn	*Penaeus monodon*	F	Gulf of Mexico		Tresaugue, 2011
Devil's tear-thumb	*Persicaria perfoliata*	E	Washington, DC		Clark, 2009

Continued

Table 20.1. Continued

Common name	Taxa	Use	Harvest location	Value	Citation
Sea lamprey	*Petromyzon marinus*	F	Michigan		Kilmas, 2012
Common reed	*Phragmites australis*	B			Glasser and Glick, 2012
Black crappie	*Pomoxis nigromaculatus*	F, R	Maine		Allen, 2011
Purslane	*Portulaca oleracea*	F	North America		Breyer, 2012; Eattheinvaders.org, 2012
Louisiana crayfish	*Procambarus clarkii*	F, R	China, Africa		Barnes and Baldridge, 2009; Lodge *et al.*, 2012
Guava	*Psidium* spp.	F	Hawaii	US$8 each[a]	Madre Chocolate, n.d.
Lionfish	*Pterois miles, Pterois volitans*	F, R	Florida, Caribbean	US$600 day^{-1} harvest	National Oceanic and Atmospheric Administration, 2010; Barbour *et al.*, 2011; O'Hara, 2012
Kudzu	*Pueraria lobata*	B, F	North America		Breyer, 2012; Glasser and Glick, 2012
Burmese python	*Python molurus bivittatus*	F, H	Florida	US$50–100 skin^{-1}, US$1200 each[a]	Griffin, 2012; Florida Fish and Wildlife Conservation Commission, n.d.
Bullfrog	*Rana catesbeiana*	F	Washington	US$13.31 lb^{-1}	Eattheinvaders.org, 2012; Kevin, 2012
Veined rapa whelk	*Rapana venosa*	F	Uruguay		Carranza, 2011
Multiflora rose	*Rosa multiflora*	E	Washington, DC		Clark, 2009
Himalayan blackberries	*Rubus armeniacus*	F	North-western USA		Mother Nature Network, 2011
Wineberry	*Rubus phoenicolasius*	F, E	Washington, DC		Clark, 2009
Curly dock	*Rumex crispus*	F	USA		Breyer, 2012; Eattheinvaders.org, 2012
Rusa deer	*Rusa timorensis*	A	Australia		Sullivan, 2009
Chinese tallow	*Triadica sebifera*	B	Global		Global Invasive Species Program, 2008; Glasser and Glick, 2012
Pink peppercorn	*Schinus molle*	F	Hawaii	US$8 each[a]	Madre Chocolate, n.d.
Brazilian pepper	*Schinus terebinthifolius*	F	Florida		Shimonski, 2011
Feral swine	*Sus scrofa*	F, R	USA		Hainds, 2011
Dandelion	*Taraxacum officinale*	F	USA		Breyer, 2012; Eattheinvaders.org, 2012
Brushtail possums	*Trichosurus vulpecula*	H	New Zealand	US$80 million export[a]	Montaque, 2000; Textiles NZ, 2009

Note: [a]Value-added final product.

USA (Glasser and Glick, 2012), and elephant grass (*Pennisetum purpureum*), a listed invasive species in Florida for which a cultivar is being developed for use in the Gulf Coast region (Glasser and Glick, 2012).

Some notable invasive species are not candidates for cultivation due to their particularly negative reputation, though opportunistic harvesting of existing populations is under consideration. For example, kudzu (*Pueraria lobata*) has had a devastating invasive history in the USA, where it is known as 'the plant that ate the south' and costs an estimated US$500 million annually in lost forest and agricultural productivity, property damage and control efforts (Glasser and Glick, 2012). Furthermore, some studies suggest that kudzu releases compounds that contribute to ozone formation (Hickman *et al.*, 2010), which could exacerbate existing climate change issues. Because it has developed massive monocultures throughout the southern USA, one popular idea is to use standing kudzu as biofuel to clear land for other biofuel or agriculture production (Sage *et al.*, 2009). Several aquatic invasive plants are also being discussed as candidates for opportunistic biofuel harvest, including common reed (*Phragmites australis*), Eurasian watermilfoil (*Myriophyllum spicatum*), hydrilla (*Hydrilla verticillata*) and water hyacinth (*Eichhornia crassipes*) (Glasser and Glick, 2012).

Finally, non-native and modified strains of algae represent the most recent trend in biofuel development. Examples of species under consideration include *Cylindrospermopsis raciborskii*, which is associated with toxic algal blooms in the Laurentian Great Lakes, and didymo (*Didymosphenia geminata*), which is spreading throughout the USA (Glasser and Glick, 2012). Compounding the health risks posed by accidental release of toxic algae, algal species could be particularly difficult to contain once released.

Fertilizer and animal feed

Two common uses for high-protein invaders are fertilizer and animal feed. The contribution of agricultural fertilizer application to climate change is significant, and decreasing the use of soluble chemical fertilizers, particularly reactive nitrogen compounds, may reduce long-term climate forcing (Pinder *et al.*, 2012). Invasive species may supply more suitable forms of fertilizer and reduce the energy burden of modern fertilizer production.

Invasive common carp (*Cyprinus carpio*) are processed into organic fertilizer that sells for AUS$25 l^{-1} in Australia (Charlie Carp, 2012). Common carp comprise up to 90% of fish populations in some Australian waters (Industry and Investment New South Wales Government, 2010), and they can reduce water quality significantly by uprooting sediment through their feeding behaviour (Bajer *et al.*, 2009). The janitor fish (*Pterygoplichthys pardalis*), a freshwater native of South America that has become an invasive predator throughout the Philippines (Hubilla *et al.*, 2007), has also been promoted for use in fishmeal and fertilizer. In 2005, a small bounty (12 cents kg^{-1}) motivated fishermen to collected 13,444 kg of janitor fish, which were then processed into 2151 kg of fishmeal (Laguna Lake Development Authority, 2005). The meal was fed to hogs and chickens as a protein supplement and used as organic fertilizer, though high levels of heavy metals has discouraged further utilization. The most economically viable way to reduce bighead carp and Asian carp near the Laurentian Great Lakes may include commercial harvest for fishmeal (Garvey *et al.*, 2012). Such harvest has motivated the construction of an Asian carp processing plant in Grafton, Illinois, which promises to create 35 jobs and generate additional income for local fishermen (Associated Press, 2012b).

Large herbivores and heavy grazers, including multiple species of deer, Asian water buffalo (*Bubalus bubalis*) and camel (*Camelus dromedarius*) in Australia and nutria (*Myocastor coypus*) in the USA, are being utilized as ingredients in pet food (Jesser *et al.*, 2008; Buskey, 2011; Mahr, 2011). Invasive plants also represent potential sources of animal feed and fertilizer. For example, the Persian silk tree

(*Albizia julibrissin*) is an Asian native that has invaded much of the USA since it was introduced intentionally in the late 1700s (Meyer, 2010). Today, the seeds from the silk tree have potential as livestock feed (Meyer, 2010). Persian silk trees and other ornamental plants that have become invaders are also touted as important nectar sources for native bees. Although the National Park Service recommends cutting Persian silk trees as a control measure (Remaley, 2009), some apiaries encourage gardeners to plant more to help honeybee populations. Importantly, the beneficiaries of silk tree persistence (i.e. beekeepers and cattle farmers) are not necessarily the same individuals who incur the cost of the invasion (i.e. owners of colonized land).

Fur and hides

The total global production of bovine and goat hides in 2009 was more than 10 million t (Mt), while international exports of all livestock hides, furs, hair and leather was more than 5 Mt, worth over US$10 billion (Food and Agriculture Organization of the United Nations, 2012a). Many species now considered invasive were originally introduced intentionally to promote the fur trade. For example, brushtail possums (*Trichosurus vulpecula*) were first introduced in New Zealand (Clout and Barlow, 1982), and nutria (Carter and Leonard, 2002) and American mink (*Neovison vison*; Macdonald and Harrington, 2003) have been introduced to most continents for their furs. Other species, such as goats in Australia (Forsyth *et al.*, 2009) and Burmese pythons (*Python molurus bivittatus*) in the USA (Willson *et al.*, 2011), were introduced as livestock and pets, respectively, but have since become recognized for the value of their hides (Griffin, 2012).

Brushtail possums were introduced to New Zealand from Australia in the late 19th century and are now notorious for extensive negative impacts resulting from browsing damage on trees (Cowan, 1992; Duncan *et al.*, 2011). Brushtail possums also act as a reservoir for bovine tuberculosis affecting cattle production (Caley *et al.*, 1999). New Zealand now spends about US$8 million annually on control efforts (Parkes *et al.*, 2006). Commercial trapping for possum pelts could also reduce the population sizes and influence the age structure of their populations (Clout and Barlow, 1982), and is about as cost-effective as aerial poisoning (Warburton and Thomson, 2002). Furthermore, some social benefit is recouped from harvest. Pelts are used for a wide range of garments and upholstery; the fur is interwoven with local wool and commonly marketed as an environmentally friendly fur fashion product. Brushtail possum exports now amount to NZ$100 million annually (Textiles NZ, 2009), though the exact contribution of these harvests to conservation goals is unknown.

Recreational hunting

Recreational hunting has enjoyed a surge in worldwide popularity in recent decades due to increases in wealth and leisure time (Naevdal *et al.*, 2012). As in the fur trade, many current invasive species were first introduced intentionally for hunting purposes. For example, both European rabbits (*Oryctolagus cuniculus*) and red foxes (*Vulpes vulpes*) were introduced to Australia in the mid-1800s by British colonists (Myers *et al.*, 1994; Saunders and McLeod, 2007). Today, the negative impacts of rabbits in Australia are far-reaching, ranging from overgrazing and loss of plant biodiversity to the support of introduced predators including red foxes (Cooke, 2012). The cost of these introductions is difficult to measure at a continental scale, but in 1998 the Victoria Department of Primary Industries estimated the annual economic damage from rabbits in Victoria at AUS$360 million annually (Bloomfield, 2010). Foxes rank among the greatest predatory threats to native Australian animals and livestock, and their overall damages are estimated to exceed AUS$200 million each year (Saunders and McLeod, 2007).

Nevertheless, where invasive species have become popular targets, hunting may serve simultaneously as a form of population control and boost local economies. For example, hunters pay in excess of AUS$10,000 for the opportunity to safari hunt the introduced megaherbivore, banteng (*Bos javanicus*; Bradshaw and Brook, 2007). Sport fishing derbies have promoted population control efforts for Indo-Pacific lionfish (*Pterois volitans*) in the Caribbean, as well as Asian carp in the Mississippi River (http://originalredneckfishingtournament.com/) and northern snakehead (*Channa argus*) in the north-eastern USA (potomacsnakehead.com). However, the popularity of hunting invasive species can have unintended consequences if species become economically or culturally valued to the extent that sportsmen favour invasive populations (Nuñez *et al.*, 2012) such as exotic salmonid populations, which are carefully maintained in Patagonia (Pascual *et al.*, 2009).

Feral pigs (*Sus scrofa*) provide a more complex example where hunting represents both a conservation boon and burden. Feral pigs cause an estimated US$800 million in damages annually in the USA alone through their voracious appetites for vegetation and small animals, destruction of natural habitat and cropland and spread of disease (Pimentel *et al.*, 2005). Nevertheless, feral pigs have been introduced repeatedly around the globe as a food source, and they remain a popular hunting target today (Hainds, 2011). In some areas, such as the Hawaiian Islands, the popularity of hunting feral pigs has led them to become integrated with the local culture, and social pressure may inhibit eradication, regardless of the negative impacts of the invasion (Nuñez and Simberloff, 2005). Importantly, models based on costs and profits in California predicted that larger feral pig population levels would be tolerated when hunting was allowed compared to when it was illegal (Zivin *et al.*, 2000). Feral pigs also provide an example of the contrast between the opportunistic harvest of invasive species and cultivation in states such as Michigan, Missouri and Tennessee, where confrontations between state governments and ranches that raise 'feral' pigs for hunting are under way.

Timber and pulp

Forestry products such as timber, pulp fibre, firewood and charcoal are common uses for woody invasive plants around the world. Common species include the tree of heaven (*Ailanthus altissima*), wattle (*Acacia* spp.), Norway maple (*Acer platanoides*) and eucalyptus (*Eucalyptus* spp.). These trees are all planted and cultivated because they produce high-value forestry products quickly; however, they have gained invasive status in multiple countries.

Introduced to South Africa in the 1800s from Australia, various *Acacia* species were originally cultivated on plantations, for timber and tannins. The economic benefits from these species have decreased and they are no longer cultivated, but they have become some of the most aggressive invasive trees in South Africa. Rural communities use the trees for building materials and firewood, and the leaves are composted for fertilizer. However, due to the fast-growing nature of the trees, surface water and biodiversity have declined. Surface water shortages are expected to increase as climate change causes more severe droughts and irregular precipitation patterns, increasing the urgency of removing invasive wattle (Shewmake, 2008). The costs of invasive wattle in South Africa are estimated at US$1.4 billion year^{-1} lost in surface water supply and US$34 million year^{-1} in biodiversity loss, including a reduction in tea exports, ecotourism and flower and grass sales (Pejchar and Mooney, 2009). The Working for Water programme leads efforts to mitigate some costs of wattle by paying people in rural areas to harvest the trees; however, it is unclear whether the beneficiaries of the Working for Water programme are the rural communities themselves (De Neergaard *et al.*, 2005; Traynor *et al.*, 2008). The benefits of harvesting wattle include estimates of US$300 million year^{-1} in forestry, US$22 million in building materials, US$143 million in firewood and charcoal and US$1.6 billion in other value-added wood products (Pejchar and Mooney, 2009).

Elsewhere in Africa, a unique project uses pulp from invasive water hyacinth

(*E. crassipes*), which covered over 20,000 ha of Lake Victoria at its peak in 1998. Pulp from water hyacinth is being used in the production of feminine hygiene products (ZanaAfrica, 2009). The harvest contributes to the population control of water hyacinth and provides social benefits as a source of sanitation napkins for Kenyan schoolgirls.

Education and awareness

Lesson plans which engage nature represent an effective tool for the promotion of environmental awareness and stewardship (Lamberton, 1994). Furthermore, new and creative uses for pest species have garnered considerable publicity, maximizing the audience for such lessons (e.g. Gorman, 2011). Therefore, the opportunity to increase public education and awareness represents an important product of invasive species harvest. Education and awareness campaigns not only may benefit efforts to identify and control existing invasive species; increased awareness can also promote attitudes and behaviours which limit new species introductions, representing a critical and cost-effective contribution to invasive species management (Leung *et al.*, 2002).

Creating and displaying artwork presents opportunities for increasing invasive species education and awareness. For example, artists in Washington, DC, have developed a variety of art supplies from many local invasive plants, including paper from garlic mustard (*Alliaria petiolata*), paper mulberry (*Broussonetia papyrifera*) and oriental bittersweet (*Celastrus orbiculatus*), paint-brushes from multiflora rose (*Rosa multiflora*) and porcelainberry (*Ampelopsis brevipedunculata*), and dyes from white mulberry (*Morus alba*) and wineberry (*Rubus phoenicolasius*) (Weeks, 2011). In addition to harvesting invasive species for art supplies, artists also actively teach others about invasive weeds when sharing their work (e.g. http://alienweeds.com/).

The tourism and entertainment industries provide additional opportunities for education and outreach using invasive species. For example, in 2009 alone, camels (*C. dromedarius*) cost AUS$10 million in damages in Australia, and the Australian government spent AUS$16 million to cull them. Governmental agencies have responded with a campaign to increase awareness of camel invasion, including selling camel burgers, using camels as an ingredient in pet food and hosting camel rides (Borrell, 2009; Mahr, 2011). Invasive wild horses (*Equus ferus caballus*) are also a problem in Australia, and to educate the public and reduce population sizes, wild horses are caught and used for show ponies, farm work and trail rides. Tourists pay AUS$1500 to go on a horse-catching tour (Hampson and Zabek, 2011). Similarly, in the USA, 'pony penning' has become a popular summer tourist attraction at Chincoteague Island in Virginia, where young ponies from the nearby Assateague Island wild horse herds sell for an average of US$1400 per pony (Chincoteague, 2012).

Food and nutrition

Humankind has succeeded repeatedly in depleting previously abundant populations of fish (Worm *et al.*, 2009) and other edible taxa, so the consumption of pest species as a control strategy has attracted considerable attention. Furthermore, confrontations and opportunities for global food security in the face of biological invasions, rising carbon dioxide levels and other drivers of global change (e.g. Ziska *et al.*, 2012) presently represent popular areas of research. Large- and small-scale 'experiments' premised upon eating invasive species as part of management and control strategies are already under way. For example, the US National Oceanic and Atmospheric Administration (2010) is leading the 'Eat Lionfish' campaign to combat lionfish, which threaten reef communities through voracious predation and represent a human health concern due to their venomous spines (Whitfield *et al.*, 2002), throughout their invaded range along the US Atlantic coast and in the Caribbean. Demand for lionfish appears to have grown by highlighting its invasive status and negative impacts on local biodiversity and recreation (Barbour *et al.*, 2011). Lionfish, tilapia, Chinese mitten crab, rusty crayfish and other invasive species have also appeared in popular sustainable seafood guides as a way to raise public awareness (Food and Water Watch, 2011). Such campaigns are not only limited to invasive fish species. For example, marketing invasive grey squirrels (*Sciurus carolinensis*) in the UK also successfully increased awareness of the negative impacts of the species (Davies, 2008). Cookbooks (e.g. Franke, 2007; Cera, 2009) and websites such as eattheinvaders.org and invasivore.org (maintained by the authors) now provide recipes for cooking a taxonomically diverse assortment of invasive species.

In the Midwestern USA, managers are considering commercial harvest and consumption as an option to control invasive Asian carp (*Hypophthalmichthys molitrix* and *Hypophthalmichthys nobilis*; Asian Carp Regional Coordinating Committee, 2010; Charlebois *et al.*, 2010). The density of Asian carp on the Illinois River is estimated at 2800 fish per river mile, with annual total biomass production in excess of 3 Mt year^{-1} (Garvey *et al.*, 2012). There is concern that these primarily planktivorous fish will become established in the Laurentian Great Lakes and devastate the US$4 billion commercial and recreational fishing industries by preventing primary production resources from entering higher trophic levels (Cooke and Hill, 2010). However, the global market for Asian carp is extensive, representing the single largest aquaculture production sector in 2010, with culture and capture production at nearly 7 Mt (Food and Agriculture Organization of the United Nations, 2010, 2012b). To apply these markets to the Asian carp threat in the Great Lakes, Illinois recently negotiated the sale of over 22,000 t of Asian carp to China, with a value around US$20 million (Food and Agriculture Organization of the United Nations, 2010) and creating 180 jobs (Asian Carp Regional Coordinating Committee, 2010), while also promoting the use of Asian carp to feed families in need (Illinois Department of Natural Resources, 2012). Managers hope that intensive harvesting of Asian carp at the invasion front nearest Lake Michigan will reduce recruitment of these species into the Great Lakes and prevent their establishment (Charlebois *et al.*, 2010).

Synthesis

Unprecedented global changes in climate and biodiversity require creative management actions. We have provided the first comprehensive list of ways people are adapting to biological invasions through harvest (Table 20.1). In compiling this list, important trends and lessons have emerged. First and foremost, there exists growing recognition that different elements of global change, such as the spread of invasive species and changes in global climate, are related through direct and indirect feedbacks (Dukes and Mooney, 1999). For example, climate change and its effects (e.g. increased disturbance and floods, increased minimum winter temperatures) may exacerbate the spread of some invasive weeds promoted by biofuel interests (Glasser and Glick, 2012). Because of such interrelationships, the management of biological invasions could influence climate in addition to biodiversity (Pyke *et al.*, 2008). With the exception of invasive species harvest for biofuel production (e.g. Sage *et al.*, 2009), our search uncovered sparse examples of invasive species harvest that directly considered the repercussions for climate change. We recommend that climate effects become an explicit consideration of invasive species harvest plans moving forward.

We have demonstrated that a variety of motivations for invasive species harvest exist, and invasive species represent a considerable economic resource. Biofuel production, animal feed, fertilizer, furs, hides, recreation, timber, pulp, educational and outreach opportunities and human consumption all represent creative and potentially beneficial adaptations to biological invasions. We have also identified a spectrum of possibilities for invasive species harvest, ranging from intensive control of wild populations (e.g. Asian carp) to cultivation (e.g. various biofuel crops). Included between these extremes, other categories can be identified, including occasional opportunistic harvests (e.g. goats) and put-and-take harvests (e.g. trout). Additionally, the harvest of many species differs between locations. Intensive control and opportunistic harvests describe the collection of invasive species already present in an ecosystem. In the case of edible invasive species, these types of harvest are epitomized by the slogan, 'If you can't beat 'em, eat 'em!', which has emerged as a common byline in the popular press (e.g. Schaper, 2006). The potential benefits from opportunistic harvest include invasive species population reduction and opportunities for public education and awareness campaigns, which could, in turn, prevent future invasions (Nuñez *et al.*, 2012). Opportunistic harvest can represent a welcome form of human adaptation to the presence of invasive species by either generating benefits or decreasing costs through population and impact reduction.

In contrast to the opportunistic harvest of invasive species is the intentional cultivation of invasive species. Although cultivated crops do not represent invasive populations per se, they do represent a potential pathway toward invasion (Mack, 2000). Our list has shown that in many cases, such as biofuel development (Glasser and Glick, 2012), the cultivation of invasive species is already occurring, often without a clear understanding of the potential impacts on biodiversity and climate change. Therefore, when considering cultivation, potential harvesters need to examine all the potential positive and negative impacts of introduction.

The costs of invasion and benefits from harvest can be difficult to quantify, and this difficulty is compounded by the fact that costs and benefits are not always incurred by the same individuals or groups, generating the potential for environmental injustice similar to other environmental change adaptation strategies (e.g. Roberts and Parks, 2007). The tension between beekeepers and natural resource managers over spotted knapweed (*Centaurea maculosa*) harvest provides one example. Spotted knapweed was introduced in North America in the 1800s and is now an aggressive invader that outcompetes native plants through allelopathy (Bais *et al.*, 2003). However, the Michigan Beekeeper Association has estimated that the honey produced from spotted knapweed yields US\$15 million year^{-1} in profits, comprising a sizeable portion of the US\$4 billion year^{-1} honey industry in the USA (Jarrell, 2011). In this case, beekeepers reap only benefits from spotted knapweed, so they have a positive outlook on the invasion, whereas landowners and other natural resource managers solely incur the cost of the invasion in the form of native plant community biodiversity loss.

Although values are complex and difficult to quantify, invasive species represent an increasingly important economic resource as ecosystems and climates change. Increasing the consideration of harvest as a management strategy will likely heighten the economic importance of many species. However, there is still much to learn about the impact of harvest on invasive species control and public perceptions, as well as interrelated elements of global change such as the changing climate. Moving forward, we recommend that managers, ecologists, economists and local stakeholders work together to ensure the successful and responsible application of current ongoing programmes and collect data to inform future efforts.

Acknowledgements

This work was supported by the GLOBES IGERT (National Science Foundation Grant #DGE-0504495 'Global Linkages of Biology, the Environment, and Society' awarded to the University of Notre Dame). Thanks to Sean Hoban, Amy Klegarth, Elizabeth Miller, Sheina Sim and Sarah Winnikoff for helpful discussions about consuming invasive species. Jeffrey Dukes and an anonymous reviewer improved an earlier draft of this manuscript.

References

Allen, K. (2011) Crappies are invasive, but they're great fun, food. *Portland Press Herald*, 5 June (http://www.pressherald.com/life/outdoors/crappies-are-invasive-but-theyre-great-fun-food_2011-06-05.html, accessed 6 February 2014).

Asian Carp Regional Coordinating Committee (2010) *2011* Asian carp control strategy framework (http://www.asiancarp.us/documents/2011Framework.pdf, accessed 7 February 2014).

Associated Press (2012a) Puerto Rico's iguanas to be killed; meat will be exported. *Huffington Post*, 3 February (http://www.huffingtonpost.com/2012/02/03/puerto-rico-iguanas-killed-exported_n_1252990.html, accessed 6 February 2014).

Associated Press (2012b) Asian carp processing plant to be built in Grafton. *State Journal-Register*, 11 June (http://www.sj-r.com/x2067834548/Asian-carp-processing-plant-to-be-built-in-Grafton, accessed 7 February 2014).

Australian Government Department of Sustainability, Environment, Water, Population and Communities (2010) Camel fact sheet (http://nrmonline.nrm.gov.au/downloads/mql:2849/PDF, accessed 7 February 2014).

Australian Government Department of Sustainability, Environment, Water, Population and Communities (2011) Feral horse (*Equus caballus*) and feral donkey (*Equus asinus*) (http://www.environment.gov.au/system/files/resources/b32a088c-cd31-4b24-8a7c-70e1880508b5/files/feral-horse.pdf, accessed 7 February 2014).

Bais, H.P., Vepachedu, R., Gilroy, S., Callaway, R.M. and Vivanco, J.M. (2003) Allelopathy and exotic plant invasion: from molecules and genes to species interactions. *Science* 301, 1377–1380.

Bajer, P.G., Sullivan, G. and Sorenson, P.W. (2009) Effects of a rapidly increasing population of common carp on vegetative cover and waterfowl in a recently restored Midwestern shallow lake. *Hydrobiologia* 632, 235–245.

Barbour, A.B., Allen, M.S., Frazer, T.K. and Sherman, K.D. (2011) Evaluating the potential efficacy of invasive lionfish (*Pterois volitans*) removals. *PLoS ONE* 6, e19666.

Barnes, M.A. and Baldridge, A.K. (2009) Louisiana crayfish: good, bad, and delicious. *LiveScience* 3 April (http://www.livescience.com/environment/090403-bts-crayfish.html, accessed 6 February 2014).

Barney, J.N. and DiTomaso, J.M. (2008) Nonnative species and bioenergy: are we cultivating the next invader? *BioScience* 58, 64–70.

Blankenship, K. (2011) Blue catfish boom threatens region's river ecosystems. *Bay Journal* (http://www.bayjournal.com/article/blue_catfish_boom_threatens_regions_river_ecosystems, accessed 7 February 2014).

Bloomfield, T. (2010) *Rabbits and Their Impact*. Department of Primary Industries, Victoria, Australia.

Borrell, B. (2009) Camel burgers: Australia plans to shoot 650,000 camels. *Scientific American News Blog*, 11 August 2009 (http://www.scientificamerican.com/blog/post.cfm?id=camel-burgers-australia-plans-to-sh-2009-08-11, accessed 6 February 2014).

Bradshaw, C.J.A. and Brook, B.W. (2007) Ecological-economic models of sustainable harvest for an endangered but exotic mega-herbivore in northern Australia. *Natural Resource Modeling* 20, 129–156.

Breyer, M. (2012) 5 invasive plants you can eat. *Treehugger*, 14 May (http://www.mnn.com/food/healthy-eating/stories/5-invasive-plants-you-can-eat, accessed 6 February 2014).

Buskey, N. (2011) Dog treats, fur coats could help spell nutria's doom. *Daily Comet*, 12 September (http://www.dailycomet.com/article/20110912/ARTICLES/110919924/1212, accessed 6 February 2014).

Butchart, S.H.M., *et al.* (2010) Global biodiversity: indicators of recent declines. *Science* 328, 1164–1168.

Caley, P., Hickling, G.J., Cowan, P.E. and Pfeiffer, D.U. (1999) Effects of sustained control of brushtail possums on levels of *Mycobacterium bovis* infection in cattle and brushtail possum populations from Hohotaka, New Zealand. *New Zealand Veterinary Journal* 47, 133–142.

Canonico, G.C., Arthington, A., McCrary, J.K. and Thieme, M.L. (2005) The effects of introduced tilapias on native biodiversity. *Aquatic Conservation: Marine and Freshwater Ecosystems* 15, 463–483.

Cardinale, B.J., *et al.* (2012) Biodiversity loss and its impact on humanity. *Nature* 486, 59–67.

Carpbusters (2011) Southern Iowa Carpfest (http://carpbusters.com/carp1.html, accessed 6 February 2014).

Carranza, A. (2011) Proyecto Rapana venosa en Uruguay. *Diario Colonia* (http://diariocolonia.com/informacion-general/noticias/proyecto-rapana-venosa-en-uruguay/, accessed 6 February 2014).

Carter, J. and Leonard, B.P. (2002) A review of the literature on the worldwide distribution, spread of, and efforts to eradicate the coypu (*Myocastor coypus*). *Wildlife Society Bulletin* 30, 162–175.

Cera, G. (2009) *The Iguana Cookbook: Save Florida, Eat an Iguana*, self published.

Charlebois, P., Parks, S., TePas, K. and Peterson, M. (2010) Asian Carp Marketing Summit Proceedings (http://www.iisgcp.org/catalog/downlds_09/asian_carp_mktg_summit_proceedings.pdf, accessed 6 February 2014).

Charlie Carp (2012) Charlie Carp (http://www.charliecarp.com, accessed 6 February 2014).

Chee, Y.E. (2004) An ecological perspective on the valuation of ecosystem services. *Biological Conservation* 120, 549–565.

Chincoteague (2012) Chincoteague Wild Ponies (http://www.chincoteague.com/ponies.html, accessed 6 February 2014).

Clark, P. (2009) The Weeds of Burleith (http://www.alienweeds.com/localweeds.html, accessed 6 February 2014).

Clout, M.N. and Barlow, N.D. (1982) Exploitation of brushtail possum populations in theory and practice. *New Zealand Journal of Ecology* 5, 29–35.

Cooke, B.D. (2012) Rabbits: manageable environmental pests or participants in new Australian ecosystems? *Wildlife Research* 39, 279–289.

Cooke, S.L. and Hill, W.R. (2010) Can filter-feeding Asian carp invade the Laurentian Great Lakes? A bioenergetic modelling exercise. *Freshwater Biology* 55, 2138–2152.

Cowan, P.E. (1992) The eradication of introduced Australian brushtail possums, *Trichosurus vulpecula*, from Kapiti Island, a New Zealand nature reserve. *Biological Conservation* 61, 217–226.

Crawford, S.S. and Muir, A.M. (2007) Global introductions of salmon and trout in the genus *Oncorhynchus*: 1870–2007. *Reviews in Fish Biology and Fisheries* 18, 313–344.

Daszak, P., Cunningham, A.A. and Hyatt, A.D. (2000) Emerging infectious diseases of wildlife: threats to biodiversity and human health. *Science* 287, 443–449.

Davies, C. (2008) The ultimate ethical meal: a grey squirrel. *The Observer*, 10 May (http://www.guardian.co.uk/lifeandstyle/2008/may/11/recipes.foodanddrink, accessed 6 February 2014).

Davis, M.A., *et al.* (2011) Don't judge species on their origins. *Nature* 474, 153–154.

de Lange, C. (2011) A crayfish a day keeps the invader at bay. *New Scientist*, 20 June (http://www.newscientist.com/blogs/culturelab/2011/06/a-crayfish-a-day-keeps-the-signal-at-bay.html, accessed 6 February 2014).

De Neergaard, A., Saarnek, C., Hill, T., Khanyile, M., Berzosa, A. and Birch-Thomsen, T. (2005) Australian wattle species in the Drakensberg Region of South Africa – an invasive alien or a natural resource? *Agricultural Systems* 85, 216–233.

Delucchi, M.A. (2010) Impacts of biofuels on climate change, water use, and land use. *Annals of the New York Academy of Sciences* 1195, 28–45.

Dukes, J.S. and Mooney, H.A. (1999) Does global change increase the success of biological invaders? *Trends in Ecology and Evolution* 14, 135–139.

Duncan, R.P., Holland, E.P., Pech, R.P., Barron, M., Nugent, G. and Parkes, J.P. (2011) The relationship between possum density and browse damage on kamahi in New Zealand forests. *Austral Ecology* 36, 858–869.

Eattheinvaders.org (2012) Eat the invaders (http://eattheinvaders.org/, accessed 6 February 2014).

Fisheries and Oceans Canada (2009) Varnish Clams (http://www.dfo-mpo.gc.ca/csas/Csas/status/1999/C6-13e.pdf, accessed 7 February 2014).

Florida Fish and Wildlife Conservation Commission (n.d.) Removing pythons in Florida: Python Removal Program (http://myfwc.com/license/wildlife/nonnative-species/python-permit-program/, accessed 6 February 2014).

Food and Agriculture Organization of the United Nations (2010) 2010 FAO Yearbook of Fishery and Aquaculture Statistics (ftp://ftp.fao.org/FI/CDrom/CD_yearbook_2010/navigation/index_intro_e.htm, accessed 6 February 2014).

Food and Agriculture Organization of the United Nations (2012a) (http://faostat3.fao.org/home/index.html, accessed 6 February 2014).

Food and Agriculture Organization of the United Nations (2012b) The State of World Fisheries and Aquaculture 2012 (http://www.fao.org/docrep/016/i2727e/i2727e.pdf, accessed 6 February 2014).

Food and Water Watch (2011) Smart Seafood Glossary (http://documents.foodandwaterwatch.org/doc/SmartSeafoodGlossary2012.pdf, accessed 6 February 2014).

Forsyth, D.M., Parkes, J.P., Woolnough, A.P., Pickles, G., Collins, M. and Gordon, I. (2009) Environmental and economic factors determine the number of feral goats commercially harvested in Western Australia. *Journal of Applied Ecology* 46, 101–109.

Franke, J.M. (2007) *The Invasive Species Cookbook: Conservation through Gastronomy.* Bradford Street Press, Wauwatosa, Wisconsin.

Garvey, J.E., Sass, G.G., Trushenski, J., Glover, D., Charlebois, P.M., Levengood, J., Roth, B.,

Whitledge, G., Small, B.C., Tripp, S.J. and Secchi, S. (2012) Fishing down the bighead and silver carps: reducing the risk of invasion to the Great Lakes (http://www.asiancarp.us/documents/CARP2011.pdf, accessed 6 February 2014).

Glasser, A. and Glick, P. (2012) *Growing Risk: Addressing the Invasive Potential of Bioenergy Feedstocks*. National Wildlife Federation, Washington, DC.

Global Invasive Species Program (2008) Biofuel crops and the use of non-native species: mitigating the risks of invasion (http://www.sprep.org/att/IRC/eCOPIES/Global/155.pdf, accessed 6 February 2014).

Godfray, H.C.J., Beddington, J.R., Crute, I.R., Haddad, L., Lawrence, D., Muir, J.F., Pretty, J., Robinson, S., Thomas, S.M. and Toulmin, C. (2010) Food security: the challenge of feeding 9 billion people. *Science* 327, 812–818.

Gonzalez, E.R. (2012) Silver Springs monkeys are spreading out as the population grows, wildlife officials say. *Orlando Sentinel*, 8 January (http://articles.orlandosentinel.com/2012-01-08/news/os-central-florida-monkeys-20120108_1_silver-springs-monkeys-tish-hennessey-rhesus, accessed 6 February 2014).

Gorman, J. (2011) A diet for an invaded planet: invasive species. *New York Times*, 2 January (http://www.nytimes.com/2011/01/02/weekinreview/02gorman.html, accessed 6 February 2014).

Griffin, J. (2012) Craftsmen seek profit in Everglades python boom. *Sun Sentinel*, 20 February (http://articles.sun-sentinel.com/2012-02-20/business/fl-python-boots-20120217_1_python-population-python-hunter-snake-skin, accessed 6 February 2014).

Gumert, M.D., Fuentes, A. and Jones-Engel, L. (2011) *Monkeys on the Edge: Ecology and Management of Long-tailed Macaques and their Interface with Humans*. Cambridge University Press, Cambridge, UK.

Hainds, M.J. (2011) *Year of the Pig*. University of Alabama Press, Tuscaloosa, Alabama.

Hampson, B. and Zabek, M. (2011) Brumby & Wildlife Expeditions Australia Brumby Research Unit Program for 2011 (http://www.wildhorseresearch.com/Documents/Tours/BrumbyTours_program & cost.pdf, accessed 6 February 2014).

Hawaii Department of Land and Natural Resources (2013) DLNR announces the 2014 Lanai axis deer hunting season (http://dlnr.hawaii.gov/blog/2013/10/25/nr13-136/, accessed 7 February 2014).

HeadCount (2011) A sustainable answer to invasive species ... put them on the menu (http://www.headcount.org/a-sustainable-answer-to-invasive-species-put-them-on-the-menu/, accessed 6 February 2014).

Hickman, J.E., Wu, S.L., Mickley, L.J. and Lerdau, M.T. (2010) Kudzu (*Pueraria montana*) invasion doubles emissions of nitric oxide and increases ozone pollution. *Proceedings of the National Academy of Sciences* 107, 10115–10119.

Holmes, D. (2001) The green crab invasion: a global perspective, with lessons from Washington state. Master's Thesis, Evergreen State College, Olympia, Washington.

Hubilla, M., Kis, F. and Primavera, J. (2007) Janitor fish *Pterygoplichthys disjunctivus* in the Agusan Marsh: a threat to freshwater biodiversity. *Journal of Environmental Science and Management* 10, 10–23.

Illinois Department of Natural Resources (2012) Target hunger now! (http://www.dnr.illinois.gov/Pages/TargetHungerNow.aspx, accessed 6 February 2014).

Industry & Investment New South Wales Government (2010) NSW Control Plan for the Noxious Fish Carp (*Cyprinus carpio*) (http://www.dpi.nsw.gov.au/__data/assets/pdf_file/0010/306010/nsw-control-plan-for-carp.pdf, accessed 6 February 2014).

Invasive.org (2011) Autumn olive *Elaeagnau umbellata* Thunb (http://www.invasive.org/browse/subinfo.cfm?sub=3021, accessed 6 February 2014).

Jarrell, A. (2011) Invasive species: scourge to some, revenue to others. *Great Lakes Echo*, 13 June 2012 (http://greatlakesecho.org/2011/06/13/invasive-species-scourge-to-some-revenue-to-others, accessed 6 February 2014).

Jesser, P., Markula, A. and Csurhes, S. (2008) Pest animal risk assessment: water buffalo *Bubalus bubalis* (http://www.daff.qld.gov.au/documents/Biosecurity_EnvironmentalPests/IPA-Water-Buffalo-Risk-Assessment.pdf, accessed 6 February 2014).

Keller, R.P., Lodge, D.M., Lewis, M.A. and Shogren, J.F. (eds) (2009) *Bioeconomics of Invasive Species: Integrating Ecology, Economics, and Management*. Oxford University Press US, New York.

Kevin, H. (2012) Pond-draining stymies bullfrogs. *Chinook Observer*, 7 February (http://www.chinookobserver.com/news/pond-draining-stymies-bullfrogs/article_cb4542fa-51f0-11e1-97b4-001871e3ce6c.html?mode=print, accessed 6 February 2014).

Kilmas, L. (2012) True, freaky sea lampreys will become pie for Queen Elizibeth. *The Blaze*, 27 April (http://www.theblaze.com/stories/what-great-lakes-invasive-species-will-be-made-into-a-pie-fit-for-the-queen/, accessed 6 February 2014).

Klein, D. (2012a) Episode 85: Goose Dilemma. *The Perennial Plate* (http://www.theperennialplate.com/episodes/2012/01/episode-85-goose-dilemma/, accessed 6 February 2014).

Klein, D. (2012b) Episode 98: Fire Starter. *The Perennial Plate* (http://www.theperennialplate.com/episodes/2012/04/episode-98-fire-starter/, accessed 6 February 2014).

Laguna Lake Development Authority (2005) Benefiting from the dreaded janitor fish (http://wbi.worldbank.org/wbdm/idea/benefiting-dreaded-janitor-fish, accessed 7 February 2014).

Lamberton, K. (1994) Yucca! Teaching the essentials of a wild food feast. *American Biology Teacher* 56, 230–232.

Leung, B., Lodge, D.M., Finnoff, D., Shogren, J.F., Lewis, M. and Lamberti, G. (2002) An ounce of prevention or a pound of cure: bioeconomic risk analysis of invasive species. *Proceedings of the Royal Society B* 269, 2407–2413.

Lewandowski, I., Scurlock, J.M.O., Lindvall, E. and Christou, M. (2003) The development and current status of perennial rhizomatous grasses as energy crops in the US and Europe. *Biomass and Bioenergy* 25, 335–361.

Lodge, D.M., Williams, S., MacIsaac, H.J., Hayes, K.R., Leung, B., Reichard, S., Mack, R.N., Moyle, P.B., Smith, M., Andow, D.A., Carlton, J.T. and McMichael, A. (2006) Biological invasions: recommendations for US policy and management. *Ecological Applications* 16, 2035–2054.

Lodge, D.M., *et al.* (2012) Global introductions of crayfishes: evaluating impact of species invasions on ecosystem services. *Annual Review of Ecology, Evolution, and Systematics* 43, 449–472.

Low, T. and Booth, C. (2008) *The Weedy Truth About Biofuels.* The Invasive Species Council, Fairfield, Victoria, Australia.

Macdonald, D.W. and Harrington, L.A. (2003) The American mink: the triumph and tragedy of adaptation out of context. *New Zealand Journal of Zoology* 30, 421–441.

Mack, R.N. (2000) Cultivation fosters plant naturalization by reducing environmental stochasticity. *Biological Invasions* 2, 111–122.

McKinnell, S., Thomson, A.J., Black, E.A., Wing, L., Guthrie, C.M. III, Koerner, J.F. and Helle, J.H. (1997) Atlantic salmon in the North Pacific. *Aquaculture Research* 28, 145–157.

Madre Chocolate (n.d.) (http://madrechocolate.com, accessed 6 February 2014).

Mahr, K. (2011) Australia: killing camels for carbon credits? *Time Science Ecocentric Blog*, 10 June 2011 (http://ecocentric.blogs.time.com/2011/06/10/australia-killing-camels-for-carbon-credits/, accessed 6 February 2014).

Mardon, S. (2012) Fishers get chance to combat invasive species at New Orleans City Park fishing rodeo Saturday. *Nola.com*, 28 March (http://blog.nola.com/new_orleans/2012/03/fishers_get_chance_to_combat_i.html, accessed 6 February 2014).

Maryland Department of Natural Resources (2012) Help control the spread of snakehead fish: catch a snakehead and win a prize worth up to $200 (http://www.dnr.state.md.us/dnrnews/pressrelease2012/032812.asp, accessed 6 February 2014).

Meacham, P. (n.d.) *Nuttalia obscurata* (purple varnish clam, varnish clam, or dark mahogany clam) (http://wdfw.wa.gov/ais/nuttalia_obscurata/, accessed 6 February 2014).

Meyer, R. (2010) *Albizia julibrissin.* US Department of Agriculture, Forest Service, Rocky Mountain Research Station, Fire Sciences Laboratory (http://www.fs.fed.us/database/feis/plants/tree/albjul/all.html, accessed 6 February 2014).

Millennium Ecosystem Assessment (2005) *Ecosystems and Human Well-being: Synthesis.* Island Press, Washington, DC.

Minnesota Sea Grant (2009) Rusty crayfish: a nasty invader (http://www.seagrant.umn.edu/ais/rustycrayfish_invader, accessed 6 February 2014).

Montaque, T.L. (2000) *The Brushtail Possum: Biology, Impact and Management of an Introduced Marsupial.* Manaaki Whenua Press, Lincoln, New Zealand.

Mother Nature Network (2011) 6 edible invasive species recipes (http://www.mnn.com/food/recipes/photos/6-edible-invasive-species-recipes, accessed 6 February 2014).

Myers, K., Parer, I., Wood, D. and Cooke, B.D. (1994) The rabbit in Australia. In: Thompson, H.V. and King, C.M. (eds) *The European Rabbit: The History and Biology of a Successful Colonizer.* Oxford University Press, Oxford, UK, pp. 108–157.

Naevdal, E., Olaussen, J.O. and Skonhoft, A. (2012) A bioeconomic model of trophy hunting. *Ecological Economics* 73, 194–205.

National Oceanic and Atmospheric Administration (2010) Eat Lionfish (http://www.ccfhr.noaa.gov/docs/EatLionfishPullCard.pdf, accessed 6 February 2014).

Navigant Research (2011) Global biofuels market value to double to $185 billion by 2021 (http://www.navigantresearch.com/, accessed 7 February 2014).

Nuñez, M.A. and Simberloff, D. (2005) Invasive

species and the cultural keystone species concept. *Ecology and Society* 10, r4 (http://www.ecologyandsociety.org/vol10/iss1/resp4/, accessed 7 February 2014).

Nuñez, M.A., Kuebbing, S., Dimarco, R.D. and Simberloff, D. (2012) Invasive species: to eat or not to eat, that is the question. *Conservation Letters* doi:10.1111/j.1755-263X.2012.00250.x.

Odeh, I. and Tan, D. (2007) Expanding biofuel production in Australia: opportunities beyond the horizon. *Farm Policy Journal* 4, 29–39.

O'Hara, T. (2012) Fisherman puts invasive species on Keys menus. *KeysNews.com*, 19 March (http://keysnews.com/node/38618, accessed 6 February 2014).

Onishi, N. (2012) Crayfish to eat, and to clean the water. *New York Times*, 13 July (http://www.nytimes.com/2012/07/13/science/earth/at-lake-tahoe-commercial-fishing-is-approved.html, accessed 6 February 2014).

Original Redneck Fishing Tournament, The (2012) (http://originalredneckfishingtournament.com/, accessed 7 February 2014).

Osher, W. (2011) Spearfishing challenge helps with invasive species control (http://mauinow.com/2011/10/27/spearfishing-challenge-helps-with-invasive-species-control/, accessed 6 February 2014).

Owen, J. (2003) Eat the invading alien crabs, urge U.K. scientists. *National Geographic News*, 13 November (http://news.nationalgeographic.com/news/2003/11/1113_031113_mittencrabs.html, accessed 6 February 2014).

Parkes, J. (1993) Goats: designing solutions for a designer pest. *New Zealand Journal of Ecology* 17, 71–83.

Parkes, J., Henzell, R. and Pickles, G. (1996) *Managing Vertebrate Pests: Feral Goats*. Australian Government Publishing Service, Canberra.

Parkes, J.P., Robley, A., Forsyth, D.M. and Choquenot, D. (2006) Adaptive management experiments in vertebrate pest control in New Zealand and Australia. *Wildlife Society Bulletin* 34, 229–236.

Parry, W. (2011) On the menu: taking a bite out of invasive species. *LiveScience*, 25 March (http://www.livescience.com/13426-invasive-species-invasivory-recipes-diet.html, accessed 6 February 2014).

Pascual, M.A., Lancelotti, J.L., Ernst, B., Ciancio, J.E., Aedo, E. and García-Asorey, M. (2009) Scale, connectivity, and incentives in the introduction and management of non-native species: the case of exotic salmonids in Patagonia. *Frontiers in Ecology and the Environment* 7, 533–540.

Pejchar, L. and Mooney, H.A. (2009) Invasive species, ecosystem services and human well-being. *Trends in Ecology and Evolution* 24, 497–504.

Peters, B.W. (2010) Evaluating strategies for controlling invasive crayfish using human and fish predation. Master's Thesis, University of Notre Dame, Notre Dame, Indiana.

Pimentel, D., Zuniga, R. and Morrison, D. (2005) Update on the environmental and economic costs associated with alien-invasive species in the United States. *Ecological Economics* 52, 273–288.

Pinder, R.W., Davidson, E.A., Goodale, C.L., Greaver, T.L., Herrick, J.D. and Liu, L. (2012) Climate change impacts of US reactive nitrogen. *Proceedings of the National Academy of Sciences* 109, 7671–7675.

Potomac Snakehead Tournament (2012) Tournament home (http://potomacsnakehead.com/, accessed 6 February 2014).

Pyke, C.R., Thomas, R., Porter, R.D., Hellmann, J.J., Dukes, J.S., Lodge, D.M. and Chavarria, G. (2008) Current practices and future opportunities for policy on climate change and invasive species. *Conservation Biology* 22, 585–592.

Raghu, S., Anderson, R.C. and Daehler, C.C. (2006) Adding biofuels to the invasive species fire? *Science* 313, 1742.

Remaley, T. (2009) Plant Conservation Alliance's Alien Plant Working Group Least Wanted: Silk Tree (http://www.nps.gov/plants/alien/fact/alju1.htm, accessed 6 February 2014).

Roberts, J.T. and Parks, B.C. (2007) *A Climate of Injustice: Global Inequality, North–South Politics, and Climate Policy*. MIT Press, Cambridge, Massachusetts.

Ruesink, J.L., Lenihan, H.S., Trimble, A.C., Heiman, K.W., Micheli, F., Byers, J.E. and Kay, M.C. (2005) Introduction of non-native oysters: ecosystem effects and restoration implications. *Annual Review of Ecology, Evolution, and Systematics* 36, 643–689.

Sage, R.F., Coiner, H.A., Way, D.A., Runion, G.B., Prior, S.A., Torbert, H.A., Sicher, R. and Ziska, L. (2009) Kudzu [*Pueraria montana* (Lour.) Merr. Variety *lobata*]: a new source of carbohydrate for bioethanol production. *Biomass and Bioenergy* 33, 57–61.

Sala, O.E., *et al.* (2000) Global biodiversity scenarios for the year 2100. *Science* 287, 1770–1775.

Saunders, G. and McLeod, L. (2007) *Improving Fox Management Strategies in Australia*. Bureau of Rural Sciences, Canberra.

Schaper, D. (2006) Asian carp: can't beat them? Eat them. *National Public Radio*, 12 July (http://www.npr.org/templates/story/story.php?storyId=5542199, accessed 6 February 2014).

Schlaepfer, M.A., Sax, D.F. and Olden, J.D. (2011) The potential conservation value of non-native species. *Conservation Biology* 25, 428–437.

Shewmake, S. (2008) *Vulnerability and the Impact of Climate Change in South Africa's Limpopo River Basin*. International Food Policy Research Institute, Washington, DC.

Shimonski, J. (2011) A tasty and zesty foe. *Biscayne Times*, April (http://www.biscaynetimes.com/index.php?option=com_content&view=article&id=844:a-tasty-and-zesty-foe&catid=43:columnists-your-garden&Itemid=196, accessed 6 February 2014).

Skonberg, D.I. and Perkins, B.L. (2002) Nutrient composition of green crab (*Carcinus maenus*) leg meat and claw meat. *Food Chemistry* 77, 401–404.

Spinney, L. (2010) Dreampond revisited. *Nature* 466, 174–175.

State of Queensland, The, Department of Employment, Economic Development and Innovation (2010) Feral goat *Capra hirus* (http://www.daff.qld.gov.au/documents/Biosecurity_EnvironmentalPests/IPA-Feral-Goat-PA18.pdf, accessed 6 February 2014).

Strong, T. (2011) Trout added to, fought from Virginia waterways. *Daily Progress*, 31 March (http://www2.dailyprogress.com/news/2011/mar/31/trout-added-fought-virginia-waterways-ar-942173/, accessed 6 February 2014).

Sullivan, B. (2009) Deer management plan for Royal National Park: 2008–2009 Annual Report (http://www.environment.nsw.gov.au/resources/pestsweeds/RoyalNationalParkDeerManagementAnnualReport2009.pdf, accessed 6 February 2014).

Textiles NZ (2009) More possum fur needed to support export growth (http://www.scoop.co.nz/stories/BU0912/S00094.htm, accessed 6 February 2014).

Traynor, C.H., Hill, T., Ndela, Z. and Tshabalala, P. (2008) What'll we do with wattle? The dualistic nature of *Acacia mearnsii* as both a resource and an alien invasive species, Swaziland. *Alternation* 15, 180–205.

Tresaugue, M. (2011) Giant shrimp raises big concern as it invades the Gulf. *Houston Chronicle*, 24 December (http://www.chron.com/news/houston-texas/article/Asian-tiger-prawns-invade-the-Gulf-of-Mexico-2424242.php, accessed 6 February 2014).

United Nations Environment Programme (2009) Climate Change Science Compendium (http://www.unep.org/pdf/ccScienceCompendium2009/cc_ScienceCompendium2009_full_en.pdf, accessed 6 February 2014).

US Geological Survey (2000) Nutria, eating Louisiana's coast (http://www.nwrc.usgs.gov/factshts/020-00.pdf, accessed 6 February 2014).

Virginia Department of Forestry (2009) Control and utilization of tree-of-heaven: a guide for Virginia landowners (http://www.dof.virginia.gov/health/print/Control-And-Utilization-Of-TOH.pdf, accessed 7 February 2014).

Wanjek, C. (2012) 5 garden weeds you can eat! *LiveScience*, 10 July (http://www.livescience.com/21496-5-edible-garden-weeds.html, accessed 6 February 2014).

Warburton, B. and Thomson, C. (2002) Comparison of three methods for maintaining possums at low density. *Science for Conservation* 189, 1–20.

Weeks, L. (2011) The art of war on invasive species. *National Public Radio*, 28 February (http://www.npr.org/2011/02/28/134054004/the-art-of-war-on-invasive-species, accessed 6 February 2014).

Whitfield, P.E., Gardner, T., Vives, S.P., Gilligan, M.R., Courtenay, W.R. Jr, Ray, G.C. and Hare, J.A. (2002) Biological invasion of the Indo-Pacific lionfish *Pterois volitans* along the Atlantic coast of North America. *Marine Ecology Progress Series* 235, 289–297.

Willson, J.D., Dorcas, M.E. and Snow, R.W. (2011) Identifying plausible scenarios for the establishment of invasive Burmese pythons (*Python molurus*) in southern Florida. *Biological Invasions* 13, 1493–1504.

Woomer, P.L., Muzira, R., Bwamiki, D., Mutetikka, D., Amoding, A. and Bekunda, M.A. (2000) Biological management of water hyacinth waste in Uganda. *Biological Agriculture and Horticulture* 17, 181–196.

Worm, B., *et al.* (2009) Rebuilding global fisheries. *Science* 325, 578–585.

Wright, J. (2011) How to make sea buckthorn fizz. *The Guardian Word of Mouth Blog*, 13 July (http://www.guardian.co.uk/lifeandstyle/wordofmouth/2011/jul/13/how-make-sea-buckthorn-fizz, accessed 6 February 2014).

ZanaAfrica (2009) ZanaAfrica (http://www.zanaafrica.org/, 7 February 2014).

Ziska, L.H., Bunce, J.A., Shimono, H., Gealy, D.R., Baker, J.T., Newton, P.C.D., Reynolds, M.P., Jagadish, K.S.V., Zhu, C., Howden, M., and Wilson, L.T. (2012) Food security and climate change: on the potential to adapt global crop production by active selection to rising carbon dioxide. *Proceedings of the Royal Society B* 279, 4097–4105.

Zivin, J., Hueth, B.M. and Zilberman, D. (2000) Managing a multiple-use resource: the case of feral pig management in California rangeland. *Journal of Environmental Economics and Management* 39, 189–204.

Index